Powers of 2, 3, 4, 5

k	2^k	3^k	4^k	[illegible]
2	4	9	16	[illegible]
3	8	27	64	[illegible]
4	16	81	256	625
5	32	243	1 024	3 125
6	64	729	4 096	15 625
7	128	2 187	16 384	78 125
8	256	6 561	65 536	390 625
9	512	19 683	262 144	1 953 125
10	1 024	59 049	1 048 576	9 765 625
11	2 048	177 147	4 194 304	48 828 125
12	4 096	531 441	16 777 216	244 140 625
13	8 192	1 594 323	67 108 864	1 220 703 125
14	16 384	4 782 969	268 435 456	6 103 515 625
15	32 768	14 348 907	1 073 741 824	30 517 578 125

Roots of 2, 3, 4, 5

k	$\sqrt[k]{2}$	$\sqrt[k]{3}$	$\sqrt[k]{4}$	$\sqrt[k]{5}$
2	1.414 213 562	1.732 050 808	2.000 000 000	2.236 067 977
3	1.259 921 050	1.442 249 570	1.587 401 052	1.709 975 947
4	1.189 207 115	1.316 074 013	1.414 213 262	1.495 348 781
5	1.148 698 355	1.245 730 940	1.319 507 911	1.379 729 662

Powers of 6, 7, 8, 9

k	6^k	7^k	8^k	9^k
2	36	49	64	81
3	216	343	512	729
4	1 296	2 401	4 096	6 561
5	7 776	16 807	32 768	59 049
6	46 656	117 649	262 144	531 441
7	279 936	823 543	2 097 152	4 782 469
8	1 679 616	5 764 801	16 777 216	43 046 721
9	10 077 696	40 353 607	134 217 728	387 420 489
10	60 466 176	282 475 249	1 073 741 824	3 486 784 401
11	362 797 056	1 977 326 743	8 589 934 592	31 381 059 609
12	2 176 782 336	13 841 287 201	68 719 476 736	282 429 536 481
13	13 060 694 016	96 889 010 407	549 755 813 888	2 541 865 828 329
14	78 364 164 096	678 223 072 849	4 398 046 511 104	22 876 792 454 961
15	470 184 984 576	4 747 561 509 943	35 184 372 088 832	205 891 132 094 649

Roots of 6, 7, 8, 9

k	$\sqrt[k]{6}$	$\sqrt[k]{7}$	$\sqrt[k]{8}$	$\sqrt[k]{9}$
2	2.449 489 743	2.645 751 311	2.828 427 125	3.000 000 000
3	1.817 120 593	1.912 931 183	2.000 000 000	2.080 083 823
4	1.565 084 580	1.626 576 562	1.681 792 831	1.732 050 808
5	1.430 969 081	1.475 773 162	1.515 716 567	1.551 845 574

Technology Mathematics Handbook

Technology Mathematics Handbook

DEFINITIONS ▪ FORMULAS ▪ GRAPHS
SYSTEMS OF UNITS ▪ PROCEDURES
CONVERSION TABLES
NUMERICAL TABLES

Jan J. Tuma, Ph.D.
Engineering Consultant
Boulder, Colorado

McGraw-Hill Book Company

New York St. Louis San Francisco Auckland Düsseldorf
Johannesburg Kuala Lumpur London Mexico Montreal
New Delhi Panama Paris São Paulo Singapore
Sydney Tokyo Toronto

Library of Congress Cataloging in Publication Data

Tuma, Jan J
Technology mathematics handbook.

Includes index.
1. Engineering mathematics—Handbooks, manuals, etc.
I. Title.
TA332.T86 510 74-26962
ISBN 0-07-065431-X

1234567890 VHVH 784321098765

The editors for this book were Harold B. Crawford and Ruth Weine and its production was supervised by Teresa F. Leaden.

It was printed and bound by Von Hoffmann Press.

To my father Joseph,
and my son Peter

CONTENTS

APPENDIX B. CONVERSION TABLES

PREFACE

This Handbook presents in one volume a concise summary of the major definitions, formulas, graphs, tables, and examples of elementary and intermediate mathematics. It places emphasis on technological applications and was prepared to serve as a desk-top reference book for aeronautical, architectural, civil, mechanical, chemical, industrial, electrical, and construction technologists and the practicing engineers in these fields.

The content of the book is grouped into four parts, each related to a particular type of technical calculations.

Part I - Elementary mathematics (Chapters 1 to 5) covers arithmetic, algebra, plane geometry, space geometry, and plane trigonometry.

Part II - Intermediate mathematics (Chapters 6 to 11) presents analytical geometry, differential calculus, sequences, series, integral calculus, matrices, determinants, and vectors.

Part III - Numerical procedures (Chapter 12 and Appendix A) gives a comprehensive outline of operations with decimal and complex numbers, applications of tables of numerical constants and elementary functions, calculations of interest and annuities, and various practical approximations.

Part IV - Conversion procedures (Chapter 13 and Appendix B) introduces the systems of units of measure (FPS system, SI system) with emphasis on their definitions, classification, and conversion.

The form of presentation has many special characteristics allowing easy and rapid location of the desired information and permitting the indexing of this information.

1. Each statement in the book is a coded sentence designated by the position number and key word.
2. The related sentences form logical sequences and their lengths allow speed reading.
3. The extensive index of all key words (of all sentences) given in the last part of the book offers the possibility of using this handbook as a dictionary of technology mathematics.
4. All formulas are presented in general symbols and their applications are illustrated by examples where this is desirable.
5. The application of tables of numerical coefficients is described in step-by-step procedures in Chapter 12, with cross-references given below each table.

In the preparation of this book free use was made of the pertinent material from my earlier book, "Engineering Mathematics Handbook," McGraw-Hill, New York, 1970. Although this handbook overlaps in part the Engineering Mathematics Handbook, their respective levels and objectives are so different that instead of competing with each other, they form a complementary set which many users may find useful in their professional work.

Boulder, Colorado *Jan J. Tuma*

1
ARITHMETIC

1.01 DEFINITIONS AND NOTATIONS

(1) Definitions

(a) Arithmetic is the systematic study of fundamental operations with real numbers and of the use of these operations in solving practical problems.

(b) Real numbers are:
(α) The *natural numbers* (Sec. 1.01–1*c*).
(β) The *rational numbers* (Sec. 1.10–1*d*).
(γ) The *irrational numbers* (Sec. 1.10–1*e*).

(c) Natural numbers (also called the arabic numbers or positive integers) are the symbols arrived at by counting such as 1, 2, 3, 4, . . . , where the three dots mean "and so on."

(d) Four fundamental operations of arithmetic are:
(α) The *addition* (Sec. 1.02–1).
(β) The *subtraction* (Sec. 1.02–2).
(γ) The *multiplication* (Sec. 1.02–3).
(δ) The *division* (Sec. 1.02–4).

(2) Symbols of Relationship

The following symbols define the relationship of two numbers:

$=$ or $::$	Equals	$\ddagger$ or $\neq$	Does not equal
$>$	Greater than	$<$	Less than
$\geq$	Greater than or equal	$\leq$	Less than or equal
$\equiv$	Identical	$\simeq$	Approximately equal
$\ngtr$	Not greater than	$\nless$	Not less than

(3) Symbols of Aggregation

The symbols of grouping (aggregation) are:

( )	Parentheses	{ }	Braces
[ ]	Brackets	—	Vinculum

(4) Signs of Operations

The signs of operations are:

$+$	Plus or positive	$-$	Minus or negative
$\pm$	Plus or minus, positive or negative	$\mp$	Minus or plus, negative or positive
$\times$ or $\cdot$	Multiplied by	$\div$ or :	Divided by
a^n	nth power of a	$\sqrt[n]{a}$	nth root of a
log, $\log_{10}$	Common logarithm, or Briggs's logarithm	ln, $\log_e$	Natural logarithm, or Napier's logarithm

1.02 FOUR FUNDAMENTAL OPERATIONS

(1) Addition

(a) Addition is the operation of finding the sum of two or more numbers.

example:

$2+3+4=9$

where 9 is the *sum* and 2, 3, 4 are the *terms* of the sum.

(b) Order of terms in addition may be changed without affecting the sum (commutative law).

examples:

$2+3+4=9 \qquad 3+4+2=9 \qquad 4+2+3=9$

(c) Grouping of terms in addition may be changed without affecting the sum (associative law).

examples:

$(2+3)+4=5+4=9 \qquad 2+(3+4)=2+7=9$

(2) Subtraction

(a) Subtraction is the operation of finding the difference of two numbers.

example:

$9-4=5$

where 5 is the *difference*, 9 is the *minuend*, and 4 is the *subtrahend*.

(b) Difference of two equal numbers is zero.

example:

$9-9=0$

(3) Multiplication

(a) Multiplication is the operation of finding the product of two or more numbers.

example:

$2 \times 3 \times 4 = 24$

where 24 is the *product* and 2, 3, 4 are the *factors* of the product.

(b) Order of factors in multiplication may be changed without affecting the product (commutative law).

examples:

$2 \times 3 \times 4 = 24 \qquad 3 \times 4 \times 2 = 24 \qquad 4 \times 2 \times 3 = 24$

(c) Grouping of factors in multiplication may be changed without affecting the product (associative law).

examples:

$(2 \times 3) \times 4 = 6 \times 4 = 24 \qquad 2 \times (3 \times 4) = 2 \times 12 = 24$

(4) Division

(a) Division is the operation of finding the quotient of two numbers.

example:

$24 : 8 = 3$

where 3 is the *quotient*, 24 is the *dividend*, and 8 is the *divisor*.

(b) Quotient of two equal numbers is 1.

example:

$24 : 24 = 1$

(5) Even, Odd, and Prime Numbers

(a) Even number is an integer divisible by 2.

example:

2, 4, 6, . . . are even numbers.

(b) Odd number is an integer not divisible by 2.

example:

1, 3, 5, . . . are odd numbers.

(c) Prime number is an integer divisible only by 1 and itself.

example:

1, 2, 3, 5, 7, . . . are prime numbers.

(6) Factoring

(a) Every nonprime number greater than 1 can be expressed as a product of prime numbers.

examples:

$30 = 2 \times 3 \times 5 \qquad 60 = 2 \times 2 \times 3 \times 5$

(b) Highest common factor (HCF) of a given set of numbers is the largest number that is a factor of all the numbers.

example:

$$\underbrace{24 = 2 \times 2 \times 2 \times 3 \qquad 60 = 2 \times 2 \times 3 \times 5 \qquad 84 = 2 \times 2 \times 3 \times 7}_{\text{HCF} = 2 \times 2 \times 3 = 12}$$

where HCF is the product of the prime factors that are common to all the numbers of the set.

(c) Lowest common multiple (LCM) of a given set of numbers is the smallest number that has each of the given numbers as a factor.

example:

$$\underbrace{24 = 2 \times 2 \times 2 \times 3 \qquad 60 = 2 \times 2 \times 3 \times 5 \qquad 84 = 2 \times 2 \times 3 \times 7}_{\text{LCM} = 2 \times 2 \times 2 \times 3 \times 5 \times 7 = 840}$$

where LCM is the product of all the different prime factors of the given numbers, each taken the greatest number of times that it occurs in one of the numbers.

1.03 SIGNED NUMBERS

(1) Graphical Representation

(a) Real numbers may be represented by points on a straight line as shown in Fig. 1.03–1, where the distance between two adjacent points is constant and equals 1.

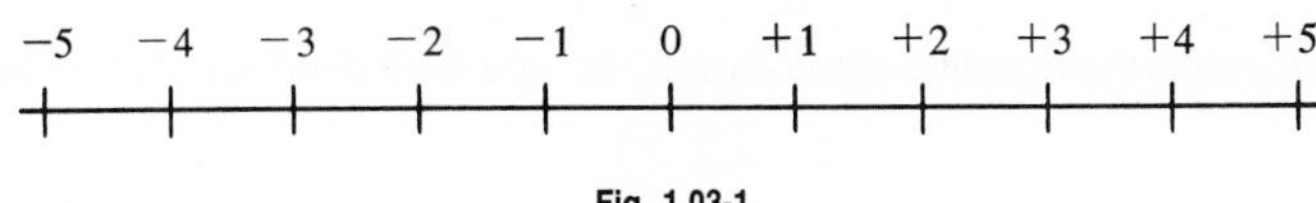

Fig. 1.03-1

(b) Positive numbers $+1, +2, +3, \ldots$ are then associated with the points on the right side of the origin designated by 0 (zero).

(c) Negative numbers $-1, -2, -3, \ldots$ are then associated with the points on the left side of the origin.

(d) Positive and negative numbers are called the *signed numbers.* Zero in arithmetic has no sign, and all unsigned numbers are assumed to be positive numbers.

(e) Absolute value of a number is its numerical value regardless of sign and is designated by two vertical lines surrounding the signed number.

examples:

$|+5| = 5 \qquad |-5| = 5 \qquad |+5| = |-5|$

which means the absolute value of a positive number equals the absolute value of the negative number and vice versa.

(2) Addition and Subtraction

(a) Sum of two numbers of like signs equals the sum of their absolute values prefixed by their sign.

examples:

$(+3)+(+5) = +8 \qquad (-3)+(-5) = -8$

(b) Sum of two numbers of unlike signs equals the difference of their absolute values prefixed by the sign of the number having the larger absolute value.

examples:

$(+3)+(-5) = -2 \qquad (-3)+(+5) = +2$

(c) Difference of the signed numbers equals their sum in which the sign of subtrahend (Sec. 1.02–2*a*) was reversed.

examples:

$$(+12)-(+3) = (+12)+(-3) = +9$$

$$(+12)-(-3) = (+12)+(+3) = +15$$

$$(-12)-(+3) = (-12)+(-3) = -15$$

$$(-12)-(-3) = (-12)+(+3) = -9$$

(3) Multiplication and Division

(a) Product of two numbers of like signs equals the positive product of their absolute values.

examples:

$(+12)\times(+3) = +(12\times 3) = +36 \qquad (-12)\times(-3) = +(12\times 3) = +36$

(b) Product of two numbers of unlike signs equals the negative product of their absolute values.

examples:

$(+12)\times(-3) = -(12\times 3) = -36 \qquad (-12)\times(+3) = -(12\times 3) = -36$

(c) Quotient of two numbers of like signs equals the positive quotient of their absolute values.

examples:

$(+12):(+3) = +(12:3) = +4 \qquad (-12):(-3) = +(12:3) = +4$

(d) Quotient of two numbers of unlike signs equals the negative quotient of their absolute values.

examples:

$(+12):(-3) = -(12:3) = -4 \qquad (-12):(+3) = -(12:3) = -4$

1.04 OPERATIONS WITH ZERO

(1) Addition and Subtraction

(a) Sum of any number N and zero equals N.

examples:

$5+0=5 \qquad 0+5=5$

(b) Difference of any number N and zero equals N.

examples:

$5-0=5$ but $0-5=-5$

(2) Multiplication and Division

(a) Product of any number N and zero equals zero.

examples:

$5\times 0=0 \qquad 0\times 5=0$

(b) Quotient of any number N and zero is meaningless, but a quotient of zero and N is zero.

example:

$0:5=0$

1.05 SIMPLE FRACTIONS

(1) Definitions

(a) Fraction is an indicated division. The part of the fraction above the vinculum is called the *numerator N* and the part of the fraction below the vinculum is called the *denominator D.*

example:

$$\frac{N}{D} = \text{fraction}$$

(b) Proper and improper fraction. When $N < D$ (N is less than D), the fraction is termed a proper fraction. When $N > D$ (N is greater than D), the fraction is termed an improper fraction.

examples:

$\frac{2}{3}$ = proper fraction $\qquad \frac{5}{3}$ = improper fraction

(c) When N and D are integers, the fraction is termed a *simple fraction.* When N or D or both are fractions, the fraction is termed a *complex fraction.* A sum of an integer and a fraction is a *mixed number.*

(2) Principles Used in Operations

(a) Enlarging a fraction by multiplying both its numerator and its denominator by the same number (except zero) does not change the value of the fraction.

example:

$$\frac{4}{6}=\frac{4\times 2}{6\times 2}=\frac{8}{12}$$

(b) Reduction of a fraction by dividing both its numerator and its denominator by the same number (except zero) does not change the value of the fraction.

example:

$$\frac{4}{6}=\frac{4:2}{6:2}=\frac{2}{3}$$

(c) Change in sign of both numerator and denominator of a fraction does not change the sign of the fraction.

examples:

$$\frac{-2}{-3}=\frac{2}{3} \qquad \frac{-2}{3}=\frac{2}{-3}=-\frac{2}{3}$$

(d) Change in sign of either numerator or denominator of a fraction changes the sign of the fraction.

examples:

$$\frac{2}{3}\neq\frac{-2}{3} \qquad \frac{2}{3}\neq\frac{2}{-3} \qquad \frac{2}{3}\neq-\frac{2}{3}$$

(e) Lowest common denominator (LCD) of two or more fractions is the lowest common multiple (LCM) (Sec. 1.02–6*c*) of their denominators.

example:

LCD of $\frac{1}{3}$, $\frac{3}{6}$, $\frac{7}{9}$ is 18 and their transformation to this denominator is accomplished by rule (a) of this section as

$$\frac{1}{3}=\frac{1\times 6}{3\times 6}=\frac{6}{18} \qquad \frac{3}{6}=\frac{3\times 3}{6\times 3}=\frac{9}{18} \qquad \frac{7}{9}=\frac{7\times 2}{9\times 2}=\frac{14}{18}$$

(3) Addition and Subtraction

(a) Fractions of common denominators can be added or subtracted.

(b) Sum or difference of two fractions with a common denominator equals a fraction whose numerator is the sum or difference of their numerators and the denominator is their common denominator.

examples:

$$\frac{6}{18}+\frac{9}{18}=\frac{6+9}{18}=\frac{15}{18}=\frac{5}{6} \qquad \frac{6}{18}-\frac{9}{18}=\frac{6-9}{18}=-\frac{3}{18}=-\frac{1}{6}$$

(c) Sum or difference of two fractions of different denominators is found by transforming the given fractions to their lowest common denominator.

examples:

$$\frac{2}{3}+\frac{4}{5}=\frac{2\times 5}{3\times 5}+\frac{4\times 3}{5\times 3}=\frac{10+12}{15}=\frac{22}{15}$$

$$\frac{2}{3}-\frac{4}{5}=\frac{2\times 5}{3\times 5}-\frac{4\times 3}{5\times 3}=\frac{10-12}{15}=-\frac{2}{15}$$

(d) Sum or difference of an integer and a fraction is found by transforming the integer into a fraction of the same denominator.

examples:

$$2+\frac{1}{3}=\frac{2\times 3}{1\times 3}+\frac{1}{3}=\frac{6+1}{3}=\frac{7}{3}$$

$$2-\frac{1}{3}=\frac{2\times 3}{1\times 3}-\frac{1}{3}=\frac{6-1}{3}=\frac{5}{3}$$

(e) Transformation of a mixed number into a fraction is found by applying rule (d) of this section.

example:

$$3\frac{1}{16}=3+\frac{1}{16}=\frac{3\times 16}{1\times 16}+\frac{1}{16}=\frac{48+1}{16}=\frac{49}{16}$$

(4) Multiplication and Division

(a) Product of two or more fractions is the product of their numerators divided by the product of their denominators. In this process, reduction (Sec. 1.05–2*b*) should be used where possible to reduce the amount of multiplication.

examples:

$$\frac{2}{3}\times\frac{5}{7}=\frac{2\times 5}{3\times 7}=\frac{10}{21}\qquad \frac{2}{3}\times\frac{3}{2}=\frac{2\times 3}{3\times 2}=\frac{1}{1}=1$$

$$\frac{2}{3}\times 5=\frac{2\times 5}{3}=\frac{10}{3}\qquad 2\times\frac{5}{7}=\frac{2\times 5}{7}=\frac{10}{7}$$

(b) Reciprocal of a number is 1 divided by this number. Reciprocal of a fraction is a fraction obtained by interchanging the numerator and the denominator in the given fraction.

examples:

Reciprocal of 5 is $\frac{1}{5}$.

Reciprocal of $\frac{2}{3}$ is $\frac{3}{2}$.

(c) Product of a number and its reciprocal equals 1. Product of a fraction and its reciprocal also equals 1.

examples:

$$5\times\frac{1}{5}=\frac{5}{1}\times\frac{1}{5}=\frac{5\times 1}{1\times 5}=\frac{1}{1}=1 \qquad \frac{2}{3}\times\frac{3}{2}=\frac{2\times 3}{3\times 2}=\frac{1}{1}=1$$

(d) Quotient of two or more fractions is the product of the first fraction with the reciprocals of the remaining fractions. In this process reduction (Sec. 1.05–2*b*) should be used where possible to reduce the amount of multiplication.

examples:

$$\frac{2}{3}:\frac{5}{7}=\frac{2}{3}\times\frac{7}{5}=\frac{2\times 7}{3\times 5}=\frac{14}{15} \qquad \frac{2}{3}:\frac{3}{2}=\frac{2}{3}\times\frac{2}{3}=\frac{2\times 2}{3\times 3}=\frac{4}{9}$$

$$\frac{2}{3}:5=\frac{2}{3}\times\frac{1}{5}=\frac{2\times 1}{3\times 5}=\frac{2}{15} \qquad 2:\frac{5}{7}=2\times\frac{7}{5}=\frac{2\times 7}{5}=\frac{14}{5}$$

1.06 MIXED NUMBERS

(1) Definitions

(a) Mixed number is a sum of an integer and a fraction (Sec. 1.05–1*c*).

examples:

$$2+\tfrac{2}{3}=2\tfrac{2}{3} \qquad \tfrac{1}{4}+7=7\tfrac{1}{4}$$

(b) Every mixed number can be converted to a simple improper fraction.

examples:

$$2\frac{2}{3}=\frac{2}{1}+\frac{2}{3}=\frac{2\times 3}{1\times 3}+\frac{2}{3}=\frac{6+2}{3}=\frac{8}{3}$$

$$7\frac{1}{4}=\frac{7}{1}+\frac{1}{4}=\frac{7\times 4}{1\times 4}+\frac{1}{4}=\frac{28+1}{4}=\frac{29}{4}$$

(c) Every simple improper fraction can be converted to a mixed number.

example:

$$\frac{15}{2}=\frac{14+1}{2}=7+\frac{1}{2}=7\frac{1}{2}$$

where 15 was resolved into a highest multiple of the denominator (which is 14) and the remainder (which is 1).

(2) Addition and Subtraction

(a) Sum of mixed numbers is the sum of their integers and of their fractions.

example:

$$2\frac{1}{3}+5\frac{1}{4}=2+5+\frac{1}{3}+\frac{1}{4}=7+\frac{4+3}{12}=7\frac{7}{12}$$

(b) Difference of two mixed numbers is the difference of their integers and of their fractions.

example:

$$5\frac{2}{3}-2\frac{1}{4}=5-2+\frac{2}{3}-\frac{1}{4}=3+\frac{8-3}{12}=3\frac{5}{12}$$

(c) In general however it is more convenient to convert the mixed numbers to improper fractions and apply rules of Sec. 1.05–3*b* and *c*.

examples:

$$2\frac{1}{3}+5\frac{1}{4}=\frac{7}{3}+\frac{21}{4}=\frac{28+63}{12}=\frac{91}{12}=7\frac{7}{12}$$

$$5\frac{2}{3}-2\frac{1}{4}=\frac{17}{3}-\frac{9}{4}=\frac{68-27}{12}=\frac{41}{12}=3\frac{5}{12}$$

(3) Multiplication and Division

(a) Product of two mixed numbers is the product of their improper fraction equivalents.

example:

$$2\frac{1}{3}\times 5\frac{1}{4}=\frac{7}{3}\times\frac{21}{4}=\frac{7}{1}\times\frac{7}{4}=\frac{49}{4}=12\frac{1}{4}$$

(b) Quotient of two mixed numbers is the quotient of their improper fraction equivalents.

example:

$$2\frac{1}{3}:5\frac{1}{4}=\frac{7}{3}:\frac{21}{4}=\frac{7}{3}\times\frac{4}{21}=\frac{1}{3}\times\frac{4}{3}=\frac{4}{9}$$

1.07 COMPLEX, COMPOUND, AND CONTINUED FRACTIONS

(1) Definitions

(a) Complex fraction is a fraction whose numerator or denominator or both are fractions.

examples:

$\frac{\frac{2}{3}}{5}, \frac{7}{\frac{11}{13}}, \frac{\frac{3}{5}}{\frac{7}{11}}$ are complex fractions.

(b) Compound fraction is a fraction whose numerator, or denominator, or both are mixed numbers.

examples:

$\frac{1\frac{2}{3}}{5}, \frac{7}{2\frac{11}{13}}, \frac{2\frac{3}{5}}{13\frac{7}{11}}$ are compound fractions.

(c) Continued fraction is an integer plus fraction whose denominator is an integer plus fraction and so on.

example:

$2+\cfrac{1}{3+\cfrac{1}{5+\cfrac{1}{7+\frac{1}{2}}}}$ is a continued fraction.

(2) Reductions

(a) Every complex, compound, and continued fraction can be reduced to a simple fraction.

(b) Reduction of a complex fraction is the quotient of its numerator and of its denominator.

examples:

$$\frac{\frac{2}{3}}{5} = \frac{\frac{2}{3}}{\frac{5}{1}} = \frac{2}{3} : \frac{5}{1} = \frac{2}{3} \times \frac{1}{5} = \frac{2}{15} \qquad \frac{7}{\frac{11}{13}} = \frac{\frac{7}{1}}{\frac{11}{13}} = \frac{7}{1} : \frac{11}{13} = \frac{7}{1} \times \frac{13}{11} = \frac{91}{11}$$

$$\frac{\frac{3}{5}}{\frac{7}{11}} = \frac{3}{5} : \frac{7}{11} = \frac{3}{5} \times \frac{11}{7} = \frac{33}{35}$$

(c) Reduction of a compound fraction is the quotient of its numerator and of its denominator, each converted to an improper fraction.

examples:

$$\frac{1\frac{2}{3}}{5} = \frac{\frac{5}{3}}{\frac{5}{1}} = \frac{5}{3} : \frac{5}{1} = \frac{5}{3} \times \frac{1}{5} = \frac{1}{3} \qquad \frac{7}{2\frac{11}{13}} = \frac{\frac{7}{1}}{\frac{37}{13}} = \frac{7}{1} : \frac{37}{13} = \frac{7}{1} \times \frac{13}{37} = \frac{91}{37}$$

$$\frac{2\frac{3}{5}}{13\frac{7}{11}} = \frac{\frac{13}{5}}{\frac{150}{11}} = \frac{13}{5} : \frac{150}{11} = \frac{13}{5} \times \frac{11}{150} = \frac{143}{750}$$

(d) Reduction of a continued fraction is found by successive reduction of compound fractions in its denominator.

example:

$$2+\cfrac{1}{3+\cfrac{1}{7+\frac{1}{2}}} = 2+\cfrac{1}{3+\cfrac{1}{\frac{15}{2}}} = 2+\cfrac{1}{3+\frac{2}{15}} = 2+\cfrac{1}{\frac{47}{15}} = 2+\frac{15}{47} = \frac{109}{47}$$

1.08 AGGREGATIONS

(1) Definitions

(a) Parentheses ( ), **brackets** [ ], **braces** { }, and the **vinculum** —— are symbols of aggregation introduced to designate groups of terms which should be treated as one quantity.

(b) Plus sign in front of a quantity indicates that the quantity is to be multiplied by +1.

example:

$$+(7-10+3-5) = +(10-15) = 10-15 = -5$$

(c) Minus sign in front of a quantity indicates that the quantity is to be multiplied by −1.

example:

$$-(7-10+3-5) = -(10-15) = -10+15 = +5$$

where + in front of 5 is usually omitted.

(d) Multiplication sign between two quantities indicates that the first quantity is to be multiplied by the second one and vice versa.

examples:

$$(7-8+3-5)\times(3-6+5) = -3\times 2 = -6$$

$$(2-8)\times(1+\tfrac{1}{3}) = -6\times\tfrac{4}{3} = -8$$

(e) Division sign between two quantities indicates that the first quantity is to be divided by the second one but not vice versa.

examples:

$$(7-8+3-5):(3-6+5) = -3:2 = -\tfrac{3}{2}$$

$$(2-8):(1+\tfrac{1}{3}) = -6:\tfrac{4}{3} = -\tfrac{9}{2}$$

(2) Composite Aggregations

(a) Composite aggregation is a group of terms, some or all of which are aggregations. Usually parentheses ( ) are used within brackets [ ] and brackets within braces { }.

(b) Process of removing symbols of complex aggregation begins with the innermost symbols (parentheses) first and terminates with removal of the outermost symbols (braces).

examples:

$$-[2(5-3)+3(-7-5)] = -[4-36] = +32$$

$$-\{-3[4-5(7-8)]+1\} = -\{-3[4+5]+1\} = -\{-27+1\} = +26$$

1.09 EXPONENTS

(1) Definitions

(a) Positive integral exponent. If n is a positive integer and A is a signed number, then

$$A^n = \underbrace{A \times A \times A \times \cdots \times A}_{n \text{ times}}$$

where A^n is the *nth power* of A, n is the *exponent,* and A is the *base.*

If n is an *even number* and A is a *positive number,* then A^n is a *positive number.*
If n is an *even number* and A is a *negative number,* then A^n is a *positive number.*
If n is an *odd number* and A is a *positive number,* then A^n is a *positive number.*
If n is an *odd number* and A is a *negative number,* then A^n is a *negative number*

If $n = 0$, then $A^n = 1$; if $n = 1$, then $A^1 = A$; if $n = 2$, then A^2 is called the square of A; and if $n = 3$, then A^3 is called the cube of A.

examples:

$(+2)^3 = (+2) \times (+2) \times (+2) = +8$ $\qquad$ $(+\frac{2}{3})^3 = (+\frac{2}{3}) \times (+\frac{2}{3}) \times (+\frac{2}{3}) = +\frac{8}{27}$

$(-2)^3 = (-2) \times (-2) \times (-2) = -8$ $\qquad$ $(-\frac{2}{3})^3 = (-\frac{2}{3}) \times (-\frac{2}{3}) \times (-\frac{2}{3}) = -\frac{8}{27}$

$(-2)^4 = (-2) \times (-2) \times (-2) \times (-2) = +16$ $\qquad$ $(-\frac{2}{3})^4 = (-\frac{2}{3}) \times (-\frac{2}{3}) \times (-\frac{2}{3}) \times (-\frac{2}{3}) = +\frac{16}{81}$

(b) Negative integral exponent. If $-n$ is a negative integer and A is a signed number, then

$$A^{-n} = \underbrace{\frac{1}{A} \times \frac{1}{A} \times \frac{1}{A} \times \cdots \times \frac{1}{A}}_{n \text{ times}} = \frac{1}{A^n}$$

where A^{-n} is the reciprocal of A^n.

(c) Sign of the exponent does not affect the sign of the base.

If $n = -1$, then $A^{-1} = 1/A$ and if $n = 0$, then $A^{-0} = 1$.

examples:

$(+2)^{-3} = (+\frac{1}{2}) \times (+\frac{1}{2}) \times (+\frac{1}{2}) = +\frac{1}{8}$ $\qquad$ $(+\frac{2}{3})^{-3} = (+\frac{3}{2}) \times (+\frac{3}{2}) \times (+\frac{3}{2}) = +\frac{27}{8}$

$(-2)^{-3} = (-\frac{1}{2}) \times (-\frac{1}{2}) \times (-\frac{1}{2}) = -\frac{1}{8}$ $\qquad$ $(-\frac{2}{3})^{-3} = (-\frac{3}{2}) \times (-\frac{3}{2}) \times (-\frac{3}{2}) = -\frac{27}{8}$

$(-2)^{-4} = (-\frac{1}{2}) \times (-\frac{1}{2}) \times (-\frac{1}{2}) \times (-\frac{1}{2}) = +\frac{1}{16}$ $\qquad$ $(-\frac{2}{3})^{-4} = (-\frac{3}{2}) \times (-\frac{3}{2}) \times (-\frac{3}{2}) \times (-\frac{3}{2}) = +\frac{81}{16}$

(2) Laws of Exponents

(a) nth power of the product of two signed numbers A and B is the product of their nth powers and vice versa.

$$(A \times B)^n = A^n \times B^n$$

examples—positive exponent:

$$[(+2)\times(-3)]^2 = [(+2)\times(+2)]\times[(-3)\times(-3)] = +36$$

$$[(-2)\times(-3)]^2 = [(-2)\times(-2)]\times[(-3)\times(-3)] = +36$$

$$[(+2)\times(-3)]^3 = [(+2)\times(+2)\times(+2)]\times[(-3)\times(-3)\times(-3)] = -216$$

$$[(-2)\times(-3)]^3 = [(-2)\times(-2)\times(-2)]\times[(-3)\times(-3)\times(-3)] = +216$$

examples—negative exponent:

$$[(+2)\times(-3)]^{-2} = \frac{1}{[(+2)\times(+2)]\times[(-3)\times(-3)]} = +\frac{1}{36} \qquad [(+2)\times(-3)]^{-3} = -\frac{1}{216}$$

$$[(-2)\times(-3)]^{-2} = \frac{1}{[(-2)\times(-2)]\times[(-3)\times(-3)]} = +\frac{1}{36} \qquad [(-2)\times(-3)]^{-3} = +\frac{1}{216}$$

(b) nth power of the quotient of two signed numbers A and B is the quotient of their nth powers and vice versa.

$$\left(\frac{A}{B}\right)^n = \frac{A^n}{B^n}$$

examples—positive exponents:

$$[(+2):(-3)]^2 = \frac{(+2)\times(+2)}{(-3)\times(-3)} = +\frac{4}{9} \qquad [(+2):(-3)]^3 = -\frac{8}{27}$$

$$[(-2):(-3)]^2 = \frac{(-2)\times(-2)}{(-3)\times(-3)} = +\frac{4}{9} \qquad [(-2):(-3)]^3 = +\frac{8}{27}$$

examples—negative exponents:

$$[(+2):(-3)]^{-2} = \frac{(-3)\times(-3)}{(+2)\times(+2)} = +\frac{9}{4} \qquad [(+2):(-3)]^{-3} = -\frac{27}{8}$$

$$[(-2):(-3)]^{-2} = \frac{(-3)\times(-3)}{(-2)\times(-2)} = +\frac{9}{4} \qquad [(-2):(-3)]^{-3} = +\frac{27}{8}$$

(c) Product of A^m **and** A^n equals the $(m+n)$th power of A.

$$A^m \times A^n = A^{m+n}$$

where m and n are signed integers.

examples:

$+3^2 \times +3^3 = +(+3)^5 = +243 \qquad +3^2 \times (-3)^3 = -(+3)^5 = -243$

$(-3)^2 \times +3^3 = +(+3)^5 = +243 \qquad (-3)^2 \times (-3)^3 = -(+3)^5 = -243$

$+3^2 \times +3^{-3} = +(+3)^{-1} = +\frac{1}{3} \qquad +3^2 \times (-3)^{-3} = -(+3)^{-1} = -\frac{1}{3}$

$(-3)^2 \times +3^{-3} = +(+3)^{-1} = +\frac{1}{3} \qquad (-3)^2 \times (-3)^3 = -(+3)^{-1} = -\frac{1}{3}$

(d) Quotient of A^m **and** A^n equals the $(m-n)$th power of A.

$$A^m : A^n = \frac{A^m}{A^n} = A^{m-n}$$

where m and n are signed integers.

examples:

$\frac{+3^2}{+3^3} = +(+3)^{-1} = +\frac{1}{3} \qquad \frac{+3^2}{(-3)^3} = -(+3)^{-1} = -\frac{1}{3}$

$\frac{(-3)^2}{+3^3} = +(+3)^{-1} = +\frac{1}{3} \qquad \frac{(-3)^2}{(-3)^3} = -(+3)^{-1} = -\frac{1}{3}$

$\frac{+3^2}{+3^{-3}} = +(+3)^5 = +243 \qquad \frac{+3^{-2}}{+3^{-3}} = +(+3)^1 = +3$

$\frac{+3^2}{(-3)^{-3}} = -(+3)^5 = -243 \qquad \frac{+3^{-2}}{(-3)^{-3}} = -(+3)^1 = -3$

(e) n**th power of** A^m is the $(m \times n)$th power of A.

$$(A^m)^n = (A^n)^m = A^{m \times n}$$

where m and n are signed integers.

examples:

$(+3^2)^2 = +3^4 = +81 \qquad (-3^2)^2 = (-3)^4 = +81$

$(+3^2)^{-2} = +3^{-4} = +\frac{1}{81} \qquad (-3^2)^{-2} = (-3)^{-4} = +\frac{1}{81}$

$(-3^3)^3 = (-3)^9 = -19{,}683 \qquad (-3^3)^{-3} = (-3)^{-9} = -\frac{1}{19{,}683}$

1.10 RADICALS

(1) Definitions

(a) Radical r is the nth root of A where A is the nth power of r.

$$r = \sqrt[n]{A} = A^{1/n}$$

where

$$A = \underbrace{r \times r \times r \times \cdots \times r}_{n \text{ times}} = r^n$$

where A is the *base* (radicand) and n is the *index of the radical* (signed integer).

(b) Square root. When $n = 2$, the radical is called the square root of A.

$$r = \sqrt{A} = A^{1/2}$$

and the index 2 is omitted in $\sqrt{A}$.

(c) Positive and negative radical. The radical r is positive if A is positive and is negative when n is odd and A negative. When n is even and A is negative then r is imaginary. Only real radicals are considered in this section.

examples:

$$\sqrt{+9} = \sqrt{(+3)\times(+3)} = +3 \qquad \sqrt{+100} = \sqrt{(+10)\times(+10)} = +10$$

$$\sqrt[3]{+8} = \sqrt[3]{(+2)\times(+2)\times(+2)} = +2 \qquad \sqrt[3]{+1{,}000} = \sqrt[3]{(+10)\times(+10)\times(+10)} = +10$$

$$\sqrt[3]{-8} = \sqrt[3]{(-2)\times(-2)\times(-2)} = -2 \qquad \sqrt[3]{-1{,}000} = \sqrt[3]{(-10)\times(-10)\times(-10)} = -10$$

(d) Rational numbers are real numbers that can be expressed in the form p/q where p and q are integers (q can be 1).

examples:

$5, \frac{1}{2}, 1\frac{1}{6}, \sqrt{4}, \sqrt[3]{1{,}000}$ are rational numbers.

(e) Irrational numbers are real numbers which cannot be expressed in the form p/q where p and q are integers.

examples:

$\sqrt{2}, \sqrt{3}, \sqrt{10}, \sqrt{\frac{1}{2}}, \sqrt[3]{5}$ are irrational numbers.

(2) Laws of Indices

(a) ***n*th root of the product of two signed numbers** A and B is the product of their nth roots and vice versa.

$$\sqrt[n]{A \times B} = \sqrt[n]{A} \times \sqrt[n]{B} = A^{1/n} \times B^{1/n}$$

examples:

$$\sqrt{+4 \times +9} = \sqrt{+2 \times +2} \times \sqrt{+3 \times +3} = +2 \times +3 = +6$$

$$2\sqrt[3]{+4} \times 5\sqrt[3]{+16} = (2 \times 5)\sqrt[3]{+4 \times +16} = 10\sqrt[3]{+64} = +40$$

(b) ***n*th root of the quotient of two signed numbers** A and B is the quotient of their nth roots and vice versa.

$$\sqrt[n]{\frac{A}{B}} = \frac{\sqrt[n]{A}}{\sqrt[n]{B}} = \frac{A^{1/n}}{B^{1/n}}$$

examples:

$$\sqrt[3]{\frac{-27}{+64}} = \frac{\sqrt[3]{-27}}{\sqrt[3]{+64}} = -\frac{3}{4} \qquad \sqrt[4]{\frac{+81}{+16}} = \frac{\sqrt[4]{+81}}{\sqrt[4]{+16}} = +\frac{3}{2}$$

(c) ***n*th power of the *m*th root of** A which is a signed number equals the mth root of the nth power of A.

$$(\sqrt[m]{A})^n = \sqrt[m]{A^n} = A^{n/m}$$

and if $m = n$,

$$(\sqrt[m]{A})^m = \sqrt[m]{A^m} = A$$

examples:

$$\sqrt[3]{27^2} = (\sqrt[3]{27})^2 = +3^2 = +9$$

$$\sqrt[3]{\left(+\frac{125}{216}\right)^2} = \left(\sqrt[3]{+\frac{125}{216}}\right)^2 = \left(+\frac{5}{6}\right)^2 = +\frac{25}{36}$$

(d) ***m*th root of the *n*th root** of A which is a signed number is the $(m \times n)$th root of A.

$$\sqrt[m]{\sqrt[n]{A}} = \sqrt[n \times m]{A} = A^{1/(m \times n)}$$

examples:

$$\sqrt[3]{\sqrt[4]{+5}} = \sqrt[12]{+5} \qquad \sqrt{\sqrt{+\tfrac{1}{2}}} = \sqrt[4]{+\tfrac{1}{2}}$$

2
ALGEBRA

2.01 DEFINITIONS AND NOTATIONS

(1) Definitions

(a) Algebra is a systematic investigation of general numbers (algebraic terms) and their relationships. *General numbers* are letter symbols (A, B, $C, \ldots$; a, b, $c, \ldots$; α, β, $\gamma, \ldots$) representing constant or variable quantities.

(b) Algebraic term is a product or quotient of one or more general numbers and of a numerical factor (which can be any natural number) with a prefixed sign (plus sign is frequently omitted).

examples:

$2a, 3x^2, -5x^3/y, -\sqrt{ab}, e^x$ are algebraic terms.

(c) Algebraic expression is a collection of one or more algebraic terms connected by the symbols of relationship (Sec. 1.01–2), and/or signs of aggregations (Sec. 1.01–3) and/or signs of operation (Sec. 1.01–4).

examples:

$ax^2 + bx + c,\ (a+b)(c+d),\ ax+b,\ a > b,\ b < c,\ c = d,\ d \neq e,$

$\dfrac{a+bx}{c+dx}, \left(\dfrac{a+bx}{c+dx}\right)^n, \sqrt[m]{\dfrac{a+bx}{c+dx}}$ are algebraic expressions.

(d) Multinomial is an algebraic expression consisting of more than one term. Special cases of multinomials are the *binomial* (two terms), the *trinomial* (three terms), and so on. A *monomial* consists of one term only.

examples:

$ax^2 + bx + c,\ ax + by + cz, \ldots$ are trinomials.
$ax + b,\ ax + by,\ 3x^2y + 2xy, \ldots$ are binomials, but they are also multinomials.

(e) Polynomial is a multinomial in which every term is integral and rational in literals.

examples:

$5x^5 + 3x^3y + x^3y,\ 4ab + 3cd + 8ef$ are polynomials, but
$5/x^5 + 3y/x^4 + x^3/y,\ 4a/b + 3c/d + 8e/f,\ 4\sqrt{ab} + 3\sqrt{cd} + 8\sqrt{ef}$ are not polynomials.

(2) Algebraic Operations

(a) Algebraic transformations involve a finite number of binary operations, governed by algebraic laws (Sec. 2.01–3) and rules of signs (Sec. 2.01–4).

(b) Four basic algebraic operations are: *addition* (Sec. 2.02–1) and its inverse, *subtraction* (Sec. 2.02–2); and *multiplication* (Sec. 2.02–3) and its inverse, *division* (Sec. 2.02–4).

(c) Three higher algebraic operations are: *involution* (Sec. 2.10–1) and its two inverses, *evolution* (Sec. 2.11–1) and finding the logarithm (Sec. 2.13–1).

(3) Algebraic Laws

Commutative law:

$a + b = b + a$

$ab = ba$

Associative law:

$a + (b + c) = (a + b) + c$

$a(bc) = (ab)c$

Distributive law:

$a(b + c) = ab + bc$

$(a + b)c = ac + bc$

Division law:

If $ab = 0$, then $a = 0$ and/or $b = 0$.

(4) Rules of Signs

Summation:	*Multiplication:*	*Division:*
$a + (+b) = a + b$	$(+a)(+b) = +ab$	$(+a):(+b) = +(a:b)$
$a + (-b) = a - b$	$(+a)(-b) = -ab$	$(+a):(-b) = -(a:b)$
$a - (+b) = a - b$	$(-a)(+b) = -ab$	$(-a):(+b) = -(a:b)$
$a - (-b) = a + b$	$(-a)(-b) = +ab$	$(-a):(-b) = +(a:b)$

2.02 ADDITION AND SUBTRACTION

(1) Addition

(a) Addition is the operation of finding the sum of two or several terms. Only like terms can be added.

example:

$$2a + 4b + 6c + 7a + 8b = 9a + 12b + 6c$$

(b) Order of terms in addition may be changed without affecting the sum (commutative law, Sec. 2.01–3).

example:

$$a + b + c = b + c + a = c + a + b = \cdots$$

(c) Grouping of terms in addition may be changed without affecting the sum (associative law, Sec. 2.01–3).

example:

$$a + (b + c) = (a + b) + c = (a + c) + b$$

(2) Subtraction

(a) Subtraction is the operation of finding the difference of two terms or of two quantities.

example:

$$(2a + 4b + 6c) - (7a + 8b) = 2a + 4b + 6c - 7a - 8b = -5a - 4b + 6c$$

(b) Difference of two equal terms is zero.

example:

$$5x - 5x = 0$$

2.03 MULTIPLICATION

(1) Basic Products

(a) Multiplication is the operation of finding the product of two or more terms.

example:

$$2ab \cdot 3bc \cdot 4cd = 24ab^2c^2d$$

where $\cdot$ is the algebraic symbol of multiplication, which shall be omitted hereafter.

(b) Order of terms in multiplication may be changed without affecting the product (commutative law, Sec. 2.01–3).

example:

$$(2ab)(3bc)(4cd) = (3bc)(4cd)(2ab) = (4cd)(2ab)(3bc)$$

(c) Grouping of terms in multiplication may be changed without affecting the product (associative law, Sec. 2.01–3).

example:

$$[(2ab)(3bc)](4cd) = (2ab)[(3bc)(4cd)]$$

(2) Special Products

(a) Multiplication by a factor.

$$(a - b + c - d)(+m) = am - bm + cm - dm$$

$$(a - b + c - d)(-m) = -am + bm - cm + dm$$

(b) Products of binomials.

$$(a + b)(c + d) = ac + bc + ad + bd$$

$$(a + b)(c - d) = ac + bc - ad - bd$$

$$(a - b)(c + d) = ac - bc + ad - bd$$

$$(a - b)(c - d) = ac - bc - ad + bd$$

$$(a + m)(a + n) = a^2 + (m + n)a + mn$$

$$(a + m)(a - n) = a^2 + (m - n)a - mn$$

$$(a + b)(a + b) = a^2 + 2ab + b^2$$

$$(a + b)(a - b) = a^2 - b^2$$

$$(a - b)(a - b) = a^2 - 2ab + b^2$$

2.04 HIGHEST COMMON FACTOR AND LOWEST COMMON MULTIPLE

(1) Definitions

(a) Factors of a given algebraic expression are two or more algebraic expressions, the product of which is the given expression.

(b) Common factor of two or more expressions is a factor of each of these expressions.

examples:

$ab + ad = a(b + d)$ a is the common factor

$a^2b^2c + ab^2c^2 = ab^2c(a + c)$ ab^2c is the common factor

(c) Prime factor is an algebraic expression divisible by no other expression than itself and 1.

example:

$a(b + c)$ a and $(b + c)$ are the prime factors

(d) Multiple of a factor is an algebraic expression divisible by this factor.

(e) Common multiple of two or more factors is an expression divisible by each of these factors.

(f) Highest common factor (HCF) of two or more expressions is the expression of the highest degree and largest numerical coefficients which is a factor of each of these expressions.

example:

$4(a + b)^8$, $2(a + b)^5$, $4(a + b)^3$ $2(a + b)^3$ is the HCF

(g) Lowest common multiple (LCM) of two or more factors is an algebraic expression of the lowest degree and smallest numerical coefficients divisible by each of these factors.

example:

$4(a + b)^8$, $2(a + b)^5$, $4(a + b)^3$ $4(a + b)^8$ is the LCM

(2) Factoring

(a) Factoring into prime factors. If the given algebraic expressions can be factored into prime factors, their HCF and LCM can be determined at once from these factors.

(b) Highest common factor is the product obtained by taking each factor to the lowest power to which it occurs in any of these expressions.

(c) Lowest common multiple is the product obtained by taking each factor to the highest power to which it occurs in any of these expressions.

(d) The process of factoring an algebraic expression is also known as *decomposition.*

2.05 DECOMPOSITION OF BINOMIALS AND TRINOMIALS

(1) General Cases

(a) Decomposition of binomials, zero remainder ($k = 1, 2, 3, \ldots$).

$$a^{2k} - b^{2k} = (a \mp b)(a^{2k-1} \pm a^{2k-2}b + a^{2k-3}b^2 \pm \cdots \pm b^{2k-1})$$

$$a^{2k+1} - b^{2k+1} = (a - b)(a^{2k} + a^{2k-1}b + a^{2k-2}b^2 + \cdots + b^{2k})$$

$$a^{2k+1} + b^{2k+1} = (a + b)(a^{2k} - a^{2k-1}b + a^{2k-2}b^2 - \cdots + b^{2k})$$

(b) Decomposition of binomials, with remainder ($k = 1, 2, 3, \ldots$).

$$a^{2k} + b^{2k} = (a \mp b)(a^{2k-1} \pm a^{2k-2}b + a^{2k-3}b^2 \pm \cdots \pm b^{2k-1}) + 2b^{2k}$$

$$a^{2k+1} - b^{2k+1} = (a + b)(a^{2k} - a^{2k-1}b + a^{2k-2}b^2 - \cdots + b^{2k}) - 2b^{2k+1}$$

$$a^{2k+1} + b^{2k+1} = (a - b)(a^{2k} + a^{2k-1}b + a^{2k-2}b^2 + \cdots + b^{2k}) + 2b^{2k+1}$$

(c) Decomposition of binomials into products of binomials.

$$a^{2k} - b^{2k} = (a^k + b^k)(a^k - b^k)$$

$$a^{2k} + b^{2k} = (a^k + b^k)(a^k - b^k) + 2b^{2k}$$

(d) Decomposition of trinomials.

$$x^2 + px + q = (x + m)(x + n)$$

where $p = m + n$ and $q = mn$.

example:

$$x^2 - 5x - 14 = (x - 7)(x + 2)$$

(2) Special Cases

(a) Decomposition of even-power binomials ($a^{2k} - a^{2k}$).

$$a^2 - b^2 = (a - b)(a + b)$$

$$a^4 - b^4 = (a - b)(a^3 + a^2b + ab^2 + b^3)$$

$$a^6 - b^6 = (a - b)(a^5 + a^4b + a^3b^2 + a^2b^3 + ab^4 + b^5)$$

$$a^4 - b^4 = (a + b)(a^3 - a^2b + ab^2 - b^3)$$

$$a^6 - b^6 = (a + b)(a^5 - a^4b + a^3b^2 - a^2b^3 + ab^4 - b^5)$$

(b) Decomposition of even-power binomials ($a^{2k} + b^{2k}$).

$$a^2 + b^2 = (a + \sqrt{2ab} + b)(a - \sqrt{2ab} + b)$$

$$a^4 + b^4 = (a^2 + ab\sqrt{2} + b^2)(a^2 - ab\sqrt{2} + b)$$

$$a^6 + b^6 = (a^3 + \sqrt{2a^3b^3} + b^3)(a^3 - \sqrt{2a^3b^3} + b^3)$$

(c) Decomposition of odd-power binomials $(a^{2k+1} - b^{2k+1})$.

$$a^3 - b^3 = (a - b)(a^2 + ab + b^2)$$

$$a^5 - b^5 = (a - b)(a^4 + a^3 b + a^2 b^2 + ab^3 + b^4)$$

(d) Decomposition of odd-power binomials $(a^{2k+1} + b^{2k+1})$.

$$a^3 + b^3 = (a + b)(a^2 - ab + b^2)$$

$$a^5 + b^5 = (a + b)(a^4 - a^3 b + a^2 b^2 - ab^3 + b^4)$$

2.06 DIVISION

(1) Basic Quotients

(a) Division is the operation of finding the quotient of two terms.

examples:

$$6a^2 b : 3ab = 2a$$

$$8(a^2 + ab) : 4a = 2(a + b)$$

(b) Quotient of two equal terms is 1.

example:

$$3abc : 3abc = 1$$

(2) General Quotients

(a) Binomials, zero remainder $(k = 1, 2, 3, \ldots)$.

$$(a^{2k} - b^{2k}) : (a \mp b) = a^{2k-1} \pm a^{2k-2} b + a^{2k-3} b^3 \pm \cdots \pm b^{2k-1}$$

$$(a^{2k+1} - b^{2k+1}) : (a - b) = a^{2k} + a^{2k-1} b + a^{2k-2} b^2 + \cdots + b^{2k}$$

$$(a^{2k+1} + b^{2k+1}) : (a + b) = a^{2k} - a^{2k-1} b + a^{2k+2} b^2 - \cdots + b^{2k}$$

$$(a^{2k} - b^{2k}) : (a^k + b^k) = a^k - b^k$$

$$(a^{2k} - b^{2k}) : (a^k - b^k) = a^k + b^k$$

(b) Binomials, with remainder $(k = 1, 2, 3, \ldots)$.

$$(a^{2k} + b^{2k}) : (a \mp b) = a^{2k-1} \pm a^{2k-2} b + a^{2k-3} b^2 \pm \cdots \pm b^{2k-1} + \frac{2b^{2k}}{a \mp b}$$

$$(a^{2k+1} - b^{2k+1}) : (a + b) = a^{2k} - a^{2k-1} b + a^{2k-2} b^2 - \cdots + b^{2k} - \frac{2b^{2k+1}}{a + b}$$

$$(a^{2k+1} + b^{2k+1}) : (a - b) = a^{2k} + a^{2k-1} b + a^{2k-2} b^2 + \cdots + b^{2k} + \frac{2b^{2k+1}}{a + b}$$

(3) Special Quotients

(a) Even-power binomials.

$$(a^2 - b^2) : (a + b) = a - b$$
$$(a^2 - b^2) : (a - b) = a + b$$

$$(a^4 - b^4) : (a + b) = a^3 - a^2b + ab^2 - b^3$$
$$(a^4 - b^4) : (a - b) = a^3 + a^2b + ab^2 + b^3$$

$$(a^6 - b^6) : (a + b) = a^5 - a^4b + a^3b^2 - a^2b^3 + ab^4 - b^5$$
$$(a^6 - b^6) : (a - b) = a^5 + a^4b + a^3b^2 + a^3b^2 + a^2b^3 + ab^4 + b^5$$

$$(a^4 - b^4) : (a^2 + b^2) = a^2 - b^2$$
$$(a^4 - b^4) : (a^2 - b^2) = a^2 + b^2$$

$$(a^6 - b^6) : (a^3 + b^3) = a^3 - b^3$$
$$(a^6 - b^6) : (a^3 - b^3) = a^3 + b^3$$

(b) Odd-power binomials.

$$(a^3 - b^3) : (a - b) = a^2 + ab + b^2$$
$$(a^3 + b^3) : (a + b) = a^2 - ab + b^2$$

$$(a^5 - b^5) : (a - b) = a^4 + a^3b + a^2b^2 + ab^3 + b^4$$
$$(a^5 + b^5) : (a + b) = a^4 - a^3b + a^2b^2 - ab^3 + b^4$$

2.07 SIMPLE FRACTIONS

(1) Definitions

(a) Algebraic fraction is an indicated division of two algebraic expressions called again the *numerator* N and *denominator* D (Sec. 1.05–1).

examples:

$\dfrac{2ab}{3b}, \dfrac{a^2 - b^2}{a + b}, \dfrac{3 + x}{x^2 + 6x + 8}$ are rational algebraic fractions.

(b) Simple and complex fractions. When N and D are integral expressions, the fractions are termed *simple fractions*. When N or D or both are simple fractions, the fraction is termed a *complex fraction*.

examples:

$\dfrac{1}{a}, \dfrac{a}{b}, \dfrac{a^2 + b^2}{c + d}$ are simple fractions. $\dfrac{1}{a/b}, \dfrac{a/b}{c}, \dfrac{(a^2 + b^2)/c}{(d + e)/f^2}$ are complex fractions.

(2) Principles Used in Operations

(a) Enlarging of fraction by multiplying both its numerator and its denominator by the same expression (except zero) does not change the value of the fraction (Sec. 1.05–2*a*).

examples:

$$\frac{a}{b} = \frac{am}{bm} \qquad \frac{x+2}{x+3} = \frac{(x+2)(x+4)}{(x+3)(x+4)} = \frac{x^2+6x+8}{x^2+7x+12}$$

(b) Reduction of fraction by dividing both its numerator and its denominator by the same expression (except zero) does not change the value of the fraction (Sec. 1.05–2*b*).

examples:

$$\frac{a\cancel{m}}{b\cancel{m}} = \frac{a}{b} \qquad \frac{x^2+6x+8}{x^2+7x+12} = \frac{(x+2)\cancel{(x+4)}}{(x+3)\cancel{(x+4)}} = \frac{x+2}{x+3}$$

where the diagonal line across a term in the numerator and across an identical term in the denominator indicates the canceling of common factors; i.e., the numerator and the denominator are divided by this factor.

(c) Change in sign of both numerator and denominator of a fraction does not change the sign of the fraction (Sec. 1.05–2*c*).

examples:

$$\frac{a}{b} = \frac{-a}{-b} \qquad \frac{-a}{b} = \frac{a}{-b} = -\frac{a}{b} \qquad \frac{-a}{c-d} = \frac{a}{d-c} = -\frac{a}{c-d}$$

(d) Change in sign of either numerator or denominator changes the sign of the fraction (Sec. 1.05–2*d*).

examples:

$$\frac{a}{b} \neq \frac{-a}{b} \qquad \frac{a}{b} \neq \frac{a}{-b} \qquad \frac{a}{b} \neq -\frac{a}{b} \qquad \frac{a-b}{c-d} \neq -\frac{a-b}{c-d}$$

(e) Lowest common denominator (LCD) of two or more fractions is the lowest common multiple (Sec. 2.04–1*g*) of their denominators (Sec. 1.05–2*e*).

example:

LCD of $\frac{1}{a}, \frac{1}{(x+2)}, \frac{1}{x^2+5x+6}$ is $a(x+2)(x+3)$.

Their transformation to this denominator accomplished by rule (a) of this section is, since $x^2+5x+6 = (x+2)(x+3)$,

$$\frac{1}{a} = \frac{(x+2)(x+3)}{a(x+2)(x+3)} \qquad \frac{1}{x+2} = \frac{a(x+3)}{a(x+2)(x+3)} \qquad \frac{1}{x^2+5x+6} = \frac{a}{a(x+2)(x+3)}$$

(3) Addition and Subtraction

(a) Algebraic fractions of common denominators can be added or subtracted (Sec. 1.05–3*a*).

(b) Sum or difference of two algebraic fractions with a *common denominator* equals a fraction whose numerator is the sum or difference of their numerators and whose denominator is their common denominator (Sec. 1.05–3*b*).

example:

$$\frac{a}{m} \pm \frac{b}{m} = \frac{a \pm b}{m}$$

(c) Sum and difference of two algebraic fractions of different denominators is found by transforming the given fractions to their LCD (Sec. 1.05–3*c*).

examples:

$$\frac{a}{m} \pm \frac{b}{n} = \frac{an \pm bm}{mn} \qquad \frac{a}{m} \pm \frac{b}{mn} = \frac{an \pm b}{mn}$$

$$\frac{a}{m} \pm \frac{b}{n} \pm \frac{c}{p} = \frac{anp \pm bmp \pm cmn}{mnp}$$

$$\frac{a}{1+x} + \frac{b}{1-x} + \frac{c}{1-x^2} = \frac{a - ax + b + bx + c}{1-x^2} = \frac{a + b + c - x(a - b)}{1-x^2}$$

(d) Sum and difference of an algebraic expression and of an algebraic fraction is found by transforming the algebraic expression to a fraction of the same denominator.

example:

$$a \pm \frac{b}{m} = \frac{am}{m} \pm \frac{b}{m} = \frac{am \pm b}{m}$$

(4) Multiplication and Division

(a) Product of two or more algebraic fractions is the product of their numerators divided by the product of their denominators (Sec. 1.05–4*a*).

examples:

$$\frac{a}{m}\frac{b}{n} = \frac{ab}{mn} \qquad \frac{x+1}{x-1}\frac{x+1}{x+2} = \frac{(x+1)(x+1)}{(x-1)(x+2)} = \frac{x^2+2x+1}{x^2+x-2}$$

$$\frac{x+1}{x-1}\frac{x+2}{x+1}\frac{x+3}{x+2} = \frac{\cancel{(x+1)}\cancel{(x+2)}(x+3)}{(x-1)\cancel{(x+1)}\cancel{(x+2)}} = \frac{x+3}{x-1}$$

where $(x+1)$ and $(x+2)$ have been canceled out (Sec. 1.05–2*b*).

(b) Product of an algebraic expression and an algebraic fraction is a fraction whose numerator is the product of the expression and the numerator of the fraction and whose denominator is the denominator of the fraction.

examples:

$$a\frac{b}{m} = \frac{ab}{m} \qquad a\frac{x+b}{m} = \frac{ax+ab}{m}$$

(c) Reciprocal of an algebraic expression is 1 divided by this expression. *Reciprocal of an algebraic fraction* is a fraction obtained by interchanging numerator and denominator in the given fraction (Sec. 1.05–*b*).

examples:

Reciprocal of a is $\frac{1}{a}$. Reciprocal of $\frac{a}{b}$ is $\frac{b}{a}$.

(d) Product of an algebraic expression and its reciprocal is 1. **Product of an algebraic fraction and its reciprocal** is also 1 (Sec. 1.05–4*c*).

examples:

$$a\frac{1}{a} = 1 \qquad \frac{a}{b}\frac{b}{a} = 1$$

(e) Quotient of two or more algebraic fractions is the product of the first fraction and the reciprocals of the remaining fractions (Sec. 1.05–4*d*).

examples:

$$\frac{a}{m} : \frac{b}{n} = \frac{a}{m}\frac{n}{b} = \frac{an}{bm} \qquad \left(\frac{a}{m} : \frac{b}{n}\right) : \frac{c}{p} = \frac{a}{m}\frac{n}{b}\frac{p}{c} = \frac{anp}{bcm}$$

where the parentheses indicate that the whole quantity $a/m : b/n$ is to be divided by c/p.

(f) Quotient of an algebraic fraction and an algebraic expression is the fraction whose numerator was divided by the expression or whose denominator was multiplied by the same expression.

examples:

$$\frac{a}{m} : b = \frac{a:b}{m} \qquad \text{or} \qquad \frac{a}{m} : b = \frac{a}{bm}$$

$$\frac{a^2-b^2}{m} : (a-b) = \frac{a^2-b^2}{m(a-b)} = \frac{(a+b)\cancel{(a-b)}}{m\cancel{(a-b)}} = \frac{a+b}{m}$$

where $(a-b)$ has been canceled out.

2.08 COMPLEX, COMPOUND, AND CONTINUED FRACTIONS

(1) Complex Fractions

(a) Definition of complex algebraic fraction is identical to that of complex fraction in arithmetic (Sec. 1.07–1a).

(b) Reduction procedure of a complex algebraic fraction follows the outline of Sec. 1.07–1b as illustrated symbolically below.

examples:

$$\frac{\frac{a}{b}}{\frac{c}{d}} = \frac{ad}{bc} \qquad \frac{\frac{am}{bn}}{\frac{cm}{dn}} = \frac{ad\cancel{mn}}{bc\cancel{mn}} = \frac{ad}{bc}$$

$$\frac{\frac{a}{b}}{c} = \frac{\frac{a}{b}}{\frac{c}{1}} = \frac{a}{bc} \qquad \frac{a}{\frac{c}{d}} = \frac{\frac{a}{1}}{\frac{c}{d}} = \frac{ad}{c}$$

(c) Operations. Once the complex algebraic fraction is reduced to a simple algebraic fraction, the prescribed operations are performed according to the rules of Secs. 2.07–3 and 2.07–4. The same applies in cases of compound and continued fractions.

examples:

$$\frac{a/m}{b/n} \pm \frac{c/p}{d/q} = \frac{an}{bm} \pm \frac{cq}{dp} = \frac{adnp \pm bcmq}{bdnp}$$

$$\frac{a/m}{b/n}\frac{c/p}{d/q} = \frac{an}{bm}\frac{cq}{dp} = \frac{acnq}{bdmp} \qquad \frac{a/m}{b/n} : \frac{c/p}{d/q} = \frac{an}{bm} : \frac{cq}{dp} = \frac{adnp}{bcmq}$$

(2) Compound Fractions

(a) Definition of a compound algebraic fraction is identical to that of arithmetic compound fraction (Sec. 1.07–1b).

(b) Reduction procedure of compound algebraic fraction follows the outline of Sec. 1.07–2c as illustrated symbolically below.

examples:

$$\frac{a \pm b/m}{c \pm d/n} = \frac{(am \pm b)/m}{(cn \pm d)/n} = \frac{n(am \pm b)}{m(cn \pm d)}$$

$$\frac{a \pm b/m}{c} = \frac{(am \pm b)/m}{c/1} = \frac{am \pm b}{cm}$$

$$\frac{a}{c \pm d/n} = \frac{a/1}{(cn \pm d)/n} = \frac{an}{cn \pm d}$$

(3) Continued Fractions

(a) Definition of continued algebraic fraction is identical to that of arithmetic compound fraction (Sec. 1.07–1c).

(b) Reduction procedure of continued algebraic fraction follows the outline of Sec. 1.07–2d as illustrated symbolically below.

example:

$$a+\cfrac{1}{b+\cfrac{1}{c+1/d}} = a+\cfrac{1}{b+\cfrac{1/1}{(cd+1)/d}} = a+\cfrac{1}{b+\cfrac{d}{1+cd}}$$

$$= a+\cfrac{\frac{1}{1}}{\cfrac{b(1+cd)+d}{1+cd}} = a+\frac{1+cd}{d+b(1+cd)}$$

$$= \frac{ad+ab(1+cd)+1+cd}{d+b(1+cd)} = \frac{ad+(1+ab)(1+cd)}{d+b(1+cd)}$$

2.09 OPERATIONS WITH ZERO

(1) Addition and Subtraction

(a) Sum of an algebraic expression and zero equals the expression.

examples:

$a+0=a \qquad 0+a=a$

(b) Difference of an algebraic expression and zero equals the expression.

examples:

$a-0=a$ but $0-a=-a$

(2) Multiplication and Division

(a) Product of an algebraic expression and zero equals zero.

examples:

$a\cdot 0=0 \qquad 0\cdot a=0$

(b) Quotient of an algebraic expression and zero is undefined but a quotient of zero and of an algebraic expression is zero.

examples:

$a:0=$ indeterminate $\qquad 0:a=0$

2.10 EXPONENTS

(1) Definitions

(a) Positive integral exponent. If n is a positive integer and a is an algebraic expression, then

$$a^n = \underbrace{a \cdot a \cdot a \cdot \cdots \cdot a}_{n \text{ times}}$$

is said to be the *nth power* of a, n is the *exponent* of the power, and a is the *base.*

(b) Negative integral exponent. If $-n$ is a negative integer and a is an algebraic expression, then

$$a^{-n} = \underbrace{\frac{1}{a} \cdot \frac{1}{a} \cdot \frac{1}{a} \cdot \cdots \cdot \frac{1}{a}}_{n \text{ times}} = \frac{1}{a^n}$$

is said to be the *nth power* of the *reciprocal* of a.

(c) Rules of signs. If $2n$ is an even integer and $2n+1$ is an odd integer, then

$$(\pm a)^{2n} = a^{2n} \qquad (\pm a)^{2n+1} = \pm a^{2n+1}$$

(d) Special cases. If $a \neq 0$, then

$$a^0 = 1 \qquad a^1 = a \qquad \frac{1}{a^0} = 1 \qquad \frac{1}{a^1} = \frac{1}{a}$$

(2) Laws of Exponents (m, n = integers)

(a) Addition and subtraction (p, q = algebraic expressions).

$$pa^m \pm qa^m = (p \pm q)a^m$$

$$pa^{-m} \pm qa^{-m} = (p \pm q)a^{-m} = \frac{p \pm q}{a^m}$$

(b) Multiplication and division.

$$a^m a^n = a^{m+n} \qquad a^m : a^n = a^{m-n} \qquad a^m a^{-n} = a^{m-n} \qquad a^m : a^{-n} = a^{m+n}$$

$$a^{-m} a^n = a^{-m+n} \qquad a^{-m} : a^n = a^{-m-n} \qquad a^{-m} a^{-n} = a^{-m-n} \qquad a^{-m} : a^{-n} = a^{-m+n}$$

(c) Powers of products and quotients.

$$(ab)^m = a^m b^m \qquad (a : b)^m = a^m : b^m$$

$$(ab)^{-m} = \frac{1}{a^m b^m} \qquad (a : b)^{-m} = b^m : a^m$$

(d) Powers of powers and fractions.

$$(a^m)^n = (a^n)^m = a^{mn} \qquad \left(\frac{a}{b}\right)^m = \frac{a^m}{b^m}$$

$$(a^m)^{-n} = (a^{-m})^n = \frac{1}{a^{mn}} \qquad \left(\frac{a}{b}\right)^{-m} = \frac{b^m}{a^m}$$

(3) Powers of Binomials and Trinomials

(a) Binomials (see also Sec. 2.19–1).

$$(a \pm b)^2 = a^2 \pm 2ab + b^2$$

$$(a \pm b)^3 = a^3 \pm 3a^2b + 3ab^2 \pm b^3$$

$$(a \pm b)^4 = a^4 \pm 4a^3b + 6a^2b^2 \pm 4ab^3 + b^4$$

$$(a \pm b)^5 = a^5 \pm 5a^4b + 10a^3b^2 \pm 10a^2b^3 + 5ab^4 \pm b^5$$

$$(a \pm b)^6 = a^6 + 6a^5b + 15a^4b^2 \pm 20a^3b^3 + 15a^2b^4 \pm 6ab^5 + b^6$$

(b) Trinomials.

$$(a \pm b + c)^2 = a^2 + b^2 + c^2 \pm 2ab \pm 2bc + 2ac$$

$$(a \pm b + c)^3 = a^3 \pm b^3 + c^3 + 3a^2(\pm b + c) + 3b^2(a + c) + 3c^2(a \pm b) \pm 6abc$$

2.11 RADICALS

(1) Definitions

(a) Positive integral index. If n is a positive integer and a is an algebraic expression, then

$$r = \sqrt[n]{a} = a^{1/n}$$

is said to be the *radical* (*nth root*) of a, which must satisfy the relationship

$$a = \underbrace{r \cdot r \cdot r \cdot \cdots \cdot r}_{n \text{ times}} = r^n$$

where a is the *radicand* (*base*) and n is the *index* of the radical.

(b) Square root. When $n = 2$, the *radical* of a is called a square root of a,

$$r = \sqrt{a}$$

and the index 2 is omitted; that is, $\sqrt[2]{a} = \sqrt{a}$.

(c) Negative integral index. If $-n$ is a negative integer and a is again an algebraic expression, then

$$\frac{1}{r} = \sqrt[-n]{a} = \frac{1}{\sqrt[n]{a}} = \frac{1}{a^{1/n}}$$

is said to be the *reciprocal of the radical* (*nth root*) of a, which must satisfy the relationship

$$\frac{1}{a} = \underbrace{\frac{1}{r} \cdot \frac{1}{r} \cdot \frac{1}{r} \cdot \cdots \cdot \frac{1}{r}}_{n \text{ times}} = \frac{1}{r^n}$$

where a is again the *radicand* (*base*) and $-n$ is the *negative index* of the radical.

(d) Rational number. When a in r is the nth power of a real number, then r is a rational number.

(e) Irrational number. When a in r is not the nth power of a real number, then r is an irrational number.

(f) Surdic expression (surd) consists of a rational part and an irrational part. Two surds are *conjugates* of each other if they differ in sign.

examples:

$a + \sqrt{b}$ is a surdic expression.

$a + \sqrt{b},\ a - \sqrt{b}$ are conjugate surdic expressions.

(g) Rules of signs. If $2n$ is an even integer and $2n + 1$ is an odd integer, then

$$\sqrt[2n]{+a^{2n}} = \pm a \qquad \sqrt[2n+1]{+a^{2n+1}} = +a \qquad \sqrt[2n+1]{-a^{2n+1}} = -a$$

(2) Laws of Indices (m, n = integers)

(a) Addition and subtraction (p, q = algebraic expressions).

$$p\sqrt[m]{a} \pm q\sqrt[m]{a} = (p \pm q)\sqrt[m]{a} \qquad \frac{p}{\sqrt[m]{a}} \pm \frac{q}{\sqrt[m]{a}} = \frac{p \pm q}{\sqrt[m]{a}}$$

(b) Multiplication and division.

$$\sqrt[m]{a}\sqrt[n]{a} = a^{(m+n)/mn} \qquad \sqrt[m]{a} : \sqrt[n]{a} = a^{(n-m)/mn}$$

examples:

$$\sqrt[3]{x}\sqrt[5]{x} = x^{1/3+1/5} = x^{8/15} = \sqrt[15]{x^8} \qquad \sqrt[3]{x} : \sqrt[5]{x} = x^{1/3-1/5} = x^{2/15} = \sqrt[15]{x^2}$$

(c) Radical of product, quotient, and fraction.

$$\sqrt[m]{ab} = \sqrt[m]{a}\sqrt[m]{b} \qquad \sqrt[m]{a:b} = \sqrt[m]{a} : \sqrt[m]{b} \qquad \sqrt[m]{\frac{a}{b}} = \frac{\sqrt[m]{a}}{\sqrt[m]{b}}$$

examples:

$$\sqrt[3]{8x^3} = \sqrt[3]{2^3}\sqrt[3]{x^3} = 2x \qquad \sqrt{x^2 : y^2} = \sqrt{x^2} : \sqrt{y^2} = x : y$$

$$\sqrt{\frac{64x^2y^2}{25u^2v^2}} = \frac{\sqrt{64x^2y^2}}{\sqrt{25u^2v^2}} = \frac{8xy}{5uv}$$

(d) Power of radical and radical of radical.

$$(\sqrt[m]{a})^n = \sqrt[m]{a^n} = a^{n/m} \qquad \sqrt[m]{\sqrt[n]{a}} = \sqrt[mn]{a} = a^{1/mn}$$

examples:

$$\sqrt[3]{x^6} = x^{6/3} = x^2 \qquad \sqrt[3]{\sqrt[5]{x}} = \sqrt[15]{x}$$

(e) Rationalization of denominator.

$$\frac{1}{\sqrt[m]{a}} = \frac{\sqrt[m]{a^{m-1}}}{a} \qquad \frac{1}{\sqrt[m]{a^n}} = \frac{\sqrt[m]{a^{m-n}}}{a}$$

examples:

$$\frac{1}{2\sqrt{a}} = \frac{\sqrt{a}}{2\sqrt{a}\sqrt{a}} = \frac{\sqrt{a}}{2a} \qquad \frac{1}{2\sqrt[3]{a}} = \frac{\sqrt[3]{a^2}}{2\sqrt[3]{a}\sqrt[3]{a^2}} = \frac{\sqrt[3]{a^2}}{2a}$$

(3) Operations with Surdic Expressions

(a) Elementary surds.

$$(a+\sqrt{b})+(a-\sqrt{b}) = 2a \qquad (a+\sqrt{b})(a-\sqrt{b}) = a^2-b$$

$$(a+\sqrt{b})-(a-\sqrt{b}) = 2\sqrt{b} \qquad (a+\sqrt{b}):(a-\sqrt{b}) = \frac{a^2+b+2a\sqrt{b}}{a^2-b}$$

(b) Higher surds.

$$\sqrt{a+\sqrt{b}}\pm\sqrt{a-\sqrt{b}} = \sqrt{2(a\pm\sqrt{a^2-b})}$$

$$\sqrt{a\pm\sqrt{b}} = \sqrt{\frac{a+\sqrt{a^2-b}}{2}} \pm \sqrt{\frac{a-\sqrt{a^2-b}}{2}}$$

(c) Surdic fractions.

$$\frac{c}{a\pm\sqrt{b}} = \frac{c(a\mp\sqrt{b})}{a^2-b} \qquad \frac{c}{\sqrt{a}\pm\sqrt{b}} = \frac{c(\sqrt{a}\mp\sqrt{b})}{a-b}$$

2.12 IMAGINARY AND COMPLEX NUMBERS

(1) Definitions

(a) Second root of a negative algebraic expression (number) defined as

$$\sqrt{-b^2} = b\sqrt{-1} = bi$$

is called *imaginary algebraic expression (number)*.

(b) Imaginary unity (unit of imaginaries)

$$i = \sqrt{-1}$$

is the basis of imaginary numbers.

(c) Complex algebraic expression (number) consists of the real part and the imaginary part,

$$p = a + bi$$

(d) If $a = 0$, the complex expression becomes an imaginary expression and *if* $b = 0$, the complex expression becomes a real expression.

(e) Conjugate complex expression (number) is a pair of binomic complex terms differing in sign only,

$$p = a + bi \qquad q = a - bi$$

(f) Complex surd is the square root of a complex expression,

$$\sqrt{a + bi}$$

(g) Conjugate complex surd is a pair of binomic complex surds differing in sign only,

$$\sqrt{a + bi} \qquad \sqrt{a - bi}$$

(2) Rules of Signs ($n = 0, 1, 2, \ldots$)

(a) Basic cases.

$$(i)^0 = +1 \qquad (i)^1 = +i \qquad (i)^2 = -1 \qquad (i)^3 = -i$$

(b) General cases.

$$(i)^{4n} = +1 \qquad (i)^{4n+1} = +i \qquad (i)^{4n+2} = -1 \qquad (i)^{4n+3} = -i$$

(3) Operations with Complex Numbers*

(a) Addition and subtraction

$$(a + bi) + (c + di) = (a + c) + (b + d)i \qquad (a + bi) - (c + di) = (a - c) + (b - d)i$$

$$(a + bi) + (a - bi) = 2a \qquad (a + bi) - (a - bi) = 2bi$$

(b) Multiplication and division

$$(a + bi)(c + di) = (ac - bd) + (ad + bc)i \qquad (a + bi)(a - bi) = a^2 + b^2$$

$$(a + bi)(a + bi) = a^2 + 2abi - b^2 \qquad (a - bi)(a - bi) = a^2 - 2abi - b^2$$

$$(a + bi) : (c + di) = \frac{(ac + bd) + (bc - ad)i}{c^2 + d^2} \qquad \frac{1}{a + bi} = \frac{a - bi}{a^2 + b^2}$$

$$(a + bi) : (a - bi) = \frac{a^2 + 2abi - b^2}{a^2 + b^2} \qquad \frac{1}{a - bi} = \frac{a + bi}{a^2 + b^2}$$

(c) Resolutions

$$a^2 + b^2 = (a + bi)(a - bi)$$

$$a^3 + b^3 = (a + b)\left(a + \frac{1 - i\sqrt{3}}{2} b\right)\left(a + \frac{1 + i\sqrt{3}}{2} b\right)$$

$$a^4 + b^4 = \left[a + \frac{\sqrt{2}}{2}(1 + i)b\right]\left[a - \frac{\sqrt{2}}{2}(1 + i)b\right]\left[a + \frac{\sqrt{2}}{2}(1 - i)b\right]\left[a - \frac{\sqrt{2}}{2}(1 - i)b\right]$$

*For numerical applications refer to Sec. 12.09.

2.13 LOGARITHMS

(1) Definitions

(a) If $a^c = b$, then c is the *logarithm* of b to the base a.

(b) Logarithms to the base 10 *are called common or Briggs's logarithms and are denoted as* $\log x$ or $\log_{10} x$.

(c) Logarithms to the base e = **2.71828** are called natural or Napier's logarithms and are denoted as $\ln x$ or $\log_e x$.

(d) Transformation relationships between $\log x$ and $\ln x$ are:

$\log x = \dfrac{\ln x}{\ln 10} = (0.43429\ldots) \ln x$	$\ln x = \dfrac{\log x}{\log e} = (2.30258\ldots) \log x$

(2) Basic Formulas

$10^0 = 1$	$\log 1 = 0$	$e^0 = 1$	$\ln 1 = 0$
$10^1 = 10$	$\log 10 = 1$	$e^1 = e$	$\ln e = 1$
$10^2 = 100$	$\log 100 = 2$	$e^2 = e^2$	$\ln e^2 = 2$
$10^{-1} = \dfrac{1}{10}$	$\log \dfrac{1}{10} = -1$	$e^{-1} = \dfrac{1}{e}$	$\ln \dfrac{1}{e} = -1$
$10^{-2} = \dfrac{1}{100}$	$\log \dfrac{1}{100} = -2$	$e^{-2} = \dfrac{1}{e^2}$	$\ln \dfrac{1}{e^2} = -2$

(3) Basic Operations*

$\log xy = \log x + \log y$	$\ln xy = \ln x + \ln y$
$\log \dfrac{x}{y} = \log x - \log y$	$\ln \dfrac{x}{y} = \ln x - \ln y$
$\log x^k = k \log x$	$\ln x^k = k \ln x$
$\log \sqrt[k]{x} = \dfrac{1}{k} \log x$	$\ln \sqrt[k]{x} = \dfrac{1}{k} \ln x$
$\log 10^k = k$	$\ln e^k = k$

(4) Composite Operations*

$\log \dfrac{x}{yz} = \log x - \log y - \log z$	$\ln \dfrac{x}{yz} = \ln x - \ln y - \ln z$
$\log \dfrac{x^m}{y^n z^p} = m \log x - n \log y - p \log z$	$\ln \dfrac{x^m}{y^n z^p} = m \ln x - n \ln y - p \ln z$
$\log \dfrac{\sqrt[m]{x}}{\sqrt[n]{y}\sqrt[p]{z}} = \dfrac{1}{m} \log x - \dfrac{1}{n} \log y - \dfrac{1}{p} \log z$	$\ln \dfrac{\sqrt[m]{x}}{\sqrt[n]{y}\sqrt[p]{z}} = \dfrac{1}{m} \ln x - \dfrac{1}{n} \ln y - \dfrac{1}{p} \ln z$

*For numerical applications refer to Sec. 12.05.

2.14 LINEAR EQUATIONS IN ONE UNKNOWN

(1) Definitions

(a) Equation is a statement of equality of two expressions.

(b) Equations are classified as identities, conditional equations, and functional equations.

(c) Identity is the equality of two constant terms.

examples:

$A = B \qquad 1+2=3$

(d) Conditional equations contain one or several unknowns.

examples:

$ax + b = 0 \qquad ax + by + cz + d = 0$

where x, y, z are the unknowns.

(e) Functional equations (functions) state the relationship between two or several variables.

examples:

$y = ax + b \qquad y = a_0 + a_1x_1 + a_2x_2 + \cdots$

where y changes as x changes or as $x_1, x_2, \ldots$ change.

(f) Algebraic equation of nth degree in the unknown x is

$$a_0 + a_1x + a_2x^2 + \cdots + a_{n-1}x^{n-1} + a_nx^n = 0$$

in which $a_0, a_1, a_2, \ldots$ are real or complex quantities.

(g) Linear algebraic equation in one unknown is

$$ax + b = 0$$

where $a \neq 0$.

(2) Solution

(a) Roots of equation. To solve an algebraic equation for the unknown x means to find values of x (roots of equation) which satisfy this equation. The fundamental theorem of algebra states that any algebraic equation of the nth degree has n real or complex roots, if m-fold roots are counted m times.

(b) System of n algebraic equations. A set of n equations which are valid only for a certain definite set of values of the unknowns $x_1, x_2, \ldots, x_n$ is called a system of n equations. Such a set of values is called the *solution of this system.*

(3) Axioms of Solution

(a) Equivalent equations. Two equations that have the same roots are said to be equivalent equations.

(b) Addition or subtraction of equal terms. If the same term is added to or subtracted from both sides of a given equation, the new equation is equivalent to the given equation (has the same roots).

example, addition:

Given equation	$x - a = b$
Addition of a	$+a \quad +a$
Equivalent equation	$x = b + a$

This operation is equivalent to the carrying of term a with the opposite sign to the other side of the equation.

$$x - \overset{0}{\not{a}} = b + a$$

example, subtraction:

Given equation	$x + c = d$
Subtraction of c	$-c \quad -c$
	$x = d - c$

This operation is also equivalent to the carrying of term c with the opposite sign to the other side of the equation.

$$x + \overset{0}{\not{c}} = d - c$$

(c) Multiplication or division by equal terms. If both sides of an equation are multiplied or divided by the same term (not zero), the new equation is equivalent to the original equation (has the same roots).

example, multiplication:

Given equation	$\frac{x}{a} = \frac{b+c}{d}$
Multiplication by a	$\frac{x}{a} a = \frac{b+c}{d} a$
Equivalent equation	$x = \frac{a(b+c)}{d}$

This procedure is also equivalent to the carrying of term a from the denominator of one side to the numerator of the other side.

$$\frac{x}{a} = a\frac{b+c}{d}$$

But if $x/a + e = (b+c)/d$, the term e must be first carried over to the right side as

$$\frac{x}{a} = \frac{b+c}{d} - e$$

and then the fraction of the left side is cleared by rule (c) as

$$x = \left(\frac{b+c}{d} - e\right)a$$

example, division:

Given equation	$fx = \frac{g+h}{k}$
Division by f	$\frac{fx}{f} = \frac{g+h}{fk}$
Equivalent equation	$x = \frac{g+h}{fk}$

This procedure is also equivalent to the carrying of term f from the numerator of one side to the denominator of the other side.

$$fx = \frac{g+h}{fk}$$

But if $fx + m = (g+h)/k$, the term m must be first carried over to the right side as

$$fx = \frac{g+h}{k} - m$$

and then the product fx of the left side is cleared by rule (c) as

$$x = \frac{(g+h)/k - m}{f}$$

2.15 LINEAR EQUATIONS IN TWO UNKNOWNS

(1) Definitions

(a) Independent equations. Two linear equations in two unknowns x and y,

$$a_1x + b_1y = c_1$$

$$a_2x + b_2y = c_2$$

are independent if neither can be derived from the other by algebraic operations.

(b) Consistent and simultaneous equations. The two equations in (a) are said to be consistent and simultaneous if a, b, c are constants, a, b are not both zero in one equation, both c's are not zero, and they are satisfied by one pair of values (one value for x; another value for y).

(2) Methods of Solution

(a) Three methods of solution are commonly used for the solution of linear equations in two unknowns: substitution, comparison, and elimination.

(b) Substitution method. First express one of the unknowns from one of the equations. Then substitute this value in the other equation and solve for the second unknown. Finally return to the equivalent first equation and solve for the final value of the first unknown.

example:

Given $\quad a_1x + b_1y = c_1 \qquad a_2x + b_2y = c_2 \qquad$ (1), (2)

From (1), $\quad x = \dfrac{c_1 - b_1y}{a_1} \qquad$ (3)

Substitution of (3) in (2) yields

$a_2\dfrac{c_1 - b_1y}{a_1} + b_2y = c_2 \quad$ from which $\quad y = \dfrac{a_1c_2 - a_2c_1}{a_1b_2 - a_2b_1} \qquad$ (4), (5)

Then (3) in terms of (5) gives $\quad x = \dfrac{c_1b_2 - c_2b_1}{a_1b_2 - a_2b_1} \qquad$ (6)

(c) Comparison method. Express the first unknown from each equation and equate their right sides. Then solve this equation for the second unknown. Finally return to the equivalent equation (to the simplest one) and solve for the first unknown.

example:

Given $\quad a_1x + b_1y = c_1 \qquad a_2x + b_2y = c_2 \qquad$ (1), (2)

From (1) and (2) respectively,

$x = \dfrac{c_1 - b_1y}{a_1} \qquad x = \dfrac{c_2 - b_2y}{a_2} \qquad$ (3), (4)

Equality of the right sides of (3) and (4) gives

$\dfrac{c_1 - b_1y}{a_1} = \dfrac{c_2 - b_2y}{a_2} \quad$ from which $\quad y = \dfrac{a_1c_2 - a_2c_1}{a_1b_2 - a_2b_1} \qquad$ (5), (6)

Then (3) or (4) in terms of (6) yields $\quad x = \dfrac{c_1b_2 - c_2b_1}{a_1b_2 - a_2b_1} \qquad$ (7)

(d) Elimination method. Multiply or divide each equation by such factors as to make equal the coefficients of one unknown in each equation. Then add or subtract these modified equations so as to eliminate one unknown. Solve the resulting equation for the remaining unknown. Then substitute this value in the simplest one of the original equations and solve for the first unknown.

example:

Given $a_1x + b_1y = c_1 \qquad a_2x + b_2y = c_2$ (1), (2)

Divide (1) and (2) by a_1 and a_2 respectively and subtract the second equation from the first one or vice versa.

$$(1) : a_1 \text{ is} \quad x + \frac{b_1}{a_1}y = \frac{c_1}{a_1} \tag{3}$$

$$-(2) : a_2 \text{ is} \; -x - \frac{b_2}{a_2}y = -\frac{c_2}{a_2} \tag{4}$$

The sum is

$$y\left(\frac{b_1}{a_1} - \frac{b_2}{a_2}\right) = \frac{c_1}{a_1} - \frac{c_2}{a_2} \tag{5}$$

from which

$$y = \frac{a_1c_2 - a_2c_1}{a_1b_2 - a_2b_1} \tag{6}$$

Then (3) or (4) in terms of (6) yields

$$x = \frac{c_1b_2 - c_2b_1}{a_1b_2 - a_2b_1} \tag{7}$$

2.16 MEANS AND PROPORTIONS

(1) Definition of Means

If $a_1, a_2, \ldots, a_n$ are n real algebraic quantities, then their

(a) Arithmetic mean is

$$A = \frac{a_1 + a_2 + \cdots + a_n}{n}$$

(b) Geometric mean is

$$G = \sqrt[n]{a_1a_2 \ldots a_n}$$

(c) Harmonic mean is

$$H = \frac{1}{n}\left(\frac{1}{a_1} + \frac{1}{a_2} + \cdots + \frac{1}{a_n}\right)$$

(d) Their relationship is

$$A \geq G \geq H$$

provided that all a's are positive.

(2) Definitions of Proportions

(a) Simple proportion is the equality of two ratios.

$$a : b = c : d$$

(b) Successive proportion is the equality of several ratios.

$$a : b = c : d = e : f = g : h$$

(c) Arithmetic proportion is defined as

$$(a - x) : (x - b) = c : c$$

where x is the arithmetic mean of a and b, and c is an arbitrary term (also 1 but not zero).

(d) Geometric proportion is defined as

$$a : x = x : b$$

where x is the geometric mean of a and b.

(e) Harmonic proportion is defined as

$$(a - x) : (x - b) = a : b$$

where x is the harmonic mean of a and b.

(3) Transformations of Simple Proportion

If $a : b = c : d$ is a given proportion of known or unknown terms, the following operations are possible and admissible.

(a) Multiplication and division of terms.

$$am : bm = c : d \qquad \frac{a}{m} : \frac{b}{m} = c : d$$

$$am : b = cm : d \qquad \frac{a}{m} : b = \frac{c}{m} : d$$

$$a : bm = c : dm \qquad a : \frac{b}{m} = c : \frac{d}{m}$$

$$a : b = cm : dm \qquad a : b = \frac{c}{m} : \frac{d}{m}$$

(b) Sum and difference of terms.

$$(a \pm b) : b = (c \pm d) : d$$

$$a : (a \pm b) = c : (c \pm d)$$

$$(a \pm b) : (a \mp b) = (c \pm d) : (c \mp d)$$

(c) Powers and roots of terms.

$$a^n : b^n = c^n : d^n \qquad \sqrt[n]{a} : \sqrt[n]{b} = \sqrt[n]{c} : \sqrt[n]{d}$$

where m is an arbitrary factor and n is a real natural number.

2.17 QUADRATIC EQUATIONS IN ONE UNKNOWN

(1) Standard Case

(a) Quadratic equation in one unknown is an integral rational equation in which the term of highest degree in the unknown is of the second degree.

$$ax^2 + bx + c = 0$$

where a, b, c are real, $a \neq 0$, and x is the unknown.

(b) Roots. The solution of this equation is

$$x_{1,2} = \frac{-b \pm \sqrt{b^2 - 4ac}}{2a}$$

where $b^2 - 4ac$ is called the *discriminant.*

(c) Classification of roots. If

$b^2 - 4ac > 0$, the roots are real and unequal,

$b^2 - 4ac = 0$, the roots are real and equal,

$b^2 - 4ac < 0$, the roots are conjugate complex.

example, real and unequal roots:

Given $\quad 2x^2 - 4x - 16 = 0$

By formula (b), $\quad x_{1,2} = \dfrac{-(-4) \pm \sqrt{(-4)^2 - (4)(2)(-16)}}{(2)(2)} = \dfrac{4 \pm 12}{4} = 1 \pm 3$

and $\quad x_1 = 4 \qquad x_2 = -2$

example, real and equal roots:

Given $\quad 2x^2 + 8x + 8 = 0$

By formula (b), $\quad x_{1,2} = \dfrac{-(8) \pm \sqrt{(8)^2 - 4(2)(8)}}{(2)(2)} = \dfrac{-8}{4} = -2$

and $\quad x_1 = x_2 = -2$

example, conjugate complex roots:

Given $\quad 2x^2 - 3x + 4 = 0$

By formula (b), $\quad x_{1,2} = \dfrac{-(-3) \pm \sqrt{(-3)^2 - 4(2)(4)}}{(2)(2)} = \dfrac{3 \pm i\sqrt{23}}{4}$

and $\quad x_1 = \dfrac{3 + i\sqrt{23}}{4} \qquad x_2 = \dfrac{3 - i\sqrt{23}}{4}$

(2) Special Cases

(a) If $b = 0$, the reduced quadratic equation is called pure quadratic equation.

$$ax^2 + c = 0 \qquad \text{and} \qquad x_{1,2} = \pm\sqrt{-\frac{c}{a}}$$

where if $c/a > 0$, the roots are imaginary, equal but of opposite sign, and if $c/a < 0$, the roots are real, equal but of opposite sign.

(b) If $c = 0$, the standard form reduces to

$$ax^2 + bx = 0 \qquad \text{or} \qquad x(ax + b) = 0$$

and

$$x_1 = 0 \qquad x_2 = -\frac{b}{a}$$

(3) Biquadratic Equation

(a) Biquadratic equation in one unknown is an integral rational equation in which the unknowns are of the fourth and second degree.

$$ax^4 + bx^2 + c = 0$$

(b) Transformation. By the substitution $y = x^2$ this equation reduces to the quadratic equation

$$ay^2 + by + c = 0$$

(c) Roots. The solution of this transformed equation is

$$y_{1,2} = \frac{-b \pm \sqrt{b^2 - 4ac}}{2a}$$

from which in terms of $y = x^2$,

$$x_{1,2,3,4} = \pm\sqrt{\frac{-b \pm \sqrt{b^2 - 4ac}}{2a}}$$

2.18 FACTORIALS

(1) Definitions

(a) The factorial of an integer $n > 0$ is defined as

$$n! = n(n-1)\cdots 3 \cdot 2 \cdot 1$$

(b) The main property of the factorial is the relationship

$$n! = n(n-1)!$$

$$n! = n(n-1)(n-2)!$$

....

(c) By definition, $0! = 1$.

(2) Table of Factorials of 0, 1, 2, . . . , 20

n	$n!$	n	$n!$
0	1		
1	1	11	39 916 800
2	2	12	479 001 600
3	6	13	6 227 020 800
4	24	14	87 178 291 200
5	120	15	1 307 674 368 000
6	720	16	20 922 789 888 000
7	5 040	17	355 687 428 096 000
8	40 320	18	6 402 373 705 728 000
9	362 880	19	121 645 100 408 832 000
10	3 628 800	20	2 432 902 008 176 640 000

2.19 BINOMIAL THEOREM

(1) Definitions

(a) Expansion. The nth power of $(a+b)$ is given by the general Newton's formula as

$$(a+b)^n = a^n + \binom{n}{1} a^{n-1} b + \binom{n}{2} a^{n-2} b^2 + \cdots + b^n$$

where n is a positive integer and b is a positive or negative term.

(b) Binomial coefficients. The coefficients of the binomial expansion are given symbolically as

$$\binom{n}{k} = \frac{n(n-1)(n-2)\cdots(n-k+1)}{k!} = \binom{n}{n-k}$$

where $k = 0, 1, 2, \ldots$ (see also Table A.43 and Sec. 2.22).

example:

$$\binom{5}{3} = \frac{(5)(4)(3)}{(3)(2)(1)} = 10$$

(2) Properties of $\binom{n}{k}$

(a) Symmetry. The binomial coefficients in the Newton's formula are symmetrical,

$$\binom{n}{k} = \binom{n}{n-k}$$

(b) Special values.

$$\binom{n}{0} = 1 \qquad \binom{n}{1} = n \qquad \binom{n}{n-1} = n \qquad \binom{n}{n} = 1$$

2.20 COUNTING AND PERMUTATIONS

(1) Definitions (see also Secs. 2.18 and 2.19)

(a) Counting. If one thing can be done in m different ways, provided that it can be done by any of these m ways, and if another thing can be done in n ways, provided again that it can be done by any of these n ways, then the total number of ways the things can be done in the indicated order is mn. This statement is known as the principle of counting (Example, Sec. 2.20–2).

(b) Permutation is an arrangement of m elements (things). The number of all possible permutations of m different elements is

$$P_m = m(m-1)(m-2)\cdots(2)(1) = m!$$

(c) The number of all different permutations of m elements, among which there are a elements of equal value (same), is

$$P_{am} = \frac{m(m-1)(m-2)\cdots(2)(1)}{a(a-1)(a-2)\cdots(2)(1)} = \frac{m!}{a!}$$

(d) The number of all different permutations of m elements, among which there are a elements of one value (of one type) and b elements of another equal value (of another type), is

$$P_{a,b,m} = \frac{m(m-1)(m-2)\cdots(2)(1)}{a(a-1)(a-2)\cdots(2)(1)b(b-1)(b-2)\cdots(2)(1)} = \frac{m!}{a!b!}$$

(2) Examples

	Elements	m, n, a
	A_1, A_2, B_1, B_2, B_3	$m = 2,\ n = 3$
Counting Case (a)	A_1, B_1 A_2, B_1 A_1, B_2 A_2, B_2 A_1, B_3 A_2, B_3	$mn = 6$
	A, B, C	$m = 3$
Permutations Case (b)	ABC BCA CAB ACB BAC CBA	$P_3 = 3! = 6$
	A, A, C	$m = 3,\ a = 2$
Permutations Case (c)	AAC ACA CAA	${}_2P_3 = \frac{3!}{2!} = 3$

2.21 VARIATIONS AND COMBINATIONS

(1) Definitions (see also Secs. 2.19 and 2.20)

(a) A variation is an arrangement of m elements (things) into a group of k terms, without regard to their order.

(b) The number of all possible variations of m elements into groups of k elements, if each element is used any number of times but only once in one group, is

$$V_m^k = \frac{m!}{(m-k)!} = \binom{m}{k} k!$$

(c) The number of all possible variations of m elements into groups of k elements, if each element is used any number of times and repeated up to k times in one group, is

$$V_m^{k*} = m^k$$

(d) A combination is an arrangement of m elements (things) into a group of k terms, with regard to their order.

(e) The number of all possible combinations of m elements into groups of k elements, where each element is used any number of times but only once in one group, is

$$C_m^k = \frac{m!}{(m-k)!k!} = \binom{m}{k}$$

(f) The number of all possible combinations of m elements into groups of k terms, where each element is used any number of times and repeated up to k times in one group, is

$$C_m^{k*} = \frac{(m+k-1)!}{(m-1)!k!} = \binom{m+k-1}{k}$$

(2) Examples

	Elements A, B, C	$m = 3,\ k = 2$
Variation Case (b)	$AB \quad BC \quad CA$ $BA \quad CB \quad AC$	$V_m^k = \frac{3!}{1!} = 6$
Variation Case (c)	$AA \quad AB \quad AC$ $BA \quad BB \quad BC$ $CA \quad CB \quad CC$	$V_m^{k*} = 3^2 = 9$
Combination Case (e)	$AB \quad AC \quad BC$	$C_m^k = \frac{3!}{1!2!} = 3$
Combination Case (f)	$AA \quad AB \quad AC$ $BB \quad BC$ CC	$C_m^{k*} = \frac{4!}{2!2!} = 6$

2.22 PASCAL'S TRIANGLE

(1) Definition (see also Sec. 2.19)

(a) Coefficients of the binomial expansion can also be represented by a triangle of integers, where the lower number equals the sum of the two adjacent numbers above.

(b) Triangle

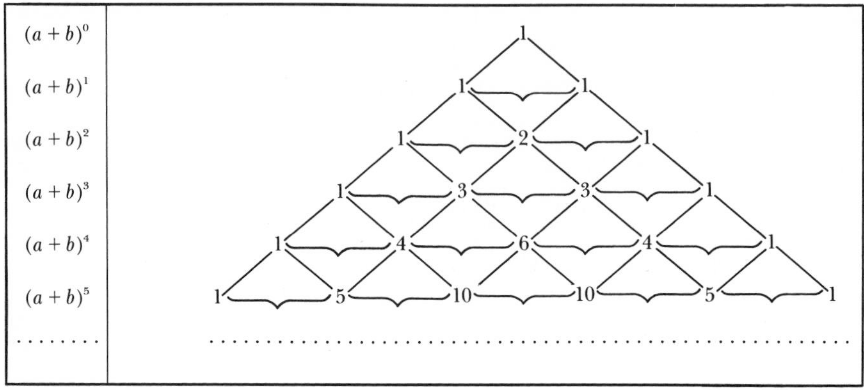

(2) Applications

$$(a \pm b)^0 = 1$$
$$(a \pm b)^1 = a \pm b$$
$$(a \pm b)^2 = a^2 \pm 2ab + b^2$$
$$(a \pm b)^3 = a^3 \pm 3a^2b + 3ab^2 \pm b^3$$
$$(a \pm b)^4 = a^4 \pm 4a^3b + 6a^2b^2 \pm 4ab^3 + b^4$$
$$(a \pm b)^5 = a^5 \pm 5a^4b + 10a^3b^2 \pm 10a^2b^3 + 5ab^4 \pm b^5$$
....

3
PLANE GEOMETRY

3.01 DEFINITIONS AND NOTATIONS

(1) Definitions

(a) Plane geometry (planimetry) is a systematic investigation of plane geometric elements and their relationships by synthetic methods (geometric constructions) and/or numerical methods.

(b) Plane geometric elements are: points, lines, angles, and segments.

(c) Geometric figure is a collection of one or several geometric elements devised from the axioms of geometry and algebra by means of geometric constructions.

(d) Axiom (postulate) is a statement (conclusion) based on experience or observation admitted to be true without proof.

(e) Proof is course of logical reasoning by which the truth or falsity of a statement is established.

(2) Symbols

$A, B, C, \ldots$ =	points, vertices	h_a, h_b, h_c =	altitudes
A =	area	m_a, m_b, m_c =	medians
C =	circumference	$2p$ =	perimeter (sum of sides)
D =	diameter	r =	inradius
R =	circumradius	r_a, r_b, r_c =	escumradii
$a, b, c, \ldots$ =	segments, sides	t_a, t_b, t_c =	internal bisectors
$e, f, \ldots$ =	diagonals	$\alpha, \beta, \gamma, \ldots$ =	angles

3.02 POINTS, LINES, AND SEGMENTS

(1) Definitions

(a) Point has no dimension and denotes position only.

(b) Straight line is the shortest distance between two points.

(c) Ray is a straight line of which one end is in the infinity.

(d) Segment is the portion of the straight line between two points.

(e) Parallel lines (parallels) do not intersect no matter how far extended (Fig. 3.02–1).

(f) Concurrent lines meet (intersect) in a point called concurrence (Fig. 3.02–2).

(g) Transversal is a line intersecting one or several other lines.

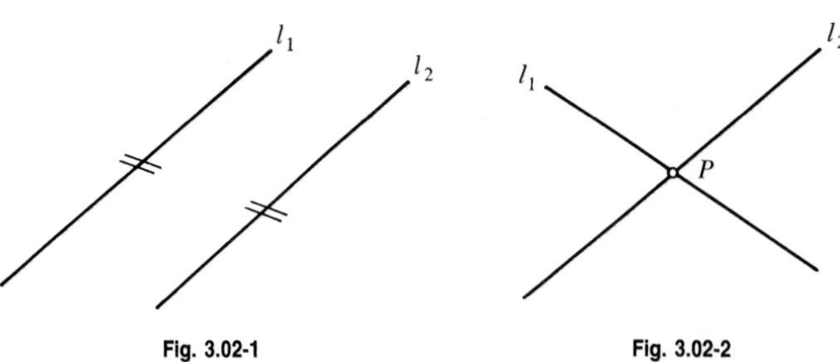

Fig. 3.02-1 Fig. 3.02-2

(2) Proportions

(a) Two parallels cut off proportional segments on any two transversals (Fig. 3.02–3).

$$a : b = c : d$$

(b) Two transversals cut off segments on two parallels proportional to the segments cut off by the parallels on the transversals (Fig. 3.02–3).

$$x : y = a : (a + b) \qquad x : y = c : (c + d)$$

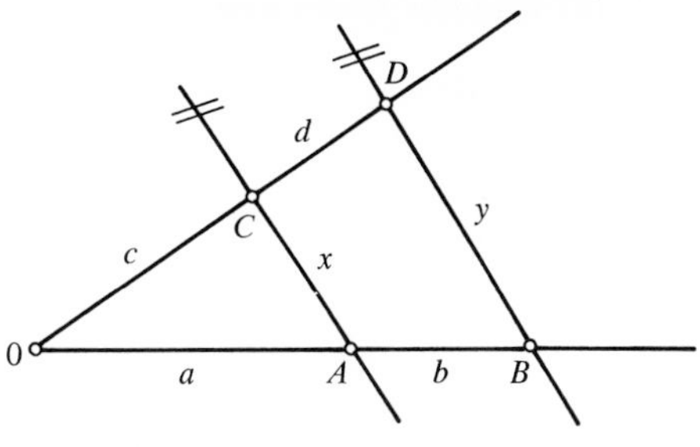

Fig. 3.02-3

(c) Harmonic division of a segment. A line segment $\overline{AB}$ is divided harmonically by the points C and D if it is divided internally by C and externally by D in the same ratio (Fig. 3.02–4).

$$a : b = d : c = m : n$$

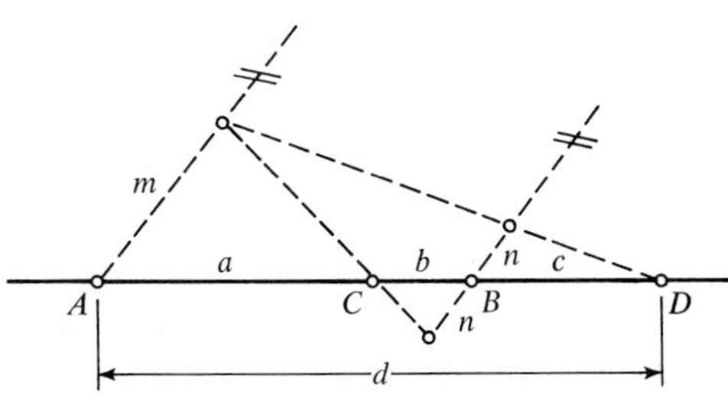

Fig. 3.02-4

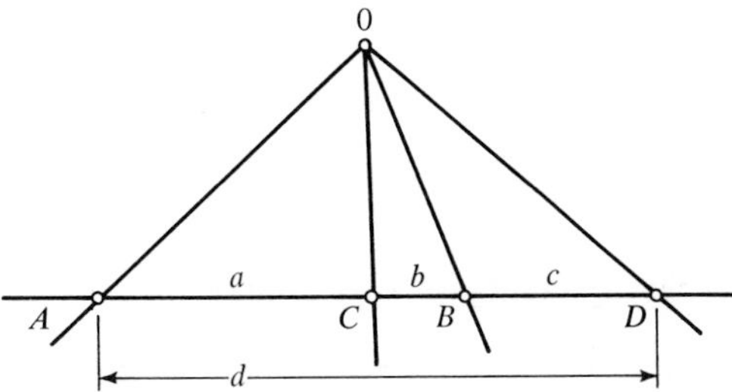

Fig. 3.02-5

(d) Harmonic set. A set of concurrent lines that divides any transversal harmonically is called a harmonic set (Fig. 3.02–5).

(e) Golden section. A line segment $\overline{AB}$ *is divided internally by the point* C in the mean and extreme ratio (Fig. 3.02–6) if its greater segment ($\overline{AC} = x$) is the *geometric mean* of the whole segment ($\overline{AB} = a$) and its smaller segment ($\overline{BC} = a - x$).

$$a : x = x : (a - x) \quad \text{or} \quad x = \frac{a}{2}(\sqrt{5} - 1) \approx \frac{3a}{5}$$

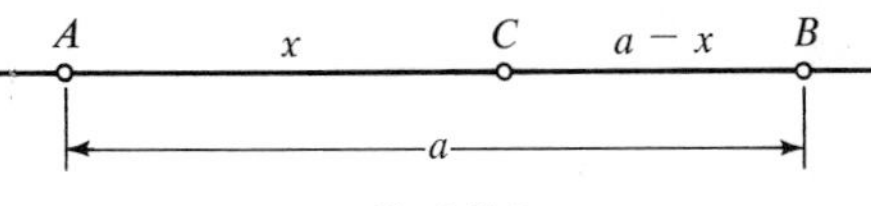

Fig. 3.02-6

3.03 ANGLES

(1) Definitions

(a) Angle. Two rays (arms) proceeding from one point (Fig. 3.03–1) divide the plane into two parts called the *internal angle* α and the *external angle* α'.

(b) Sense. The angle is *positive* if measured in the counterclockwise direction and is *negative* if measured in the clockwise direction.

(c) Complete revolution of a line is a rotation through an angle called the *perigon*.

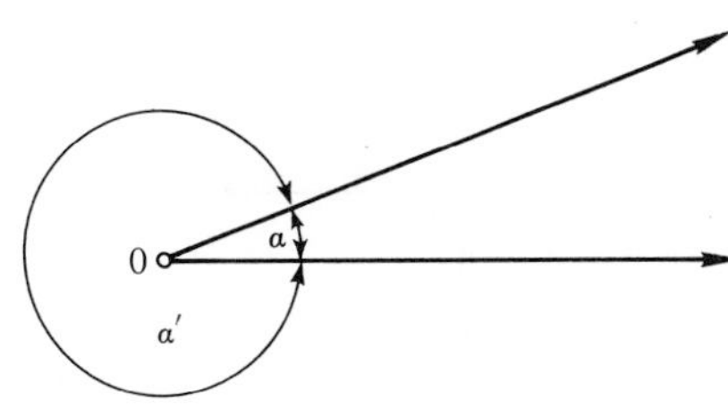

Fig. 3.03-1

(2) Angular Measures

(a) Angles are measured in *degrees, minutes,* and *seconds* or in *radians.*

(b) One degree (1°) is 1/360 of the perigon (Sec. 3.03–1*c*). *One minute* (1′) is one-sixtieth of 1 degree and *1 second* (1″) is 1/60 of 1 minute.

$$\frac{\text{Perigon}}{360} = 1° = 60' = 3{,}600''$$

(c) One radian (1 rad) is the angle subtended at the center of a circle of radius R by an arc of length R (Fig. 3.03–2).

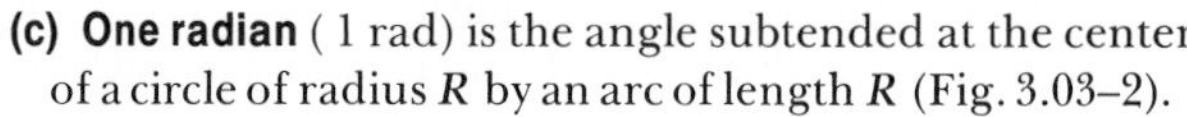

$$1 \text{ rad} = \frac{\pi}{180}$$

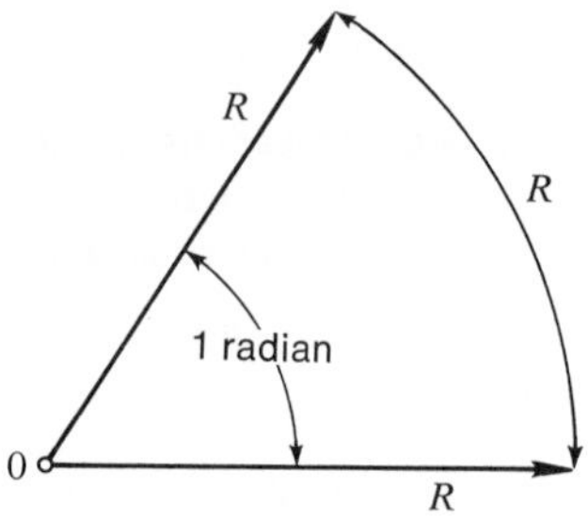

Fig. 3.03-2

(d) Relations between degrees and radians are (see also Appendix Tables):

$1° = 0.01745$ rad $\qquad$ 1 rad $= 57°17'44.8''$

(3) Classification of Angles

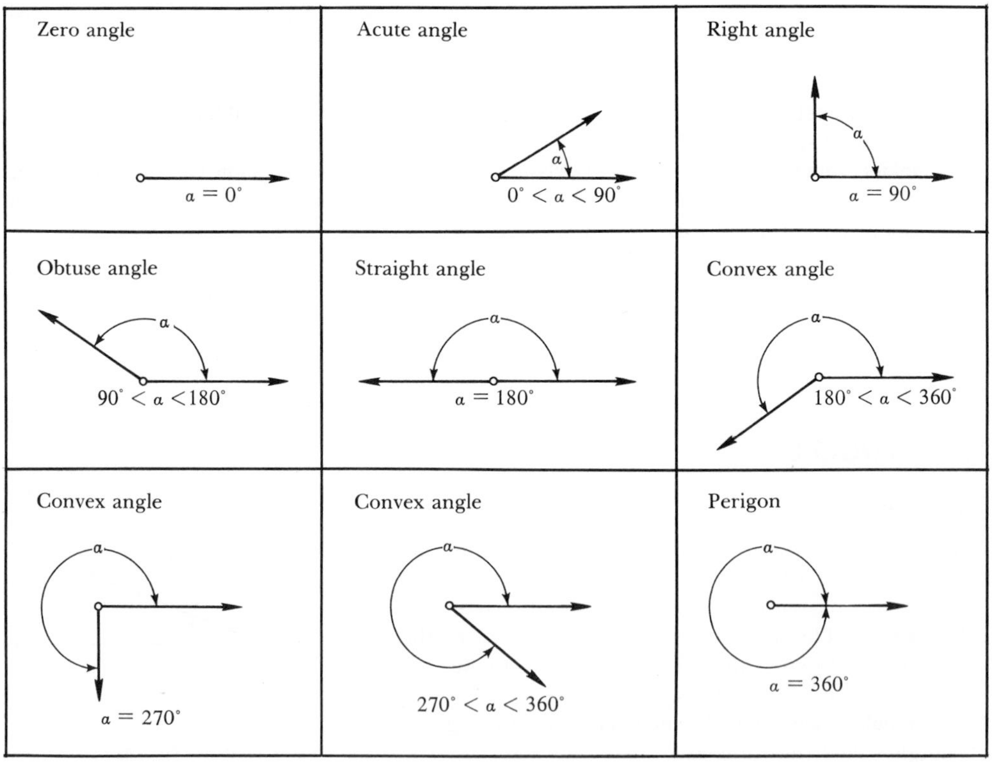

Zero angle $\alpha = 0°$	Acute angle $0° < \alpha < 90°$	Right angle $\alpha = 90°$
Obtuse angle $90° < \alpha < 180°$	Straight angle $\alpha = 180°$	Convex angle $180° < \alpha < 360°$
Convex angle $\alpha = 270°$	Convex angle $270° < \alpha < 360°$	Perigon $\alpha = 360°$

(4) Two Angles

(a) Complementary angles. Two angles (α and β) are complementary (Fig. 3.03–3) when their sum is equal to the right angle.

$$\alpha + \beta = 90° \qquad \text{or} \qquad \alpha + \beta = \frac{\pi}{2}$$

(b) Supplementary angles. Two angles (α and β) are supplementary (Fig. 3.03–4) when their sum is equal to the straight angle.

$$\alpha + \beta = 180° \qquad \text{or} \qquad \alpha + \beta = \pi$$

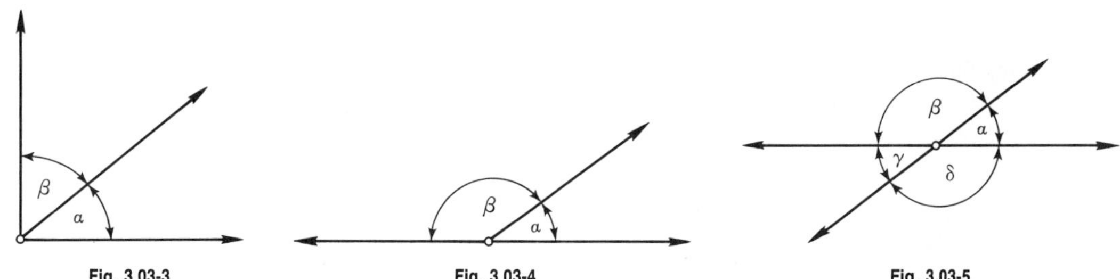

Fig. 3.03-3 Fig. 3.03-4 Fig. 3.03-5

(c) Conjugate angles. Two angles (α and α') are conjugate (Fig. 3.03–1) when their sum is equal to the perigon.

$$\alpha + \alpha' = 360° \qquad \text{or} \qquad \alpha + \alpha' = 2\pi$$

(d) Vertically opposite angles. Two angles which have a common vertex and whose arms are two straight lines (Fig. 3.03–5) are said to be vertically opposite to one another.

$$\alpha = \gamma \qquad \beta = \delta \qquad \alpha + \beta = 180° \qquad \alpha + \delta = 180° \qquad \alpha + \beta + \gamma + \delta = 360°$$

(e) Angles formed by a transversal to a set of parallels (Fig. 3.03–6) are called:

Step angles:

$$\alpha_1 = \beta_1 \qquad \alpha_2 = \beta_2$$

$$\alpha_3 = \beta_3 \qquad \alpha_4 = \beta_4$$

Alternate angles:

$$\alpha_1 = \beta_3 \qquad \alpha_2 = \beta_4$$

$$\alpha_3 = \beta_1 \qquad \alpha_4 = \beta_2$$

Opposite angles:

$$\alpha_1 + \beta_4 = 180° \qquad \alpha_2 + \beta_3 = 180°$$

$$\alpha_3 + \beta_2 = 180° \qquad \alpha_4 + \beta_1 = 180°$$

Fig. 3.03-6

3.04 TRIANGLES

(1) Definitions

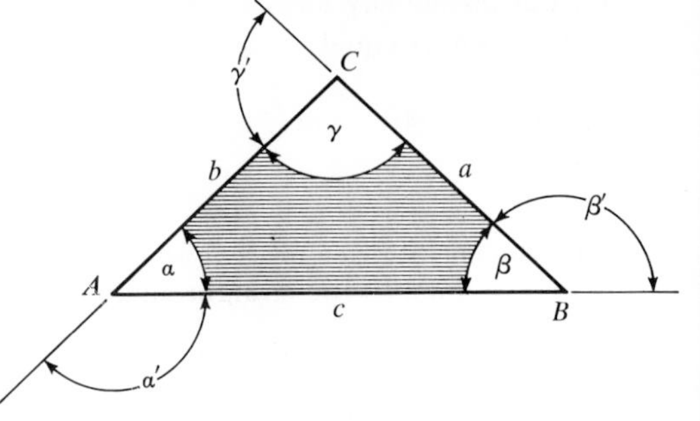

Fig. 3.04-1

(a) Triangle is a portion of a plane bounded by three segments called sides. The sum of sides is the *perimeter* $2p$.

$$2p = a + b + c$$

(b) Sum of interior angles of a triangle is 180°, and the *sum of exterior angles* is 360° (Fig. 3.04–1).

$$\alpha + \beta + \gamma = 180° \qquad \alpha' + \beta' + \gamma' = 360°$$

(c) Special cases. Triangle with three unequal sides is called a *scalene* triangle (Fig. 3.04–2). A triangle with two equal sides is called an *isosceles triangle* (Fig. 3.04–3). A triangle is called an *equilateral* triangle when its three sides are equal (Fig. 3.04–4).

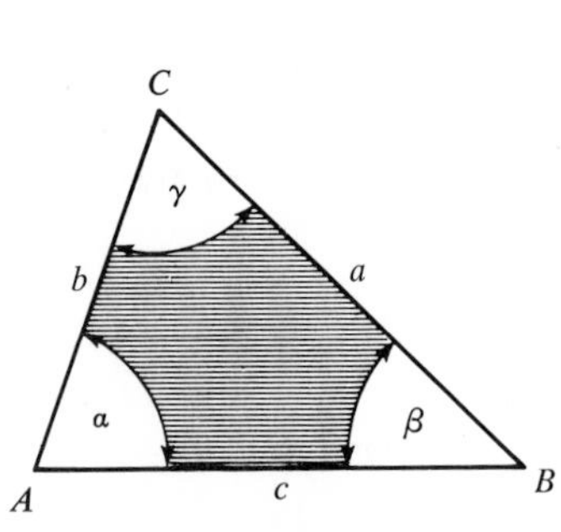

Fig. 3.04-2

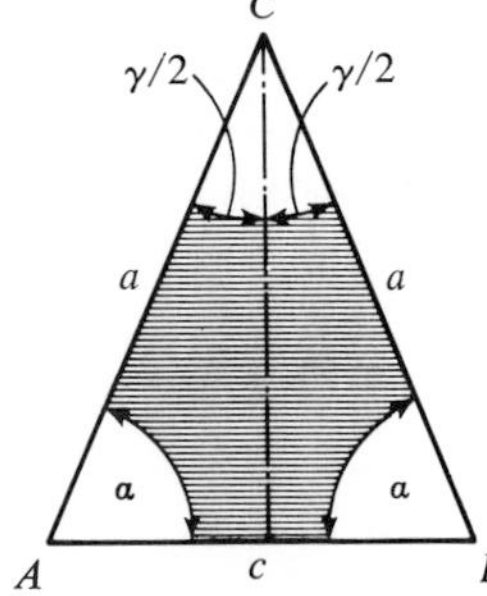

Fig. 3.04-3

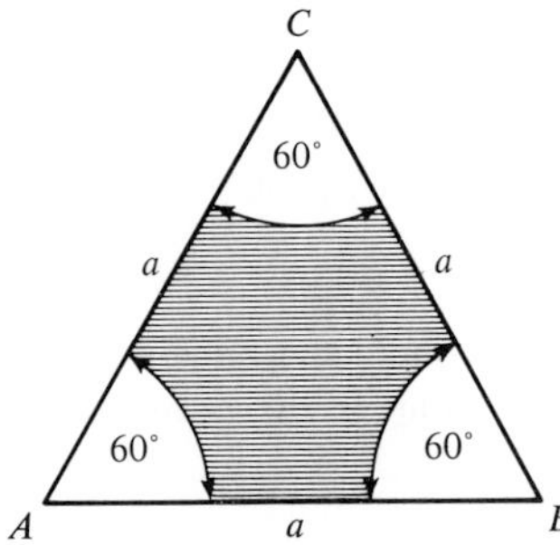

Fig. 3.04-4

(2) Properties

(a) Median m of a triangle (Fig. 3.04–5) is a segment joining the vertex with the middle point of the opposite side. Three medians intersect in one point which is the *center of gravity* of the triangle. Each median is divided by this point in the ratio 2 : 1.

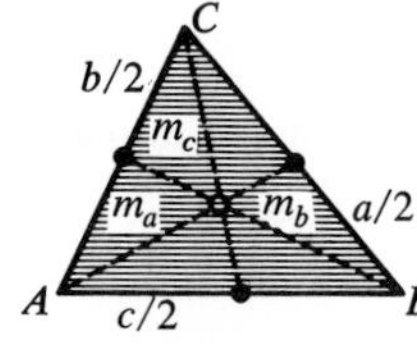

Fig. 3.04-5

$$m_a = \tfrac{1}{2}\sqrt{2(b^2 + c^2) - a^2} \qquad m_b = \tfrac{1}{2}\sqrt{2(c^2 + a^2) - b^2}$$

$$m_c = \tfrac{1}{2}\sqrt{2(a^2 + b^2) - c^2}$$

(b) Bisectors t_a, t_b, t_c (Fig. 3.04–6) halve the respective interior angle of the triangle and intersect at the incenter.

$$t_a = \frac{\sqrt{bc[(b+c)^2 - a^2]}}{b+c} \qquad t_b = \frac{\sqrt{ca[(c+a)^2 - b^2]}}{c+a} \qquad t_c = \frac{\sqrt{ab[(a+b)^2 - c^2]}}{a+b}$$

(c) Incenter is the center of the inscribed circle of radius

$$r = \frac{\sqrt{p(p-a)(p-b)(p-c)}}{p} = \frac{A}{p}$$

(d) Axes of symmetry of the respective sides (Fig. 3.04–7) intersect at the circumcenter.

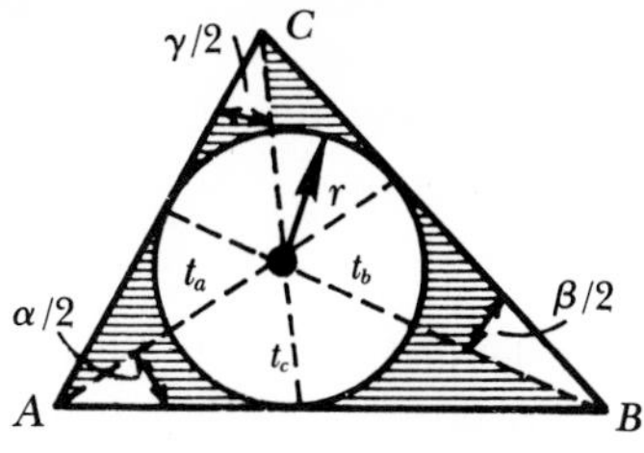

Fig. 3.04-6

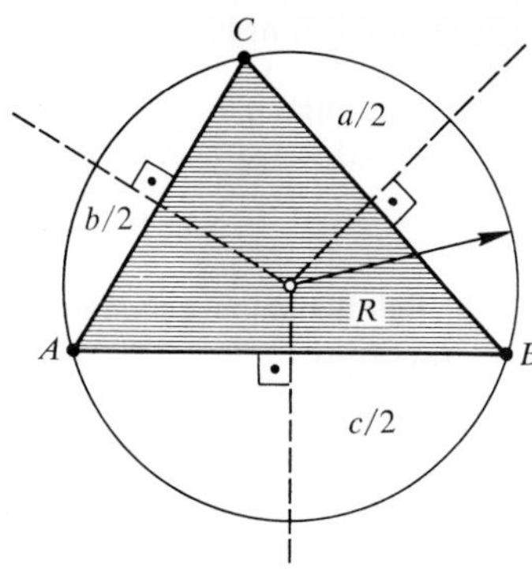

Fig. 3.04-7

(e) Circumcenter is the center of the circumscribed circle of radius

$$R = \frac{abc}{4\sqrt{p(p-a)(p-b)(p-c)}} = \frac{abc}{4A}$$

(f) Escribed circle touches the side of the triangle and the extensions of two other sides. Every triangle has three escribed circles (Fig. 3.04–8). Their radii are

$$r'_a = \frac{\sqrt{p(p-a)(p-b)(p-c)}}{p-a} = \frac{A}{p-a}$$

$$r'_b = \frac{\sqrt{p(p-a)(p-b)(p-c)}}{p-b} = \frac{A}{p-b}$$

$$r'_c = \frac{\sqrt{p(p-a)(p-b)(p-c)}}{p-c} = \frac{A}{p-c}$$

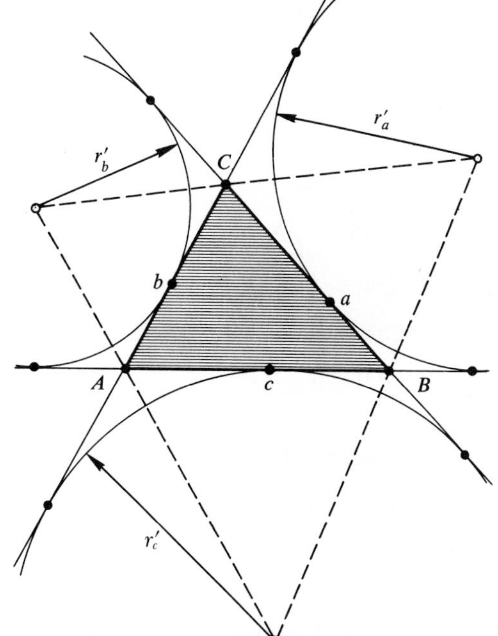

Fig. 3.04-8

(g) Altitudes of triangle h_a, h_b, h_c are perpendicular to the respective sides and they intersect at the orthocenter (Fig. 3.04–9).

$$h_c = \frac{2\sqrt{p(p-a)(p-b)(p-c)}}{c} = \frac{2A}{c}$$

$$h_a = \frac{2\sqrt{p(p-a)(p-b)(p-c)}}{a} = \frac{2A}{a}$$

$$h_b = \frac{2\sqrt{p(p-a)(p-b)(p-c)}}{b} = \frac{2A}{b}$$

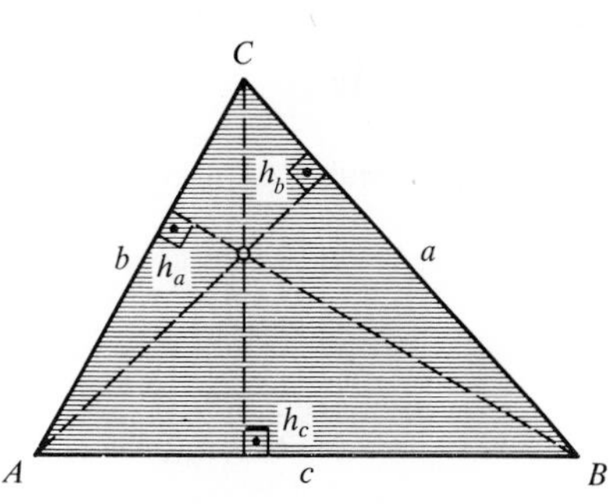

Fig. 3.04-9

(h) Area of triangle (symbols, Sec. 3.01–2).

$$A = \frac{ah_a}{2} = \frac{bh_b}{2} = \frac{ch_c}{2} = rp = \sqrt{p(p-a)(p-b)(p-c)}$$

(3) Right Triangle

(a) Right triangle has one right angle. The side opposite to it is the *hypotenuse* and the other two sides are *legs* (Fig. 3.04–10).

$$\gamma = 90° \qquad \alpha + \beta = 90°$$

(b) Relationships between the sides of the right triangle are defined by the Pythagorean theorem.

$$a^2 + b^2 = c^2$$

(c) Three geometric means relate a, b, c, x, y, and h.

$b : c = x : b$ or $b = \sqrt{cx}$ (Fig. 3.04–11)

$a : c = y : a$ or $a = \sqrt{cy}$ (Fig. 3.04–12)

$h : x = y : h$ or $h = \sqrt{xy}$ (Fig. 3.04–13)

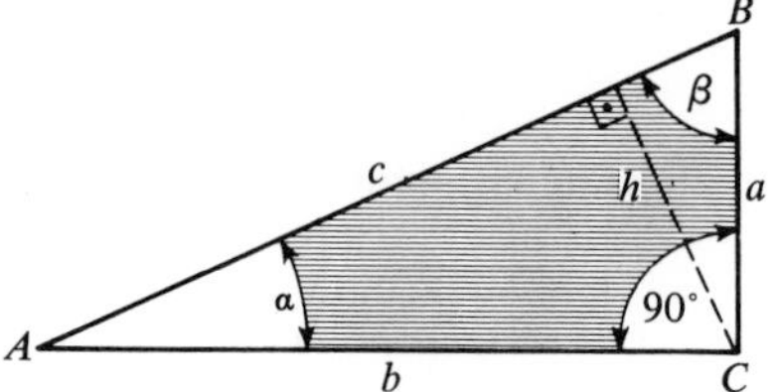

Fig. 3.04-10

(d) Formulas (symbols, Sec. 3.01–2).

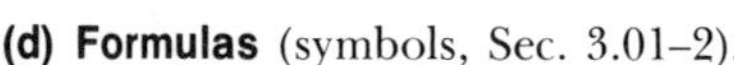

$$A = \frac{ab}{2} \qquad h = \frac{ab}{c} \qquad r = \frac{a+b-c}{2} \qquad R = \frac{c}{2}$$

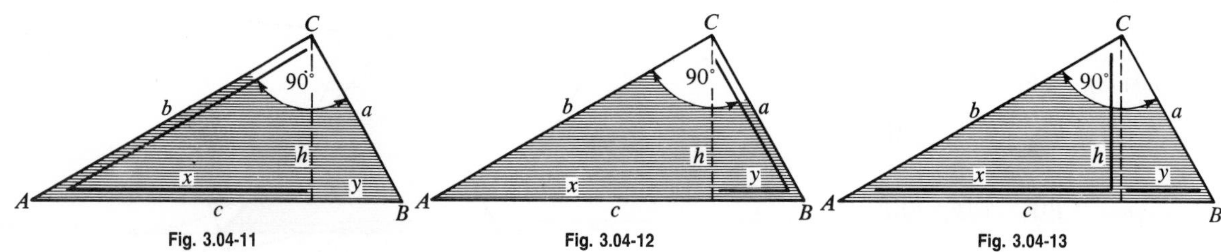

Fig. 3.04-11 Fig. 3.04-12 Fig. 3.04-13

(4) Isosceles Triangle

(a) Isosceles triangle has two equal sides and two equal angles (Fig. 3.04–14)

$$a = b \qquad \alpha = \beta \qquad \alpha + \frac{\gamma}{2} = 90°$$

(b) Relationships between sides of the isosceles triangle and its altitude are defined by the Pythagorean theorem.

$$a^2 - \left(\frac{c}{2}\right)^2 = h^2$$

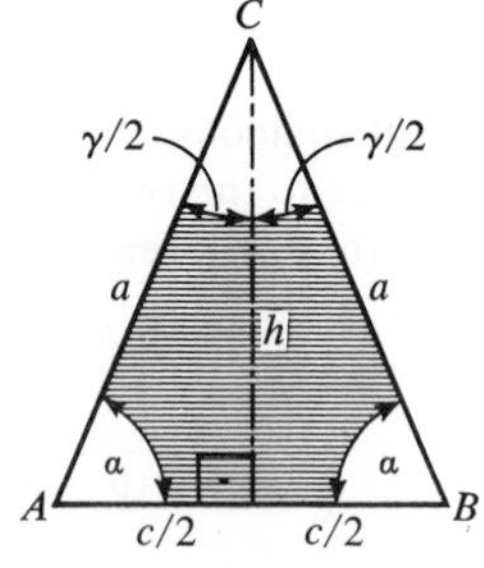

Fig. 3.04-14

(c) Formulas (symbols, Sec. 3.01–2).

$$A = \frac{ch}{2} \qquad h = \sqrt{a^2 - \left(\frac{c}{2}\right)^2} \qquad r = \frac{c}{2}\sqrt{\frac{2a-c}{2a+c}}$$

$$R = \frac{a^2}{\sqrt{(2a+c)(2a-c)}}$$

(4) Equilateral Triangle

(a) Equilateral triangle has three equal sides and three equal angles (Fig. 3.04–15).

$$a = b = c \qquad \alpha = \beta = \gamma = 60°$$

(b) Formulas (symbols, Sec. 3.01–2).

$$A = \frac{a^2}{2}\sqrt{3} \qquad h = \frac{a}{2}\sqrt{3} \qquad r = \frac{a}{6}\sqrt{3} \qquad R = \frac{a}{3}\sqrt{3}$$

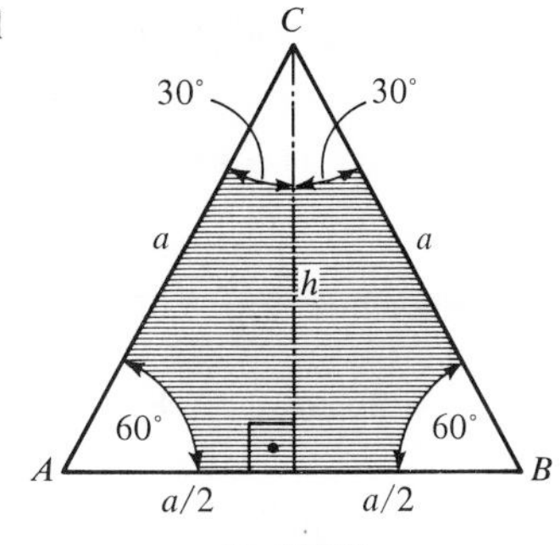

Fig. 3.04-15

3.05 QUADRILATERALS

(1) Definitions

(a) Quadrilateral is a portion of a plane bounded by four segments. The sum of interior angles of a convex quadrilateral is 360°.

(b) Quadrilaterals are classified as: *square* (Fig. 3.05–1), *rectangle* (Fig. 3.05–2), *rhombus* (Fig. 3.05–4), *rhomboid* (Fig. 3.05–5), *trapezoid* (Fig. 3.05–3), and *trapezium* (Fig. 3.05–6).

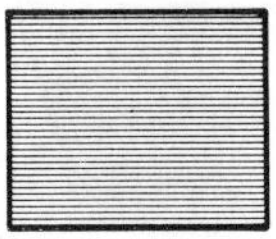

Fig. 3.05-1

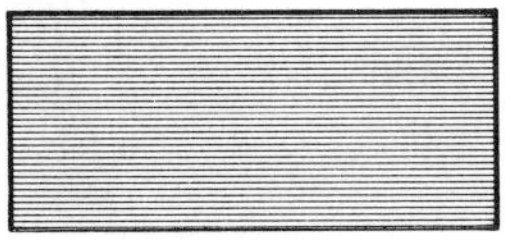

Fig. 3.05-2

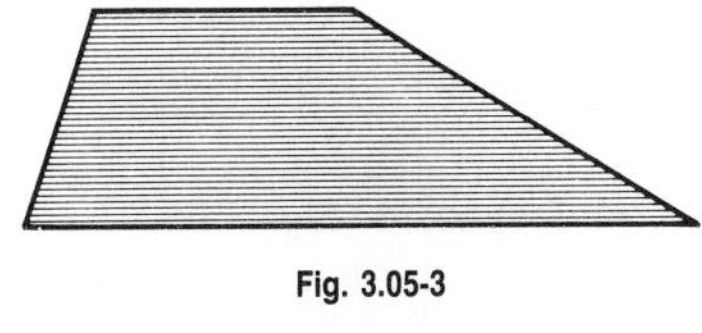

Fig. 3.05-3

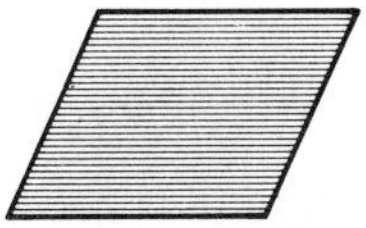

Fig. 3.05-4

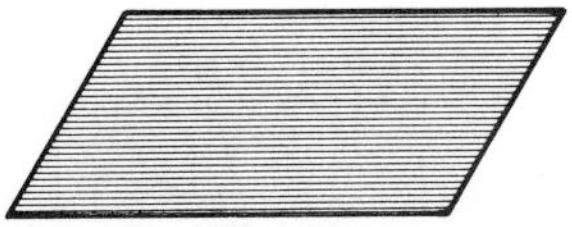

Fig. 3.05-5

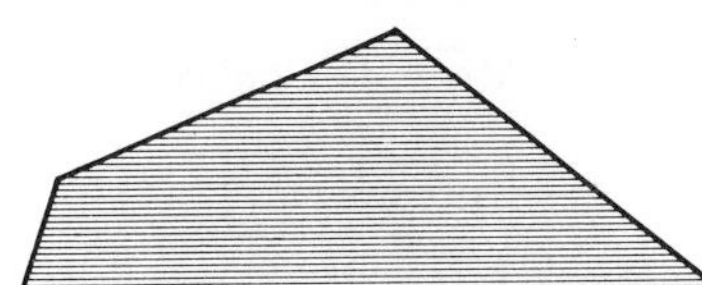

Fig. 3.05-6

(2) Particular Cases (symbols, Sec. 3.01–2).

(a) Square (Fig. 3.05–7)

$$e = a\sqrt{2} = 1.4142a \qquad R = \frac{a\sqrt{2}}{2} = 0.7071a$$

$$r = \frac{a}{2} \qquad A = a^2$$

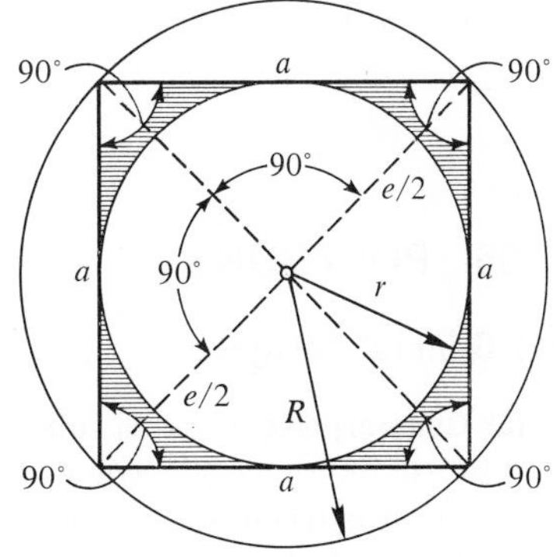

Fig. 3.05-7

(b) Rectangle (Fig. 3.05–8).

$$e = \sqrt{a^2 + b^2} \qquad R = \frac{e}{2} \qquad A = ab$$

$$\cos\omega = (1 - k^2) : (1 + k^2) \qquad \text{where } k = \frac{b}{a}$$

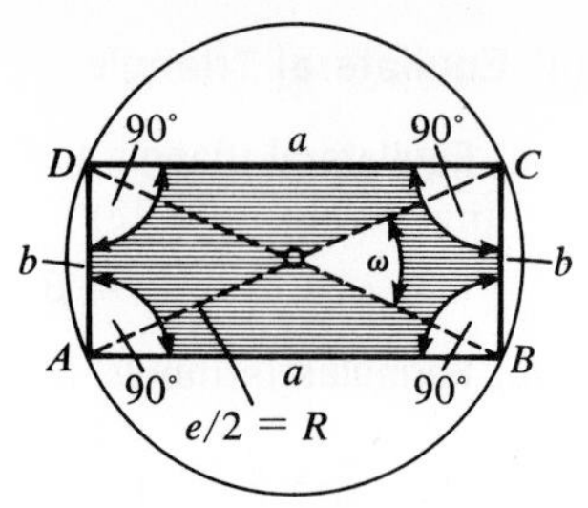

Fig. 3.05-8

(c) Rhombus (Fig. 3.05–9).

$$e = 2a\cos\frac{\alpha}{2} \qquad h = a\sin\alpha$$

$$f = 2a\sin\frac{\alpha}{2} \qquad r = \frac{a}{2}\sin\alpha \qquad e^2 + f^2 = 4a^2$$

$$A = ah = \frac{ef}{2} = a^2\sin\alpha \qquad \alpha + \beta = 180°$$

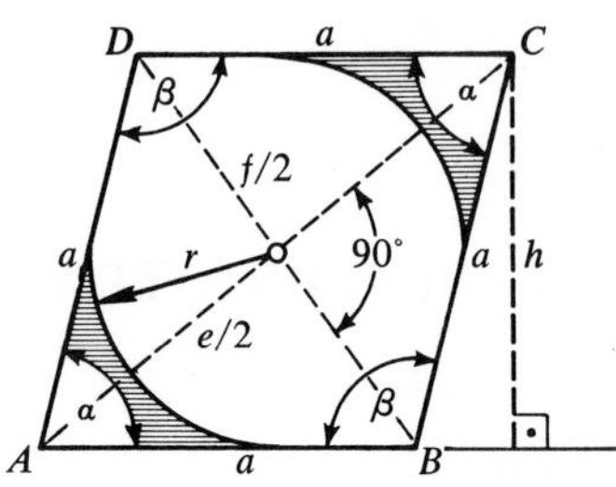

Fig. 3.05-9

(d) Rhomboid (Fig. 3.05–10).

$$e = \sqrt{a^2 + b^2 - 2ab\cos\beta} \qquad h_a = b\sin\alpha$$

$$f = \sqrt{a^2 + b^2 - 2ab\cos\alpha} \qquad h_b = a\sin\alpha$$

$$e^2 + f^2 = 2(a^2 + b^2) \qquad \cos\omega = \frac{b^2 - a^2}{4ef}$$

$$A = ah_a = bh_b = ab\sin\alpha \qquad \alpha + \beta = 180°$$

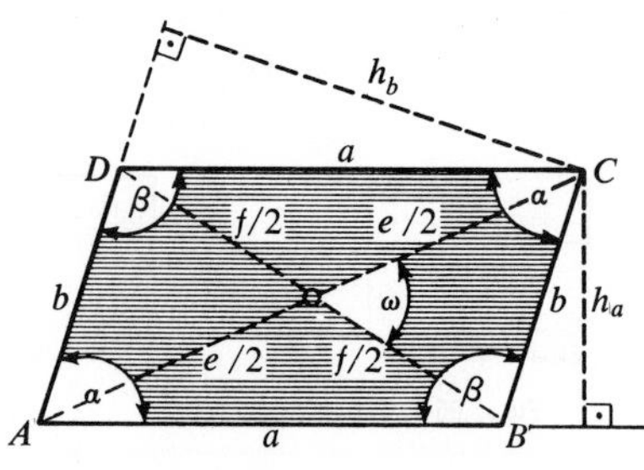

Fig. 3.05-10

(e) Trapezoid (Fig. 3.05–11).

$$e = \sqrt{a^2 + b^2 - 2ab\cos\beta} = \sqrt{c^2 + d^2 - 2cd\cos\delta}$$

$$f = \sqrt{a^2 + d^2 - 2ad\cos\alpha} = \sqrt{b^2 + c^2 - 2bc\cos\gamma}$$

$$s = \frac{a + b + c + d}{2}$$

$$h = \frac{2}{a - c}\sqrt{s(s - a + c)(s - b)(s - d)}$$

$$h = d\sin\alpha = b\sin\beta \qquad A = \frac{(a + c)h}{2}$$

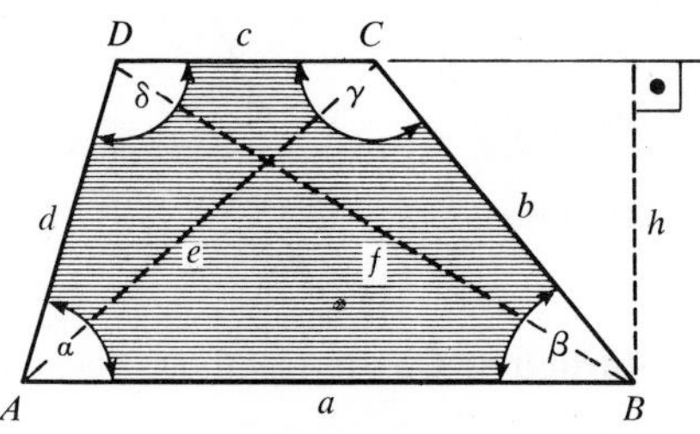

Fig. 3.05-11

3.06 POLYGONS

(1) General Polygons

(a) General polygon with n sides (Fig. 3.06–1) has the sum of interior angles equal to $(n - 2)180°$, and the sum of exterior angles equal to 360°.

(b) Area is found by decomposition into triangles (Fig. 3.06–1).

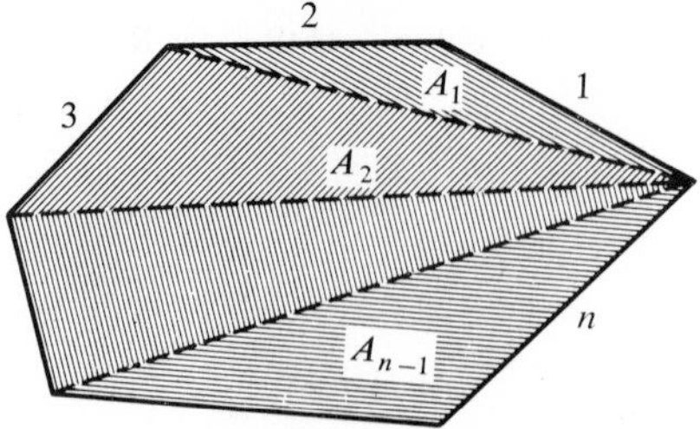

Fig. 3.06-1

(2) Regular Polygons

(a) Regular polygon has n equal sides and n equal angles (Fig. 3.06–2).

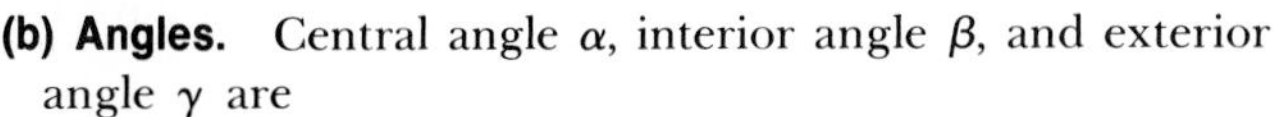

(b) Angles. Central angle α, interior angle β, and exterior angle γ are

$$\alpha = \frac{360°}{n} \qquad \beta = \frac{(n-2)180°}{n} \qquad \gamma = \frac{360°}{n}$$

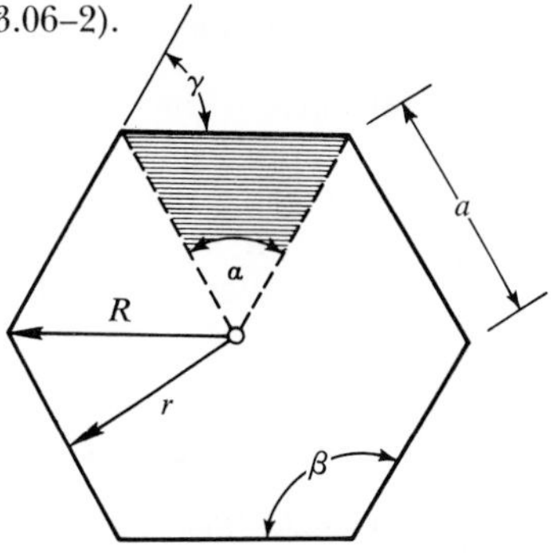

Fig. 3.06-2

(c) Formulas (symbols, Sec. 3.01–2).

$$A = \frac{na^2}{4}\cot\frac{\pi}{n} = \frac{nR^2}{2}\sin\frac{2\pi}{n} \qquad R = \frac{a}{2}\csc\frac{\pi}{n} \qquad r = \frac{a}{2}\cot\frac{\pi}{n}$$

The evaluation of these formulas for $n = 3, 4, \ldots, 48, 64$ is given in Table 3.06–1.

TABLE 3.06–1 Coefficients of Regular Polygons

n	$180°/n$	A/a^2	A/R^2	A/r^2	R/a	r/a	R/r
3	60.000	0.433013	1.299038	5.196152	0.577350	0.288675	2.000000
4	45.000	1.000000	2.000000	4.000000	0.707107	0.500000	1.414214
5	36.000	1.720477	2.377642	3.632713	0.850651	0.688191	1.236068
6	30.000	2.598076	2.598076	3.464102	1.000000	0.866025	1.154701
7	25.714* . . .	3.633914	2.736408	3.371021	1.152383	1.038261	1.109916
8	22.500	4.828427	2.828427	3.313710	1.306563	1.207107	1.082392
9	20.000	6.181825	2.892544	3.275732	1.461902	1.373739	1.064177
10	18.000	7.694208	2.938926	3.249197	1.618034	1.538842	1.051462
12	15.000	11.196154	3.000000	3.215389	1.931852	1.866025	1.035277
15	12.000	17.642362	3.050524	3.188348	2.404867	2.352314	1.022341
16	11.250	20.109363	3.061464	3.182596	2.562917	2.513670	1.019592
20	9.000	31.568769	3.090168	3.167687	3.196228	3.156877	1.012465
24	7.500	45.574519	3.105827	3.159659	3.830649	3.797877	1.008629
32	5.625	81.225378	3.121442	3.151724	5.101151	5.076586	1.004839
48	3.750	183.084812	3.132619	3.146082	7.644910	7.628533	1.002147
64	2.8125*	325.687826	3.136541	3.144114	10.190024	10.177744	1.001206

*180°/7 = 25.714 285 714 . . .° (periodic), 180°/64 = 2.8125° (finite)

examples:

Given $n = 6$, $a = 10m$, by Table 3.06–1,

$$A = 2.5981(10)^2 = 259.81m^2 \qquad R = 1.0000(10) = 10m \qquad r = 0.8660(10) = 8.6m$$

Given $n = 10$, $R = 20m$, by Table 3.06–1

$$A = 2.9389(20)^2 = 1175.56m^2$$

$$\frac{R}{a} = 1.6180 \qquad \text{from which } a = 20 : 1.6180 = 12.36m$$

$$\frac{R}{r} = 1.0515 \qquad \text{from which } r = 20 : 1.0515 = 19.02m$$

3.07 CIRCLES

(1) Definitions

(a) Circle is the part of a plane bounded by a curved line, all points of which have equal distance from a point within called the *center.*

(b) Length of the bounding line is called the *circumference C,* and the equal distance is called the *radius R.*

(c) Diameter of the circle D is the length of a chord through the center, $D = 2R$.

(2) Particular Cases (symbols, Sec. 3.01–2)

(a) Circle (Fig. 3.07–1).

$$C = 2\pi R = \pi D = \text{circumference}$$

$$A = \pi R^2 = \frac{\pi D^2}{4} = \frac{CR}{2}$$

Fig. 3.07-1

(b)* Circular sector (Fig. 3.07–2).

$$S = \frac{\pi R \alpha^\circ}{180^\circ} = R\alpha = \frac{D\alpha}{2} = \text{arc length}$$

$$A = \frac{R^2\alpha^\circ}{360^\circ} = \frac{R^2\alpha}{2}$$

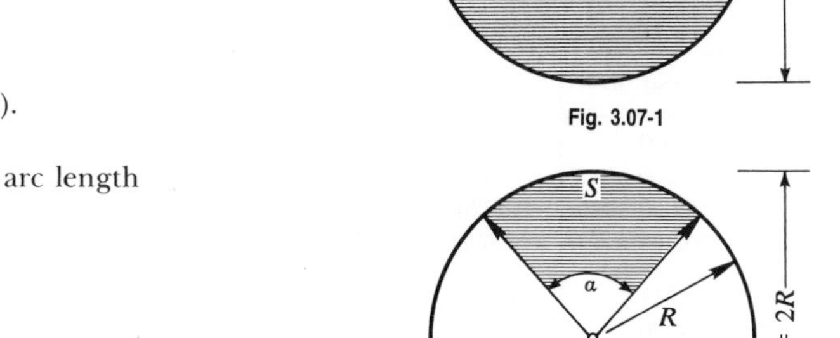

Fig. 3.07-2

(c)* Circular segment (Fig. 3.07–3).

$$l = \sqrt{R^2 - h^2} = R \sin\frac{\alpha}{2} = h \tan\frac{\alpha}{2}$$

$$h = \sqrt{R^2 - l^2} = R \cos\frac{\alpha}{2} = l \cot\frac{\alpha}{2}$$

$$v = R - h \qquad \alpha = 2\cos^{-1}\frac{h}{R}$$

$$A = \frac{R^2}{2}\left(\frac{\pi\alpha^\circ}{180} - \sin\alpha\right) = \frac{R^2}{2}(\alpha - \sin\alpha)$$

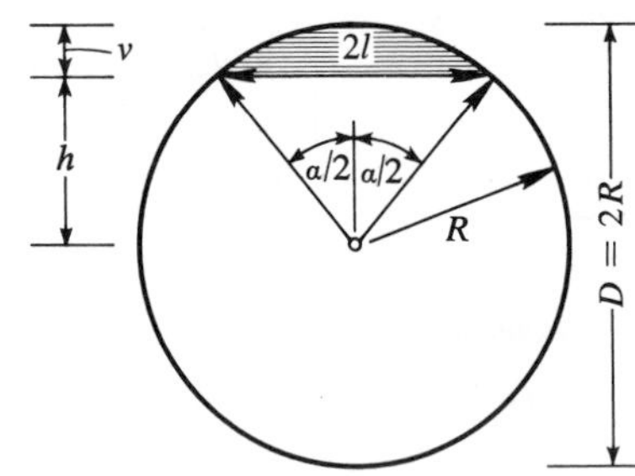

Fig. 3.07-3

(3) Constants Involving π = 3.14159†

$$\text{Arc } 1^\circ = \frac{\pi}{180} = 0.017\,453\,293 \text{ rad}$$

$$\text{Arc } 1' = \frac{\pi}{10800} = 0.000\,290\,888 \text{ rad}$$

$$\text{Arc } 1'' = \frac{\pi}{648000} = 0.000\,004\,848 \text{ rad}$$

*α° = central angle in degrees, α = central angle in radians.

†For tables of constants involving π see front endpapers and Tables A.01, A.02, A.24, and A.26.

(4) Relationships

(a) All peripheral angles belonging to the same chord with vertex on the same side of the circle are equal (Figs. 3.07–4 and 3.07–5).

$$\beta_1 = \beta_2 = \beta_3 = \cdots = \beta \qquad \delta_1 = \delta_2 = \delta_3 = \cdots = \delta$$

(b) Peripheral angle is half the size of the central angle (Figs. 3.07–4 and 3.07–5).

$$\beta = \alpha \qquad \delta = \gamma$$

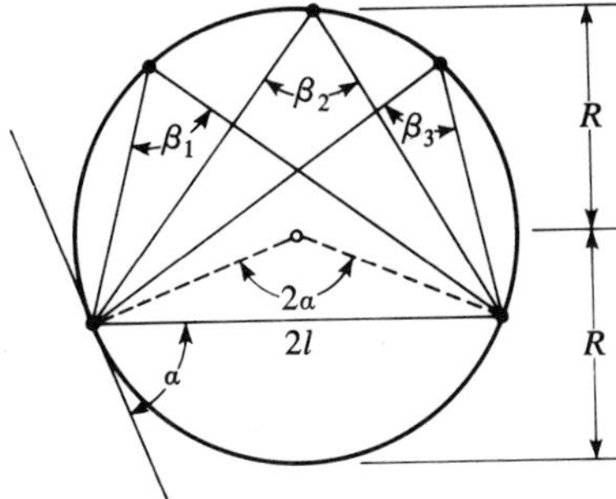

Fig. 3.07-4

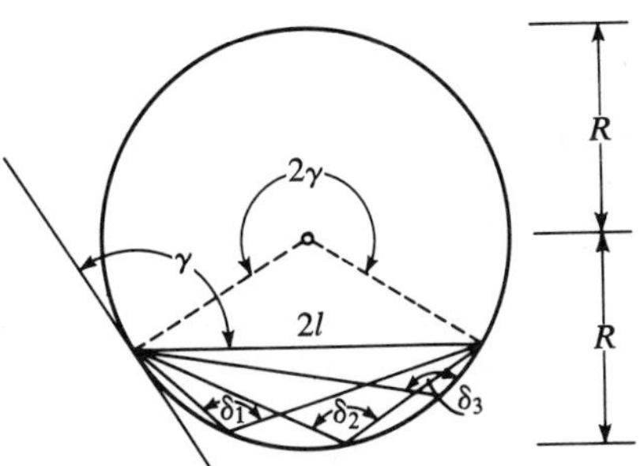

Fig. 3.07-5

(c) Sum of central angles belonging to the same chord is 360° (Fig. 3.07–5).

$$2\alpha + 2\gamma = 360° = 2\beta + 2\delta \qquad \text{or} \qquad \beta + \delta = 180°$$

(d) All peripheral angles belonging to the same chord $2l = 2R$ are right angles (Fig. 3.07–6).

$$\beta_1 = \beta_2 = \beta_3 = \cdots = 90° \qquad \delta_1 = \delta_2 = \delta_3 = \cdots = 90°$$

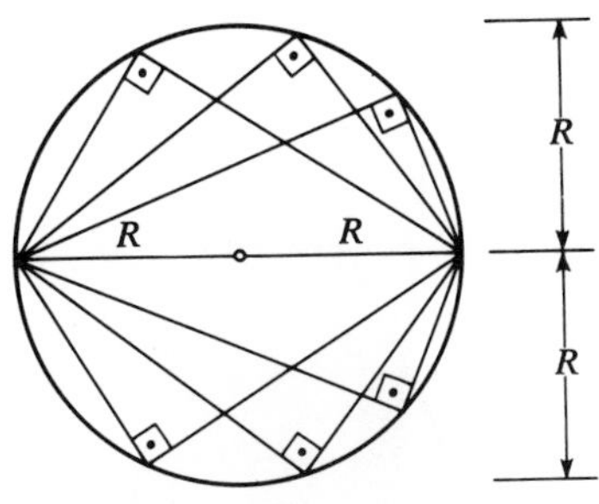

Fig. 3.07-6

(e) If two chords intersect inside a circle, the product of the segments of one equals the product of the segments of the other (Fig. 3.07–7).

$$\overline{FG} \cdot \overline{GB} = \overline{DG} \cdot \overline{GC}$$

(f) If a tangent and a secant are drawn from a point outside a circle, the tangent is the mean proportion between the secant and its external segment (Fig. 3.07–7).

$$\overline{AB} = \sqrt{\overline{AF} \cdot \overline{AE}} = \sqrt{\overline{AD} \cdot \overline{AC}}$$

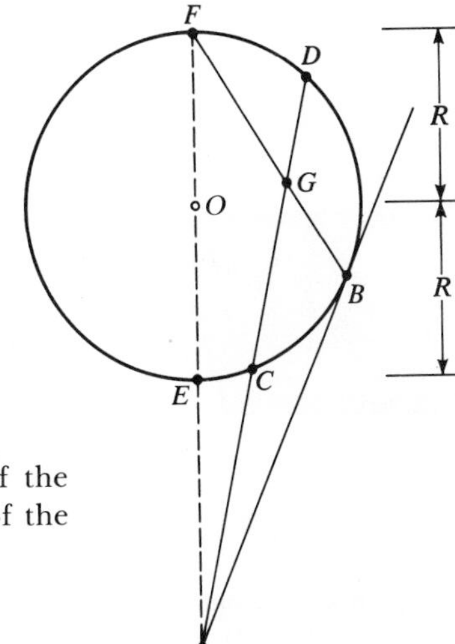

Fig. 3.07-7

3.08 CONGRUENCE, SIMILARITY, AND EQUALITY

(1) Congruence Theorems

(a) Triangle. Two triangles are congruent if they have identical:
(α) Two sides and the included angle
(β) Two angles and the included side
(γ) Two sides and the angle opposite to the larger side
(δ) Three sides

(b) Polygon. Two polygons are congruent if their corresponding sides and angles are identical.

(c) Circle. Two circles are congruent if their radii are equal.

(2) Similarity Theorems

(a) Triangle. Two triangles are similar if they are equal in:
(α) Three angles
(β) Ratio of three sides
(γ) Ratio of two sides and the included angle
(δ) Ratio of two sides and the angle opposite to the larger side

(b) Polygon. Two polygons are similar if their corresponding angles are equal and their corresponding sides are proportional.

(c) Circle. All circles are similar.

(3) Equality Theorems

(a) Triangle. Two triangles are equivalent if their areas are equal (Fig. 3.08–1) (a, h congruent).

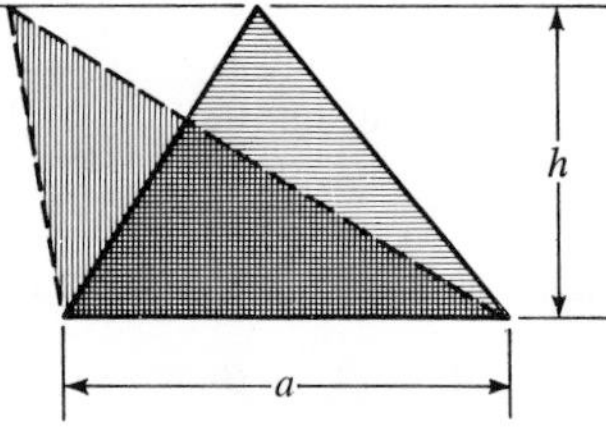

Fig. 3.08-1

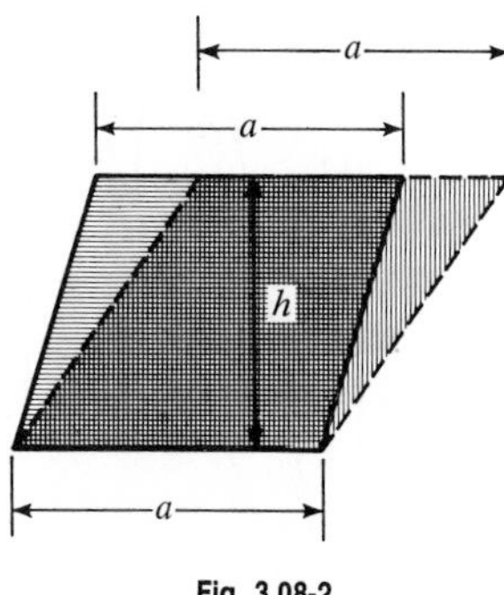

Fig. 3.08-2

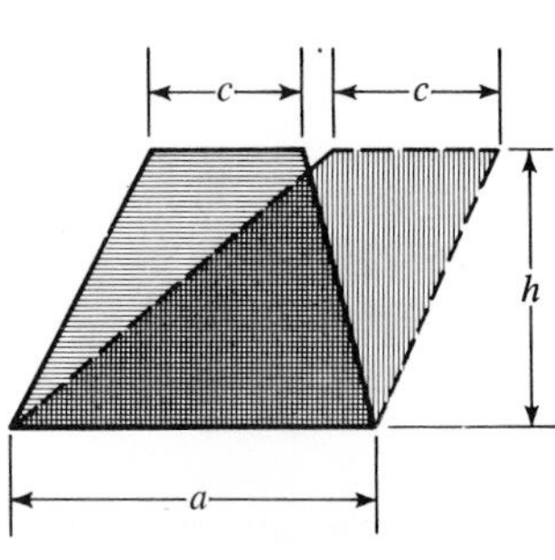

Fig. 3.08-3

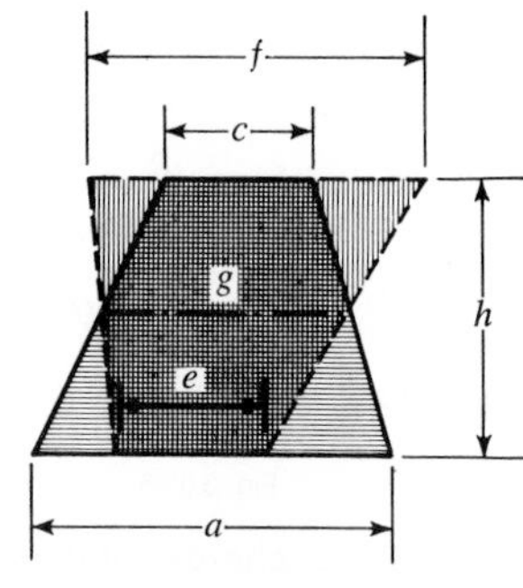

Fig. 3.08-4

(b) Parallelogram. Two parallelograms are equivalent if their areas are equal (Fig. 3.08–2) (a, h congruent).

(c) Trapezoids. Two trapezoids are equivalent if their areas are equal (Figs. 3.08–3 and 3.08–4) [a, c, h or g, h congruent, where $g = (a + c)/2 = (e + f)/2$].

4

SPACE GEOMETRY

4.01 DEFINITIONS AND NOTATIONS

(1) Definitions

(a) Space geometry (solid geometry, stereometry) is the extension of the plane geometric methods to the systematic investigation of geometric elements and their relationships in three dimensions.

(b) Space geometric elements are points, lines, angles, segments, areas, and volumes (solids), designated again by letter symbols and representing constant quantities.

(2) Symbols

A = lateral area, lateral surface	M = area of midsection
B_b, B_t = area of lower base, upper base	S = total surface
C = circumference	V = volume
D = diameter	

4.02 POINT, LINE, AND PLANE

(1) Definitions

(a) Plane is a subset of points in space such that a straight line joining any two points of this subset lies in this plane.

(b) Plane is determined by: three noncolinear points; or one straight line and a point not on the line; or two concurrent straight lines; or two parallel straight lines.

(2) Two Straight Lines

(a) Two straight lines on a plane are parallel, or concurrent.

(b) Two lines are skew in space if they do not intersect one another and do not lie in a plane.

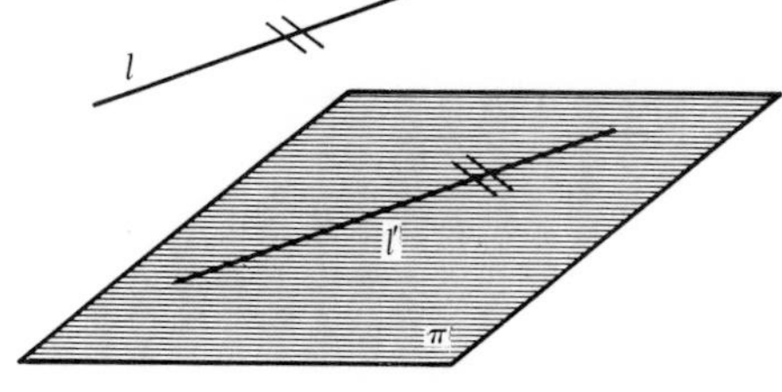

Fig. 4.02-1

(3) Straight Line and Plane

(a) Straight line l is parallel to a plane π if it is parallel to one straight line l' on that plane (Fig. 4.02–1).

(b) Straight line k is perpendicular to a plane π if it intersects that plane and is perpendicular to every straight line (l, m, n, . . . , contained in that plane) passing through the point of intersection P (Fig. 4.02–2).

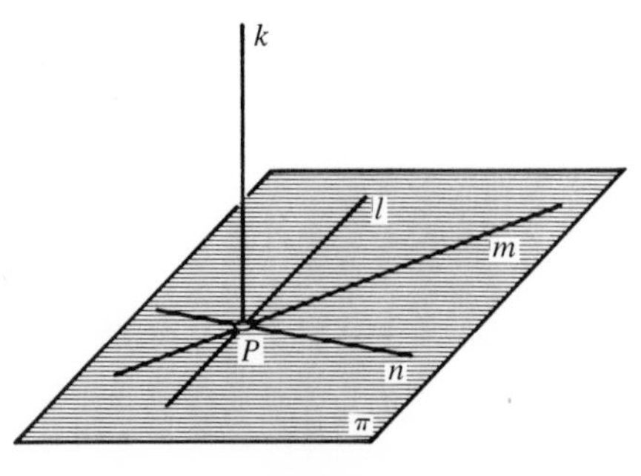

Fig. 4.02-2

(4) Point and Plane

(a) Shortest distance. A straight perpendicular line is the shortest distance d between a point A and a plane π (Fig. 4.02–3).

(b) Plane of symmetry π of a segment $\overline{AB}$ is a plane perpendicular to $\overline{AB}$ and bisecting the segment at C. $\overline{AC} = \overline{CB} = d$.

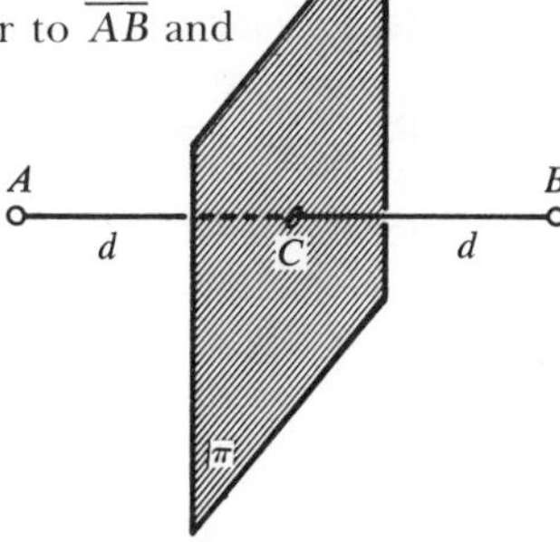

Fig. 4.02-3

4.03 TWO AND THREE PLANES

(1) Two Planes

(a) Skew planes. If two planes cut each other, their intersection is a straight line.

(b) Parallel planes. If two planes are perpendicular to the same line, they have no line of intersection and they are parallel.

(c) Dihedral angle. The opening between two intersecting planes is called the dihedral angle ω. It is measured by two perpendiculars $\overline{AB}$, $\overline{BC}$ to the line of plane intersection at B (Fig. 4.03–1).

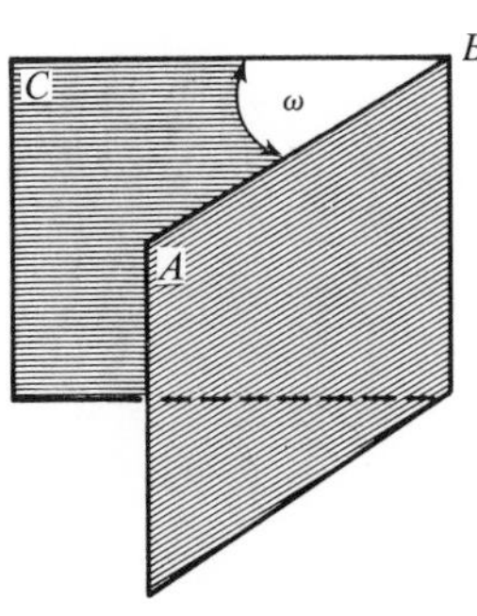

Fig. 4.03-1

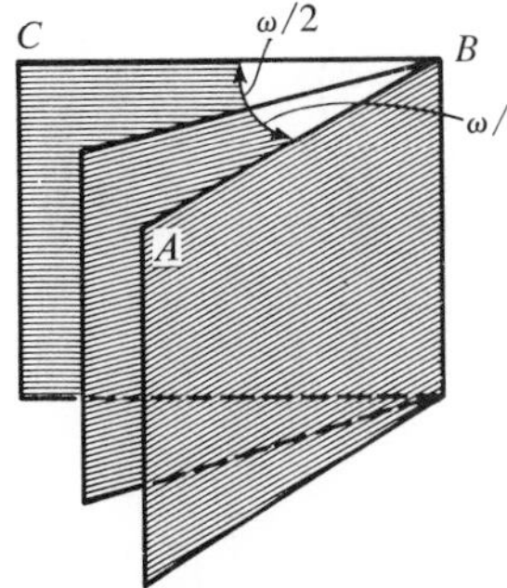

Fig. 4.03-2

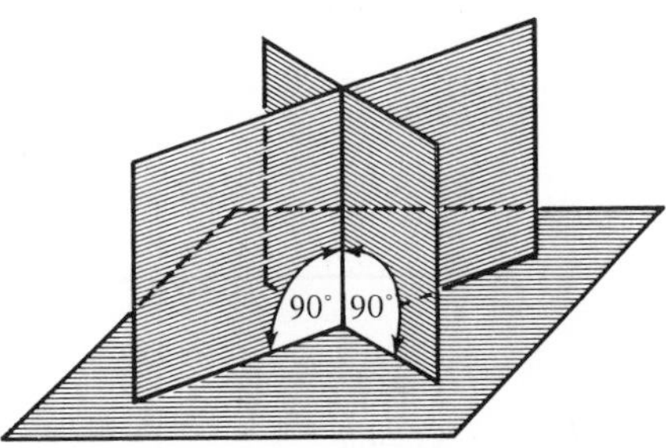

Fig. 4.03-3

(d) Normal plane. Two planes are perpendicular (normal), if their dihedral angle is a right angle.

(e) Plane of symmetry. Every point in a plane which bisects the dihedral angle is equidistant from the planes forming the angle and lies on the plane of symmetry of this angle (Fig. 4.03–2).

(2) Three Planes

(a) Two skew planes normal to a third plane. If two intersecting planes are perpendicular to a third plane, their line of intersection is also perpendicular to the third plane (Fig. 4.03–3).

(b) Two parallel planes skew to a third plane. If two parallel planes are intersected by a third plane, their lines of intersection are parallel lines (Fig. 4.03–4).

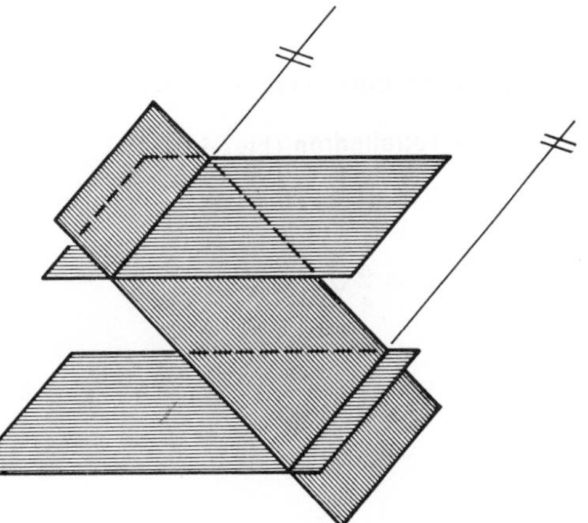

Fig. 4.03-4

4.04 POLYHEDRONS

(1) General Polyhedrons

(a) Polyhedron is a solid bounded by planes. The bounding planes are called the *faces*; the intersections of the faces, the *edges*; and the intersections of the edges, the *vertices*.

(b) Polyhedron is convex if the dihedral angles ω formed by the adjacent planes are less than 180°.

(c) Relations. If j is the number of vertices, k the number of faces, e the number of edges, and m the number of angles between the edges, then for any convex polyhedron,

$$j+k = e+2 \qquad 2e = m$$

(2) Regular Polyhedrons

(a) Polyhedron is regular if its faces are regular congruent f-sided polygons and if the same number of edges meets at each vertex.

(b) Classification. The polyhedron of 4 faces is called a *tetrahedron* (Fig. 4.04–1); of 6 faces, a *hexadron* or a *cube* (Fig. 4.04–2); of 8 faces, an *octahedron* (Fig. 4.04–3); of 12 faces, a *dodecahedron* (Fig. 4.04–4); and of 20 faces, an *icosahedron* (Fig. 4.04–5).

(c) General relationships (symbols, Sec. 4.01–2).

a = edge length, $g = m/j$

Dihedral angle	$\omega = 2\sin^{-1}\left(\cos\frac{\pi}{g}\csc\frac{\pi}{f}\right)$
Inradius	$r = \frac{a}{2}\cot\frac{\pi}{f}\tan\frac{\omega}{2}$
Circumradius	$R = \frac{a}{2}\tan\frac{\pi}{g}\tan\frac{\omega}{2}$
Surface	$S = gk\frac{a^2}{4}\cot\frac{\pi}{g}$
Volume	$V = gk\frac{a^3}{24}\cot^2\frac{\pi}{g}\tan\frac{\omega}{2}$

(3) Formulas (symbols, Sec. 4.01–2)

(a) Tetrahedron (Fig. 4.04–1).

4 triangles, 6 edges, 4 vertices, $\omega = 70°31'44''$

$$R = \frac{a}{4}\sqrt{6} \qquad r = \frac{a}{12}\sqrt{6} \qquad v = \frac{a\sqrt{6}}{3}$$

$$S = a^2\sqrt{3} = 1.7321a^2 \qquad V = \frac{a^3}{12}\sqrt{2} = 0.1179a^3$$

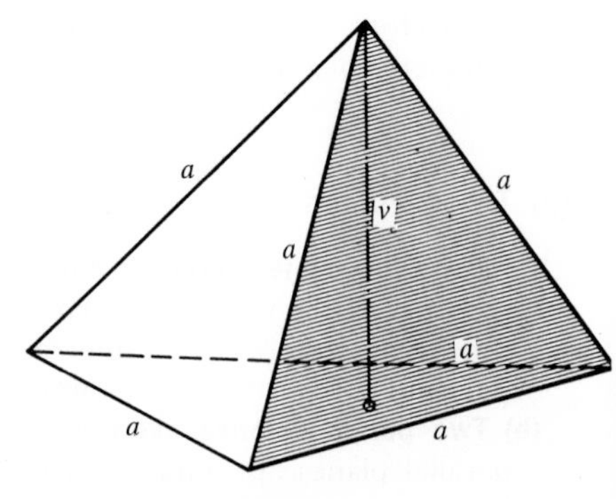

Fig. 4.04-1

(b) Cube (Fig. 4.04–2).
6 squares, 12 edges, 8 vertices, $\omega = 90°$

$$R = \frac{a}{2}\sqrt{3} \qquad r = \frac{a}{2} \qquad e = a\sqrt{3}$$

$$S = 6a^2 \qquad V = a^3$$

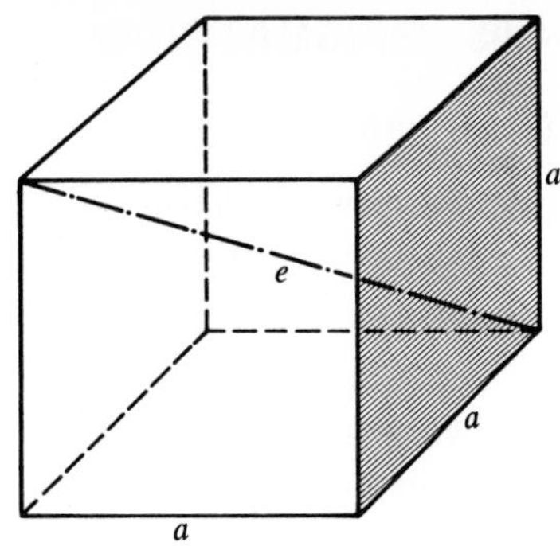

Fig. 4.04-2

(c) Octahedron (Fig. 4.04–3).
8 triangles, 12 edges, 6 vertices, $\omega = 109°28'16''$

$$R = \frac{a}{2}\sqrt{2} \qquad r = \frac{a}{6}\sqrt{6}$$

$$S = 2a^2\sqrt{3} = 3.4642a^2 \qquad V = \frac{a^3}{3}\sqrt{2} = 0.4714a^3$$

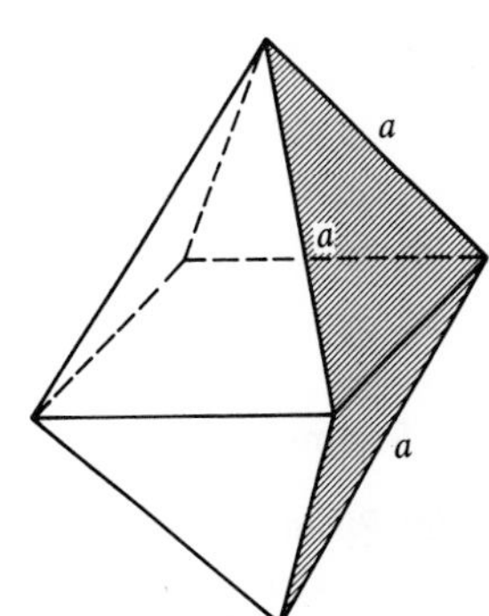

Fig. 4.04-3

(d) Dodecahedron (Fig. 4.04–4).
12 pentagons, 30 edges, 20 vertices, $\omega = 116°33'54''$

$$R = \frac{a(1+\sqrt{5})\sqrt{3}}{4} \qquad r = \frac{a}{4}\sqrt{\frac{50+22\sqrt{5}}{5}}$$

$$S = 3a^2\sqrt{5(5+2\sqrt{5})} = 20.6457a^2$$

$$V = \frac{a^3}{4}(15+7\sqrt{5}) = 7.6631a^3$$

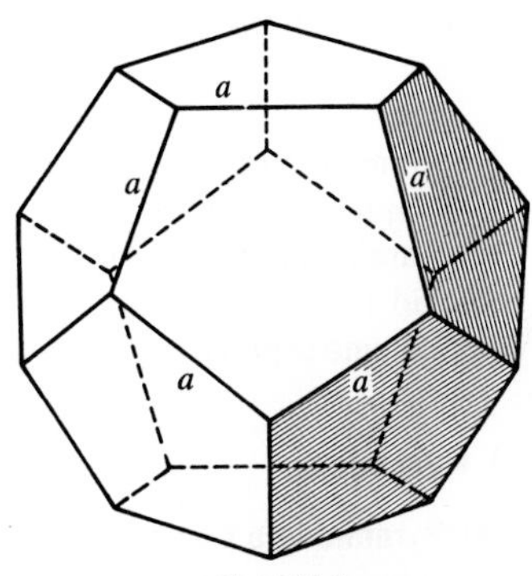

Fig. 4.04-4

(e) Icosahedron (Fig. 4.04–5).
20 triangles, 30 edges, 12 vertices, $\omega = 138°11'23''$

$$R = \frac{a}{4}\sqrt{2(5+\sqrt{5})}$$

$$r = \frac{a}{2}\sqrt{\frac{7+3\sqrt{5}}{6}}$$

$$S = 5a^2\sqrt{3} = 8.6603a^2 \qquad V = \frac{5a^3}{12}(3+\sqrt{5}) = 2.1817a^3$$

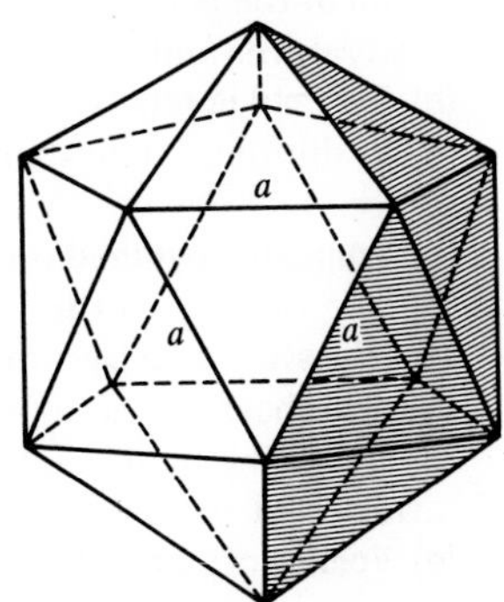

Fig. 4.04-5

4.05 PRISMATOIDS

(1) General case

(a) Prismatoid. Polyhedron (Fig. 4.05–1) is called a prismatoid if its two polygonal bases are in parallel planes, and the lateral surface is formed by triangles or trapezoids with one side common with one base and the opposite vertex or side common with other base. The altitude of a prismatoid is perpendicular to the bases. Its midsection is made by a plane perpendicular to the altitude, bisecting the altitude and all the lateral edges.

(b) Volume of a prismatoid is equal to the product of one-sixth of its altitude v times the sum of the areas of its bases B_b, B_t and four times the area of its midsection M.

$$V = \frac{(B_b + B_t + 4M)v}{6}$$

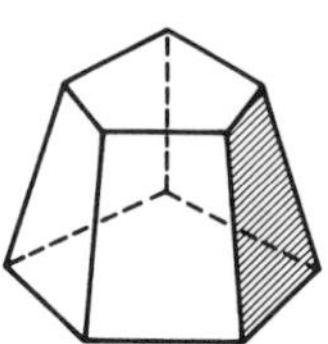
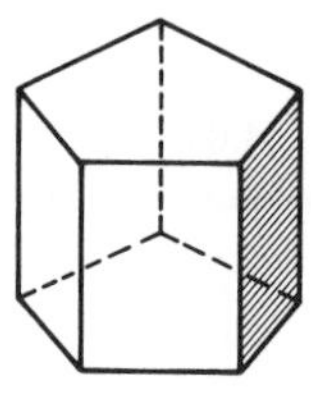
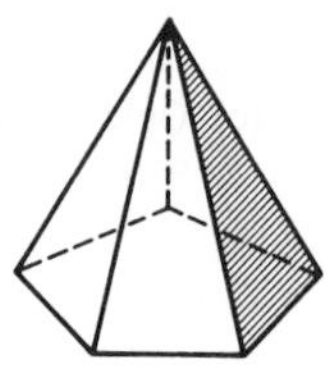

Fig. 4.05-1

(2) Prisms

(a) Prism is a polyhedron of which two bases are equal polygons in parallel planes, and the lateral faces are parallelograms. The altitude of a prism is the perpendicular distance between the bases. A right section of a prism is a section made by a plane perpendicular to the lateral edges. A *right prism* is a prism whose lateral edges are perpendicular to the bases. An *oblique prism* has lateral edges oblique to their bases. The sum of its lateral faces is called the lateral area and its surface is the sum of areas of its bases and of the lateral area.

(b) Truncated prism is the part of the prism between the base and the section made by a plane oblique to the base.

(c) Parallepiped is a prism whose bases are parallelograms. If all six faces are rectangles, the solid is called a *rectangular parallelepiped.*

(d) Volume of prism (B = area of base, v = altitude) is $V = Bv$.

(3) Pyramids

(a) Pyramid is a polyhedron of which the base is a polygon and the lateral faces are triangles having a common vertex. The intersection of the lateral faces are called the edges and the sum of the lateral faces is called the lateral area. The surface of a pyramid is the sum of its lateral area and of the area of its base. Its altitude is the distance of the vertex to its base.

(b) Pyramid is regular if its base is a regular polygon whose center coincides with the foot of the altitude. The slant height of a rectangular pyramid is the altitude of any of the lateral faces.

(c) Truncated pyramid is the portion of a pyramid between the base and the section made by a plane oblique to the base, cutting all the lateral edges.

(d) Frustum of a pyramid is the portion of a pyramid between the base and a section parallel to the base. The altitude of a frustum is the distance between the bottom and the top base. The slant height of a frustum is different for each face and is equal to altitude of the trapezoid forming the respective face.

(e) Volume of pyramid (B = area of base, v = altitude) is $V = \frac{1}{3}Bv$.

(4) Particular Cases (symbols, Sec. 4.01–2)

(a) Rectangular parallelepiped (Fig. 4.05–2).

$$e = \sqrt{a^2 + b^2 + c^2} \qquad R = \frac{\sqrt{a^2 + b^2 + c^2}}{2}$$

$$S = 2(ab + bc + ca) \qquad V = abc$$

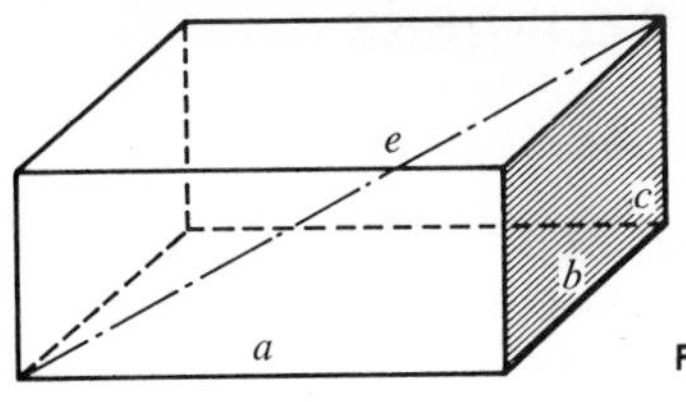

Fig. 4.05-2

(b) Prism (Fig. 4.05–3).

$2p$ = perimeter of right section

$$A = 2ph \qquad S = A + 2B \qquad V = Bv$$

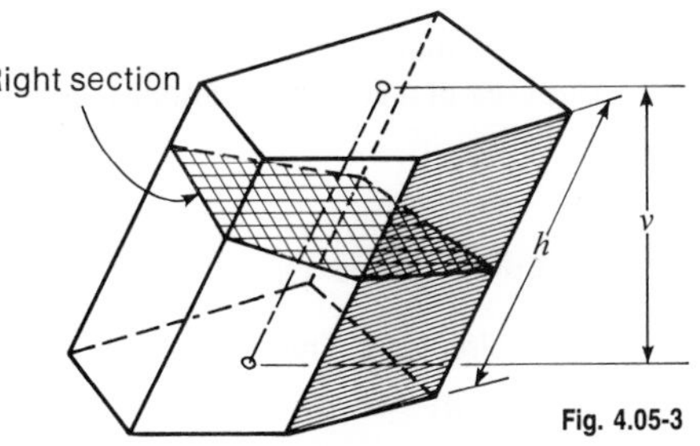

Fig. 4.05-3

(c) Right pyramid (Fig. 4.05–4).

$$A = a\sqrt{v^2 + \left(\frac{b}{2}\right)^2} + b\sqrt{v^2 + \left(\frac{a}{2}\right)^2}$$

$$B = ab \qquad S = A + B \qquad V = \frac{Bv}{3}$$

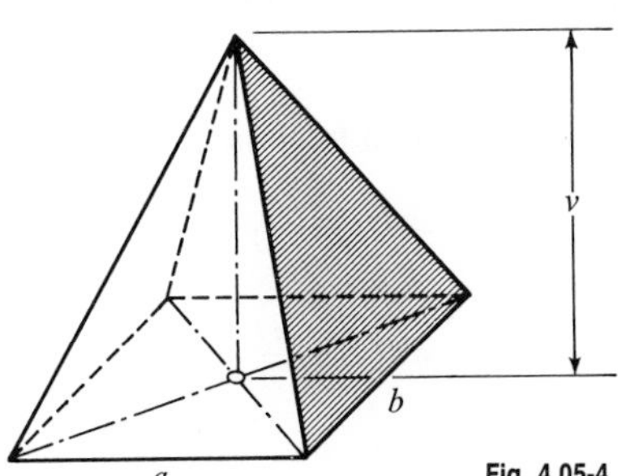

Fig. 4.05-4

(d) Frustum of right pyramid (Fig. 4.05–5).

$$A = (a_b + a_t)\sqrt{v^2 + \left(\frac{b_b - b_t}{2}\right)^2} + (b_b + b_t)\sqrt{v^2 + \left(\frac{a_b - a_t}{2}\right)^2}$$

$$S = A + a_b b_b + a_t b_t \qquad V = \frac{v}{3}(B_b + B_t + \sqrt{B_b B_t})$$

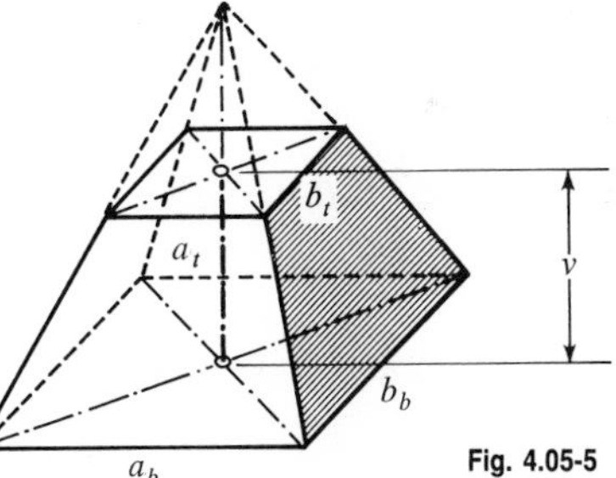

Fig. 4.05-5

(e) Right wedge (Fig. 4.05–6).

$$A = 2(a + c)\sqrt{v^2 + b^2} + 2b\sqrt{v^2 + (a - c)^2}$$

$$S = A + 4ab \qquad V = \frac{2bv}{3}(2a + c)$$

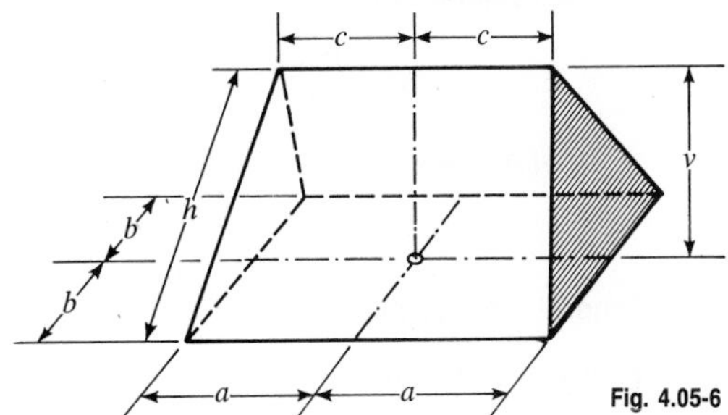

Fig. 4.05-6

4.06 CYLINDERS

(1) Definitions

(a) Cylinder is a solid bounded by a cylindrical surface and two parallel bases. The cylindrical surface is generated by a straight line translating parallel to a fixed line along a fixed plane curve. The translating line is called the generatrix and the fixed curve the directrix. The axis of a cylinder is a straight line connecting the centers of the bases. The altitude is the perpendicular distance between the bases. The axis of a right cylinder is perpendicular to the bases; the axis of an oblique cylinder is oblique to the bases.

(b) Directrix of a circular cylinder is a circle. A *right circular cylinder* may be generated by the revolution of a rectangle about one side.

(2) Particular Cases (symbols, Sec. 4.01–2)

(a) Right circular cylinder (Fig. 4.06–1).

$$A = 2\pi R v \qquad B = \pi R^2$$

$$S = 2\pi R(R+v) \qquad V = \pi R^2 v$$

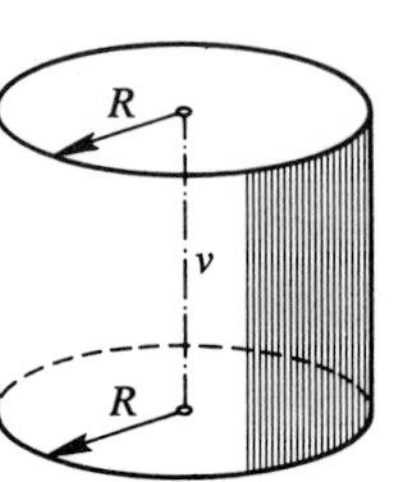

Fig. 4.06-1

(b) Truncated frustum of a cylinder (Fig. 4.06–2).

$$A = \pi R(h_1 + h_2) \qquad S = \pi R\left[h_1 + h_2 + R + \sqrt{R^2 + \left(\frac{h_2 - h_1}{2}\right)^2}\right]$$

$$V = \pi R^2 \frac{h_1 + h_2}{2} = \pi R^2 v$$

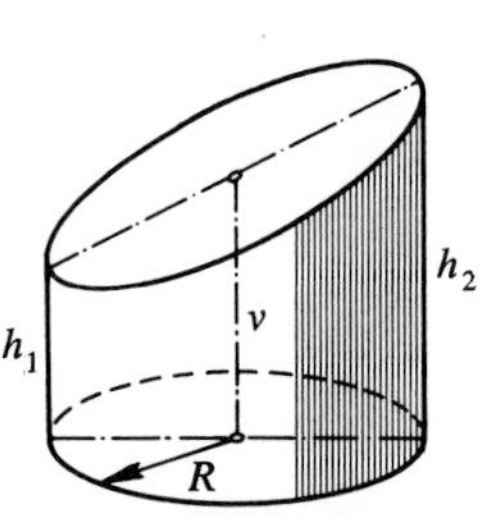

Fig. 4.06-2

(c) Ungula of a cylinder (Fig. 4.06–3).

$$A = \frac{2Rh}{b}[(b-R)\omega + a]$$

$$V = \frac{h}{3b}[a(3R^2 - a^2) + 3R^2(b-R)\omega]$$

If $a = b = R$,

$$A = 2Rh \qquad V = \tfrac{2}{3}R^2 h$$

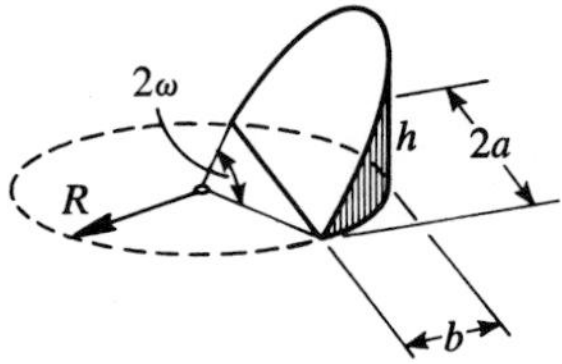

Fig. 4.06-3

(d) Hollow cylinder (Fig. 4.06–4).

$$t = R - r \qquad \rho = \frac{R+r}{2} \qquad A = 2\pi R v$$

$$B = \pi(R^2 - r^2) \qquad V = \pi v(R^2 - r^2) = 2\pi v t \rho$$

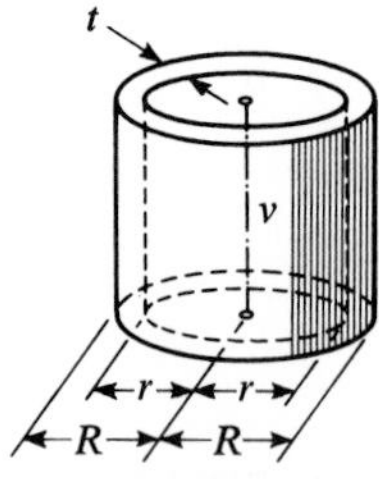

Fig. 4.06-4

(e) General circular cylinder (Fig. 4.06–5).

C = circumference of right section

$$A = Ch \qquad B = \pi R^2 \qquad V = Bv$$

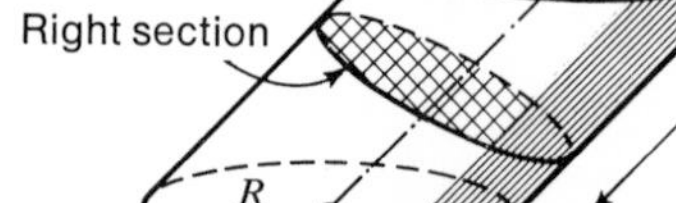

Fig. 4.06-5

4.07 CONES

(1) Definitions

(a) Cone is a solid bounded by a conical surface and a plane cutting all straight lines of this surface. The conical surface is generated by a straight line moving on a fixed plane curve and passing through a fixed point not included in the plane of the curve. The moving line is called a generatrix, the fixed curve the directrix, and the fixed point the vertex. The altitude of the cone is the perpendicular distance between the vertex and the base. The axis of the cone is a straight line connecting the vertex with the center of the base. The axis of a straight cone is perpendicular to the base, of an oblique cone oblique to the base.

(b) Directrix of a circular cone is a circle. A *right circular cone* may be generated by the revolution of a right triangle about one leg.

(2) Particular Cases (symbols, Sec. 4.01–2)

(a) Right circular cone (Fig. 4.07–1).

$$A = \pi R\sqrt{v^2+R^2} = \pi R h$$

$$B = \pi R^2 \qquad S = \pi R(R+h) \qquad V = \frac{\pi R^2 v}{3}$$

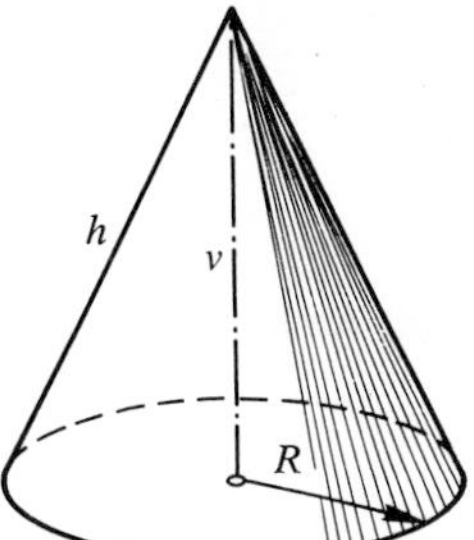

Fig. 4.07-1

(b) Frustum of a right cone (Fig. 4.07–2).

$$A = \pi(R_1+R_2)\sqrt{v^2+(R_1-R_2)^2} = \pi(R_1+R_2)h$$

$$B_b = \pi R_1^2 \qquad B_t = \pi R_2^2$$

$$S = \pi[R_1^2+(R_1+R_2)h+R_2^2] \qquad V = \frac{\pi v}{3}(R_1^2+R_1R_2+R_2^2)$$

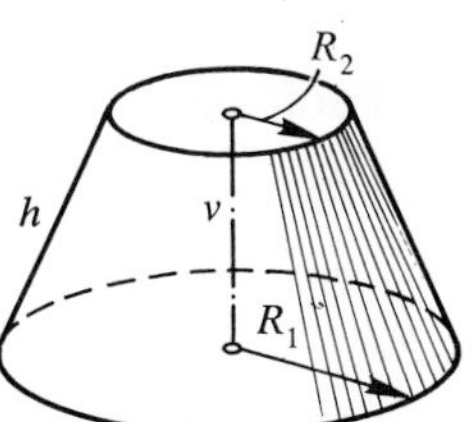

Fig. 4.07-2

(c) General circular cone (Fig. 4.07–3).

$$V = \frac{Bv}{3} = \frac{\pi R_1^2 v}{3}$$

For frustum $B_b = \pi R_1^2$, $B_t = \pi R_2^2$:

$$V = \frac{v_1}{3}(B_b + B_t + \sqrt{B_bB_t})$$

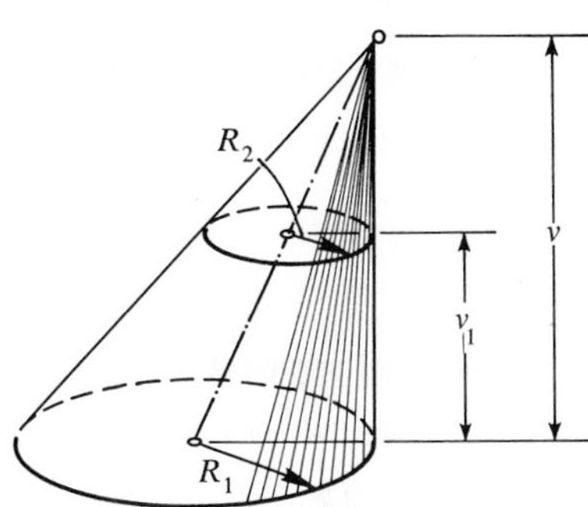

Fig. 4.07-3

4.08 SPHERES

(1) Definitions

(a) Sphere is a solid bounded by a surface all points of which are equally distant from a point within called the center. The equal distance is the radius of the sphere R.

(b) Every section of a sphere made by a plane is a *circle.* A *great circle* of a sphere is a section made by a plane which passes through the center of the sphere. Only one circle can be drawn through any three points on a sphere, and only one sphere can pass through four points not in the same plane.

(c) Spherical angle is formed by two arcs of great circles intersecting on the axis passing through the center (Fig. 4.08–1). A *spherical triangle* is spherical polygon with three sides, bounded by three great circles (Fig. 4.08–2).

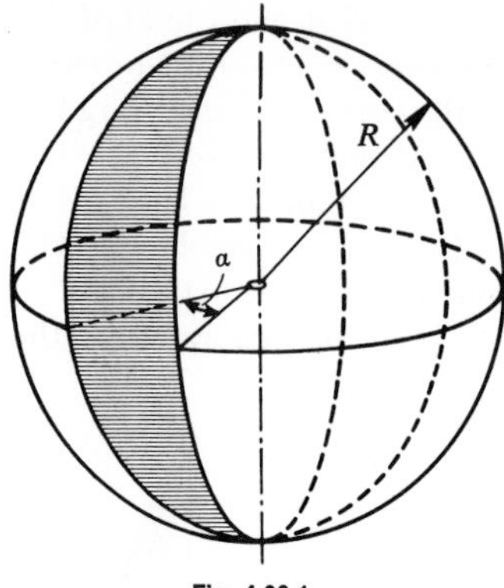

Fig. 4.08-1

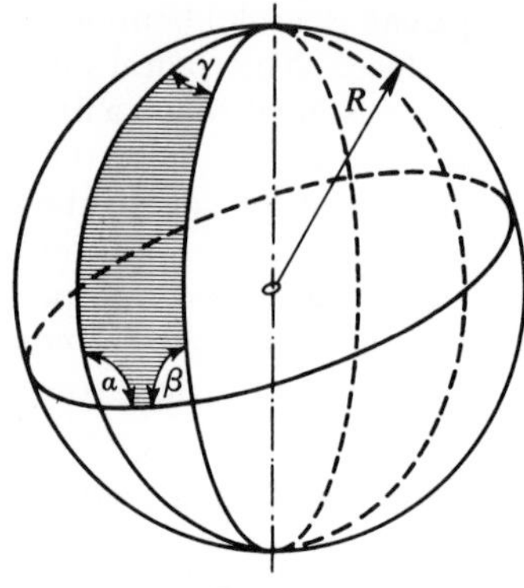

Fig. 4.08-2

(d) Areas of the spherical angle and of the *spherical triangle* are, respectively,

$$A = \frac{\pi R^2 \alpha^\circ}{90^\circ} \qquad \text{(Fig. 4.08–1)}$$

$$A = \frac{\pi R^2(\alpha^\circ + \beta^\circ + \gamma^\circ - 180^\circ)}{180^\circ} \qquad \text{(Fig. 4.08–2)}$$

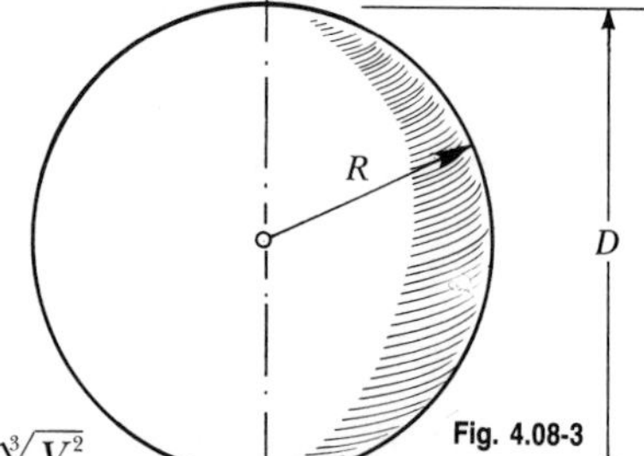

Fig. 4.08-3

(2) Particular Cases (symbols, Sec. 4.01–2)

(a) Sphere (Fig. 4.08–3).

$$S = 4\pi R^2 = 12.5664R^2 = \pi D^2 = \sqrt[3]{36\pi V^2} = 4.8362\sqrt[3]{V^2}$$

$$V = \tfrac{4}{3}\pi R^3 = 4.1888R^3 = \frac{\pi D^3}{6} = \frac{1}{6}\sqrt{\frac{S^3}{\pi}} = 0.0940\sqrt{S^3}$$

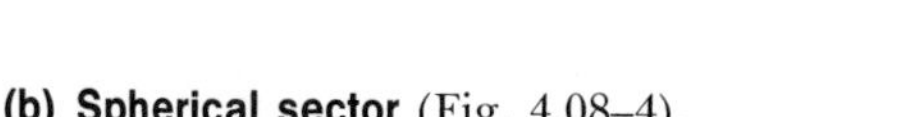

(b) Spherical sector (Fig. 4.08–4).

$$S = \pi R(2v + a) = 3.1416R(2v + a)$$

$$V = \frac{2\pi}{3}R^2 v = 2.0944R^2 v$$

Fig. 4.08-4

(c) Spherical segment (one base) (Fig. 4.08–5).

$$a = \sqrt{v(2R - v)} \qquad A = 2\pi Rv = 6.2832Rv$$

$$S = \pi v(4R - v) = 3.1416(4R - v)V$$

$$V = \frac{\pi}{3}v^2(3R - v) = 1.0472v^2(3R - v)$$

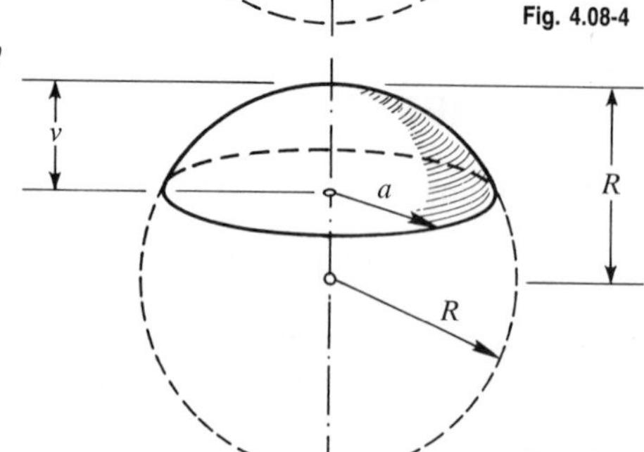

Fig. 4.08-5

(d) Spherical segment (two bases) (Fig. 4.08–6).

$$R = \sqrt{a^2 + \left(\frac{a^2 - b^2 - v^2}{2v}\right)^2}$$

$$A = 2\pi Rv = 6.2832Rv$$

$$S = \pi(2Rv + a^2 + b^2) = 3.1416(2Rv + a^2 + b^2)$$

$$V = \frac{\pi v}{6}(3a^2 + 3b^2 + v^2) = 0.5236v(3a^2 + 3b^2 + v^2)$$

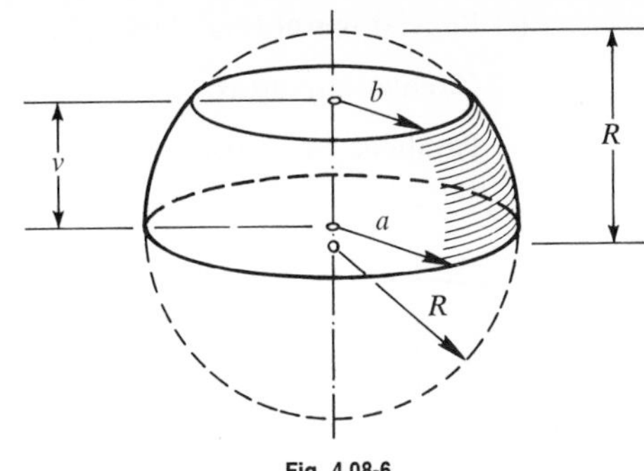

Fig. 4.08-6

4.09 GENERAL SOLIDS OF REVOLUTION

(1) Theorems of Pappus (also called Guldin's theorems)

(a) If a plane curve of length C rotates about a straight line in the plane of C, the area of the total surface generated by this rotation is equal to the product of the length of C and the length of the path made by the center of gravity of C during this rotation.

$$S = 2\pi RC$$

where R is the radius of rotation of the centroid of C.

(b) If a plane area A rotates about a straight line in the plane of A, the volume of the solid generated by this rotation is equal to the product of the area A and the length of the path made by the center of gravity of A during this rotation α.

$$V = \alpha RA$$

where R is the radius of rotation of the centroid of A and in case of a complete revolution $\alpha = 2\pi$.

(2) Particular Cases (symbols, Sec. 4.01–2)

(a) Conical ring (Fig. 4.09–1).

$$S = 2\pi R\left(v + \sqrt{R^2 - \frac{v^2}{4}}\right) = 6.2832R\left(v + \sqrt{R^2 - \frac{v^2}{4}}\right)$$

$$V = \frac{2\pi}{3}R^2 v = 2.0943R^2 v$$

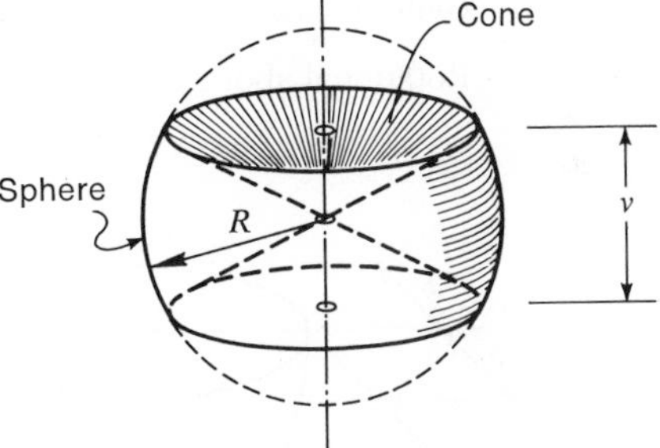

Fig. 4.09-1

(b) Torus (Fig. 4.09–2).

$$S = 4\pi^2 Rr = 39.4784Rr \qquad V = 2\pi^2 Rr^2 = 19.7392Rr^2$$

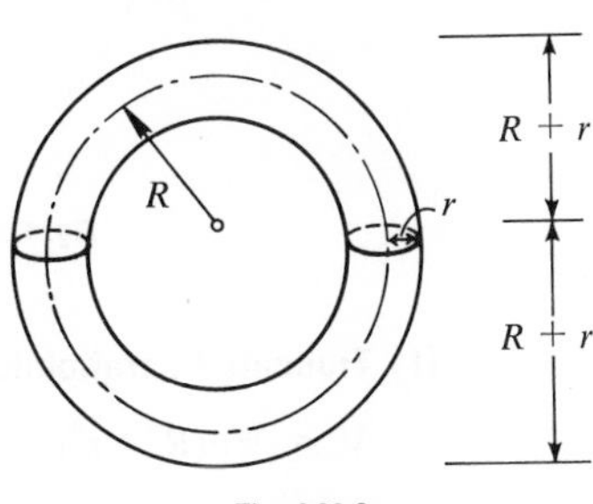

Fig. 4.09-2

(c) Circular barrel (Fig. 4.09–3).

Circular curvature: $V = \frac{1}{3}\pi v(2R^2 + r^2) = 1.0472v(2R^2 + r^2)$

Parabolic curvature: $V = \frac{1}{15}\pi v(8R^2 + 4Rr + 3r^2) = 0.2094v(8R^2 + 4Rr + 3r^2)$

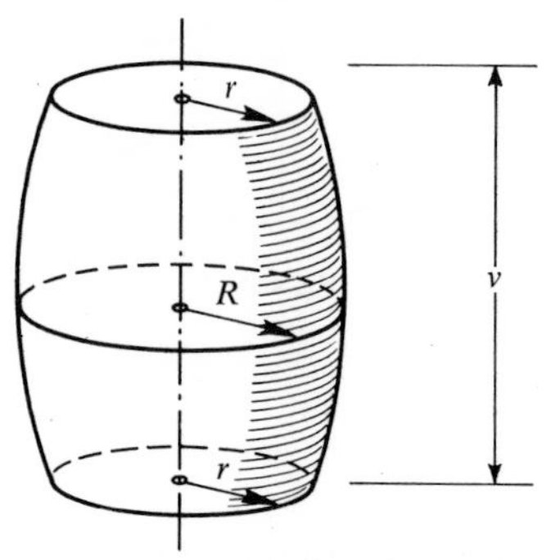

Fig. 4.09-3

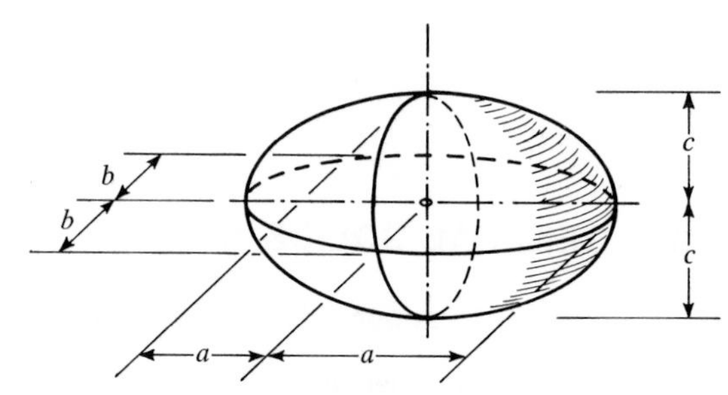

Fig. 4.09-4

(d) Ellipsoid (Fig. 4.09–4).

General case: $V = \frac{4}{3}\pi abc = 4.1888abc$

Rotational about $2a$ axis, $b = c$: $V = \frac{4}{3}\pi ab^2 = 4.1888ab^2$

Rotational about $2b$ axis, $a = c$: $V = \frac{4}{3}\pi a^2 b = 4.1888a^2 b$

(e) Paraboloid (Fig. 4.09–5).

General case: $V = \frac{1}{2}\pi abv = 1.5708abv$

Rotational about v axis, $a = b$: $V = \frac{1}{2}\pi a^2 v = 1.5708a^2 v$

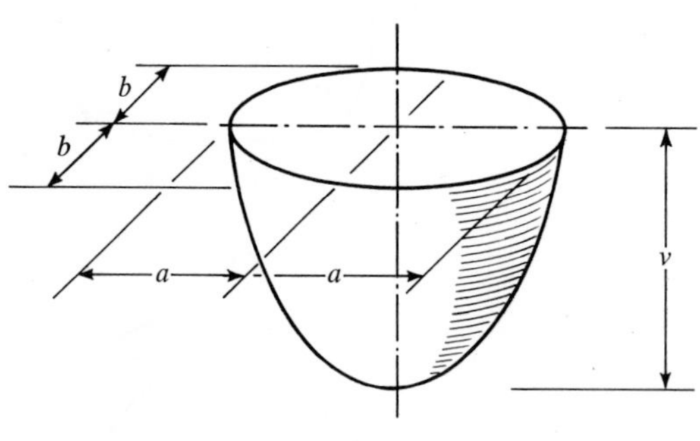

Fig. 4.09-5

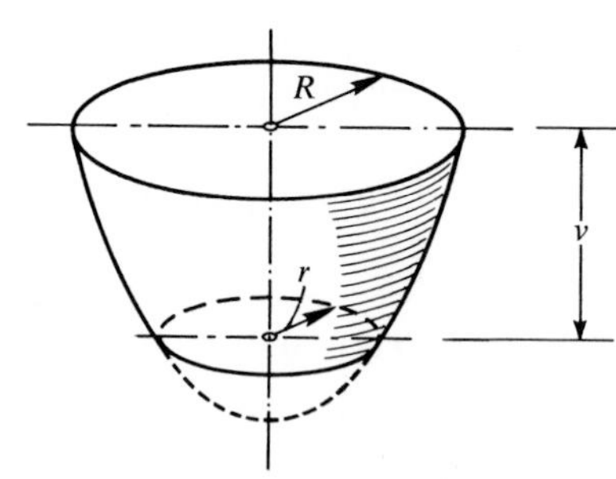

Fig. 4.09-6

(f) Truncated paraboloid of revolution (Fig. 4.09–6).

$V = \frac{1}{2}\pi v(R^2 + r^2) = 1.5708v(R^2 + r^2)$

5
PLANE TRIGONOMETRY

5.01 DEFINITIONS AND NOTATIONS

(1) Definitions

(a) Plane trigonometry describes relations between the sides and angles of plane triangles and polygons by means of trigonometric functions (Sec. 5.02–1).

(b) Basic relations. Since all plane polygons can be decomposed into a system of right triangles, the basic trigonometric relations are those derived from the properties of the right triangle.

(2) Notations

A = area R = circumradius $a, b, c, \ldots$ = sides	$\alpha, \beta, \gamma, \ldots$ = angles h_a, h_b, h_c = altitudes m_a, m_b, m_c = medians	$2p$ = perimeter r = inradius t_a, t_b, t_c = bisectors

5.02 TRIGONOMETRIC FUNCTIONS

(1) Functions and Their Argument

(a) Trigonometric functions of the angle α are defined by means of the trigonometric circle of radius $R = 1$ (Fig. 5.02–1), or for the acute angle $\alpha (\alpha \le 90°)$, by means of the right triangle (Fig. 5.02–2).

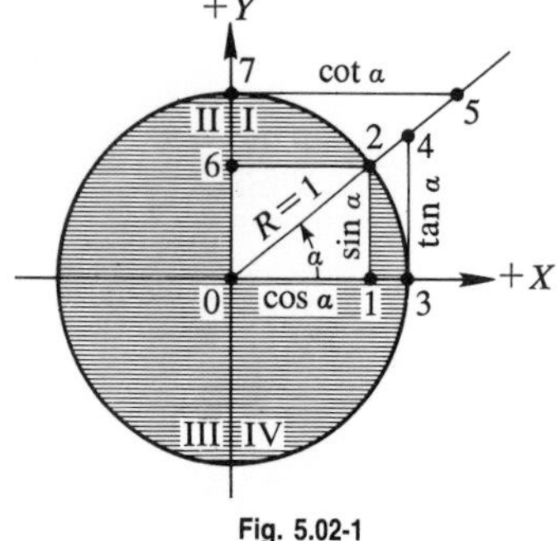

Fig. 5.02-1

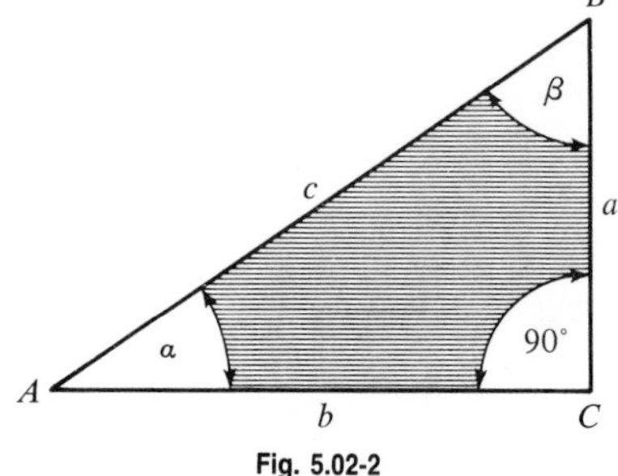

Fig. 5.02-2

sine $\alpha = \sin\alpha = \overline{12} = \dfrac{a}{c}$	cosecant $\alpha = \csc\alpha = \overline{05} = \dfrac{c}{a}$	versine $\alpha = \text{vers}\,\alpha = \overline{13} = \dfrac{c-b}{c}$
cosine $\alpha = \cos\alpha = \overline{01} = \dfrac{b}{c}$	secant $\alpha = \sec\alpha = \overline{04} = \dfrac{c}{a}$	coversine $\alpha = \text{covers}\,\alpha = \overline{67} = \dfrac{c-a}{c}$
tangent $\alpha = \tan\alpha = \overline{34} = \dfrac{a}{b}$	cotangent $\alpha = \cot\alpha = \overline{75} = \dfrac{b}{a}$	

where $\overline{02}$ is always positive but $\overline{01}$ and $\overline{12}$ are directed segments (coordinates of point 2) whose signs depend on their directions. If measured in the direction of the respective axis ($\overline{01}$ along $+X$, $\overline{12}$ along $+Y$) they are positive; otherwise they are negative.

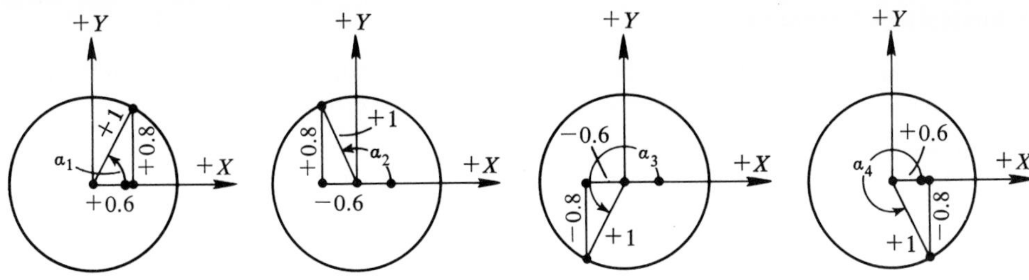

Fig. 5.02-3

examples:

For the angles α_1, α_2, α_3, and α_4 shown in Fig. 5.02–3 find their sine, cosine, and tangent.

$$\sin\alpha_1 = \frac{+0.8}{+1.0} = +0.800 \qquad \sin\alpha_2 = \frac{+0.8}{+1.0} = +0.800 \qquad \sin\alpha_3 = \frac{-0.8}{+1.0} = -0.800 \qquad \sin\alpha_4 = \frac{-0.8}{+1.0} = -0.800$$

$$\cos\alpha_1 = \frac{+0.6}{+1.0} = +0.600 \qquad \cos\alpha_2 = \frac{-0.6}{+1.0} = -0.600 \qquad \cos\alpha_3 = \frac{-0.6}{+1.0} = -0.600 \qquad \cos\alpha_4 = \frac{+0.6}{+1.0} = +0.600$$

$$\tan\alpha_1 = \frac{+0.8}{+0.6} = +1.333 \qquad \tan\alpha_2 = \frac{+0.8}{+0.6} = -1.333 \qquad \tan\alpha_3 = \frac{-0.8}{-0.6} = +1.333 \qquad \tan\alpha_4 = \frac{-0.8}{+0.6} = -1.333$$

(b) Argument of a trigonometric function is the angle α (sometimes designated as A). This angle is positive if measured in the counterclockwise direction and negative if measured in the clockwise direction.

(2) Measures of Angles

(a) Units of α are degrees, minutes, and seconds, or radians.

(b) One degree (1°) is the measure of the central angle subtended by an arc of a circle equal to 1/360 of the circumference of the circle. *One minute* (1′) is 1/60 of a degree. *One second* (1″) is 1/60 of a minute.

(c) One radian (1 rad) is the measure of the central angle subtended by an arc of a circle equal to the radius of the circle.

(d) Circumference of a circle is $2\pi R$ and subtends an angle of 360°. Then for $R = 1$ (Fig. 5.02–4),

$$2\pi \text{ rad} = 360° \qquad \text{or} \qquad \pi \text{ rad} = 180° \qquad \text{where } \pi = 3.14159\ldots$$

and

$$1 \text{ rad} = \frac{180°}{\pi} = 57.29578° \qquad 1° = \frac{\pi}{180} = 0.017453 \text{ rad}$$

Fig. 5.02-4

examples:

$$30° = 30°\frac{\pi}{180°} = \frac{\pi}{6} \text{ rad} = 0.5236 \text{ rad} \qquad \frac{4}{3} \text{ rad} = \frac{4}{3}\frac{180°}{\pi} = \frac{240°}{\pi} = 76°23'40''$$

(3) Analysis of Functions

(a) Sign of the respective trigonometric function depends on the quadrant in which the radius $\overline{01}$ of the angle α lies (Figs. 5.02–1 and 5.02–3) as shown below.

Quadrant	sin	cos	tan	cot	sec	csc	vers	covers
I	+	+	+	+	+	+	+	+
II	+	−	−	−	−	+	+	+
III	−	−	+	+	−	−	+	+
IV	−	+	−	−	+	−	+	+

(b) Graphical representation of the eight trigonometric functions is shown in Fig. 5.02–5. Their special values are given in Table 5.02–1.

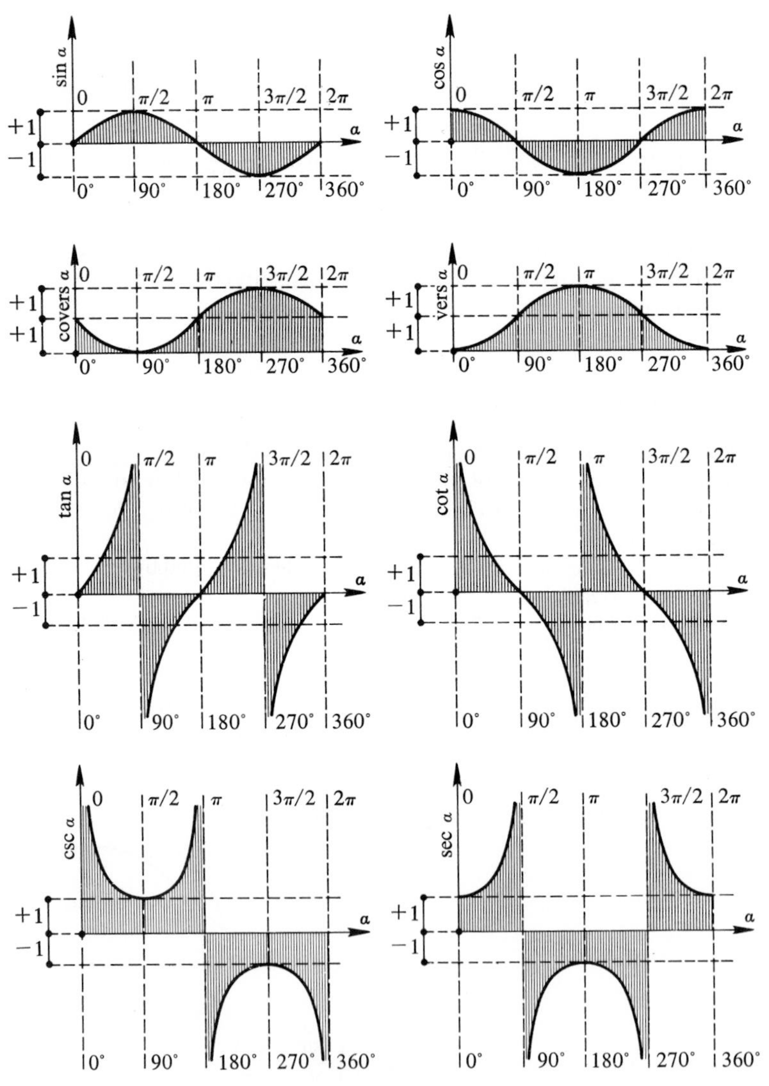

Fig. 5.02-5

TABLE 5.02–1 Special Values of Trigonometric Functions

Degrees	Radians	sin α	cos α	tan α	cot α	sec α	csc α	vers α	covers α
0	0	0	$+1$	0	$\mp\infty$	$+1$	$\mp\infty$	0	$+1$
30	$\frac{\pi}{6}$	$+\frac{1}{2}$	$+\frac{\sqrt{3}}{2}$	$+\frac{\sqrt{3}}{3}$	$+\sqrt{3}$	$+\frac{2\sqrt{3}}{3}$	$+2$	$+1-\frac{\sqrt{3}}{2}$	$+\frac{1}{2}$
45	$\frac{\pi}{4}$	$+\frac{\sqrt{2}}{2}$	$+\frac{\sqrt{2}}{2}$	$+1$	$+1$	$+\sqrt{2}$	$+\sqrt{2}$	$+1-\frac{\sqrt{2}}{2}$	$+1-\frac{\sqrt{2}}{2}$
60	$\frac{\pi}{3}$	$+\frac{\sqrt{3}}{2}$	$+\frac{1}{2}$	$+\sqrt{3}$	$+\frac{\sqrt{3}}{3}$	$+2$	$+\frac{2\sqrt{3}}{3}$	$+\frac{1}{2}$	$+1-\frac{\sqrt{3}}{2}$
90	$\frac{\pi}{2}$	$+1$	0	$\pm\infty$	0	$\pm\infty$	$+1$	$+1$	0
120	$\frac{2\pi}{3}$	$+\frac{\sqrt{3}}{2}$	$-\frac{1}{2}$	$-\sqrt{3}$	$-\frac{\sqrt{3}}{3}$	-2	$+\frac{2\sqrt{3}}{3}$	$+\frac{3}{2}$	$+1-\frac{\sqrt{3}}{2}$
135	$\frac{3\pi}{4}$	$+\frac{\sqrt{2}}{2}$	$-\frac{\sqrt{2}}{2}$	-1	-1	$-\sqrt{2}$	$+\sqrt{2}$	$+1+\frac{\sqrt{2}}{2}$	$+1-\frac{\sqrt{2}}{2}$
150	$\frac{5\pi}{6}$	$+\frac{1}{2}$	$-\frac{\sqrt{3}}{2}$	$-\frac{\sqrt{3}}{3}$	$-\sqrt{3}$	$-\frac{2\sqrt{3}}{3}$	$+2$	$+1+\frac{\sqrt{3}}{2}$	$+\frac{1}{2}$
180	π	0	-1	0	$\mp\infty$	-1	$\mp\infty$	$+2$	$+1$
210	$\frac{7\pi}{6}$	$-\frac{1}{2}$	$-\frac{\sqrt{3}}{2}$	$+\frac{\sqrt{3}}{3}$	$+\sqrt{3}$	$-\frac{2\sqrt{3}}{3}$	-2	$+1+\frac{\sqrt{3}}{2}$	$+\frac{3}{2}$
225	$\frac{5\pi}{4}$	$-\frac{\sqrt{2}}{2}$	$-\frac{\sqrt{2}}{2}$	$+1$	$+1$	$-\sqrt{2}$	$-\sqrt{2}$	$+1+\frac{\sqrt{2}}{2}$	$+1+\frac{\sqrt{2}}{2}$
240	$\frac{4\pi}{3}$	$-\frac{\sqrt{3}}{2}$	$-\frac{1}{2}$	$+\sqrt{3}$	$+\frac{\sqrt{3}}{3}$	-2	$-\frac{2\sqrt{3}}{3}$	$+\frac{3}{2}$	$+1+\frac{\sqrt{3}}{2}$
270	$\frac{3\pi}{2}$	-1	0	$\pm\infty$	0	$\mp\infty$	-1	$+1$	$+2$
300	$\frac{5\pi}{3}$	$-\frac{\sqrt{3}}{2}$	$+\frac{1}{2}$	$-\sqrt{3}$	$-\frac{\sqrt{3}}{3}$	$+2$	$-\frac{2\sqrt{3}}{3}$	$+\frac{1}{2}$	$+1+\frac{\sqrt{3}}{2}$
315	$\frac{7\pi}{4}$	$-\frac{\sqrt{2}}{2}$	$+\frac{\sqrt{2}}{2}$	-1	-1	$+\sqrt{2}$	$-\sqrt{2}$	$+1-\frac{\sqrt{2}}{2}$	$+1+\frac{\sqrt{2}}{2}$
330	$\frac{11\pi}{6}$	$-\frac{1}{2}$	$+\frac{\sqrt{3}}{2}$	$-\frac{\sqrt{3}}{3}$	$-\sqrt{3}$	$+\frac{2\sqrt{3}}{3}$	-2	$+1-\frac{\sqrt{3}}{2}$	$+\frac{3}{2}$
360	2π	0	$+1$	0	$\mp\infty$	$+1$	$\mp\infty$	0	$+1$

(4) Change in Argument

(a) Negative argument.

$\sin(-\alpha) = -\sin\alpha$	$\cos(-\alpha) = \cos\alpha$
$\tan(-\alpha) = -\tan\alpha$	$\cot(-\alpha) = -\cot\alpha$
$\csc(-\alpha) = -\csc\alpha$	$\sec(-\alpha) = \sec\alpha$
$\text{covers}(-\alpha) = 2 - \text{covers}\,\alpha$	$\text{vers}(-\alpha) = \text{vers}\,\alpha$

(b) Argument $(90° - \alpha)$.

$\sin(90° - \alpha) = \cos\alpha$	$\cos(90° - \alpha) = \sin\alpha$
$\tan(90° - \alpha) = \cot\alpha$	$\cot(90° - \alpha) = \tan\alpha$
$\csc(90° - \alpha) = \sec\alpha$	$\sec(90° - \alpha) = \csc\alpha$
$\text{covers}(90° - \alpha) = \text{vers}\,\alpha$	$\text{vers}(90° - \alpha) = \text{covers}\,\alpha$

(c) Argument $(90° < \alpha < 360°)$.

Function	$\beta = 90° + \alpha$	$\beta = 180° \pm \alpha$	$\beta = 270° \pm \alpha$	$\beta = 360° - \alpha$
$\sin\beta$	$+\cos\alpha$	$\mp\sin\alpha$	$-\cos\alpha$	$-\sin\alpha$
$\cos\beta$	$-\sin\alpha$	$-\cos\alpha$	$\pm\sin\alpha$	$+\cos\alpha$
$\tan\beta$	$-\cot\alpha$	$\pm\tan\alpha$	$\mp\cot\alpha$	$-\tan\alpha$
$\cot\beta$	$-\tan\alpha$	$\pm\cot\alpha$	$\mp\tan\alpha$	$-\cot\alpha$
$\sec\beta$	$-\csc\alpha$	$-\sec\alpha$	$\pm\csc\alpha$	$+\sec\alpha$
$\csc\beta$	$+\sec\alpha$	$\mp\csc\alpha$	$-\sec\alpha$	$-\csc\alpha$
$\text{vers}\,\beta$	$2 - \text{covers}\,\alpha$	$2 - \text{vers}\,\alpha$	$\frac{\text{covers}\,\alpha}{2 - \text{covers}\,\alpha}$	$\text{vers}\,\alpha$
$\text{covers}\,\beta$	$\text{vers}\,\alpha$	$\frac{2 - \text{covers}\,\alpha}{\text{covers}\,\alpha}$	$2 - \text{vers}\,\alpha$	$2 - \text{covers}\,\alpha$

examples:

$\sin 120° = \sin(90° + 30°) = \cos 30°$ or $\sin 120° = \sin(180° - 60°) = \sin 60°$

$\cos 190° = \cos(180° + 10°) = -\cos 10°$ or $\cos 190° = \cos(270° - 80°) = -\sin 80°$

$\tan 290° = \tan(270° + 20°) = -\cot 20°$ or $\tan 290° = \tan(360° - 70°) = -\tan 70°$

$\text{vers}\,110° = \text{vers}(90° + 20°) = 2 - \text{covers}\,20°$ or $\text{vers}\,110° = \text{vers}(180° - 70°) = 2 - \text{vers}\,70°$

(d) Argument $(n360° + \alpha)$.

$$F(n360° + \alpha) = F(\alpha)$$

where $F(\ )$ is the respective trigonometric function and $n = 1, 2, 3, \ldots$.

example:

$\tan 780° = \tan[2(360°) + 60°] = \tan 60°$

5.03 RELATIONS BETWEEN FUNCTIONS

(1) Basic Relations

(a) Formulas.

$\sin^2 \alpha + \cos^2 \alpha = 1$ $\sec^2 \alpha - \tan^2 \alpha = 1$ $\csc^2 \alpha - \cot^2 \alpha = 1$	$\sin \alpha \csc \alpha = 1$ $\cos \alpha \sec \alpha = 1$ $\tan \alpha \cot \alpha = 1$
$\sin \alpha + \text{covers}\, \alpha = 1$ $\cos \alpha + \text{vers}\, \alpha = 1$	$\tan \alpha = \dfrac{\sin \alpha}{\cos \alpha}$ $\cot \alpha = \dfrac{\cos \alpha}{\sin \alpha}$

(b) Geometric models. The basic formulas of Sec. 5.03–1*a* can be conveniently derived from Figs. 5.03–1, 5.03–2, and 5.03–3.

examples:

From Fig. 5.03–1:

$$\sin^2 \alpha + \cos^2 \alpha = 1 \qquad \sin \alpha + \text{covers}\, \alpha = 1$$

$$\cos \alpha + \text{vers}\, \alpha = 1$$

$$\tan \alpha = \frac{\sin \alpha}{\cos \alpha} \qquad \cot \alpha = \frac{\cos \alpha}{\sin \alpha}$$

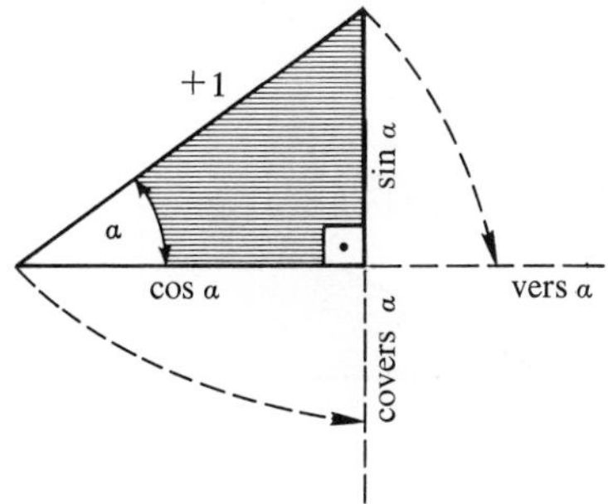

Fig. 5.03-1

From Fig. 5.03–2:

$$1 + \tan^2 \alpha = \sec^2 \alpha \qquad \cos \alpha = \frac{1}{\sec \alpha} \qquad \sin \alpha = \frac{\tan \alpha}{\sec \alpha}$$

$$\cot \alpha = \frac{1}{\tan \alpha}$$

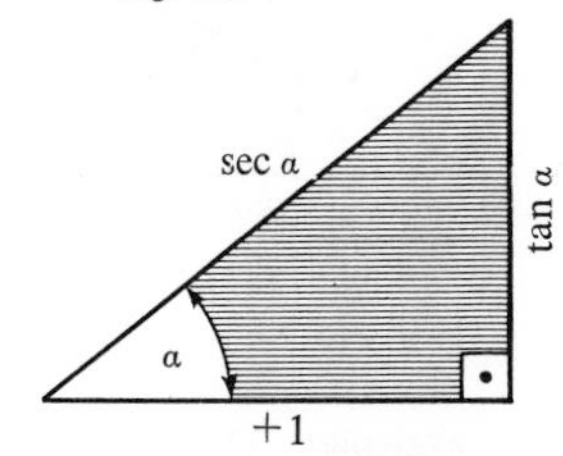

Fig. 5.03-2

From Fig. 5.03–3:

$$1 + \cot^2 \alpha = \csc^2 \alpha \qquad \sin \alpha = \frac{1}{\csc \alpha} \qquad \cos \alpha = \frac{\cot \alpha}{\csc \alpha}$$

$$\tan \alpha = \frac{1}{\cot \alpha}$$

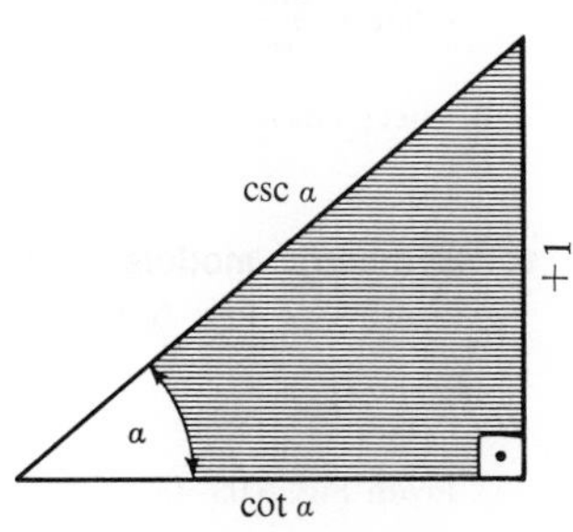

Fig. 5.03-3

(2) Transformations

(a) Direct transformations. The basic formulas of Sec. 5.03–1*a* give direct relations between $\sin\alpha$ and $\cos\alpha$, $\sin\alpha$ and $\csc\alpha$, $\cos\alpha$ and $\sec\alpha$, $\tan\alpha$ and $\sec\alpha$, $\cot\alpha$ and $\csc\alpha$, $\tan\alpha$ and $\cot\alpha$, $\sin\alpha$ and $\operatorname{covers}\alpha$, $\cos\alpha$ and $\operatorname{vers}\alpha$.

examples:

From $\sin^2\alpha+\cos^2\alpha=1$: $\quad \sin\alpha=\pm\sqrt{1-\cos^2\alpha}\quad$ or $\quad\cos\alpha=\pm\sqrt{1-\sin^2\alpha}$

From $\sin\alpha\csc\alpha=1$: $\quad \sin\alpha=\dfrac{1}{\csc\alpha}\quad$ or $\quad\csc\alpha=\dfrac{1}{\sin\alpha}$

From $\sin\alpha+\operatorname{covers}\alpha=1$: $\quad \sin\alpha=1-\operatorname{covers}\alpha\quad$ or $\quad\operatorname{covers}\alpha=1-\sin\alpha$

(b) Indirect transformations. The remaining relations cannot be derived from one basic formula, and two or more relations of Sec. 5.03–1*a* must be employed.

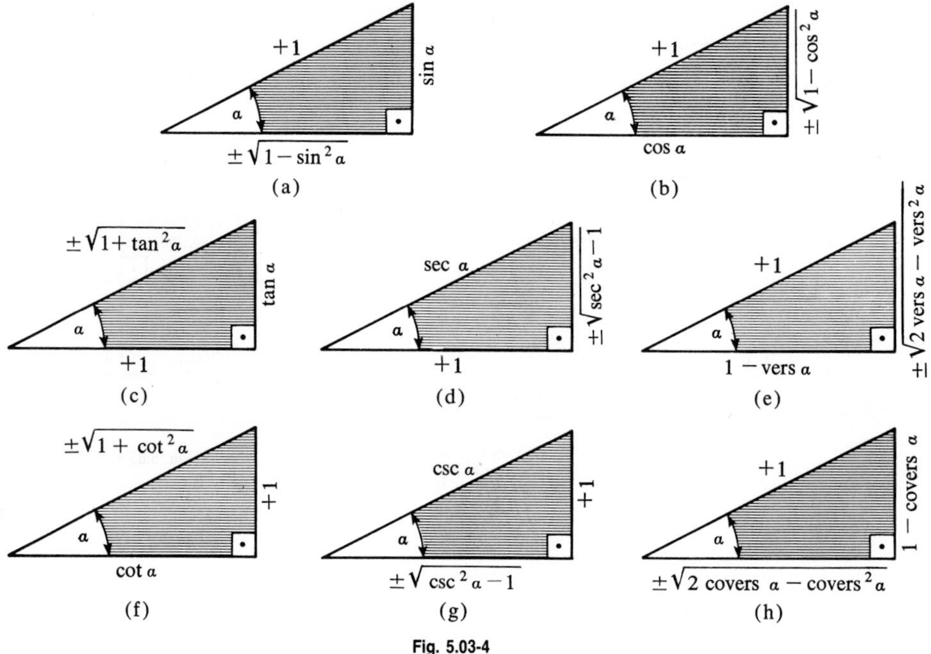

Fig. 5.03-4

example: The relation between $\tan\alpha$ and $\sin\alpha$ can be derived from

$$\tan\alpha=\frac{\sin\alpha}{\cos\alpha}$$

where $\cos\alpha=\pm\sqrt{1-\sin^2\alpha}$ as shown in the preceding example.

(c) Geometric models. The same relations can be conveniently derived from the right triangles of Fig. 5.03–4.

examples:

From Fig. 5.03–4*c*: $\quad \sin\alpha=\dfrac{\tan\alpha}{\pm\sqrt{1+\tan^2\alpha}}\qquad\cos\alpha=\dfrac{1}{\pm\sqrt{1+\tan^2\alpha}},\quad\ldots$

From Fig. 5.03–4*e*: $\quad \sin\alpha=\pm\sqrt{2\operatorname{vers}\alpha-\operatorname{vers}^2\alpha}\qquad\cos\alpha=1-\operatorname{vers}\alpha,\quad\ldots$

(d) Transformation table. A complete set of transformation relations is given in Table 5.03–1.

TABLE 5.03–1 Transformations of Trigonometric Functions*

	$\sin\alpha$	$\cos\alpha$	$\tan\alpha$	$\cot\alpha$	$\sec\alpha$	$\csc\alpha$	$\operatorname{vers}\alpha$	$\operatorname{covers}\alpha$
$\sin\alpha$	$\sin\alpha$	$\pm\sqrt{1-\cos^2\alpha}$	$\dfrac{\tan\alpha}{\pm\sqrt{1+\tan^2\alpha}}$	$\dfrac{1}{\pm\sqrt{1+\cot^2\alpha}}$	$\dfrac{\pm\sqrt{\sec^2\alpha-1}}{\sec\alpha}$	$\dfrac{1}{\csc\alpha}$	$\pm\sqrt{2\operatorname{vers}\alpha-\operatorname{vers}^2\alpha}$	$1-\operatorname{covers}\alpha$
$\cos\alpha$	$\pm\sqrt{1-\sin^2\alpha}$	$\cos\alpha$	$\dfrac{1}{\pm\sqrt{1+\tan^2\alpha}}$	$\dfrac{\cot\alpha}{\pm\sqrt{1+\cot^2\alpha}}$	$\dfrac{1}{\sec\alpha}$	$\dfrac{\pm\sqrt{\csc^2\alpha-1}}{\csc\alpha}$	$1-\operatorname{vers}\alpha$	$\pm\sqrt{2\operatorname{covers}\alpha-\operatorname{covers}^2\alpha}$
$\tan\alpha$	$\dfrac{\sin\alpha}{\pm\sqrt{1-\sin^2\alpha}}$	$\dfrac{\pm\sqrt{1-\cos^2\alpha}}{\cos\alpha}$	$\tan\alpha$	$\dfrac{1}{\cot\alpha}$	$\pm\sqrt{\sec^2\alpha-1}$	$\dfrac{1}{\pm\sqrt{\csc^2\alpha-1}}$	$\dfrac{\pm\sqrt{2\operatorname{vers}\alpha-\operatorname{vers}^2\alpha}}{1-\operatorname{vers}\alpha}$	$\dfrac{1-\operatorname{covers}\alpha}{\pm\sqrt{2\operatorname{covers}\alpha-\operatorname{covers}^2\alpha}}$
$\cot\alpha$	$\dfrac{\pm\sqrt{1-\sin^2\alpha}}{\sin\alpha}$	$\dfrac{\cos\alpha}{\pm\sqrt{1-\cos^2\alpha}}$	$\dfrac{1}{\tan\alpha}$	$\cot\alpha$	$\dfrac{1}{\pm\sqrt{\sec^2\alpha-1}}$	$\pm\sqrt{\csc^2\alpha-1}$	$\dfrac{1-\operatorname{vers}\alpha}{\pm\sqrt{2\operatorname{vers}\alpha-\operatorname{vers}^2\alpha}}$	$\dfrac{\pm\sqrt{2\operatorname{covers}\alpha-\operatorname{covers}^2\alpha}}{1-\operatorname{covers}\alpha}$
$\sec\alpha$	$\dfrac{1}{\pm\sqrt{1-\sin^2\alpha}}$	$\dfrac{1}{\cos\alpha}$	$\pm\sqrt{1+\tan^2\alpha}$	$\dfrac{\pm\sqrt{1+\cot^2\alpha}}{\cot\alpha}$	$\sec\alpha$	$\dfrac{\csc\alpha}{\pm\sqrt{\csc^2\alpha-1}}$	$\dfrac{1}{1-\operatorname{vers}\alpha}$	$\dfrac{1}{\pm\sqrt{2\operatorname{covers}\alpha-\operatorname{covers}^2\alpha}}$
$\csc\alpha$	$\dfrac{1}{\sin\alpha}$	$\dfrac{1}{\pm\sqrt{1-\cos^2\alpha}}$	$\dfrac{\pm\sqrt{1+\tan^2\alpha}}{\tan\alpha}$	$\pm\sqrt{1+\cot^2\alpha}$	$\dfrac{\sec\alpha}{\pm\sqrt{\sec^2\alpha-1}}$	$\csc\alpha$	$\dfrac{1}{\pm\sqrt{2\operatorname{vers}\alpha-\operatorname{vers}^2\alpha}}$	$\dfrac{1}{1-\operatorname{covers}\alpha}$
$\operatorname{vers}\alpha$	$1\mp\sqrt{1-\sin^2\alpha}$	$1-\cos\alpha$	$1\mp\dfrac{1}{\sqrt{1+\tan^2\alpha}}$	$1\mp\dfrac{\cot\alpha}{\sqrt{1+\cot^2\alpha}}$	$1\mp\dfrac{\sqrt{\sec^2\alpha-1}}{\sec\alpha}$	$1\mp\dfrac{\sqrt{\csc^2\alpha-1}}{\csc\alpha}$	$\operatorname{vers}\alpha$	$1\mp\sqrt{2\operatorname{covers}\alpha-\operatorname{covers}^2\alpha}$
$\operatorname{covers}\alpha$	$1-\sin\alpha$	$1\mp\sqrt{1-\cos^2\alpha}$	$1\mp\dfrac{\tan\alpha}{\sqrt{1+\tan^2\alpha}}$	$1\mp\dfrac{1}{\sqrt{1+\cot^2\alpha}}$	$1\mp\dfrac{\sqrt{\sec^2\alpha-1}}{\sec\alpha}$	$1\mp\dfrac{1}{\csc\alpha}$	$1\mp\sqrt{2\operatorname{vers}\alpha-\operatorname{vers}^2\alpha}$	$\operatorname{covers}\alpha$

*The sign is governed by the quadrant in which the argument terminates (Sec. 5.02–3a).

examples:

$\cos 60° = +\sqrt{1-\sin^2 60°}$ $\qquad \cos 150° = -\sqrt{1-\sin^2 150°}$

$\cos 220° = -\sqrt{1-\sin^2 220°}$ $\qquad \cos 320° = +\sqrt{1-\sin^2 320°}$

$\sin 60° = +\sqrt{1-\cos^2 60°}$ $\qquad \sin 150° = +\sqrt{1-\cos^2 150°}$

$\sin 220° = -\sqrt{1-\cos^2 220°}$ $\qquad \sin 320° = -\sqrt{1-\cos^2 320°}$

5.04 TRIGONOMETRIC IDENTITIES

(1) Sums

(a) $F(\alpha \pm \beta)$.

$$\sin(\alpha \pm \beta) = \sin\alpha\cos\beta \pm \sin\beta\cos\alpha \qquad \tan(\alpha \pm \beta) = \frac{\tan\alpha \pm \tan\beta}{1 \mp \tan\alpha\tan\beta}$$

$$\cos(\alpha \pm \beta) = \cos\alpha\cos\beta \mp \sin\alpha\sin\beta \qquad \cot(\alpha \pm \beta) = \frac{\cot\alpha\cot\beta \mp 1}{\cot\beta \pm \cot\alpha}$$

(b) $F(\alpha) \pm F(\beta)$.

$$\sin\alpha + \sin\beta = 2\sin\tfrac{1}{2}(\alpha+\beta)\cos\tfrac{1}{2}(\alpha-\beta) \qquad \tan\alpha \pm \tan\beta = \frac{\sin(\alpha \pm \beta)}{\cos\alpha\cos\beta}$$

$$\sin\alpha - \sin\beta = 2\sin\tfrac{1}{2}(\alpha-\beta)\cos\tfrac{1}{2}(\alpha+\beta)$$

$$\cos\alpha + \cos\beta = 2\cos\tfrac{1}{2}(\alpha+\beta)\cos\tfrac{1}{2}(\alpha-\beta) \qquad \cot\alpha \pm \cot\beta = \frac{\sin(\alpha \pm \beta)}{\sin\alpha\sin\beta}$$

$$\cos\alpha - \cos\beta = -2\sin\tfrac{1}{2}(\alpha+\beta)\sin\tfrac{1}{2}(\alpha-\beta)$$

(c) $F(\alpha) \pm G(\alpha)$.

$$\sin\alpha + \cos\alpha = \sqrt{2}\sin\left(\frac{\pi}{4}+\alpha\right) = \sqrt{2}\cos\left(\frac{\pi}{4}-\alpha\right)$$

$$\sin\alpha - \cos\alpha = -\sqrt{2}\cos\left(\frac{\pi}{4}+\alpha\right) = -\sqrt{2}\sin\left(\frac{\pi}{4}-\alpha\right)$$

$$\tan\alpha + \cot\alpha = 2\csc 2\alpha$$

$$\tan\alpha - \cot\alpha = -2\cot 2\alpha$$

(d) $F(\alpha) + F(\beta) \pm F(\gamma)$, $(\alpha + \beta + \gamma = 180°)$.

$$\sin\alpha + \sin\beta + \sin\gamma = 4\sin\frac{\alpha+\beta}{2}\cos\frac{\alpha}{2}\cos\frac{\beta}{2}$$

$$\sin\alpha + \sin\beta - \sin\gamma = 4\sin\frac{\alpha+\beta}{2}\sin\frac{\alpha}{2}\sin\frac{\beta}{2}$$

$$\cos\alpha + \cos\beta + \cos\gamma = 4\cos\frac{\alpha+\beta}{2}\sin\frac{\alpha}{2}\sin\frac{\beta}{2} + 1$$

$$\cos\alpha + \cos\beta - \cos\gamma = 4\cos\frac{\alpha+\beta}{2}\cos\frac{\alpha}{2}\cos\frac{\beta}{2} - 1$$

$$\tan\alpha + \tan\beta + \tan\gamma = -\tan(\alpha+\beta)\tan\alpha\tan\beta$$

$$\tan\alpha + \tan\beta - \tan\gamma = -\cot(\alpha+\beta)\tan\alpha\tan\beta$$

(e) $F(\alpha + \beta) \pm F(\alpha - \beta)$.

$$\sin(\alpha+\beta) + \sin(\alpha-\beta) = 2\sin\alpha\cos\beta$$

$$\sin(\alpha+\beta) - \sin(\alpha-\beta) = 2\cos\alpha\sin\beta$$

$$\cos(\alpha+\beta) + \cos(\alpha-\beta) = 2\cos\alpha\cos\beta$$

$$\cos(\alpha+\beta) - \cos(\alpha-\beta) = -2\sin\alpha\sin\beta$$

(2) Multiple Angles

For $\binom{n}{k}$ refer to Table A.43.

(a) $\sin n\alpha \ (n = 2, 3, 4, \ldots)$.

$$\sin 2\alpha = 2 \sin \alpha \cos \alpha$$

$$\sin 3\alpha = 3 \sin \alpha \cos^2 \alpha - \sin^3 \alpha = (\sin \alpha)(3 - 4 \sin^2 \alpha)$$

$$\sin 4\alpha = 4 \sin \alpha \cos^3 \alpha - 4 \sin^3 \alpha \cos \alpha = (4 \sin \alpha \cos \alpha)(2 \cos^2 \alpha - 1)$$

$$\sin 5\alpha = 5 \sin \alpha \cos^4 \alpha - 10 \sin^3 \alpha \cos^2 \alpha + \sin^5 \alpha$$

$$\sin 6\alpha = 6 \sin \alpha \cos^5 \alpha - 20 \sin^3 \alpha \cos^3 \alpha + 6 \sin^5 \alpha \cos \alpha$$

$$\sin 7\alpha = 7 \sin \alpha \cos^6 \alpha - 35 \sin^3 \alpha \cos^4 \alpha + 21 \sin^5 \alpha \cos^2 \alpha - \sin^7 \alpha$$

$$\sin n = \binom{n}{1} \sin \alpha \cos^{n-1} \alpha - \binom{n}{3} \sin^3 \alpha \cos^{n-3} \alpha + \binom{n}{5} \sin^5 \alpha \cos^{n-5} \alpha - \cdots$$

(b) $\cos n\alpha \ (n = 2, 3, \ldots)$.

$$\cos 2\alpha = \cos^2 \alpha - \sin^2 \alpha = 1 - 2 \sin^2 \alpha = 2 \cos^2 \alpha - 1$$

$$\cos 3\alpha = \cos^3 \alpha - 3 \sin^2 \alpha \cos \alpha = (\cos \alpha)(4 \cos^2 \alpha - 3)$$

$$\cos 4\alpha = \cos^4 \alpha - 6 \sin^2 \alpha \cos^2 \alpha + \sin^4 \alpha = 1 - 8 \cos^2 \alpha + 8 \cos^4 \alpha$$

$$\cos 5\alpha = \cos^5 \alpha - 10 \sin^2 \alpha \cos^3 \alpha + 5 \sin^4 \alpha \cos \alpha$$

$$\cos 6\alpha = \cos^6 \alpha - 15 \sin^2 \alpha \cos^4 \alpha + 15 \sin^4 \alpha \cos^2 \alpha - \sin^6 \alpha$$

$$\cos 7\alpha = \cos^7 \alpha - 21 \sin^2 \alpha \cos^5 \alpha + 35 \sin^4 \alpha \cos^3 \alpha - 7 \sin^6 \alpha \cos \alpha$$

$$\cos n\alpha = \cos^n \alpha - \binom{n}{2} \sin^2 \alpha \cos^{n-2} \alpha + \binom{n}{4} \sin^4 \alpha \cos^{n-4} \alpha - \cdots$$

(c) $\tan n\alpha \ (n = 2, 3, 4, \ldots)$.

$$\tan 2\alpha = \frac{2 \tan \alpha}{1 - \tan^2 \alpha} = \frac{2}{\cot \alpha - \tan \alpha}$$

$$\tan 3\alpha = \frac{3 \tan \alpha - \tan^3 \alpha}{1 - 3 \tan^2 \alpha}$$

$$\tan 4\alpha = \frac{4 \tan \alpha - 4 \tan^3 \alpha}{1 - 6 \tan^2 \alpha + \tan^4 \alpha}$$

$$\tan 5\alpha = \frac{5 \tan \alpha - 10 \tan^3 \alpha + \tan^5 \alpha}{1 - 10 \tan^2 \alpha + 5 \tan^4 \alpha}$$

$$\tan 6\alpha = \frac{6 \tan \alpha - 20 \tan^3 \alpha + 6 \tan^5 \alpha}{1 - 15 \tan^2 \alpha + 15 \tan^4 \alpha - \tan^6 \alpha}$$

$$\tan 7\alpha = \frac{7 \tan \alpha - 35 \tan^3 \alpha + 21 \tan^5 \alpha - \tan^7 \alpha}{1 - 21 \tan^2 \alpha + 35 \tan^4 \alpha - 7 \tan^6 \alpha}$$

$$\tan n\alpha = \frac{\binom{n}{1} \tan \alpha - \binom{n}{3} \tan^3 \alpha + \binom{n}{5} \tan^5 \alpha - \cdots}{1 - \binom{n}{2} \tan^2 \alpha + \binom{n}{4} \tan^4 \alpha - \binom{n}{6} \tan^6 \alpha + \cdots}$$

(d) cot $n\alpha$ $(n = 2, 3, 4, \ldots)$.

$$\cot 2\alpha = \frac{\cot^2 \alpha - 1}{2 \cot \alpha} = \frac{\cot \alpha - \tan \alpha}{2}$$

$$\cot 3\alpha = \frac{\cot^3 \alpha - 3 \cot \alpha}{3 \cot^2 \alpha - 1}$$

$$\cot 4\alpha = \frac{\cot^4 \alpha - 6 \cot^2 \alpha + 1}{4 \cot^3 \alpha - 4 \cot \alpha}$$

$$\cot 5\alpha = \frac{\cot^5 \alpha - 10 \cot^3 \alpha + 5 \cot \alpha}{5 \cot^4 \alpha - 10 \cot^2 \alpha + 1}$$

$$\cot 6\alpha = \frac{\cot^6 \alpha - 15 \cot^4 \alpha + 15 \cot^2 \alpha - 1}{6 \cot^5 \alpha - 20 \cot^3 \alpha + 6 \cot \alpha}$$

$$\cot 7\alpha = \frac{\cot^7 \alpha - 21 \cot^5 \alpha + 35 \cot^3 \alpha - \cot \alpha}{7 \cot^6 \alpha - 35 \cot^4 \alpha - 21 \cot^2 \alpha + 1}$$

$$\cot n\alpha = \frac{\cot^n \alpha - \binom{n}{2} \cot^{n-2} \alpha + \binom{n}{4} \cot^{n-4} \alpha - \cdots}{\binom{n}{1} \cot^{n-1} \alpha - \binom{n}{3} \cot^{n-3} \alpha + \binom{n}{5} \cot^{n-5} \alpha - \cdots}$$

(3) Half Angles

(a) $F(\alpha/2) = G(\alpha)$.

$$\sin \frac{\alpha}{2} = \sqrt{\frac{1 - \cos \alpha}{2}} \qquad \cos \frac{\alpha}{2} = \sqrt{\frac{1 + \cos \alpha}{2}}$$

$$\tan \frac{\alpha}{2} = \sqrt{\frac{1 - \cos \alpha}{1 + \cos \alpha}} = \frac{\sin \alpha}{1 + \cos \alpha} = \frac{1 - \cos \alpha}{\sin \alpha}$$

$$\cot \frac{\alpha}{2} = \sqrt{\frac{1 + \cos \alpha}{1 - \cos \alpha}} = \frac{1 + \cos \alpha}{\sin \alpha} = \frac{\sin \alpha}{1 - \cos \alpha}$$

(b) $F(\alpha) = H(\alpha/2)$.

$$\sin \alpha = 2 \sin \frac{\alpha}{2} \cos \frac{\alpha}{2} \qquad \cos \alpha = \cos^2 \frac{\alpha}{2} - \sin^2 \frac{\alpha}{2}$$

$$\tan \alpha = \frac{2 \tan (\alpha/2)}{1 - \tan^2 (\alpha/2)} = \frac{2 \sin (\alpha/2) \cos (\alpha/2)}{\cos^2 (\alpha/2) - \sin^2 (\alpha/2)} = \frac{2}{\cot (\alpha/2) - \tan (\alpha/2)}$$

$$\cot \alpha = \frac{\cot^2 (\alpha/2) - 1}{2 \cot (\alpha/2)} = \frac{\cos^2 (\alpha/2) - \sin^2 (\alpha/2)}{2 \sin (\alpha/2) \cos (\alpha/2)} = \frac{\cot (\alpha/2) - \tan (\alpha/2)}{2}$$

(c) $F(\alpha) = H(2\alpha)$.

$$\sin \alpha = \sqrt{\frac{1 - \cos 2\alpha}{2}} \qquad \cos \alpha = \sqrt{\frac{1 + \cos}{2}}$$

$$\tan \alpha = \sqrt{\frac{1 - \cos 2\alpha}{1 + \cos 2\alpha}} = \frac{\sin 2\alpha}{1 + \cos 2\alpha} = \frac{1 - \cos 2\alpha}{\sin 2\alpha}$$

$$\cot \alpha = \sqrt{\frac{1 + \cos 2\alpha}{1 - \cos 2\alpha}} = \frac{1 + \cos 2\alpha}{\sin 2\alpha} = \frac{\sin 2\alpha}{1 - \cos 2\alpha}$$

(4) Products

(a) $F(\alpha)F(\beta)$.

$$\sin\alpha\sin\beta = \tfrac{1}{2}\cos(\alpha-\beta) - \tfrac{1}{2}\cos(\alpha+\beta)$$

$$\cos\alpha\cos\beta = \tfrac{1}{2}\cos(\alpha-\beta) + \tfrac{1}{2}\cos(\alpha+\beta)$$

$$\tan\alpha\tan\beta = \frac{\cos(\alpha-\beta)-\cos(\alpha+\beta)}{\cos(\alpha-\beta)+\cos(\alpha+\beta)} = \frac{\tan\alpha+\tan\beta}{\cot\alpha+\cot\beta}$$

$$\cot\alpha\cot\beta = \frac{\cos(\alpha-\beta)+\cos(\alpha+\beta)}{\cos(\alpha-\beta)-\cos(\alpha+\beta)} = \frac{\cot\alpha+\tan\beta}{\tan\alpha+\cot\beta}$$

(b) $F(\alpha)G(\beta)$.

$$\sin\alpha\cos\beta = \tfrac{1}{2}\sin(\alpha-\beta) + \tfrac{1}{2}\sin(\alpha+\beta)$$

$$\tan\alpha\cot\beta = \frac{\sin(\alpha-\beta)+\sin(\alpha+\beta)}{-\sin(\alpha-\beta)+\sin(\alpha+\beta)} = \frac{\tan\alpha+\cot\beta}{\cot\alpha+\tan\beta}$$

$$\cos\alpha\sin\beta = \tfrac{1}{2}\sin(\alpha+\beta) - \tfrac{1}{2}\sin(\alpha-\beta)$$

$$\cot\alpha\tan\beta = \frac{\sin(\alpha+\beta)-\sin(\alpha-\beta)}{\sin(\alpha+\beta)+\sin(\alpha-\beta)} = \frac{\cot\alpha+\tan\beta}{\tan\alpha+\cot\beta}$$

(5) Powers

(a) $\sin^n\alpha$.

$$\sin^2\alpha = \tfrac{1}{2}(-\cos 2\alpha + 1)$$

$$\sin^3\alpha = \tfrac{1}{4}(-\sin 3\alpha + 3\sin\alpha)$$

$$\sin^4\alpha = \tfrac{1}{8}(\cos 4\alpha - 4\cos 2\alpha + 3)$$

$$\sin^5\alpha = \tfrac{1}{16}(\sin 5\alpha - 5\sin 3\alpha + 10\sin\alpha)$$

(b) $\cos^n\alpha$.

$$\cos^2\alpha = \tfrac{1}{2}(\cos 2\alpha + 1)$$

$$\cos^3\alpha = \tfrac{1}{4}(\cos 3\alpha + 3\cos\alpha)$$

$$\cos^4\alpha = \tfrac{1}{8}(\cos 4\alpha + 4\cos 2\alpha + 3)$$

$$\cos^5\alpha = \tfrac{1}{16}(\cos 5\alpha + 5\cos 3\alpha + 10\cos\alpha)$$

(c) $\tan^n\alpha$.

$$\tan^2\alpha = \frac{1-\cos 2\alpha}{1+\cos 2\alpha}$$

$$\tan^3\alpha = \frac{-\sin 3\alpha + 3\sin\alpha}{\cos 3\alpha + 3\cos\alpha}$$

$$\tan^4\alpha = \frac{\cos 4\alpha - 4\cos 2\alpha + 3}{\cos 4\alpha + 4\cos 2\alpha + 3}$$

$$\tan^5\alpha = \frac{\sin 5\alpha - 5\sin 3\alpha + 10\sin\alpha}{\cos 5\alpha + 5\cos 3\alpha + 10\cos\alpha}$$

(d) $\cot^n\alpha$.

$$\cot^2\alpha = \frac{1+\cos 2\alpha}{1-\cos 2\alpha}$$

$$\cot^3\alpha = \frac{\cos 3\alpha + 3\cos\alpha}{-\sin 3\alpha + 3\sin\alpha}$$

$$\cot^4\alpha = \frac{\cos 4\alpha + 4\cos 2\alpha + 3}{\cos 4\alpha - 4\cos 2\alpha + 3}$$

$$\cot^5\alpha = \frac{\cos 5\alpha + 5\cos 3\alpha + 10\cos\alpha}{\sin 5\alpha - 5\sin 3\alpha + 10\sin\alpha}$$

5.05 PLANE RIGHT TRIANGLE

(1) Trigonometric relationships (Fig. 5.05–1)

(a) Segments x, y **and altitude** h. The altitude h divides the area of right triangle in two right triangles from which

$$h = \sqrt{xy} = a \sin \beta = b \sin \alpha = \sqrt{ab \cos \alpha \cos \beta}$$

$$x = \sqrt{b^2 - h^2} = b \cos \alpha = b \sin \beta = h \cot \alpha = h \tan \beta$$

$$y = \sqrt{a^2 - h^2} = a \cos \alpha = b \sin \beta = h \cot \beta = h \tan \alpha$$

$$c = x + y = a \cos \beta + b \cos \alpha = \sqrt{a^2 + b^2}$$

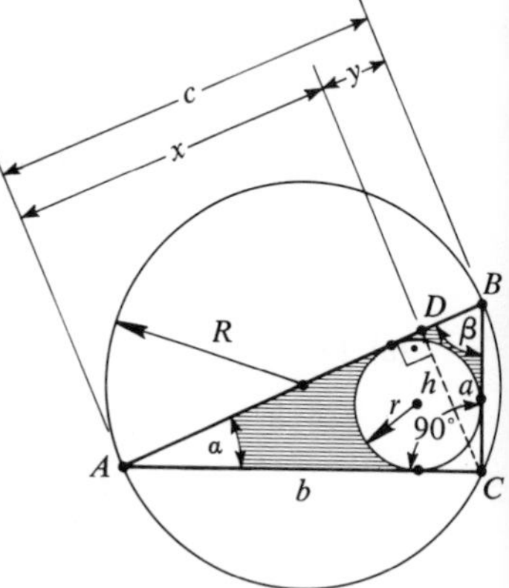

Fig. 5.05-1

(b) Radius of inscribed circle is

$$r = \frac{a + b - c}{2} = \frac{c(\sin \alpha + \sin \beta - 1)}{2} = \frac{c(\cos \alpha + \cos \beta - 1)}{2}$$

(c) Radius of circumscribed circle is

$$R = \frac{c}{2} = \frac{a \cos \beta + b \cos \alpha}{2} = \frac{a \sin \alpha + b \sin \beta}{2}$$

(d) Area of the triangle is

$$A = \frac{ab}{2} = \frac{c^2}{4} \sin 2\alpha = \frac{a^2}{2} \cot \alpha = \frac{b^2}{2} \cot \beta = \frac{a^2}{2} \tan \beta = \frac{b^2}{2} \tan \alpha$$

(2) Solutions

(a) Right triangle is uniquely determined by two sides or by one side and one angle. Consequently four basic problems can be identified in the solution of right triangle.

(b) First problem (a,b). If a and b are known in Fig. 5.05–2, then

$$c = \sqrt{a^2 + b^2} \qquad \tan \alpha = \frac{a}{b} \qquad \tan \beta = \frac{b}{a}$$

and $\alpha + \beta$ must equal 90° (which is always a useful and necessary check).

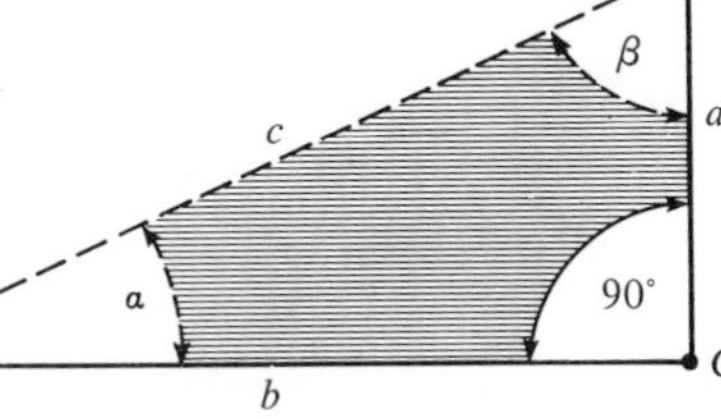

Fig. 5.05-2

(c) Second problem (a,c) **or** (b,c). If a or b and c are known in Fig. 5.05–3, then

$$b = \sqrt{c^2 - a^2} \qquad \text{or} \qquad a = \sqrt{c^2 - b^2}$$

$$\sin \alpha = \frac{a}{c} \qquad \cos \beta = \frac{a}{c}$$

or

$$\cos \alpha = \frac{b}{c} \qquad \sin \beta = \frac{b}{c}$$

In each case, $\alpha + \beta = 90°$.

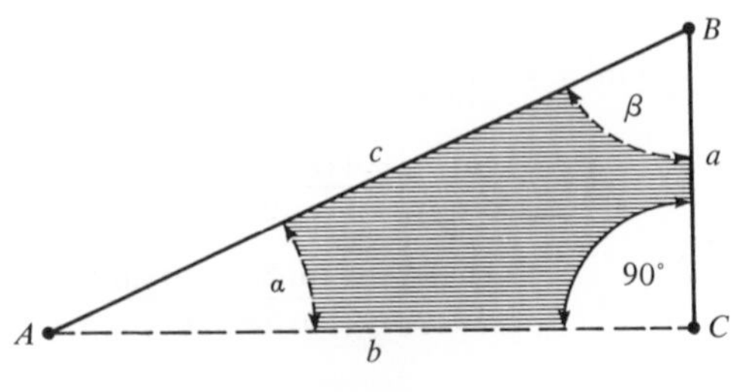

Fig. 5.05-3

(d) Third problem (a,α), (a,β), (b,α), **or** (b,β). If one angle is known in Fig. 5.05–4, then automatically the second angle is known, since $\alpha + \beta = 90°$. This then reduces the solution to finding the other leg and the hypotenuse. If a and α are known, then

from $\tan\alpha = \dfrac{a}{b}$: $\qquad b = \dfrac{a}{\tan\alpha}$

and from $\sin\alpha = \dfrac{a}{c}$: $\qquad c = \dfrac{a}{\sin\alpha}$

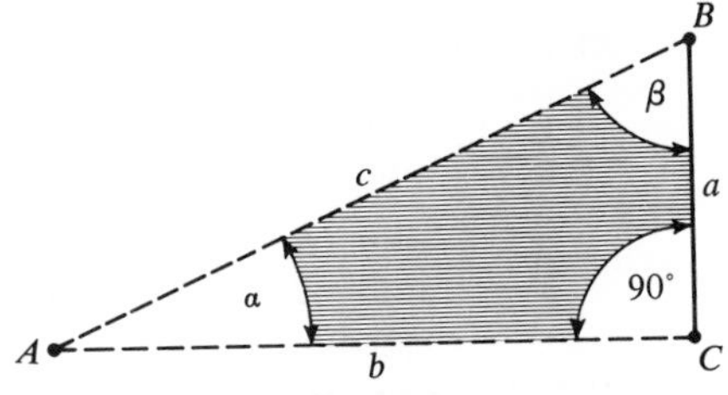

Fig. 5.05-4

(e) Fourth problem (c,α) **or** (c,β). Again if c and α are known, the solution (Fig. 5.05–5) reduces to

$$a = c\sin\alpha \qquad \text{and} \qquad b = c\cos\alpha$$

where $a^2 + b^2 = c^2$ is used as a numerical check.

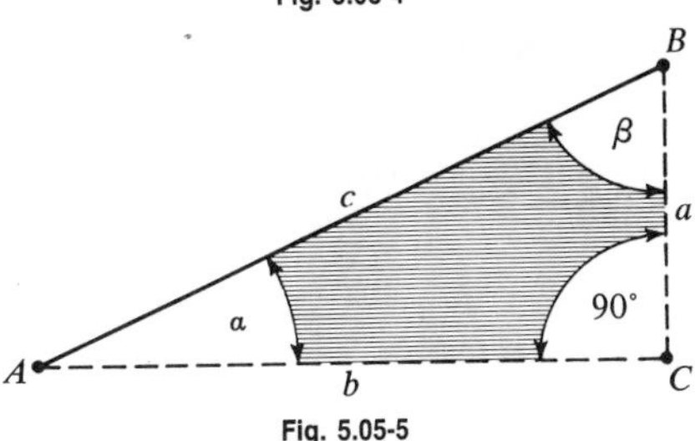

Fig. 5.05-5

5.06 PLANE OBLIQUE TRIANGLE

(1) Basic Laws (Fig. 5.06–1)

(a) Sum of any two sides of a triangle is greater than the third side. *Sum of internal angles of a triangle equals* 180°.

(b) Law of sines. In any triangle, the sides are proportional to the sines of the opposite angles.

$$\frac{a}{\sin\alpha} = \frac{b}{\sin\beta} = \frac{c}{\sin\gamma}$$

from which

$$\frac{a}{b} = \frac{\sin\alpha}{\sin\beta} \qquad \frac{b}{c} = \frac{\sin\beta}{\sin\gamma} \qquad \frac{c}{a} = \frac{\sin\gamma}{\sin\alpha}$$

Fig. 5.06-1

and also

$$\frac{a}{a+b} = \frac{\sin\alpha}{\sin\alpha + \sin\beta} \qquad \frac{b}{b+c} = \frac{\sin\beta}{\sin\beta + \sin\gamma} \qquad \frac{c}{a+c} = \frac{\sin\gamma}{\sin\alpha + \sin\gamma}$$

$$\frac{a}{a-b} = \frac{\sin\alpha}{\sin\alpha - \sin\beta} \qquad \frac{b}{b-c} = \frac{\sin\beta}{\sin\beta - \sin\gamma} \qquad \frac{c}{a-c} = \frac{\sin\gamma}{\sin\alpha - \sin\gamma}$$

$$\frac{a+b}{a-b} = \frac{\sin\alpha + \sin\beta}{\sin\alpha - \sin\beta} \qquad \frac{b+c}{b-c} = \frac{\sin\beta + \sin\gamma}{\sin\beta - \sin\gamma} \qquad \frac{a+c}{a-c} = \frac{\sin\alpha + \sin\gamma}{\sin\alpha - \sin\gamma}$$

(c) Law of cosines. In any triangle the square of any side equals the sum of the squares of the other two sides diminished by twice the product of these two sides and the cosine of the angle between them.

$$a^2 = b^2 + c^2 - 2bc\cos\alpha = (b+c)^2 - 4bc\cos^2\frac{\alpha}{2} = (b-c)^2 + 4bc\sin^2\frac{\alpha}{2}$$

$$b^2 = c^2 + a^2 - 2ca\cos\beta = (c+a)^2 - 4ca\cos^2\frac{\beta}{2} = (c-a)^2 + 4ca\sin^2\frac{\beta}{2}$$

$$c^2 = a^2 + b^2 - 2ab\cos\gamma = (a+b)^2 - 4ab\cos^2\frac{\gamma}{2} = (a-b)^2 + 4ab\sin^2\frac{\gamma}{2}$$

(d) Projection law. In any triangle,

$$a = b\cos\gamma + c\cos\beta \qquad b = c\cos\alpha + a\cos\gamma \qquad c = a\cos\beta + b\cos\alpha$$

(e) Law of tangents. In any triangle,

$$\frac{a+b}{a-b} = \frac{\tan[(\alpha+\beta)/2]}{\tan[(\alpha-\beta)/2]} = \frac{\cot(\gamma/2)}{\tan[(\alpha-\beta)/2]} \qquad \frac{b+c}{b-c} = \frac{\tan[(\beta+\gamma)/2]}{\tan[(\beta-\gamma)/2]} = \frac{\cot(\alpha/2)}{\tan[(\beta-\gamma)/2]}$$

$$\frac{a+c}{a-c} = \frac{\tan[(\alpha+\gamma)/2]}{\tan[(\alpha-\gamma)/2]} = \frac{\cot(\beta/2)}{\tan[(\alpha-\gamma)/2]}$$

(f) Mollweide's formulas. In any triangle,

$$\frac{a+b}{c} = \frac{\cos[(\alpha-\beta)/2]}{\sin(\gamma/2)} \qquad \frac{a-b}{c} = \frac{\sin[(\alpha-\beta)/2]}{\cos(\gamma/2)}$$

$$\frac{b+c}{a} = \frac{\cos[(\beta-\gamma)/2]}{\sin(\alpha/2)} \qquad \frac{b-c}{a} = \frac{\sin[(\beta-\gamma)/2]}{\cos(\alpha/2)}$$

$$\frac{c+a}{b} = \frac{\cos[(\gamma-\alpha)/2]}{\sin(\beta/2)} \qquad \frac{c-a}{b} = \frac{\sin[(\gamma-\alpha)/2]}{\cos(\beta/2)}$$

(g) Simple angle formulas $[p = (a+b+c)/2]$. In any triangle,

$$\sin\alpha = \frac{2}{bc}\sqrt{p(p-a)(p-b)(p-c)} \qquad \cos\alpha = \frac{b^2+c^2-a^2}{2bc}$$

$$\sin\beta = \frac{2}{ac}\sqrt{p(p-a)(p-b)(p-c)} \qquad \cos\beta = \frac{a^2+c^2-b^2}{2ac}$$

$$\sin\gamma = \frac{2}{ac}\sqrt{p(p-a)(p-b)(p-c)} \qquad \cos\gamma = \frac{a^2+b^2-c^2}{2ab}$$

(h) Half-angle formulas $[p = (a+b+c)/2]$. In any triangle,

$$\sin\frac{\alpha}{2} = \sqrt{\frac{(p-b)(p-c)}{bc}} \qquad \cos\frac{\alpha}{2} = \sqrt{\frac{p(p-a)}{bc}}$$

$$\sin\frac{\beta}{2} = \sqrt{\frac{(p-c)(p-a)}{ca}} \qquad \cos\frac{\beta}{2} = \sqrt{\frac{p(p-b)}{ca}}$$

$$\sin\frac{\gamma}{2} = \sqrt{\frac{(p-a)(p-b)}{ab}} \qquad \cos\frac{\gamma}{2} = \sqrt{\frac{p(p-c)}{ab}}$$

from which

$$\tan\frac{\alpha}{2} = \sqrt{\frac{(p-b)(p-c)}{p(p-a)}} \qquad \cot\frac{\alpha}{2} = \sqrt{\frac{p(p-a)}{(p-b)(p-c)}}$$

$$\tan\frac{\beta}{2} = \sqrt{\frac{(p-c)(p-a)}{p(p-b)}} \qquad \cot\frac{\beta}{2} = \sqrt{\frac{p(p-b)}{(p-c)(p-a)}}$$

$$\tan\frac{\gamma}{2} = \sqrt{\frac{(p-a)(p-b)}{p(p-c)}} \qquad \cot\frac{\gamma}{2} = \sqrt{\frac{p(p-c)}{(p-a)(p-b)}}$$

(2) Laws of Angles $(\alpha + \beta + \gamma = 180°)$

(a) Transformations. In any triangle,

$$\sin(\alpha+\beta) = \sin\gamma \qquad \cos(\alpha+\beta) = -\cos\gamma$$

$$\tan(\alpha+\beta) = -\tan\gamma \qquad \cot(\alpha+\beta) = -\cot\gamma$$

$$\sin\frac{\alpha+\beta}{2} = \cos\frac{\gamma}{2} \qquad \cos\frac{\alpha+\beta}{2} = \sin\frac{\gamma}{2}$$

$$\tan\frac{\alpha+\beta}{2} = \cot\frac{\gamma}{2} \qquad \cot\frac{\alpha+\beta}{2} = \tan\frac{\gamma}{2}$$

(b) Sums of functions of simple angle. In any triangle,

$$\sin\alpha + \sin\beta + \sin\gamma = 4\cos\frac{\alpha}{2}\cos\frac{\beta}{2}\cos\frac{\gamma}{2} \qquad \sin\alpha + \sin\beta - \sin\gamma = 4\sin\frac{\alpha}{2}\sin\frac{\beta}{2}\cos\frac{\gamma}{2}$$

$$\cos\alpha + \cos\beta + \cos\gamma = 4\sin\frac{\alpha}{2}\sin\frac{\beta}{2}\sin\frac{\gamma}{2} + 1 \qquad \cos\alpha + \cos\beta - \cos\gamma = 4\cos\frac{\alpha}{2}\cos\frac{\beta}{2}\sin\frac{\gamma}{2} - 1$$

$$\tan\alpha + \tan\beta + \tan\gamma = \tan\alpha\tan\beta\tan\gamma \qquad \tan\alpha + \tan\beta - \tan\gamma = (\tan\alpha\tan\beta - 2)(\tan\gamma)$$

(c) Sums of functions of double angles. In any triangle,

$$\sin 2\alpha + \sin 2\beta + \sin 2\gamma = 4\sin\alpha\sin\beta\sin\gamma$$

$$\cos 2\alpha + \cos 2\beta + \cos 2\gamma = -(4\cos\alpha\cos\beta\cos\gamma + 1)$$

(d) Sums of squares of functions of simple angles. In any triangle,

$$\sin^2\alpha + \sin^2\beta + \sin^2\gamma = 2(\cos\alpha\cos\beta\cos\gamma + 1)$$

$$\cos^2\alpha + \cos^2\beta + \cos^2\gamma = 1 - 2\cos\alpha\cos\beta\cos\gamma$$

(3) Altitudes and Area

(a) Altitudes h_a, h_b, h_c of any triangle (Fig. 5.06–2) are

$$h_a = b\sin\gamma = c\sin\beta$$

$$h_b = a\sin\gamma = c\sin\alpha$$

$$h_c = a\sin\beta = b\sin\alpha$$

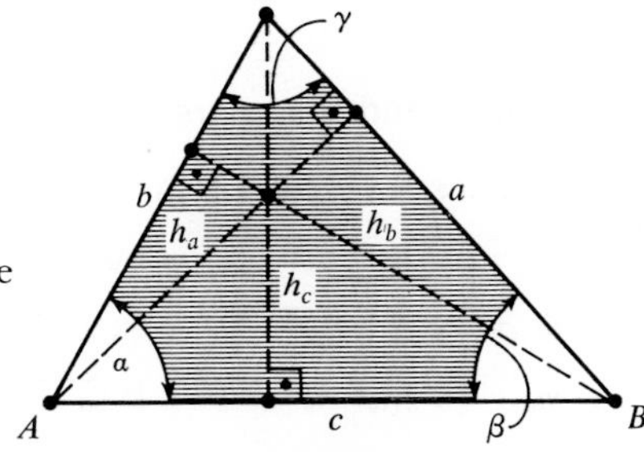

Fig. 5.06-2

(b) Area of any triangle (Fig. 5.06–2) equals half the product of its base and altitude.

$$A = \frac{ab\sin\gamma}{2} = \frac{bc\sin\alpha}{2} = \frac{ca\sin\beta}{2} = \frac{abc}{4R} = 2R^2\sin\alpha\sin\beta\sin\gamma$$

$$= \frac{a^2\sin\beta\sin\gamma}{2\sin(\beta+\gamma)} = \frac{b^2\sin\gamma\sin\alpha}{2\sin(\gamma+\alpha)} = \frac{c^2\sin\alpha\sin\beta}{2\sin(\alpha+\beta)}$$

$$= p^2\tan\frac{\alpha}{2}\tan\frac{\beta}{2}\tan\frac{\gamma}{2} = r^2\cot\frac{\alpha}{2}\cot\frac{\beta}{2}\cot\frac{\gamma}{2}$$

where R and r are respectively the radius of circumscribed and inscribed circle and $p = (a+b+c)/2$.

(4) Radii $[p = (a+b+c)/2]$

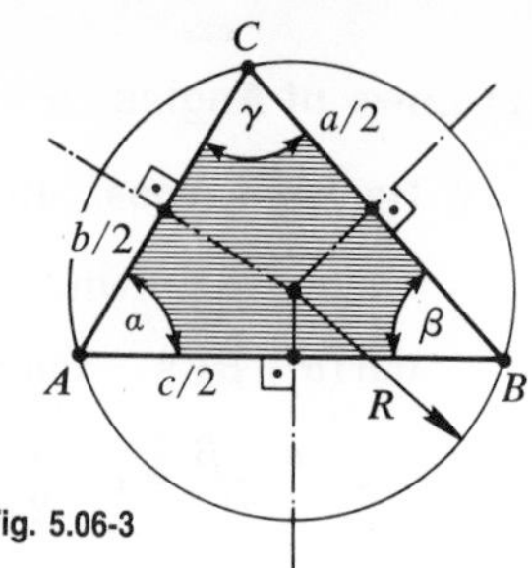

Fig. 5.06-3

(a) Circumradius R of any triangle (Fig. 5.06–3) is

$$R = \frac{a}{2\sin\alpha} = \frac{b}{2\sin\beta} = \frac{c}{2\sin\gamma} \qquad R = \frac{p}{4\cos\frac{\alpha}{2}\cos\frac{\beta}{2}\cos\frac{\gamma}{2}}$$

(b) Inradius r of any triangle (Fig. 5.06–4) is

$$r = (p-a)\tan\frac{\alpha}{2} = (p-b)\tan\frac{\beta}{2} = (p-c)\tan\frac{\gamma}{2}$$

$$r = p\tan\frac{\alpha}{2}\tan\frac{\beta}{2}\tan\frac{\gamma}{2} = 4R\sin\frac{\alpha}{2}\sin\frac{\beta}{2}\sin\frac{\gamma}{2}$$

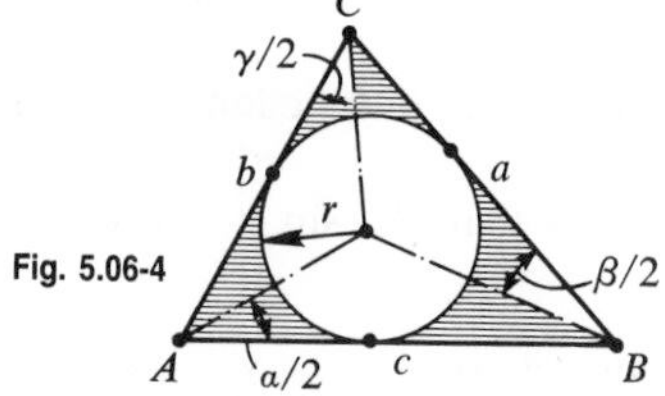

Fig. 5.06-4

(c) Escumradii r_a, r_b, r_c of any triangle (Fig. 5.06–5) are

$$r_a = p\tan\frac{\alpha}{2} = \frac{a\cos(\beta/2)\cos(\gamma/2)}{\cos(\alpha/2)}$$

$$r_b = p\tan\frac{\beta}{2} = \frac{b\cos(\gamma/2)\cos(\alpha/2)}{\cos(\beta/2)}$$

$$r_c = p\tan\frac{\gamma}{2} = \frac{c\cos(\alpha/2)\cos(\beta/2)}{\cos(\gamma/2)}$$

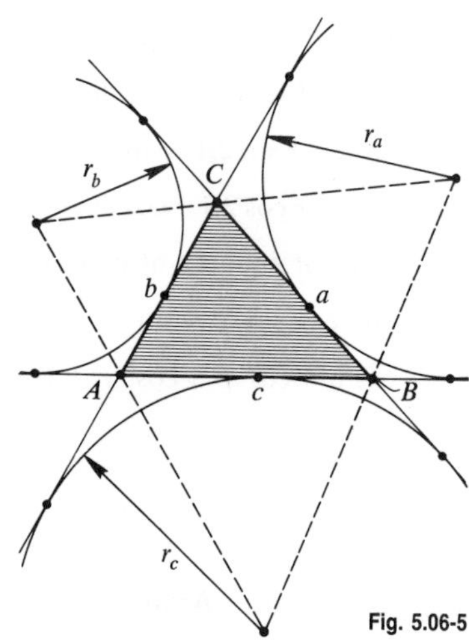

Fig. 5.06-5

(5) Medians and Bisectors

(a) Medians m_a, m_b, m_c of any triangle (Fig. 5.06–6) are

$$m_a = \tfrac{1}{2}\sqrt{2(b^2+c^2)-a^2} = \tfrac{1}{2}\sqrt{b^2+c^2+2bc\cos\alpha}$$

$$m_b = \tfrac{1}{2}\sqrt{2(c^2+a^2)-b^2} = \tfrac{1}{2}\sqrt{c^2+a^2+2ca\cos\beta}$$

$$m_c = \tfrac{1}{2}\sqrt{2(a^2+b^2)-c^2} = \tfrac{1}{2}\sqrt{a^2+b^2+2ab\cos\gamma}$$

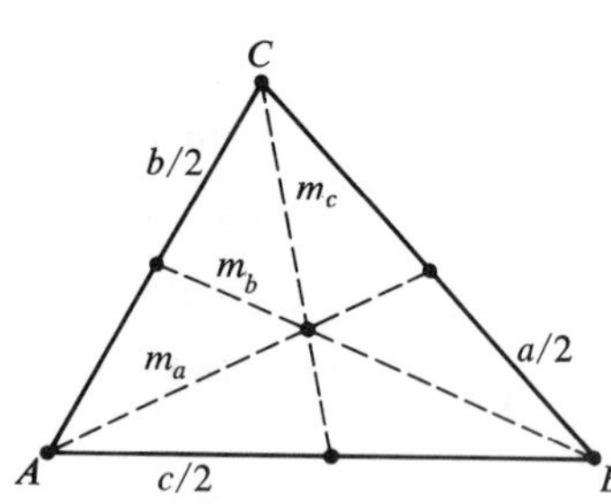

Fig. 5.06-6

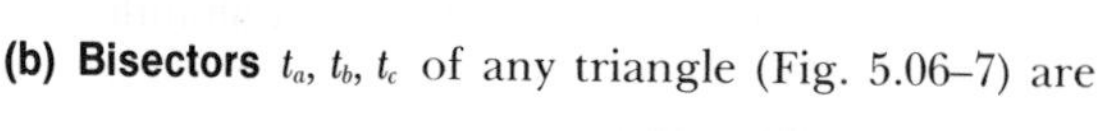

(b) Bisectors t_a, t_b, t_c of any triangle (Fig. 5.06–7) are

$$t_a = \frac{\sqrt{bc[(b+c)^2-a^2]}}{b+c} = \frac{2bc\cos(\alpha/2)}{b+c}$$

$$t_b = \frac{\sqrt{ca[(c+a)^2-b^2]}}{c+a} = \frac{2ca\cos(\beta/2)}{c+a}$$

$$t_c = \frac{\sqrt{ab[(a+b)^2-c^2]}}{a+b} = \frac{2ab\cos(\gamma/2)}{a+b}$$

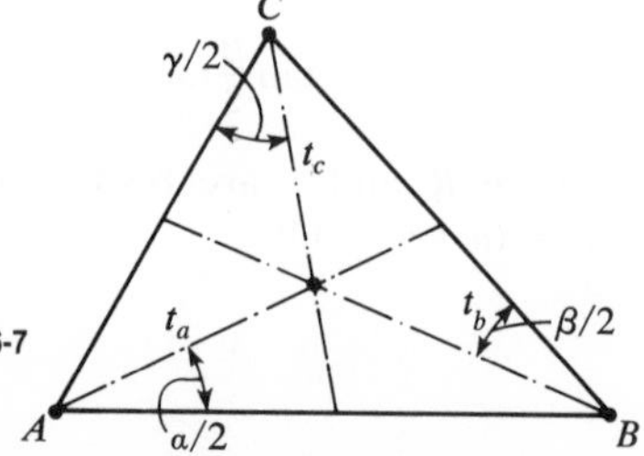

Fig. 5.06-7

(6) Solutions

(a) Any oblique triangle is uniquely determined by:

Three sides
Two sides and the angle between them
Two sides and the angle opposite the greater side
One side and two adjacent angles
One side, one adjacent angle, and one opposite angle

(b) First problem (a, b, c)**.** If a, b, c are known in Fig. 5.06–8, then the angles α, β, γ are computed from the laws of cosines as

$$\cos\alpha = \frac{b^2 + c^2 - a^2}{2bc} \quad \text{or} \quad \cos\frac{\alpha}{2} = \sqrt{\frac{p(p-a)}{bc}}$$

$$\cos\beta = \frac{a^2 + c^2 - b^2}{2ac} \quad \text{or} \quad \cos\frac{\beta}{2} = \sqrt{\frac{p(p-b)}{ac}}$$

$$\cos\gamma = \frac{a^2 + b^2 - c^2}{2ab} \quad \text{or} \quad \cos\frac{\gamma}{2} = \sqrt{\frac{p(p-b)}{ab}}$$

The results must satisfy $\alpha + \beta + \gamma = 180°$.

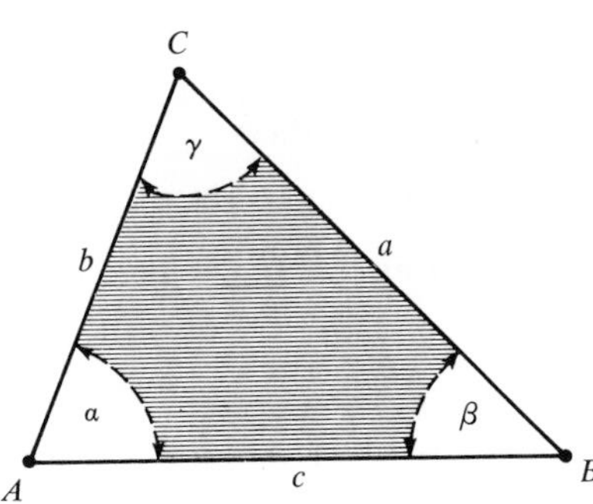

Fig. 5.06-8

(c) Second problem (a, b, γ). If a, b, γ are known in Fig. 5.06–9, then

$$\frac{\alpha + \beta}{2} = 90° - \frac{\gamma}{2}$$

and $(\alpha - \beta)/2$ is computed from the law of tangents as

$$\tan\frac{\alpha - \beta}{2} = \frac{a - b}{a + b}\cot\frac{\gamma}{2}$$

The sum and difference of these values yield

$$\alpha = \frac{\alpha + \beta}{2} + \frac{\alpha - \beta}{2} \qquad \beta = \frac{\alpha + \beta}{2} - \frac{\alpha - \beta}{2}$$

which again must satisfy $\alpha + \beta + \gamma = 180°$.

Finally, by the law of cosines,

$$c = \sqrt{a^2 + b^2 - 2ab\cos\gamma}$$

Fig. 5.06-9

(d) Third problem (a, b, α). If a, b, α are known in Fig. 5.06–10, then from the law of sines,

$$\sin\beta = \frac{b\sin\alpha}{a} \qquad \gamma = 180° - (\alpha + \beta)$$

Finally, from the law of sines,

$$c = \frac{a\sin\gamma}{\sin\alpha}$$

The results must satisfy $c = \sqrt{a^2 + b^2 - 2ab\cos\gamma}$.

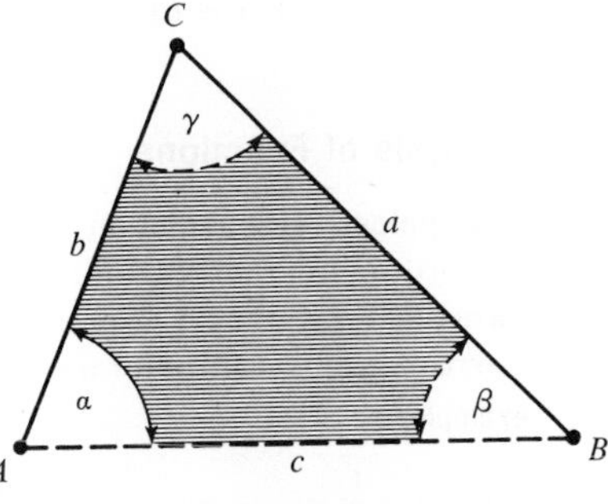

Fig. 5.06-10

(e) Fourth problem (a, β, γ). If a, β, γ are known in Fig. 5.06–11, then

$$\alpha = 180° - (\beta + \gamma)$$

and b and c are computed from the law of sines as

$$b = \frac{a \sin \beta}{\sin \alpha} = \frac{a \sin \beta}{\sin (\beta + \gamma)} \qquad c = \frac{a \sin \gamma}{\sin \alpha} = \frac{a \sin \gamma}{\sin (\beta + \gamma)}$$

The results must satisfy $a = \sqrt{b^2 + c^2 - 2bc \cos \alpha}$.

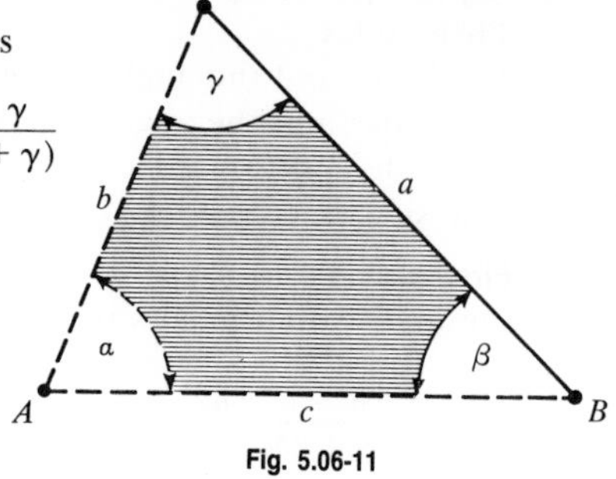

Fig. 5.06-11

(f) Fifth problem (a, α, β). If a, α, β are known in Fig. 5.06–12, then

$$\gamma = 180° - (\alpha + \beta)$$

and b and c are computed from the laws of sines as

$$b = \frac{a \sin \beta}{\sin \alpha} \qquad c = \frac{a \sin \gamma}{\sin \alpha} = \frac{a \sin (\alpha + \beta)}{\sin \alpha}$$

The results must satisfy $a = \sqrt{b^2 + c^2 - 2bc \cos \gamma}$.

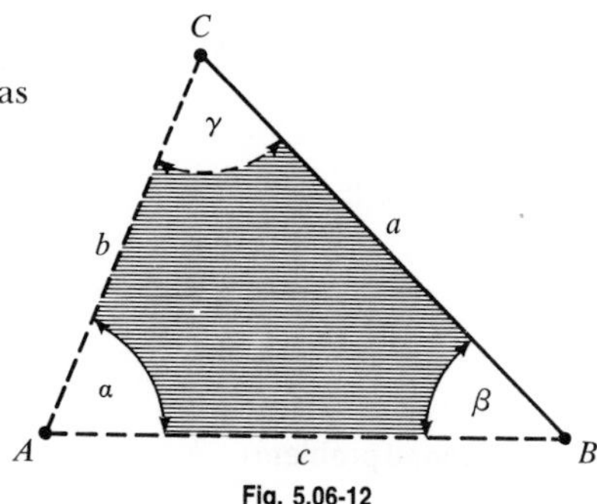

Fig. 5.06-12

5.07 INVERSE TRIGONOMETRIC FUNCTIONS

(1) Definitions and Notations

(a) Inverse function of $y = F(\alpha)$ is $\alpha = F^{-1}(y)$.

(b) Inverse trigonometric functions are defined as follows. If

$y = \sin \alpha$	then	$\alpha = \arcsin y = \sin^{-1} y$
$y = \cos \alpha$	then	$\alpha = \arccos y = \cos^{-1} y$
$y = \tan \alpha$	then	$\alpha = \arctan y = \tan^{-1} y$
$y = \cot \alpha$	then	$\alpha = \text{arccot}\, y = \cot^{-1} y$
$y = \sec \alpha$	then	$\alpha = \text{arcsec}\, y = \sec^{-1} y$
$y = \csc \alpha$	then	$\alpha = \text{arccsc}\, y = \text{cosc}^{-1} y$

which means that α (in radians) is the arc of an angle of which the trigonometric function is y.

(2) Analysis of Functions

(a) Graphical representation of inverse trigonometric functions is shown in Fig. 5.07–1. According to these graphs, the inverse trigonometric functions are not single-valued functions (as the trigonometric functions) but multiple-valued functions, each value corresponding to one branch of the function.

examples:

For $y = \frac{1}{2}$: $\quad \sin^{-1} y = \frac{\pi}{6}, \frac{5\pi}{6}, \frac{13\pi}{6}, \frac{17\pi}{6}, \ldots \qquad \cos^{-1} y = \frac{\pi}{3}, \frac{5\pi}{3}, \frac{7\pi}{3}, \frac{11\pi}{3}, \ldots$

(b) First branch of the inverse function is called the principal branch (shown by solid lines in Fig. 5.07–1).

(c) Principal values of an inverse trigonometric function are the y values of the first branch.

(d) Limits of principal values (limits of the first branch) of six inverse trigonometric functions are:

For $y \geq 0$		For $y < 0$	
$0 \leq \sin^{-1} y \leq \frac{\pi}{2}$	$0 < \cot^{-1} y \leq \frac{\pi}{2}$	$-\frac{\pi}{2} \leq \sin^{-1} y < 0$	$\frac{\pi}{2} < \cot^{-1} y < \pi$
$0 \leq \cos^{-1} y \leq \frac{\pi}{2}$	$0 \leq \sec^{-1} y < \frac{\pi}{2}$	$\frac{\pi}{2} < \cos^{-1} y \leq \pi$	$\frac{\pi}{2} < \sec^{-1} y \leq \pi$
$0 \leq \tan^{-1} y < \frac{\pi}{2}$	$0 < \csc^{-1} y \leq \frac{\pi}{2}$	$-\frac{\pi}{2} < \tan^{-1} y < 0$	$-\frac{\pi}{2} \leq \csc^{-1} y < 0$

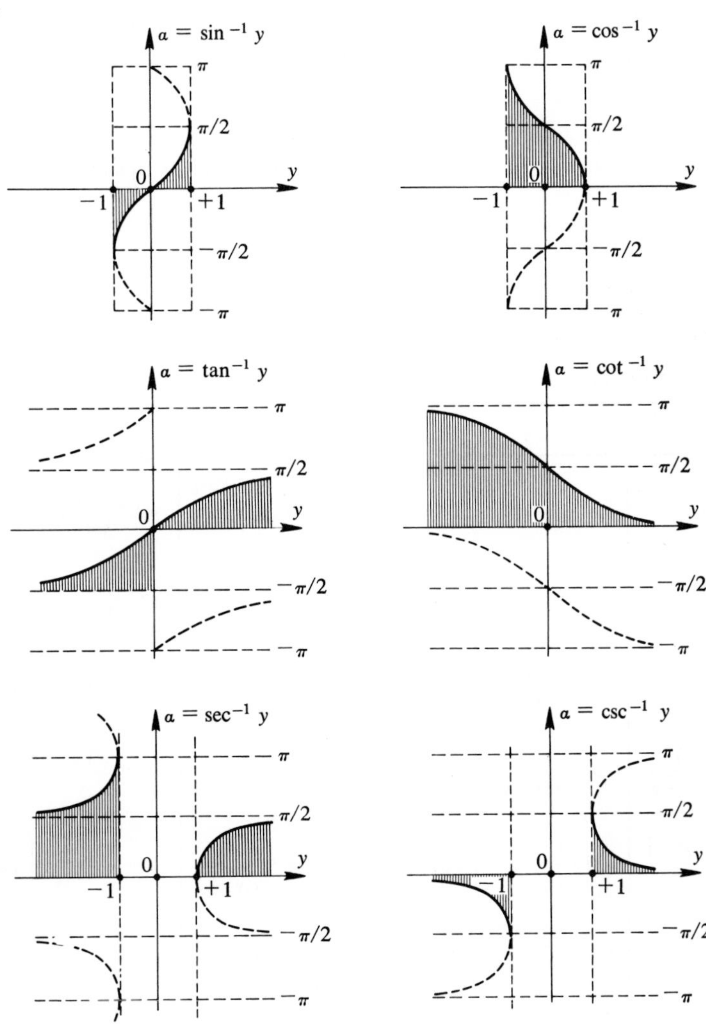

Fig. 5.07-1

(3) Relations between Functions

(a) Functions of simple arguments.

$\sin^{-1} y + \cos^{-1} y = \dfrac{\pi}{2}$	$\csc^{-1} y = \sin^{-1} \dfrac{1}{y}$	$\sin^{-1}(-y) = -\sin^{-1} y$	$\cos^{-1}(-y) = \pi - \cos^{-1} y$
$\tan^{-1} y + \cot^{-1} y = \dfrac{\pi}{2}$	$\sec^{-1} y = \cos^{-1} \dfrac{1}{y}$	$\tan^{-1}(-y) = -\tan^{-1} y$	$\cot^{-1}(-y) = \pi - \cot^{-1} y$
$\sec^{-1} y + \csc^{-1} y = \dfrac{\pi}{2}$	$\tan^{-1} y = \cot^{-1} \dfrac{1}{y}$	$\sec^{-1}(-y) = \pi - \sec^{-1} y$	$\csc^{-1}(-y) = -\csc^{-1} y$

(b) Transformation relations between the principal values of six inverse trigonometric functions for $y > 0$ are given in Table 5.07–1.

examples:

Derive $\sin^{-1} y = \cos^{-1} \sqrt{1-y^2}$.

From Table 5.03–1, $\sin\alpha = \sqrt{1-\cos^2\alpha}$. Placing each side equal to y,

$$y = \sin\alpha \quad \text{and} \quad y = \sqrt{1-\cos^2\alpha} \quad \text{or} \quad \cos\alpha = \sqrt{1-y^2}$$

and their inverses are $\alpha = \sin^{-1}\alpha$ and $\alpha = \cos^{-1}\sqrt{1-y^2}$

Derive $\tan^{-1} y = \sin^{-1} y/\sqrt{1+y^2}$.

From Table 5.03–1, $\tan\alpha = (\sin\alpha)/\sqrt{1-\sin^2\alpha}$. Placing each side equal to y,

$$y = \tan\alpha \quad \text{and} \quad y = \frac{\sin\alpha}{\sqrt{1-\sin^2\alpha}} \quad \text{or} \quad \frac{y}{\sqrt{1+y^2}} = \sin\alpha$$

Inverting, $\alpha = \tan^{-1} y$ and $\alpha = \sin^{-1}\dfrac{y}{\sqrt{1+y^2}}$

TABLE 5.07–1 Transformations of Inverse Trigonometric Functions

$y > 0$	$\sin^{-1} y$	$\cos^{-1} y$	$\tan^{-1} y$	$\cot^{-1} y$	$\sec^{-1} y$	$\csc^{-1} y$
$\sin^{-1} y$	$\sin^{-1} y$	$\cos^{-1}\sqrt{1-y^2}$	$\tan^{-1}\dfrac{y}{\sqrt{1-y^2}}$	$\cot^{-1}\dfrac{\sqrt{1-y^2}}{y}$	$\sec^{-1}\dfrac{1}{\sqrt{1-y^2}}$	$\csc^{-1}\dfrac{1}{y}$
$\cos^{-1} y$	$\sin^{-1}\sqrt{1-y^2}$	$\cos^{-1} y$	$\tan^{-1}\dfrac{\sqrt{1-y^2}}{y}$	$\cot^{-1}\dfrac{y}{\sqrt{1-y^2}}$	$\sec^{-1}\dfrac{1}{y}$	$\csc^{-1}\dfrac{1}{\sqrt{1-y^2}}$
$\tan^{-1} y$	$\sin^{-1}\dfrac{y}{\sqrt{1+y^2}}$	$\cos^{-1}\dfrac{1}{\sqrt{1+y^2}}$	$\tan^{-1} y$	$\cot^{-1}\dfrac{1}{y}$	$\sec^{-1}\sqrt{1+y^2}$	$\csc^{-1}\dfrac{\sqrt{1+y^2}}{y}$
$\cot^{-1} y$	$\sin^{-1}\dfrac{1}{\sqrt{1+y^2}}$	$\cos^{-1}\dfrac{y}{\sqrt{1+y^2}}$	$\tan^{-1}\dfrac{1}{y}$	$\cot^{-1} y$	$\sec^{-1}\dfrac{\sqrt{1+y^2}}{y}$	$\csc^{-1}\sqrt{1+y^2}$
$\sec^{-1} y$	$\sin^{-1}\dfrac{\sqrt{y^2-1}}{y}$	$\cos^{-1}\dfrac{1}{y}$	$\tan^{-1}\sqrt{y^2-1}$	$\cot^{-1}\dfrac{1}{\sqrt{y^2-1}}$	$\sec^{-1} y$	$\csc^{-1}\dfrac{y}{\sqrt{y^2-1}}$
$\csc^{-1} y$	$\sin^{-1}\dfrac{1}{y}$	$\cos^{-1}\dfrac{\sqrt{y^2-1}}{y}$	$\tan^{-1}\dfrac{1}{\sqrt{y^2-1}}$	$\cot^{-1}\sqrt{y^2-1}$	$\sec^{-1}\dfrac{y}{\sqrt{y^2-1}}$	$\csc^{-1} y$

6
PLANE ANALYTIC GEOMETRY

6.01 DEFINITIONS AND NOTATIONS

(1) Definitions

(a) Analytic geometry is a systematic investigation of geometric elements, figures, surfaces, and solids and their relationships by analytical methods.

(b) System of coordinates. The basis of this investigation is a system of coordinates: a set of rules relating the geometric elements to numbers and vice versa.

(c) Plane and space analytic geometry. Since the object of investigation may be a planar or a nonplanar system, terms plane and space analytic geometry are used respectively. Only planar systems are considered in this chapter.

(d) Classification. Two most common systems of plane coordinates are the rectangular (cartesian) coordinate system (Sec. 6.01–2*a*) and the polar coordinate system (Sec. 6.01–3*a*).

(2) Cartesian Coordinates

(a) Notations. A point P is given by two mutually perpendicular distances x, y (coordinates) measured from two mutually perpendicular axes X, Y (coordinate axes), which intersect at the origin 0 (Fig. 6.01–1) and define the coordinate plane.

(b) Quadrants. The coordinate axes divide the coordinate plane into four quadrants (I, II, III, IV) and assign directions to all coordinates (Fig. 6.01–2).

(c) Sign of x. The coordinate x (called abscissa) measured along the X axis is positive to the right and negative to the left of the Y axis.

(d) Sign of y. The coordinate y (called ordinate) measured along the Y axis is positive above and negative below the X axis.

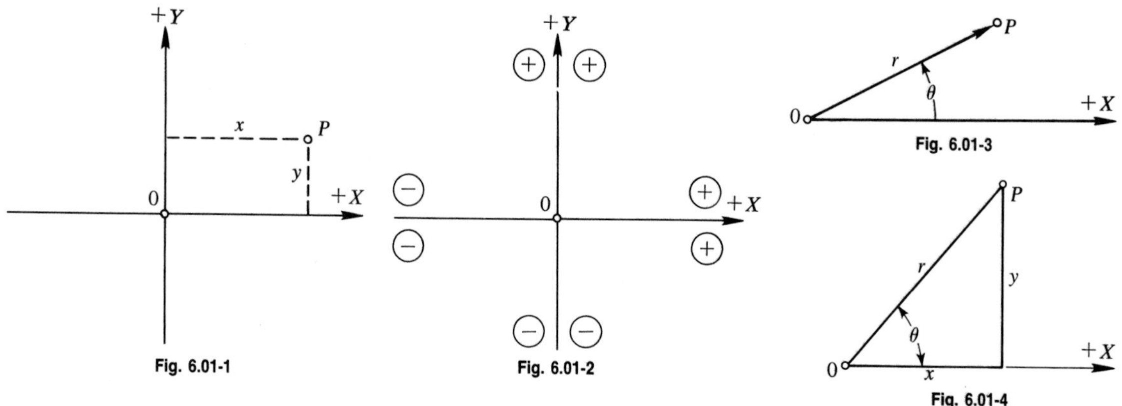

Fig. 6.01-1

Fig. 6.01-2

Fig. 6.01-3

Fig. 6.01-4

(3) Polar Coordinates

(a) Notations. A point P is given by two polar coordinates associated with a fixed axis X (polar axis) and a fixed point 0 on this axis (pole). The first coordinate is the distance from 0 to P called the radius r and the second coordinate is the position angle θ measured from the X axis to the radius r (Fig. 6.01–3).

(b) Signs of r and θ. The radius r is always positive. The angle θ is positive if measured in the counterclockwise direction and negative if measured in the opposite direction.

(c) Relations between the cartesian and polar coordinates of the point P derived from the right triangle of Fig. 6.01–4 are

$$x = r\cos\theta \qquad y = r\sin\theta \qquad r = \sqrt{x^2+y^2} \qquad \theta = \tan^{-1}\frac{y}{x}$$

where $\sin\theta = \dfrac{y}{\sqrt{x^2+y^2}} \qquad \cos\theta = \dfrac{x}{\sqrt{x^2+y^2}} \qquad \tan\theta = \dfrac{y}{x}$

6.02 POINTS

(1) Two Points

(a) Cartesian coordinates (Fig. 6.02–1). The distance of two points $P_1(x_1,y_1)$, $P_2(x_2,y_2)$ is

$$d = \overline{P_1P_2} = \sqrt{(x_2-x_1)^2+(y_2-y_1)^2}$$

The direction tangent and cosines of $\overline{P_1P_2}$ are

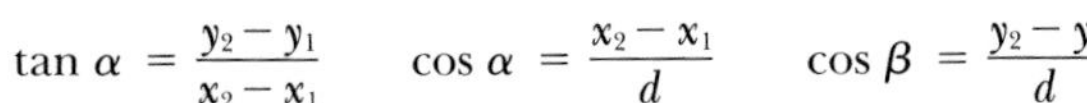

$$\tan\alpha = \frac{y_2-y_1}{x_2-x_1} \qquad \cos\alpha = \frac{x_2-x_1}{d} \qquad \cos\beta = \frac{y_2-y_1}{d}$$

Fig. 6.02-1

(b) Polar coordinates (Fig. 6.02–2). The distance of two points $P_1(r_1,\theta_1)$, $P_2(r_2,\theta_2)$ is

$$d = \overline{P_1P_2} = \sqrt{r_1^2+r_2^2-2r_1r_2\cos(\theta_2-\theta_1)}$$

The direction tangent and cosines of $\overline{P_1P_2}$ are

$$\tan\alpha = \frac{r_2\sin\theta_2-r_1\sin\theta_1}{r_2\cos\theta_2-r_1\cos\theta_1}$$

$$\cos\alpha = \frac{r_2\cos\theta_2-r_1\cos\theta_2}{d} \qquad \cos\beta = \frac{r_2\sin\theta_2-r_1\sin\theta_1}{d}$$

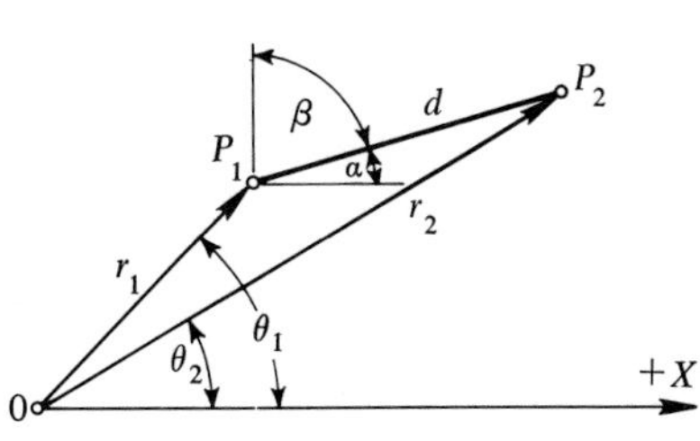

Fig. 6.02-2

(c) Coordinates of a point P_3 dividing the segment $\overline{P_1P_2}$ in a given ratio $m:n$ (Fig. 6.02–3) are

$$x_3 = \frac{nx_1+mx_2}{m+n} \qquad y_3 = \frac{ny_1+my_2}{m+n}$$

(d) Coordinates of a midpoint C of the segment $\overline{P_1P_2}$ are

$$x_C = \frac{x_1+x_2}{2} \qquad y_C = \frac{y_1+y_2}{2}$$

which are also the coordinates of the centroid of $\overline{P_1P_2}$.

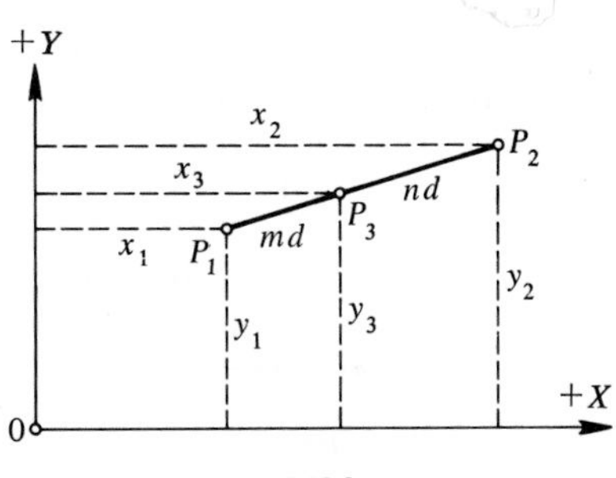

Fig. 6.02-3

(2) System of Points

(a) Area of a triangle given by vertices $P_1(x_1,y_1)$, $P_2(x_2,y_2)$, $P_3(x_3,y_3)$ is

$$A = \tfrac{1}{2}[x_1(y_2-y_3)+x_2(y_3-y_1)+x_3(y_1-y_2)]$$

where A is positive if the vertices are numbered counterclockwise and negative if numbered in the opposite direction.

(b) Coordinates of centroid of the triangle $P_1P_2P_3$ are

$$x_C = \frac{x_1+x_2+x_3}{3} \qquad y_C = \frac{y_1+y_2+y_3}{3}$$

(c) Area of a convex polygon given by vertices $P_1(x_1,y_1)$, $P_2(x_2,y_2)$, . . . , $P_n(x_n,y_n)$ is

$$A = \tfrac{1}{2}[(x_1-x_2)(y_1+y_2)+(x_2-x_3)(y_2+y_3)+\cdots+(x_n-x_1)(y_n+y_1)]$$

where A is again positive if the vertices are numbered counterclockwise and negative if numbered in the opposite direction.

6.03 STRAIGHT LINE

(1) Equations of Straight Line

(a) Every linear equation in x and y represents a straight line.

(b) Four forms of this equation are available: direction form, intercept form, normal form, and general form.

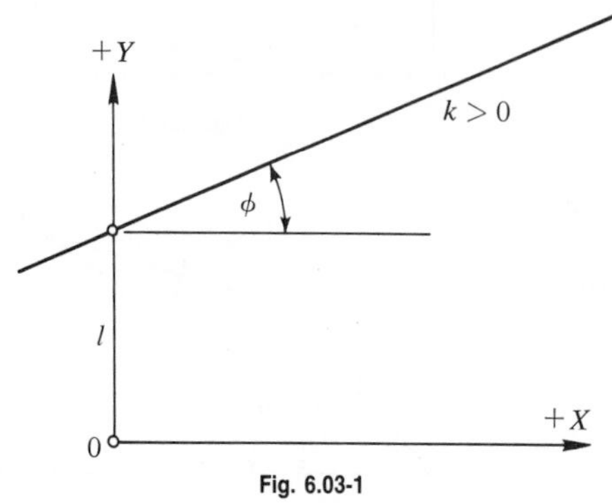

Fig. 6.03-1

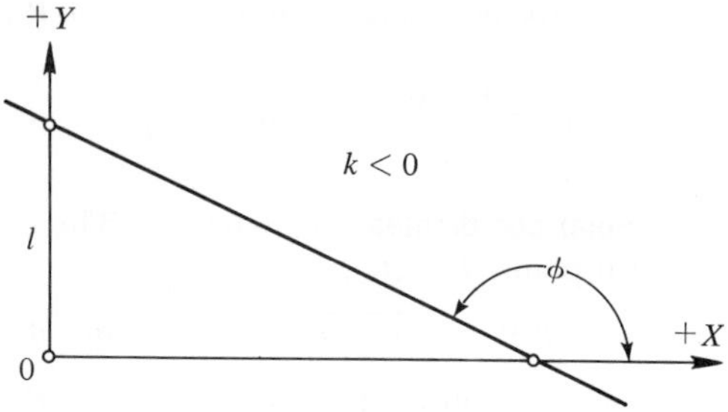

Fig. 6.03-2

(c) Direction form is

$$y = kx + l$$

where $k = \tan\phi$ is called the slope and l is the intercept on the Y axis. If $k > 0$, the line is rising (Fig. 6.03–1). If $k < 0$, the line is falling (Fig. 6.03–2).

(d) Intercept form (Fig. 6.03–3) is

$$\frac{x}{a} + \frac{y}{b} = 1$$

where a, b are the intercepts on the X and Y axis respectively and $\tan\phi = -b/a$.

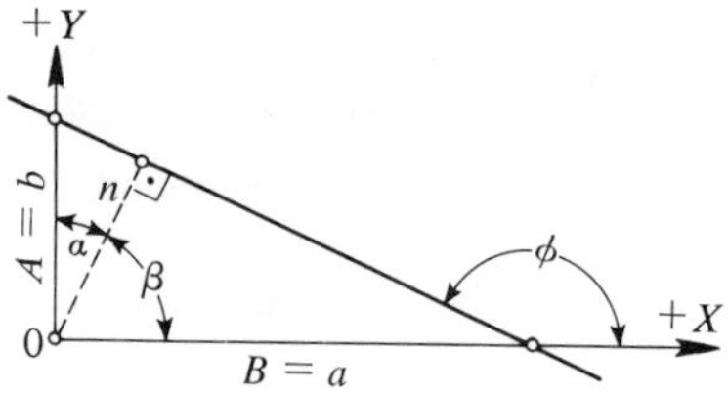

(e) General form (Fig. 6.03–3) is

$$Ax + By + C = 0$$

where $A = b$, $B = a$, $C = -ab$, and $\tan\phi = -A/B$.

Fig. 6.03-3

(f) Normal form in terms of α, β, n (Fig. 6.03–3) is

$$x\cos\beta + y\cos\alpha = n$$

where β is the angle of n with the $+X$ axis, $\alpha = 90° - \beta$, and n is the normal distance of the line to 0.

(g) Normal form in terms of A, B, C (Fig. 6.03–3) is

$$\frac{Ax + By + C}{\pm\sqrt{A^2 + B^2}} = 0$$

where the sign of the radical is opposite to the sign of C.

(h) Polar form (Fig. 6.03–4) is

$$r = \frac{n}{\cos(\theta - \beta)}$$

where n and β are the same as in (f) above.

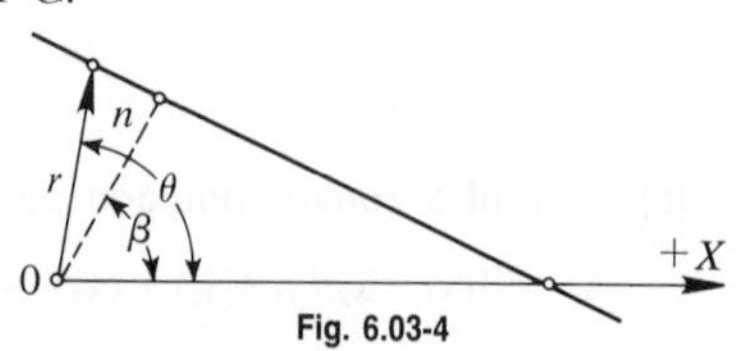

Fig. 6.03-4

(2) Special Cases of $Ax + By + C = 0$

(a) If $A = 0$, then $By + C = 0$ defines a line parallel to the X axis at $d = -C/B$ (Fig. 6.03–5).
(b) If $B = 0$, then $Ax + C = 0$ defines a line parallel to the Y axis at $d = -C/A$ (Fig. 6.03–6).
(c) If $C = 0$, then $Ax + By = 0$ defines a line passing through the origin 0 in a direction $k = -A/B$ (Fig. 6.03–7).
(d) If $A = 0$, $B = 1$, $C = 0$, then $y = 0$ defines the X axis.
(e) If $A = 1$, $B = 0$, $C = 0$, then $x = 0$ defines the Y axis.

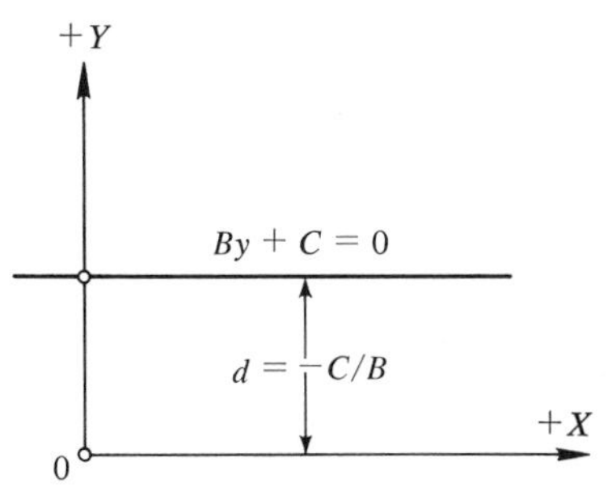

Fig. 6.03-5

Fig. 6.03-6

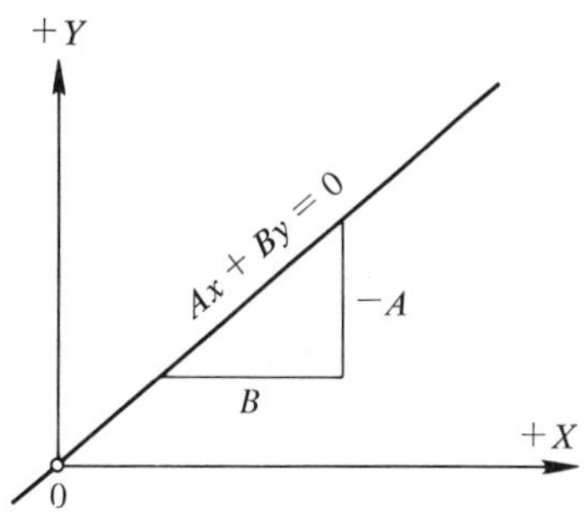

Fig. 6.03-7

(3) Points and Straight Lines

(a) Straight line passing through a point P_i in a given direction k (Fig. 6.03–8) is defined by

$$y - y_i = k(x - x_i)$$

where x_i, y_i are the coordinates of P_i and $k = \tan \phi$.

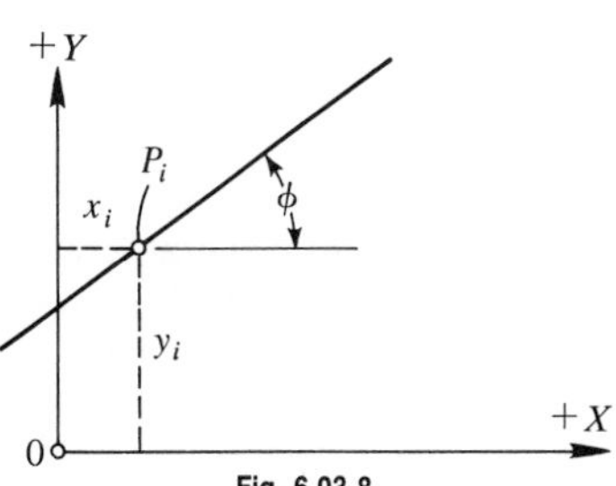

Fig. 6.03-8

(b) Straight line passing through two points (Fig. 6.03–9) is defined by

$$\frac{y - y_i}{x - x_i} = \frac{y_j - y_i}{x_j - x_i}$$

where (x_i, y_i), (x_j, y_j) are the coordinates of P_i, P_j respectively and its direction is

$$k = \tan \phi = \frac{y_j - y_i}{x_j - x_i}$$

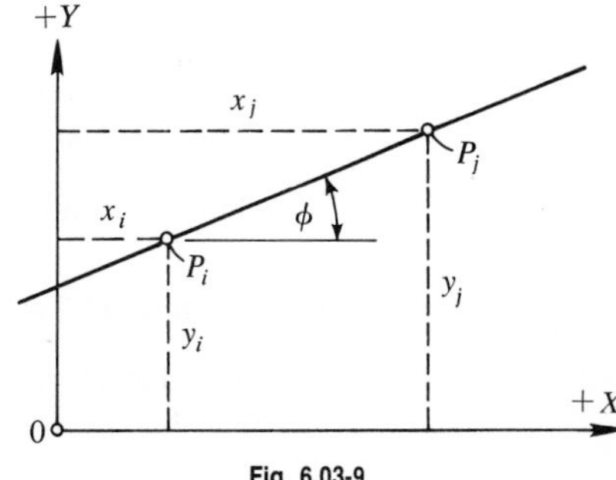

Fig. 6.03-9

(c) Distance from a point $P_i(x_i, y_i)$ to the line $Ax + By + C = 0$ (Fig. 6.03–10) is

$$d_i = \frac{Ax_i + By_i + C}{\pm\sqrt{A^2 + B^2}}$$

where the sign of the denominator is opposite to the sign of C.

(d) Distance from the origin 0 to the line $Ax + By + C = 0$ (Fig. 6.03–10) is

$$d_0 = \frac{C}{\pm\sqrt{A^2 + B^2}}$$

which is a special case of (c).

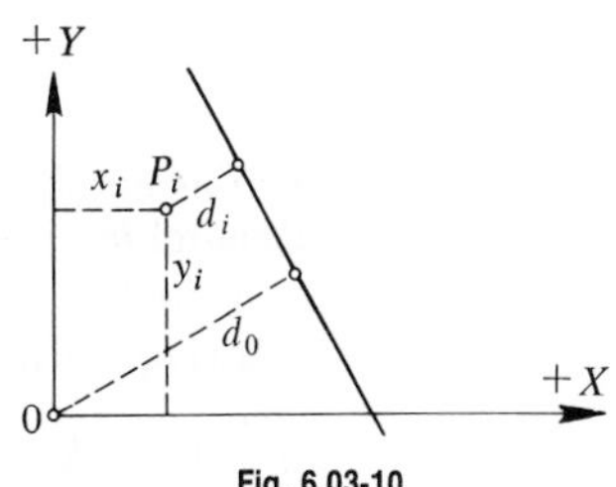

Fig. 6.03-10

6.04 TWO STRAIGHT LINES

(1) Relationships—Direction Forms

Two lines defined by

$y = k_1 x + l_1$ and $y = k_2 x + l_2$

are:

(a) Concurrent, if $k_1 \neq k_2$.

(b) Parallel, if $k_1 = k_2$, $l_1 \neq l_2$.

(c) Collinear, if $k_1 = k_2$, $l_1 = l_2$.

(d) Normal, if $k_1 = -1/k_2$.

(2) Relationships—General Forms

Two lines defined by

$A_1 x + B_1 y + C_1 = 0$ and $A_2 x + B_2 y + C_2 = 0$

are:

(a) Concurrent, if $A_1/B_1 \neq A_2/B_2$

(b) Parallel, if $A_1/B_1 = A_2/B_2$, $C_1 \neq C_2$.

(c) Collinear, if $A_1/B_1 = A_2/B_2$, $C_1 = C_2$.

(d) Normal, if $A_1/B_1 = -B_2/A_2$.

(3) Intersection of Two Lines

(a) Direction form. If the condition of Sec. 6.04–1*a* is satisfied, then the coordinates of the point of intersection i (Fig. 6.04–1) are

$$x_i = \frac{l_2 - l_1}{k_1 - k_2} \qquad y_i = \frac{k_1 l_2 - k_2 l_1}{k_1 - k_2}$$

and the angle ω between these two lines is given by

$$\tan \omega = \left| \frac{k_2 - k_1}{1 + k_1 k_2} \right|$$

taken as an absolute value.

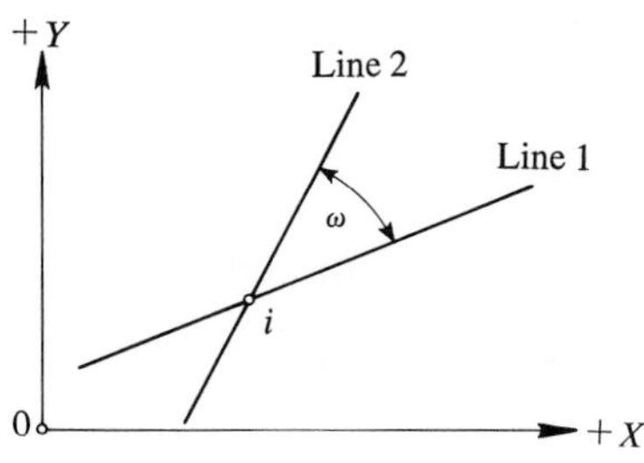

Fig. 6.04-1

(b) General form. If the condition of Sec. 6.04–2*a* is satisfied, then the coordinates of the point of intersection i are

$$x_i = \frac{B_1 C_2 - B_2 C_1}{A_1 B_2 - A_2 B_1} \qquad y_i = \frac{C_1 A_2 - C_2 A_1}{A_1 B_2 - A_2 B_1}$$

and the angle ω between these two lines is given by

$$\tan \omega = \left| \frac{A_1 B_2 - A_2 B_1}{A_1 A_2 + B_1 B_2} \right|$$

taken as an absolute value.

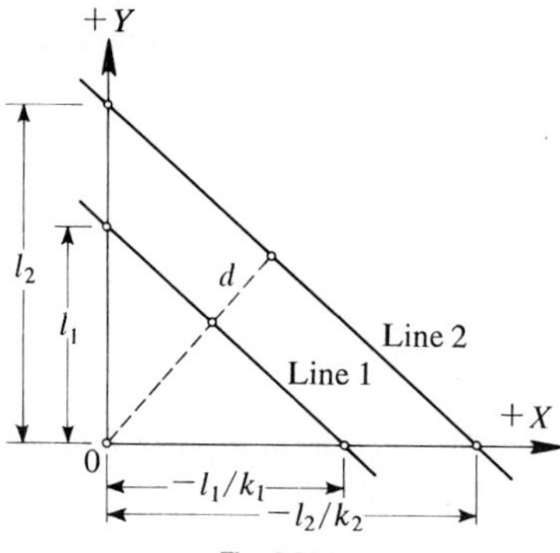

Fig. 6.04-2

(4) Distance of Two Parallel Lines

(a) Direction form. If the condition of Sec. 6.04–1*b* is satisfied, then the distance of two parallel lines (Fig. 6.04–2) is

$$d = \left| \frac{l_2 - l_1}{\pm\sqrt{1 + k^2}} \right|$$

taken as an absolute value.

(b) General form. If the condition of Sec. 6.04–2*b* is satisfied, then the distance of two parallel lines (Fig. 6.04–3) is

$$d = \left| \frac{C_2}{\pm\sqrt{A_2^2 + B_2^2}} - \frac{C_1}{\pm\sqrt{A_1^2 + B_1^2}} \right|$$

where the sign of each radical is opposite to that of C and d is taken as an absolute value.

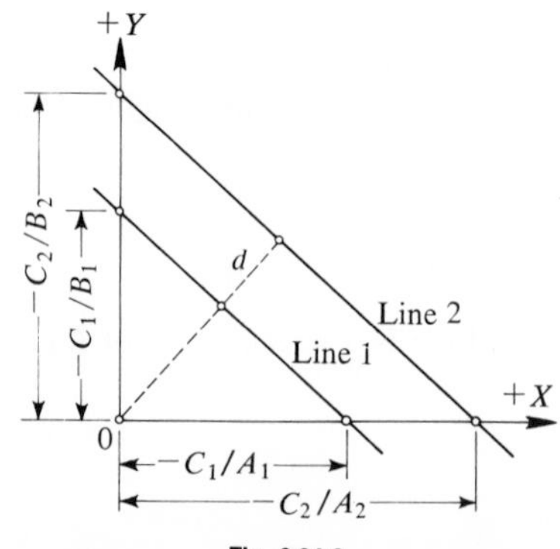

Fig. 6.04-3

6.05 TRANSFORMATION OF COORDINATES

(1) Definitions

(a) Transformation is a substitution of a function of one or more variables for a given variable.

(b) Three types of transformations are used in the analytic geometry: transformation of one system of coordinates to another (Sec. 6.01–3*c*), translation of a coordinate system (Sec. 6.05–2*a*), and rotation of a coordinate system (Sec. 6.05–2*b*).

(c) Translation. When the initial coordinate axes X, Y are moved into a new parallel position, the transformation is called the translation.

(d) Rotation. When the initial coordinate axes X, Y are rotated about their fixed origin 0 to a new position, the transformation is called the rotation.

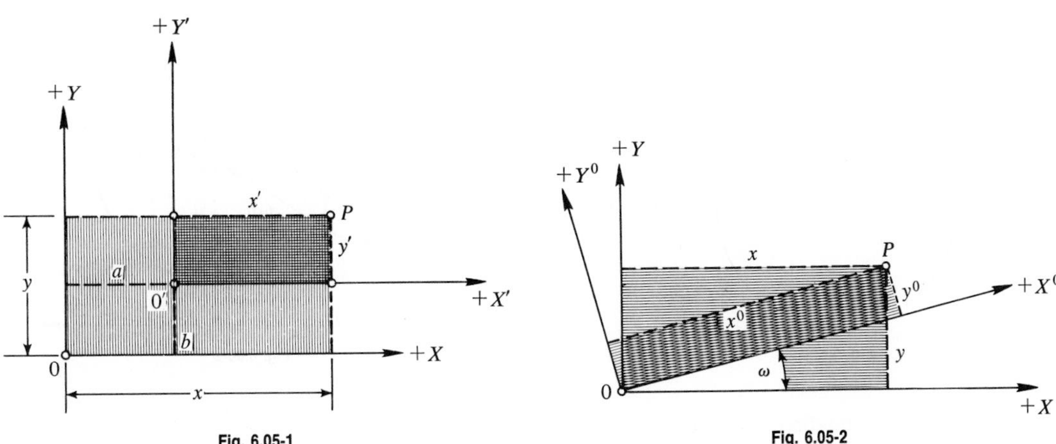

Fig. 6.05-1

Fig. 6.05-2

(2) Relations

(a) Translation. The transformation equations of translation (Fig. 6.05–1) are

$$x = x' + a \qquad y = y' + b$$

$$x' = x - a \qquad y' = y - b$$

where x, y are the coordinates of P in the X, Y system (initial axes), x', y' are the coordinates of P in the X', Y' system (translated axes) and a, b are the coordinates of the new origin $0'$.

(b) Rotation. The transformation equations of rotation (Fig. 6.05–2) are

$$x = x^\circ \cos\omega - y^\circ \sin\omega \qquad y = x^\circ \sin\omega + y^\circ \cos\omega$$

$$x^\circ = x\cos\omega + y\sin\omega \qquad y^\circ = -x\sin\omega + y\cos\omega$$

where x, y are the coordinates of P in the X, Y system (initial system), x°, y° are the coordinates of P in the X°, Y° system (rotated axes), and ω is the angle of rotation (positive in the counterclockwise direction).

6.06 CIRCLE

(1) Basic Equations

(a) Center at 0 (Fig. 6.06–1). Equation of a circle of radius R with center at the origin 0 is

$$x^2 + y^2 = R^2$$

Parametrically, $x = R \cos \tau$ and $y = R \sin \tau$, where τ is the position angle.

(b) Center at M (Fig. 6.06–2). Equation of a circle of radius R with center at an arbitrary point $M(x_M, y_M)$ is

$$(x - x_M)^2 + (y - y_M)^2 = R^2$$

Parametrically, $x = x_M + R \cos \tau$, $y = y_M + R \sin \tau$, where τ is again the position angle.

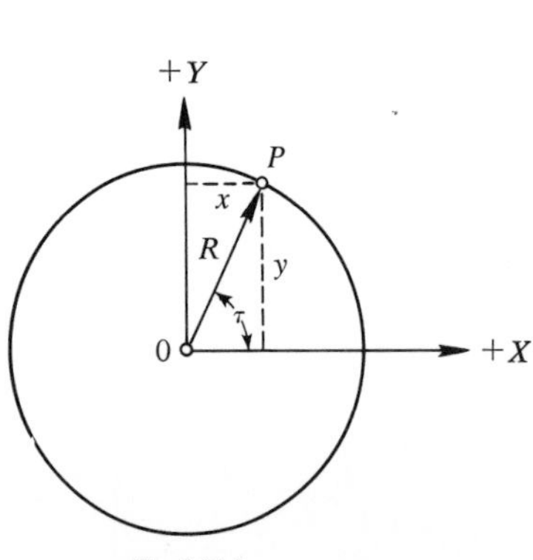

Fig. 6.06-1

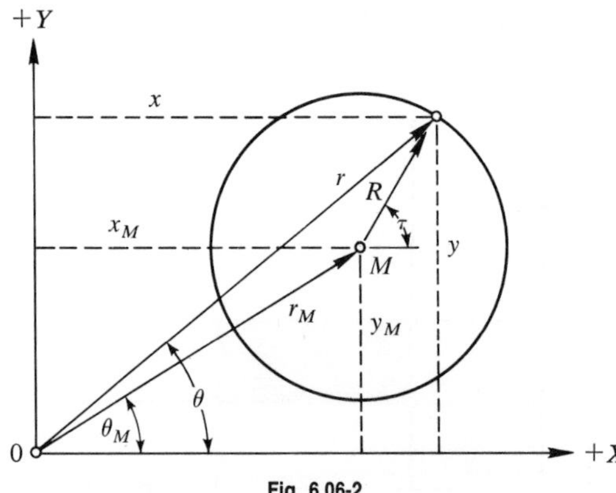

Fig. 6.06-2

(c) Polar equation of the circle of Fig. 6.06–2 is

$$r^2 - 2r_M r \cos(\theta - \theta_M) + r_M^2 = R^2$$

where r, θ are the polar coordinates of the circle and r_M, θ_M are the polar coordinates of its center M.

(d) General equation of the circle of Fig. 6.06–2 is

$$Ax^2 + Ay^2 + 2Dx + 2Ey + F = 0$$

which can be reduced (dividing each term by A) to

$$x^2 + y^2 + 2dx + 2ey + f = 0$$

from which the coordinates of the center are $x_M = -d$, $y_M = -e$, and the radius is $R = \sqrt{d^2 + e^2 - f}$.

example:

If $5x^2 + 5y^2 - 20x - 30y - 15 = 0$

then $x^2 + y^2 - 4x - 6y - 3 = 0$

where $d = -2$, $e = -3$, $f = -3$

and $x_M = 2$, $y_M = 3$, $R = \sqrt{4 + 9 + 3} = 4$

(2) Circle and a Straight Line

(a) Conditions. If $x^2 + y^2 = R^2$ and $y = kx + l$ are respectively equations of a circle and a straight line, the coordinates of their points of intersection are (Fig. 6.06–3)

$$x_{1,2} = \frac{-kl \pm \sqrt{\Delta}}{1+k^2} \qquad y_{1,2} = \frac{l \pm k\sqrt{\Delta}}{1+k^2}$$

where $\Delta = R^2 + R^2k^2 - l^2$.
If $\Delta > 0$, the line intersects the circle at two points $P_1(x_1,y_1)$, $P_2(x_2,y_2)$.
If $\Delta = 0$, the line is a tangent to the circle at $P_1(x_1,y_1) \equiv P_2(x_2,y_2)$ and $x_1 = x_2$, $y_1 = y_2$.
If $\Delta < 0$, the line does not intersect the circle and $x_{1,2}$, $y_{1,2}$ are conjugate complex numbers (coordinates of imaginary points).

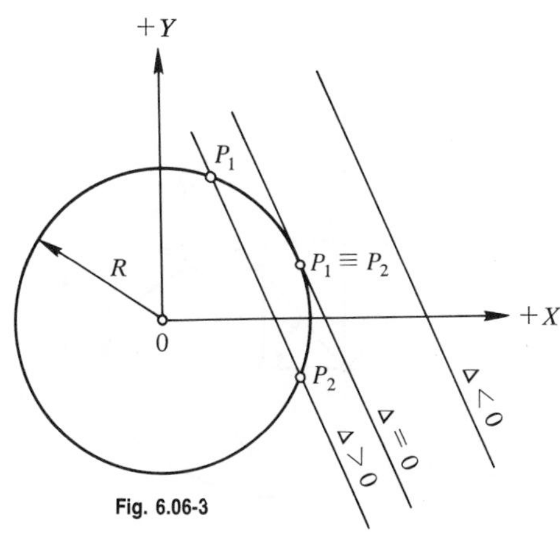

Fig. 6.06-3

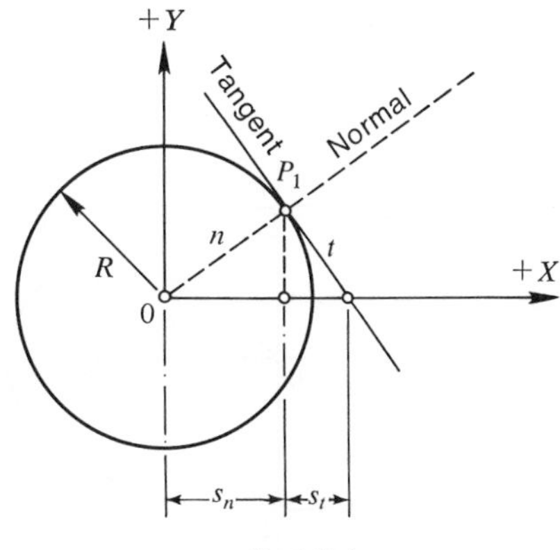

Fig. 6.06-4

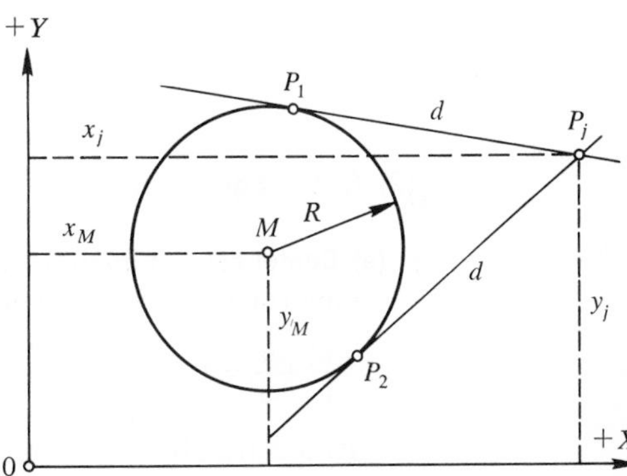

Fig. 6.06-5

(b) Tangent and normal (Fig. 6.06–4).

Equation of tangent at $P_1(x_1,y_2)$: $xx_1 + yy_1 = R^2$
Equation of normal at $P_1(x_1,y_{11})$: $yx_1 - xy_1 = 0$

Length of tangent: $t = \left|\frac{Ry_1}{x_1}\right|$

Length of normal: $n = R$

Length of subtangent: $s_t = \left|\frac{y_1^2}{x_1}\right|$

Length of subnormal: $s_n = |x_1|$

(c) Length of tangent between P_j and the point of contact P_1 or P_2 (Fig. 6.06–5) is

$$d = \sqrt{(x_j - x_M)^2 + (y_j - y_M)^2 - R^2}$$

6.07 ELLIPSE

(1) Definitions

(a) Ellipse is the locus of points whose sum of distances from two fixed points F_1 and F_2, called foci, is constant and equals $2a$ (Fig. 6.07–1); that is, $\overline{F_1P}+\overline{F_2P} = 2a$.

(b) Axes of ellipse are the major axis $\overline{AB} = 2a$ and the minor axis $\overline{CD} = 2b$.

(c) Linear eccentricity, $\overline{F_10} = \overline{0F_2} = e = \sqrt{a^2 - b^2}$, is the absolute value of the x coordinate of the focus F_1 and F_2; $\overline{F_1F_2} = 2e$.

(d) Numerical eccentricity is $\epsilon = e/a < 1$.

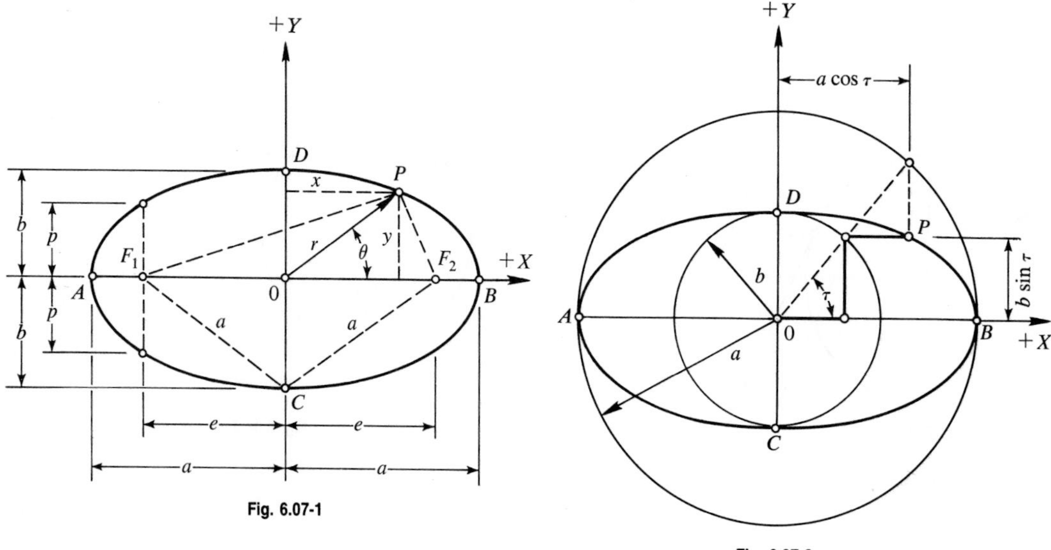

Fig. 6.07-1

Fig. 6.07-2

(2) Basic Equations

(a) Center at 0. Equation of an ellipse whose major axis $\overline{AB} = 2a$ coincides with the X axis, minor axis $\overline{CD} = 2b$ coincides with the Y axis (Fig. 6.07–1), is

$$\frac{x^2}{a^2}+\frac{y^2}{b^2}=1$$

Parametrically, $x = a \cos \tau$ and $y = b \sin \tau$, and τ is the angle defined in Fig. 6.07–2. In polar coordinates (pole at 0),

$$r^2 = \frac{a^2b^2}{a^2 \sin^2 \theta + b^2 \cos^2 \theta}$$

where r, θ are the polar coordinates (Fig. 6.07–1).

(b) Center at M (Fig. 6.07–3). Equation of an ellipse whose major and minor axes are parallel to the X and Y axes respectively at the center $M(x_M, y_M)$ is

$$\left(\frac{x - x_M}{a}\right)^2 + \left(\frac{y - y_M}{b}\right)^2 = 1$$

Parametrically, $x = x_M + a \cos \tau$, $y = y_M + b \sin \tau$.

(c) General equation of the ellipse of Fig. 6.07–3 is

$$Ax^2 + Cy^2 + 2Dx + 2Ey + F = 0$$

from which $x_M = -D/A$, $y_M = -E/C$, and the semiaxes are

$$a = \sqrt{\frac{CD^2 + AE^2 - ACF}{A^2C}} \qquad b = \sqrt{\frac{CD^2 + AE^2 - ACF}{AC^2}}$$

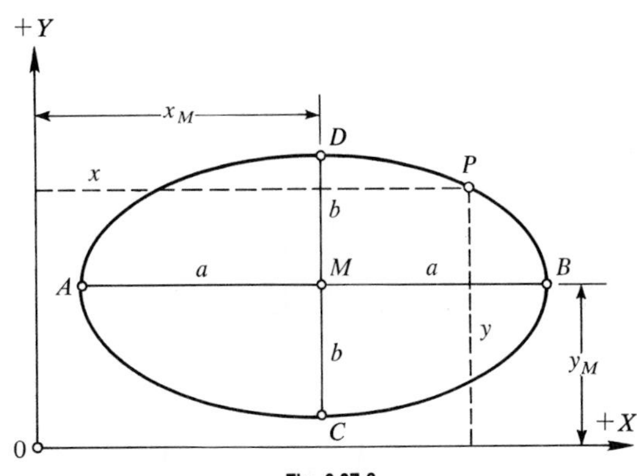

Fig. 6.07-3

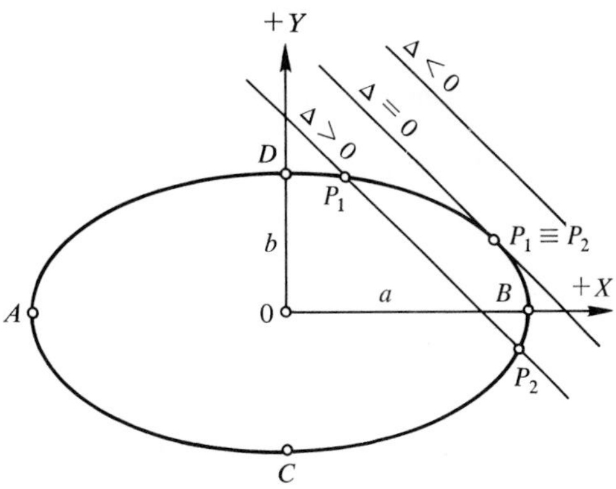

Fig. 6.07-4

(3) Ellipse and a Straight Line

(a) Conditions. If $x^2/a^2 + y^2/b^2 = 1$ and $y = kx + l$ are respectively equations of an ellipse and a straight line, the coordinates of their points of intersection (Fig. 6.07–4) are

$$x_{1,2} = \frac{-a^2kl \pm ab\sqrt{\Delta}}{b^2 + a^2k^2} \qquad y_{1,2} = \frac{b^2l \pm abk\sqrt{\Delta}}{b^2 + a^2k^2}$$

where $\Delta = b^2 + a^2k^2 - l^2$.

(b) Tangent and normal (Fig. 6.07–5).

Equation of tangent at $P_1(x_1, y_1)$: $\quad \dfrac{xx_1}{a^2} + \dfrac{yy_1}{b^2} = 1$

Equation of normal at $P_1(x_1, y_1)$: $\quad \dfrac{y - y_1}{a^2 y_1} - \dfrac{x - x_1}{b^2 x_1} = 0$

Length of tangent: $\quad t = \sqrt{y_1^2 + \left(\dfrac{a^2 - x_1^2}{x_1}\right)^2}$

Length of normal: $\quad n = \dfrac{b\sqrt{a^4 - e^2x_1^2}}{a^2}$

Length of subtangent: $\quad s_t = \left|\dfrac{a^2 - x_1^2}{x_1}\right|$

Length of subnormal: $\quad s_n = \left|\dfrac{b^2x_1}{a^2}\right|$

where $e^2 = a^2 - b^2$.

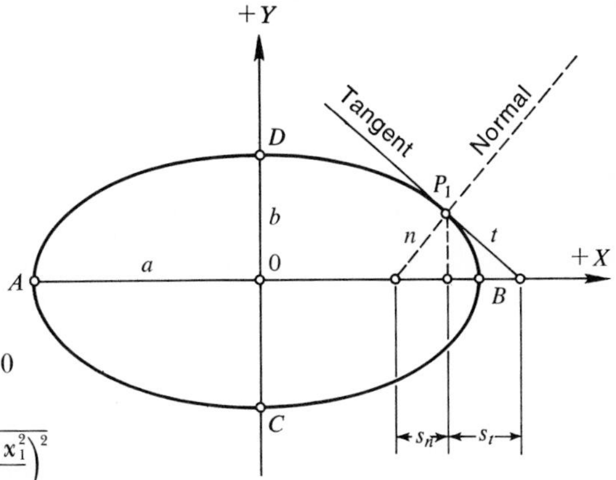

Fig. 6.07-5

6.08 HYPERBOLA

(1) Definitions

(a) Hyperbola is the locus of points whose difference of distances from two fixed points F_1, F_2, called foci, is constant and equals $2a$ (Fig. 6.08–1); that is, $\overline{F_1P} - \overline{F_2P} = 2a$.

(b) Axes of hyperbola are the major axis $\overline{AB} = 2a$ and the minor axis $\overline{CD} = 2b$. Asymptotes of a hyperbola are two tangents through 0 whose points of contact are at an infinite distance from 0.

(c) Linear eccentricity, $\overline{F_1 0} = \overline{0F_2} = e = \sqrt{a^2 + b^2}$, is the absolute value of the x coordinate of the focus F_1 and F_2; that is, $\overline{F_1F_2} = 2e$.

(d) Numerical eccentricity is $\epsilon = e/a > 1$.

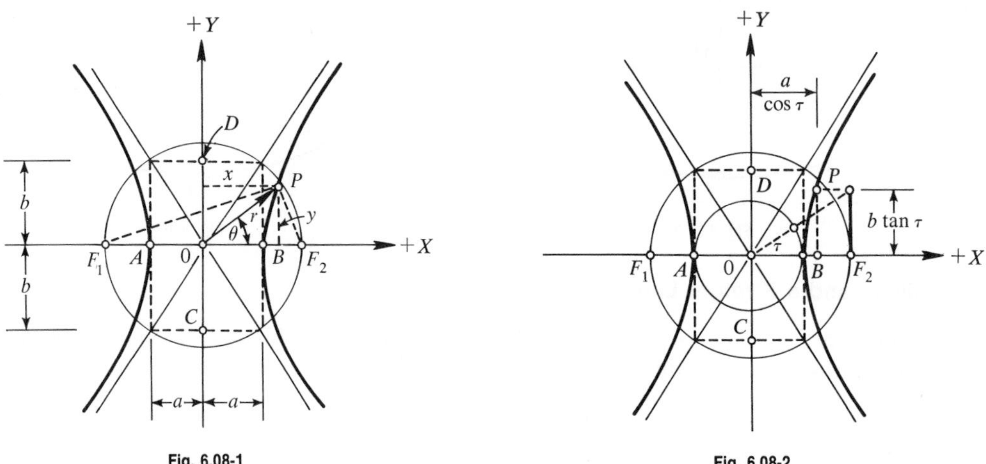

Fig. 6.08-1

Fig. 6.08-2

(2) Basic Equations

(a) Center at 0. Equation of a hyperbola whose major axis $\overline{AB} = 2a$ coincides with the X axis and minor axis coincides with the Y axis (Fig. 6.08–1) is

$$\frac{x^2}{a^2} - \frac{y^2}{b^2} = 1$$

Parametrically, $x = a/(\cos \tau)$ and $y = \pm b \tan \tau$, where τ is the angle defined in Fig. 6.08–2. In polar coordinates (pole at 0),

$$r^2 = \frac{a^2b^2}{b^2 \cos^2 \theta - a^2 \sin^2 \theta}$$

where r, θ are the polar coordinates (Fig. 6.08–1).

(b) Center at M, (Fig. 6.08–3). Equation of a hyperbola whose major and minor axes are parallel to X and Y axes respectively at the center $M(x_M, y_M)$ is

$$\left(\frac{x - x_M}{a}\right)^2 - \left(\frac{y - y_M}{a}\right)^2 = 1$$

Parametrically, $x = x_M + a/(\cos \tau)$, $y = y_M \pm b \tan \tau$.

(c) General equation of the hyperbola of Fig. 6.08–3 is

$$Ax^2 - Cy^2 + 2Dx + 2Ey + F = 0$$

from which $x_M = -D/A$, $y_M = -E/C$, and the semiaxes are

$$a = \sqrt{\frac{CD^2 - AE^2 - ACF}{A^2C}} \qquad b = \sqrt{\frac{CD^2 - AE^2 - ACF}{AC^2}}$$

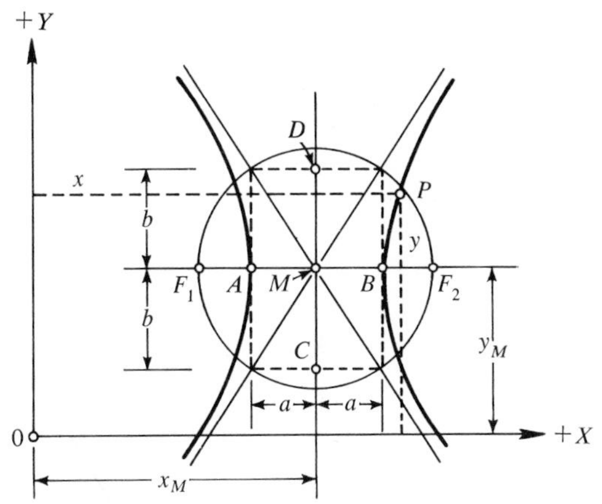

Fig. 6.08-3

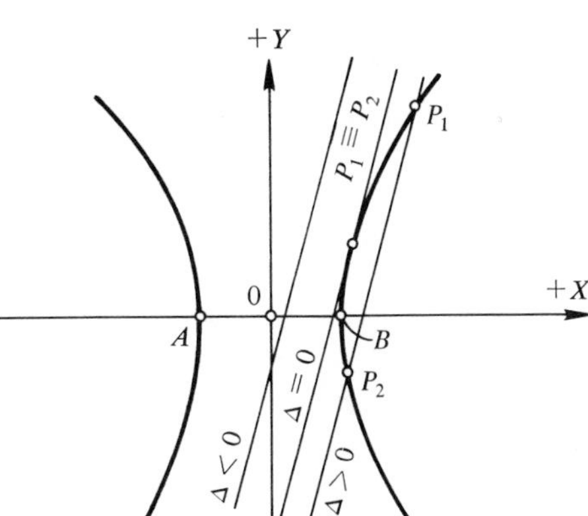

Fig. 6.08-4

(3) Hyperbola and a Straight Line

(a) Conditions. If $x^2/a^2 - y^2/b^2 = 1$ and $y = kx + l$ are respectively equations of a hyperbola and of a straight line, the coordinates of their points of intersection (Fig. 6.08–4) are

$$x_{1,2} = \frac{a^2kl \pm ab\sqrt{\Delta}}{b^2 - a^2k^2} \qquad y_{1,2} = \frac{b^2l \pm abk\sqrt{\Delta}}{b^2 - a^2k^2}$$

where $\Delta = b^2 - a^2k^2 + l^2$.

(b) Tangent and normal (Fig. 6.08–5).

Equation of tangent at $P_1(x_1,y_1)$: $\quad \dfrac{xx_1}{a^2} - \dfrac{yy_1}{b^2} = 1$

Equation of normal at $P_1(x_1,y_1)$: $\quad \dfrac{y - y_1}{a^2y_1} + \dfrac{x - x_1}{b^2x_1} = 0$

Length of tangent: $\quad t = \sqrt{y_1^2 + \left(\dfrac{x_1^2 - a^2}{x_1}\right)^2}$

Length of normal: $\quad n = \dfrac{b}{a^2}\sqrt{e^2x_1^2 - a^4}$

Length of subtangent $\quad s_t = \left|\dfrac{x_1^2 - a^2}{x_1}\right|$

Length of subnormal: $\quad s_n = \left|\dfrac{b^2x_1}{a^2}\right|$

where $e^2 = a^2 + b^2$.

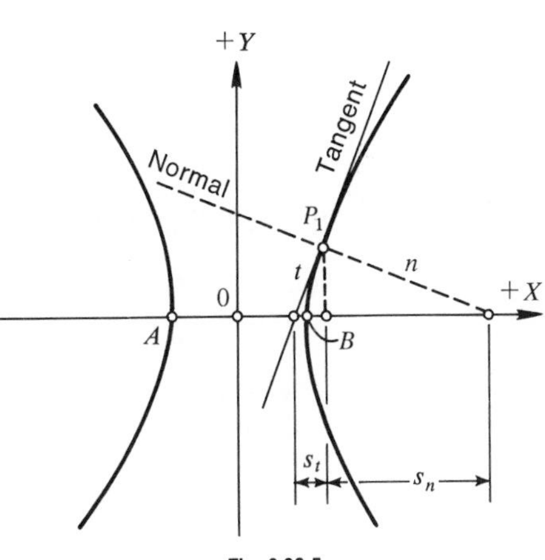

Fig. 6.08-5

6.09 PARABOLA

(1) Definitions

(a) Parabola is the locus of points whose distance from a fixed line (directrix) is equal to their distance from a fixed point F (focus) (Fig. 6.09–1); that is, $\overline{EP} = \overline{FP}$.

(b) Axis of parabolas in Figs. 6.09–1 and 6.09–2 is the X axis and Y axis respectively with the vertex at 0 bisecting the distance p (parameter) between the directrix d and the focus F, $\overline{D0} = p/2$, $\overline{0F} = p/2$.

(c) Linear eccentricity is $\overline{0F} = e = p/2$ is the x coordinate (or y coordinate) of the focus F.

(d) Numerical eccentricity is $\epsilon = e/(p/2) = 1$.

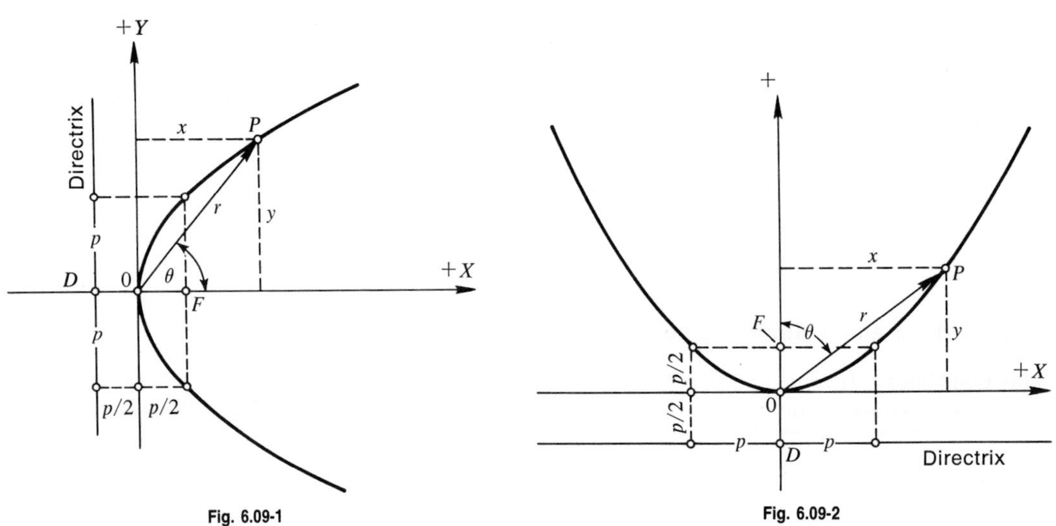

Fig. 6.09-1

Fig. 6.09-2

(2) Basic Equations

(a) Vertex at 0. Equation of a parabola whose axis is the X axis with vertex at 0 (Fig. 6.09–1) is

$$y^2 = 2px$$

which is open to the right if $p > 0$ and is open to the left if $p < 0$. If the axis of the parabola is the Y axis with the vertex at 0 (Fig. 6.09–2), the equation is

$$x^2 = 2py$$

If $p > 0$, the parabola is open upward; if $p < 0$, the parabola is open downward. Parametrically, $x = (p/2)\tau^2$, $y = p\tau$ (Fig. 6.09–1), and $y = (p/2)\tau^2$, $x = p\tau$ (Fig. 6.09–2).

In polar coordinates (pole at 0),

$$r = 2p \cos\theta\,(1 + \cot^2\theta)$$

where r, θ are the polar coordinates (Figs. 6.09–1 and 6.09–2) and θ is always measured from the axis of the parabola.

(b) Vertex at A. Equation of a parabola whose axis is parallel to the X axis with vertex at $A(x_A, y_A)$ (Fig. 6.09–3) is

$$(y - y_A)^2 = 2p(x - x_A)$$

which is open to the right if $p > 0$ and is open to the left is $p < 0$. If the axis of the parabola is the Y axis with vertex at $A(x_A, y_A)$ (Fig. 6.09–4), the equation is

$$(x - x_A)^2 = 2p(y - y_A)$$

which is open upward if $p > 0$ and is open downward if $p < 0$. Parametrically, $x = x_A + (p/2)\tau^2$, $y = y_A + p\tau$ (Fig. 6.09–3), and $x = x_A + p\tau$, $y = y_A + (p/2)\tau^2$ (Fig. 6.09–4).

(c) General equation of the parabola of Fig. 6.09–3 is

$$Cy^2 + 2Dx + 2Ey + F = 0$$

from which $x_A = (E^2 - CF)/2CD$, $y_A = -(E/C)$, and the parameter

$$p = -\frac{D}{C}$$

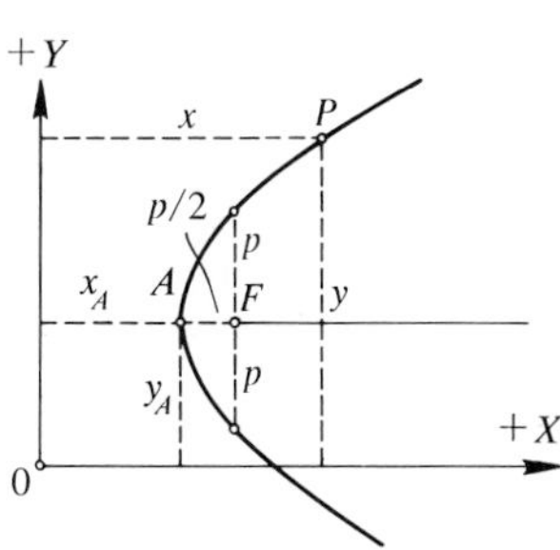

Fig. 6.09-3

(d) General equation of the parabola of Fig. 6.09–4 is

$$Ax^2 + 2Dx + 2Ey + F = 0$$

From which $x_A = -D/A$, $y_A = (D^2 - AF)/2AE$, and the parameter

$$p = -\frac{E}{A}$$

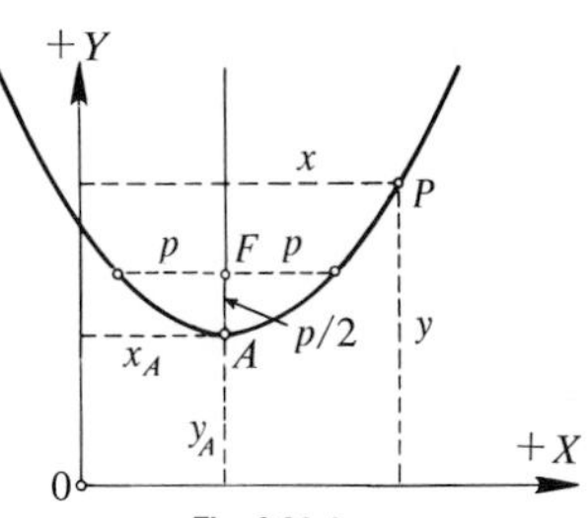

Fig. 6.09-4

(3) Parabola and a Straight Line

(a) Conditions. If $y^2 = 2px$ and $y = kx + l$ are respectively equations of a parabola and of a straight line, the coordinates of their point of intersection (Fig. 6.09–5) are

$$x_{1,2} = \frac{p - kl \pm \sqrt{\Delta}}{k^2} \qquad y_{1,2} = \frac{p \pm \sqrt{\Delta}}{k}$$

where

$$\Delta = p(p - 2kl)$$

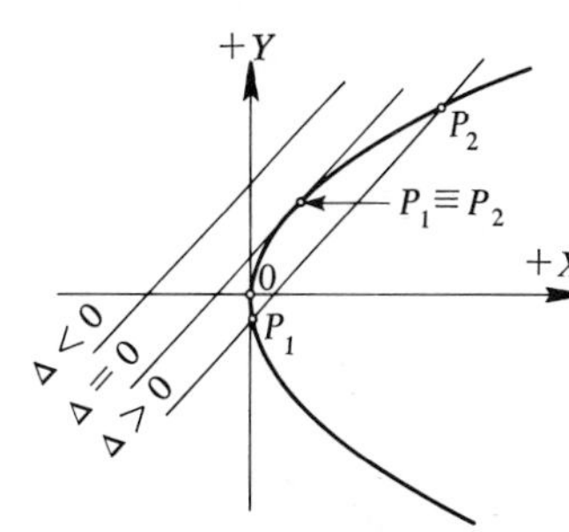

Fig. 6.09-5

(b) Tangent and normal (Fig. 6.09–6).

Equation of tangent at $P_1(x_1, y_1)$:	$yy_1 = p(x + x_1)$
Equation of normal at $P_1(x_1, y_1)$:	$p(y - y_1) + y_1(x - x_1) = 0$
Length of tangent:	$t = \sqrt{y_1^2 + 4x_1^2}$
Length of normal:	$n = \sqrt{y_1^2 + p^2}$
Length of subtangent:	$s_t = 2x_1$
Length of subnormal:	$s_n = p$

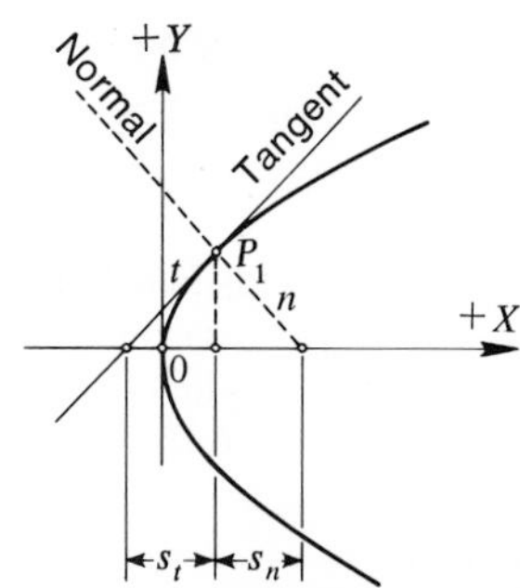

Fig. 6.09-6

6.10 GENERAL ALGEBRAIC FUNCTIONS

(1) Definitions

(a) Algebraic curve is one whose equation is of the form $f(x,y) = 0$, where $f(x,y)$ is a rational or irrational algebraic expression in x and y.

(b) Classification. They are classified as rational integral functions, rational fractional functions, irrational integral functions, and irrational fractional functions.

(2) Rational Integral Functions

(a) Function $y = Ax + B$ defines a straight line introduced in Sec. 6.03.

(b) Function $y = Ax^2 + Bx + C$ defines a quadratic parabola with vertex at $x_V = -(B/2A)$, $y_V = (4AC - B^2)/4A$, and with its axis parallel to the Y axis (Fig. 6.10–1).

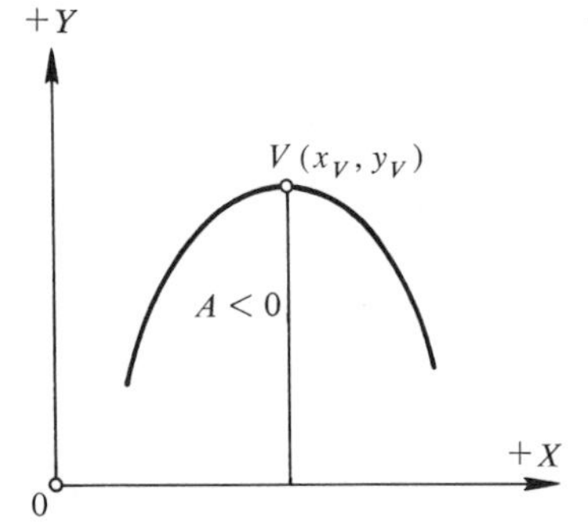

Fig. 6.10-1

(c) Function $y = Ax^3 + Bx^2 + Cx + D$ defines the cubic parabola whose shape depends on the values of A and $\Delta = 3AC - B^2$. Three typical cubic parabolas are shown in Fig. 6.10–2.

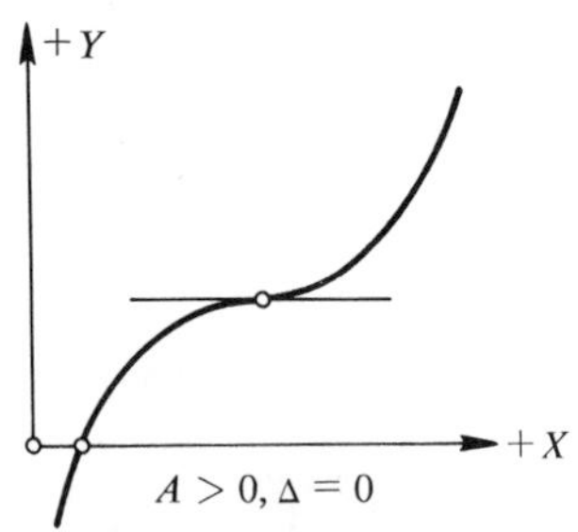

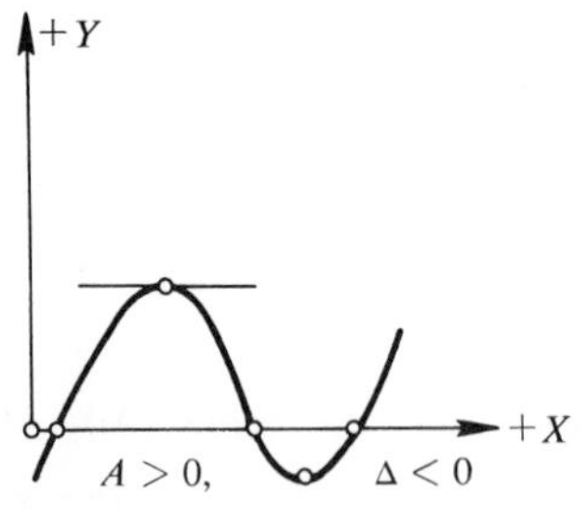

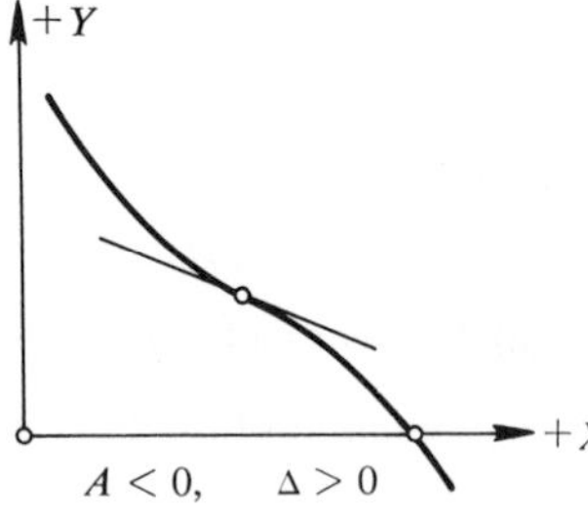

Fig. 6.10-2

(d) Function $y = Ax^n$, where n is a positive integer, defines a parabola of the nth degree. If n is even, the curve is symmetrical with respect to the Y axis and its vertex is at *0*. If n is odd, the curve is antisymmetrical with respect to the Y axis and its inflection point is at *0* (Fig. 6.10–3).

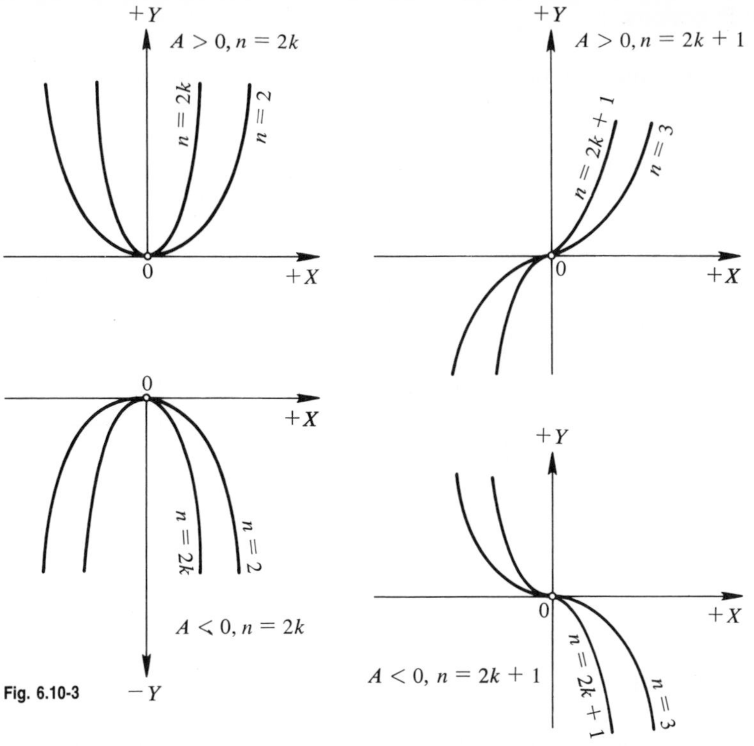

Fig. 6.10-3

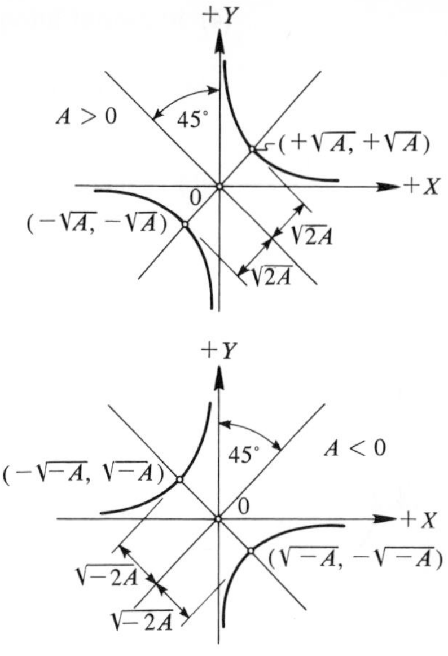

Fig. 6.10-4

(3) Rational Fractional Functions

(a) Function $y = A/x$ defines the asymptotic hyperbola whose asymptotes are the X and Y axes (Fig. 6.10–4).

(b) Function $y = 1/(Ax^2 + Bx + C)$ defines the third-degree reciprocal curve which is symmetrical with respect to an axis parallel to the Y axis, and its shape depends on $\Delta = 4AC - B^2$ (Fig. 6.10–5).

(c) Function $y = A/x^n$ defines the nth-degree asymptotic curve. If n is even, the curve is symmetric with respect to the Y axis. If n is odd, the curve is antisymmetric with respect to the Y axis (Fig. 6.10–6).

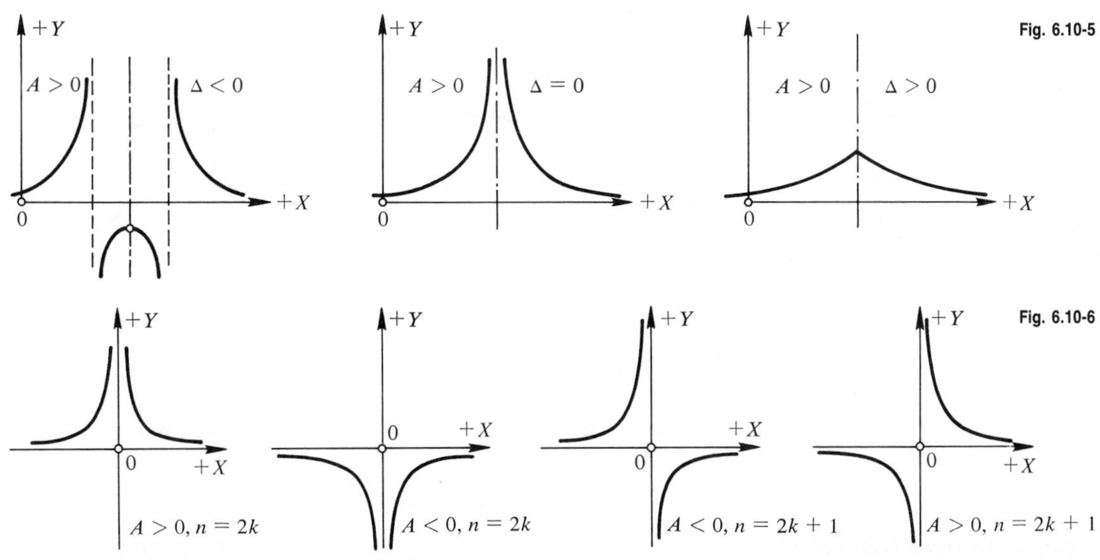

Fig. 6.10-5

Fig. 6.10-6

(4) Irrational Integral Functions

(a) Function $y = \sqrt{Ax + B}$ defines a second-degree parabola which is symmetric with respect to the X axis (Fig. 6.10–7).

(b) Function $y = \sqrt{Ax^2 + Bx + C}$ defines the hyperbola if $A > 0$ and the ellipse if $A < 0$ (Fig. 6.10–8). The orientation of the principal axis is governed by the sign of $\Delta = 4AC - B^2$.

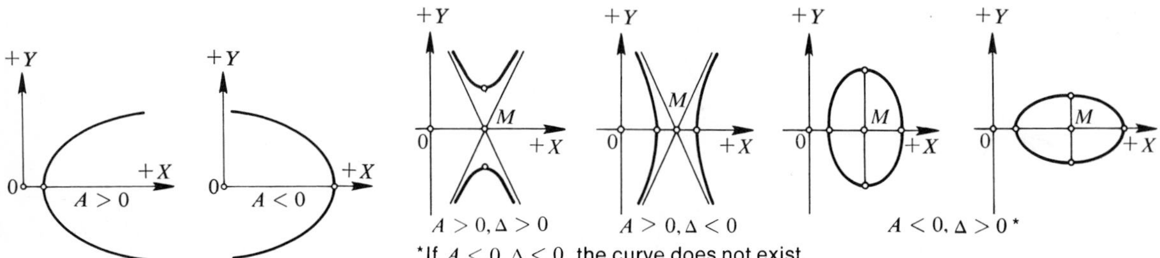

*If $A < 0$, $\Delta < 0$, the curve does not exist.

Fig. 6.10-7

Fig. 6.10-8

(5) Special Irrational Functions

(a) Function $y = \sqrt[n]{x}$ is the exponential inverse of $y = x^n$. Their graphs are symmetrical with respect to $x = y$ (Figs. 6.10–9 and 6.10–10).

(b) Functions $y = \sqrt{x^3}$ **and** $x = \sqrt{y^3}$ are the semicubic parabolas. Their graphs are symmetrical with respect to $x = y$ (Fig. 6.10–11).

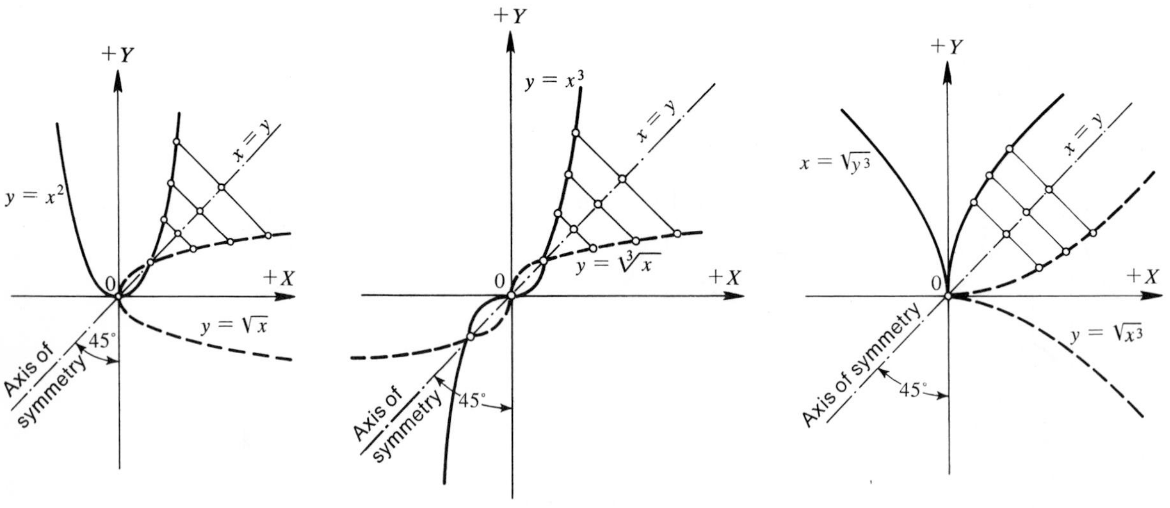

Fig. 6.10-9

Fig. 6.10-10

Fig. 6.10-11

6.11 SPECIAL ALGEBRAIC FUNCTIONS

(1) Functions of Third Degree

(a) Witch of Agnesi (Fig. 6.11–1) is defined in the cartesian coordinates by

$$y = \frac{8R^3}{x^2 + 4R^2}$$

and parametrically by

$$x = 2R \cot \phi \qquad \text{and} \qquad y = R(1 - \cos 2\phi)$$

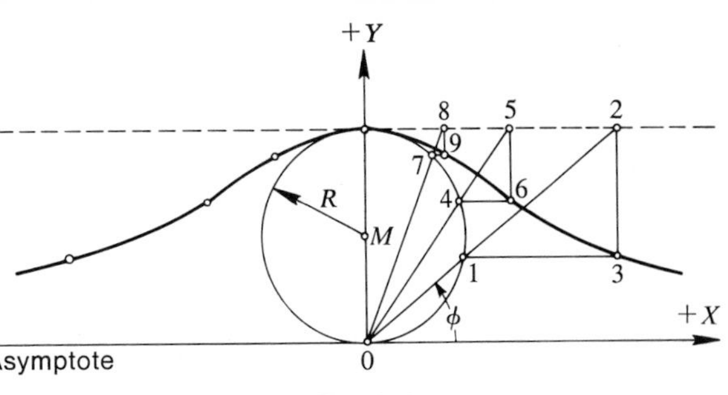

Fig. 6.11-1

Its graphical construction is shown in Fig. 6.11–1 and follows the sequence of numbers $1, 2, 3, \ldots$. The curve is asymptotic to the X axis and the area between the curve and its asymptote is $4\pi R^2$.

(b) Cissoid of Diocles (Fig. 6.11–2) is defined in the cartesian coordinates by

$$y^2 = \frac{x^3}{2R - x}$$

in polar coordinates by

$$r = 2R \sin \theta \tan \theta$$

and parametrically by

$$x = \frac{2Rt^2}{1+t^2} \qquad y = \frac{2Rt^3}{1+t^2}$$

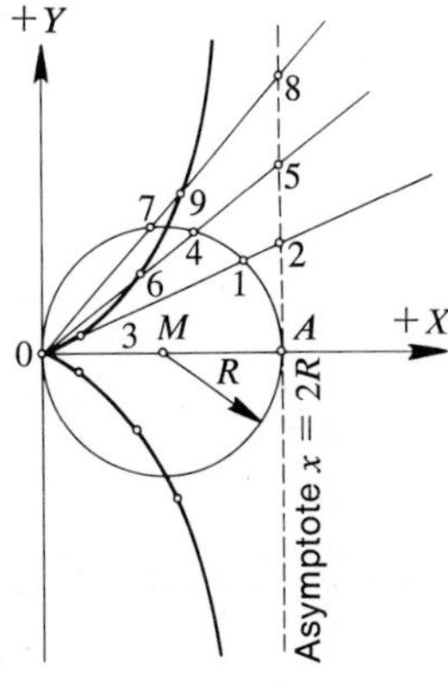

Fig. 6.11-2

Its graphical construction is shown in Fig. 6.11–2, where $\overline{12} = \overline{03}$, $\overline{45} = \overline{06}$, $\overline{78} = \overline{09}, \ldots$. The curve is asymptotic to the line $x = 2R$ and the area between the curve and its asymptote is $3\pi R^2$.

(c) Folium of Descartes (Fig. 6.11–3) is a cissoid of the ellipse $x^2 - xy + y^2 - a(x + y) = 0$ with regard to the straight line $x + y = -a$, defined in the cartesian coordinates by

$$x^3 + y^3 - 3axy = 0$$

in the polar coordinates by

$$r = \frac{3a \sin \theta \cos \theta}{\sin^3 \theta + \cos^3 \theta}$$

and in parametric form by

$$x = \frac{3at}{1+t^3} \qquad y = \frac{3at^2}{1+t^3}$$

Its graphical construction is shown in Fig. 6.11–3, where $\overline{01} = \overline{23}$, $\overline{04} = \overline{56}, \ldots$. The curve is asymptotic to $x + y = -a$ and symmetrical with respect to $x = y$, with vertex $V(3a/2, 3a/2)$. The area of the loop is $3a^2/2$ and the area between the curve and its asymptote is also $3a^2/2$.

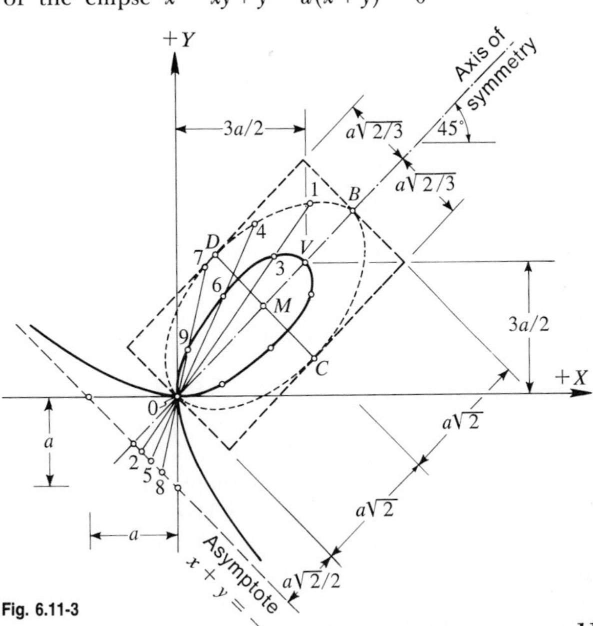

Fig. 6.11-3

(d) Strofoid (Fig. 6.11–4) is defined in the cartesian coordinates by

$$y^2 = x^2 \frac{a - x}{a + x}$$

in the polar coordinates by

$$r = a \frac{\cos 2\theta}{\cos \theta}$$

and parametrically by

$$x = \frac{a(1 - t^2)}{1 + t^2} \qquad y = \frac{at(1 - t^2)}{1 + t^2}$$

Strofoid of Fig. 6.11–4 is the locus of points P_1 and P_2 for which $\overline{0Q} = \overline{QP_1} = \overline{QP_2}$. The curve is asymptotic to $x = -a$, and symmetrical with respect to the X axis with vertex $A\,(a,0)$. The area of the loop is $a^2(1 - \pi/4)$, and the area between the curve and its asymptote is $a^2(1 + \pi/4)$.

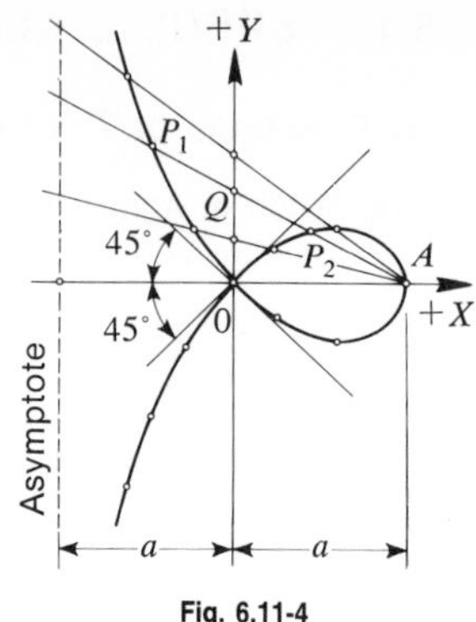

Fig. 6.11-4

(2) Functions of Fourth Degree

(a) Conchoid of Nicomedes (Fig. 6.11–5) is defined in the cartesian coordinates by

$$(x^2 + y^2)(x - a)^2 - b^2x^2 = 0$$

in the polar coordinates by

$$r = \frac{a}{\cos \theta} \pm b$$

and parametrically by

$$x = a + b \cos \phi$$

$$y = a \tan \phi + b \sin \phi$$

Geometrically, it is the locus of points P_1 and P_2, for which $\overline{0P_1} = \overline{0Q} - b$ and $\overline{0P_2} = \overline{0Q} + b$. The curve has two branches symmetrical with respect to the X axis and asymptotic to $x = a$. The area between the curve and its asymptote is ∞.

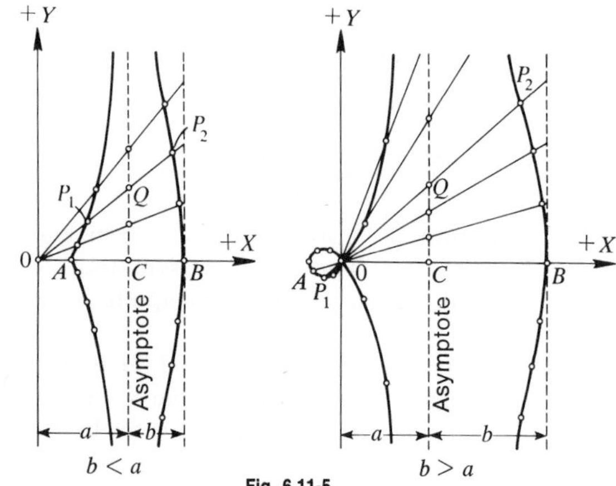

Fig. 6.11-5

(b) Pascal's snails (Figs. 6.11–6 and 6.11–7) are defined in the cartesian coordinates by

$$(x^2 + y^2 - ax^2) - b^2(x^2 + y^2) = 0$$

in the polar coordinates by

$$r = a \cos \theta \pm b$$

and parametrically by

$$x = a \cos^2 \phi + b \cos \phi \qquad y = a \cos \phi \sin \phi + b \sin \phi$$

Geometrically, its points satisfy $\overline{0P_1} = \overline{0Q} + b$ and $P_2 = \overline{0Q} - b$. The curve is symmetrical with respect to the X axis and its four typical shapes, depending on the relation of a to b, are shown in Figs. 6.11–6 and 6.11–7. For the curves without the loop the area is $\pi(b^2 + a^2/2)$. The special case of $a = b$ is called the *cardioid*.

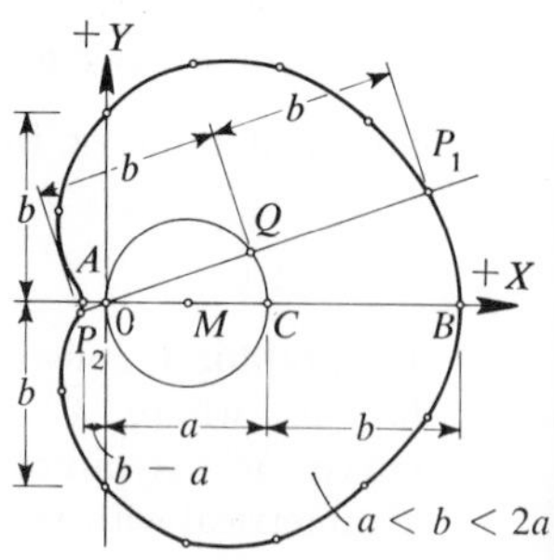

Fig. 6.11-6

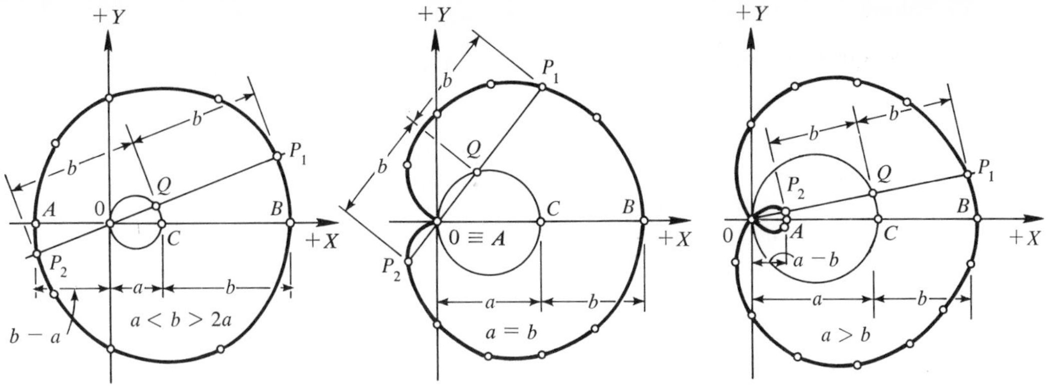

Fig. 6.11-7

(c) Ovals of Cassini (Fig. 6.11–8) are defined in the cartesian coordinates by

$$(x^2+y^2)^2 - 2e^2(x^2-y^2) = a^4 - e^4$$

and in the polar coordinates by

$$r^2 = e^2\cos 2\theta \pm \sqrt{e^4\cos^2 2\theta + (a^4-e^4)}$$

Geometrically, their points satisfy the relation $\overline{F_1P}\cdot\overline{F_2P} = a^2$, where F_1 and F_2 are two fixed points. Their shapes depend on the relation of a to e.

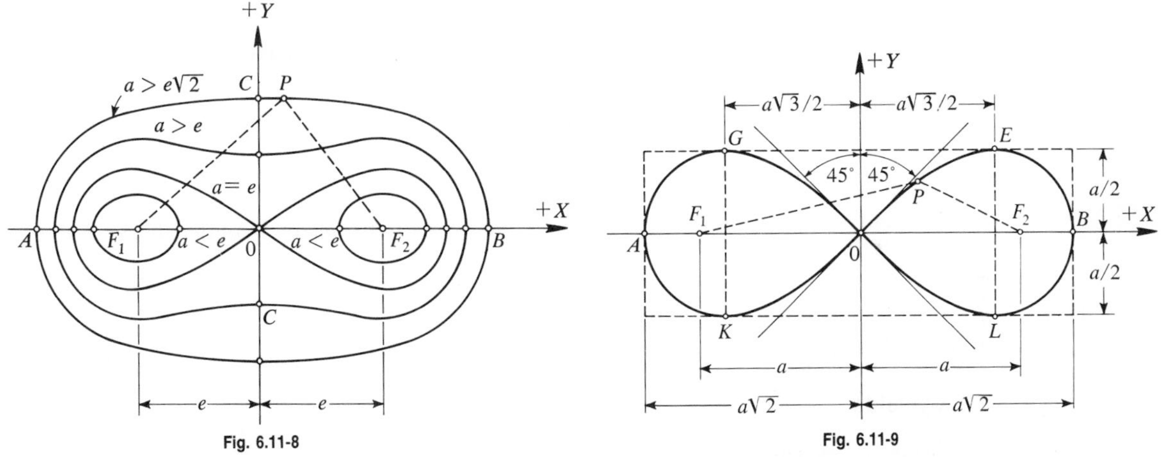

Fig. 6.11-8

Fig. 6.11-9

(d) Lemniscate of Bernoulli (Fig. 6.11–9) is a special case of the oval of Cassini, $a = e$, defined in the cartesian coordinates by

$$(x^2+y^2)^2 = 2a^2(x^2-y^2)$$

in the polar coordinates by

$$r^2 = 2a^2\cos 2\theta$$

and parametrically

$$x = \frac{at\sqrt{2}(1+t^2)}{1+t^4} \qquad y = \frac{at\sqrt{2}(1-t^2)}{1+t^4}$$

The curve is symmetrical with respect to the X and Y axes and the area of one of its loops is a^2.

(3) Spirals

(a) Spiral of Archimedes (Fig. 6.11–10) is defined in the polar coordinates by

$$r = a\theta$$

The curve is generated by the point P moving with a constant linear velocity v along a straight line which rotates with a constant angular velocity ω about the fixed origin 0.

The length of the arc of the curve is $(a/2)(\theta\sqrt{\theta^2+1}+\sinh^{-1}\theta)$ which for large θ is approximately $a\theta^2/2$. The area of a sector bounded by θ_1 and θ_2 is $a^2(\theta_2^3-\theta_1^3)/6$.

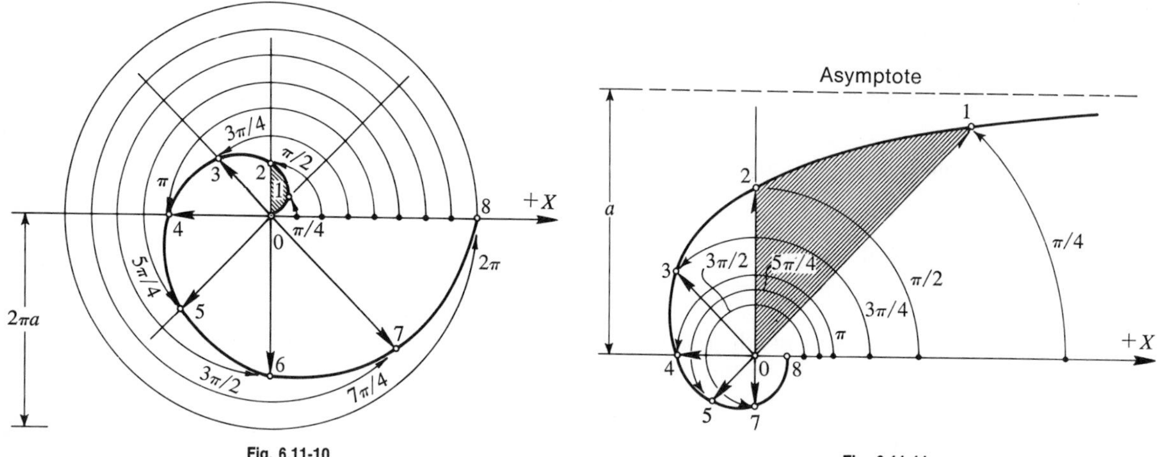

Fig. 6.11-10

Fig. 6.11-11

(b) Hyperbolic spiral (Fig. 6.11–11) is defined in the polar coordinates by

$$r = \frac{a}{\theta}$$

The curve is asymptotic to $y = a$, and the area of the sector bounded by θ_1 and θ_2 is $(a^2/2)(1/\theta_1 - 1/\theta_2)$.

6.12 CYCLIC FUNCTIONS

(1) Cycloids

(a) Ordinary cycloid (Fig. 6.12–1) is defined in the cartesian coordinates by

$$x = a\cos^{-1}\frac{a-y}{a} - \sqrt{y(2a-y)}$$

and parametrically by

$$x = a(\phi - \sin\phi) \qquad y = a(1-\cos\phi)$$

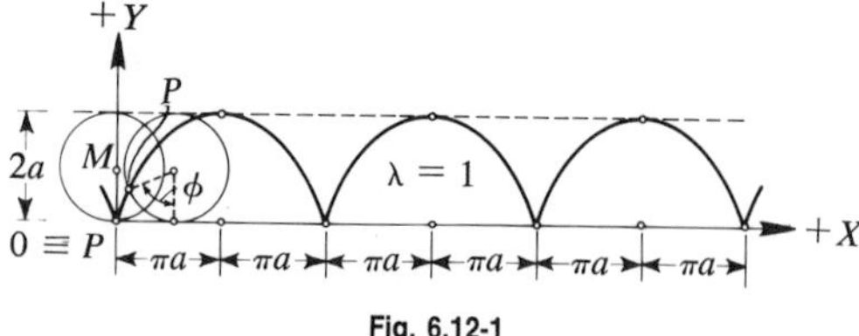

Fig. 6.12-1

This curve is generated by a fixed point P on a circle of radius a which rolls without slipping along the X axis. The curve is periodic (repeats) with the period $2\pi a$. The length of the curve between two cusps is $8a$, and the area between one full arch of the curve and the X axis is $3\pi a^2$.

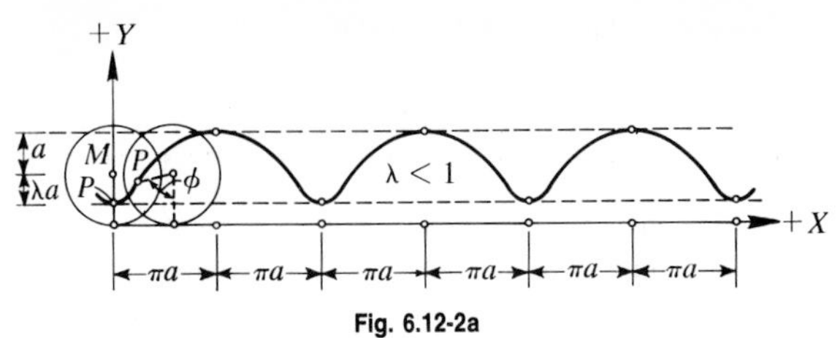

Fig. 6.12-2a

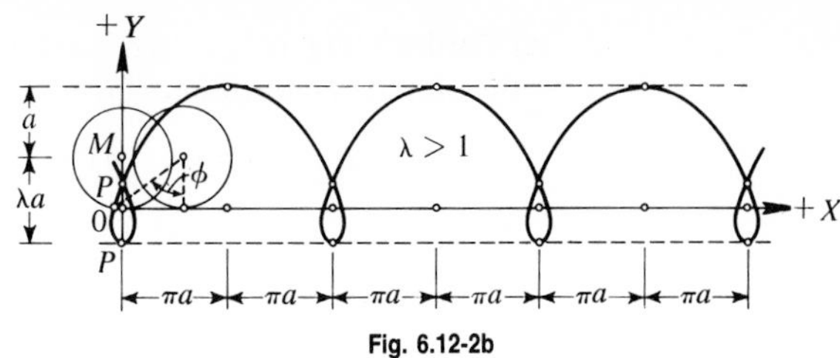

Fig. 6.12-2b

(b) Trochoid (Fig. 6.12–2) is defined parametrically by

$$x = a(\phi - \lambda \sin \phi) \qquad y = a(1 - \lambda \cos \phi)$$

This curve is generated by a fixed point P at distance λa from the center of a circle of radius a which rolls without slipping along the X axis.

If $\lambda < 1$, the curve (Fig. 6.12–2a) is called the *curtate cycloid.* If $\lambda > 1$, the curve (Fig. 6.12–2b) is called the *prolate cycloid.*

(2) Epicycloids

(a) Ordinary epicycloid (Fig. 6.12–3) is defined parametrically by

$$x = (a + b) \cos \phi - b \cos \frac{a + b}{b} \phi \qquad y = (a + b) \sin \phi - b \sin \frac{a + b}{b} \phi$$

This curve is generated by a fixed point P on a circle of radius b which rolls without slipping on the outside of a fixed circle of radius a.

If $a/b = N$ is an integer, the curve consists of N equal branches. The length of the arc of each branch is $8(a + b)/N$, and the area of one sector is $(b\pi/a)(a + b)(a + 2b)$. If N is a fraction, the branches cross one another.

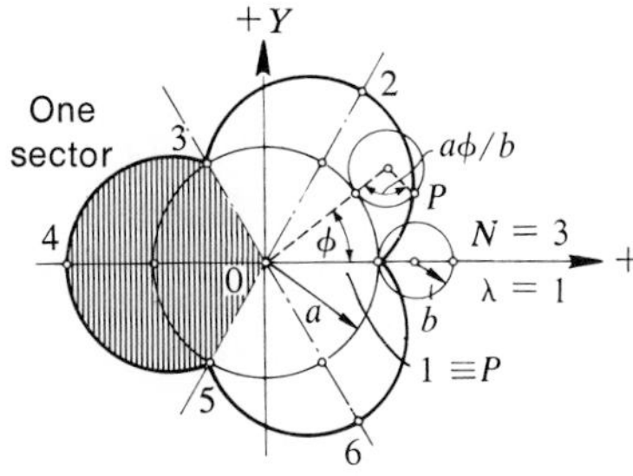

Fig. 6.12-3

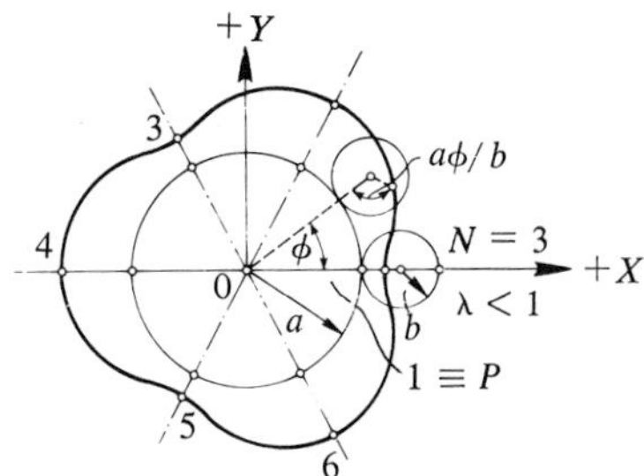

Fig. 6.12-4a

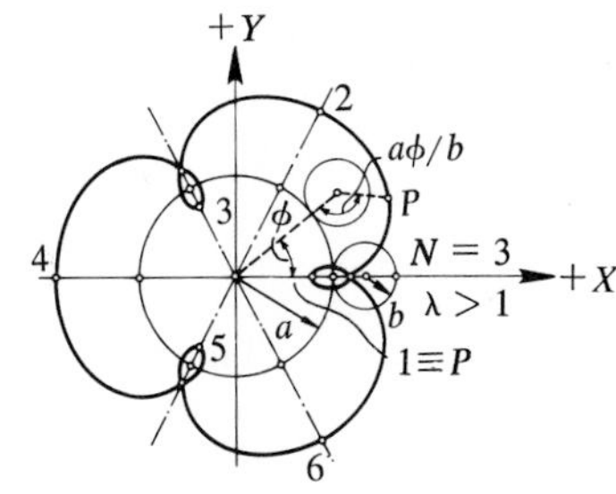

Fig. 6.12-4b

(b) Epitrochoid (Fig. 6.12–4) is defined parametrically by

$$x = (a + b) \cos \phi - \lambda b \cos \frac{a + b}{b} \phi \qquad y = (a + b) \sin \phi - \lambda b \sin \frac{a + b}{b} \phi$$

This curve is generated by a fixed point P at a distance λb from the center of a circle of radius b which rolls without slipping on the outside of a fixed circle of radius a.

If $a/b = N$ is an integer, the curve consists of N equal branches. If N is a fraction, the branches cross one another.

If $\lambda < 1$, the curve (Fig. 6.12–4a) is called the *curtate epicycloid.* If $\lambda > 1$, the curve (Fig. 6.12–4b) is called the *prolate epicycloid.*

(c) Cardioid (Fig. 6.12–5) is a special case of epicycloid ($a = b$) defined in the cartesian coordinates by

$$(x^2 + y^2 - a^2)^2 = 4a^2[(x-a)^2 + y^2]$$

in the polar coordinates by

$$r = 2a(1 - \cos\theta)$$

and parametrically by

$$x = a(2\cos\phi - \cos 2\phi) \qquad y = a(2\sin\phi - \sin 2\phi)$$

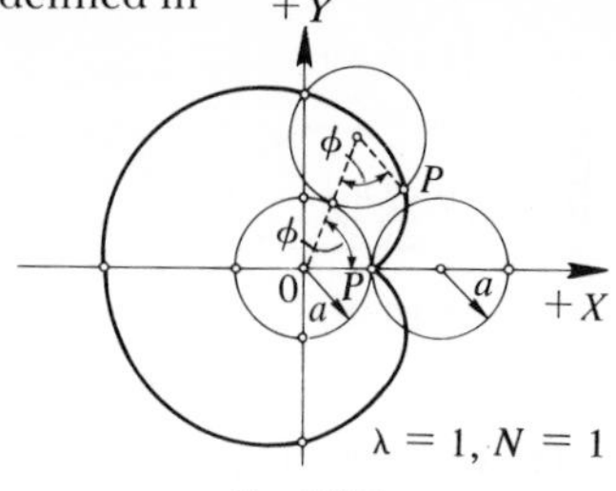

Fig. 6.12-5

The length of the curve is $16a$ and the area between the curve and the fixed circle is $6\pi a^2$.

(3) Hypocycloids

(a) Ordinary hypocycloid (Fig. 6.12–6) is defined parametrically by

$$x = (a-b)\cos\phi + b\cos\frac{a-b}{b}\phi$$

$$y = (a-b)\sin\phi - b\sin\frac{a-b}{b}\phi$$

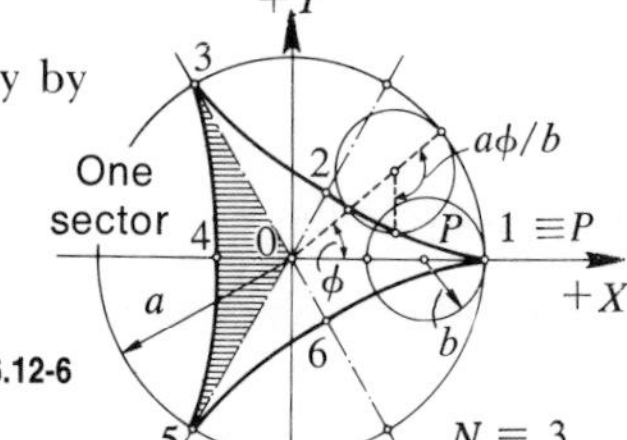

Fig. 6.12-6

This curve is generated by a fixed point P on a circle of radius b which rolls without slipping on the inside of a fixed circle of radius a.

If $a/b = N$ is an integer, the curve consists of N equal branches. The length of the arc of each branch is $8(a-b)/N$ and the area of one sector is $(b\pi/a)(a-b)(a-2b)$.

For $N = 2$ the hypocycloid reduces to the straight line (horizontal diameter of the fixed circle).

(b) Hypotrochoid (Fig. 6.12–7) is defined parametrically by

$$x = (a-b)\cos\phi + \lambda b\cos\frac{a-b}{b}\phi$$

$$y = (a-b)\sin\phi - \lambda b\sin\frac{a-b}{b}\phi$$

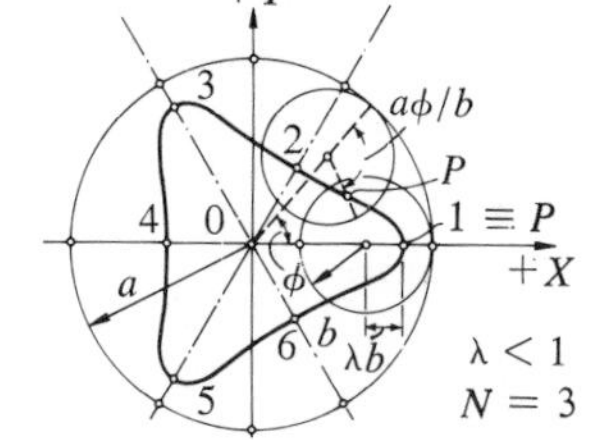

Fig. 6.12-7a

The curve is generated by a fixed point P at a distance λb from the center of a circle of radius b which rolls without slipping on the inside of a fixed circle of radius a.

If $a/b = N$ is an integer, the curve consists of N equal branches.

If N is a fraction, the branches cross one another.

If $\lambda < 1$, the curve (Fig. 6.12–7a) is called a *curtate hypocycloid*.

If $\lambda > 1$, the curve (Fig. 6.12–7b) is called a *prolate hypocycloid*.

Fig. 6.12-7b

(c) Astroid (Fig. 6.12–8) is a special case of hypocycloid ($b = a/4$) defined in the cartesian coordinates by

$$x^{2/3} + y^{2/3} = a^{2/3}$$

or

$$(x^2 + y^2 - a^2)^3 + 27a^2x^2y^2 = 0$$

and parametrically by

$$x = a\cos^3\phi \qquad y = a\sin^3\phi$$

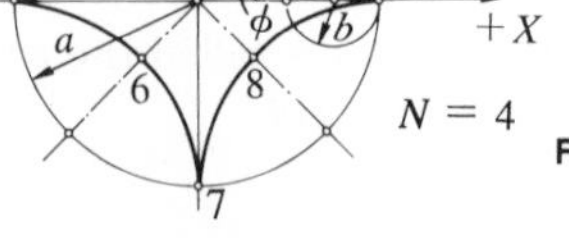

Fig. 6.12-8

The length of the curve is $6a$ and the area between the curve and the circumference of the fixed circle is $\frac{5}{8}\pi a^2$.

6.13 EXPONENTIAL AND LOGARITHMIC FUNCTIONS

(1) Simple Functions

(a) Function $y = A^x = e^{Bx}$ (where $B = \ln A$) defines the ordinary exponential curves of Fig. 6.13–1.

(b) Function $y = A^{-x} = e^{-Bx}$ (where $B = \ln A$) defines the reciprocal exponential curves of Fig. 6.13–2.

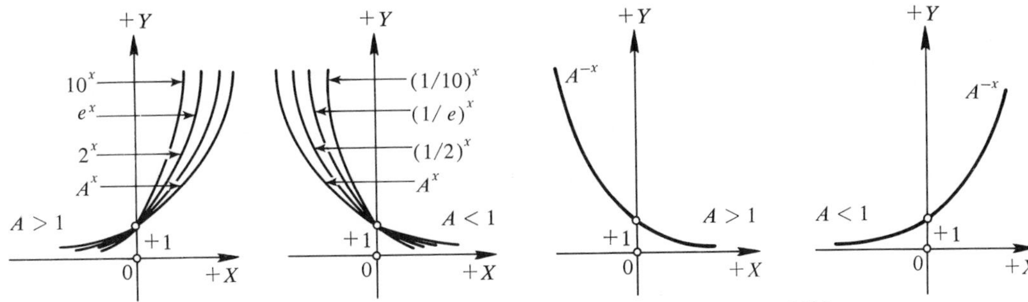

Fig. 6.13-1

Fig. 6.13-2

(c) Function $y = e^{-(Bx)^2}$ defines the distribution curve of Fig. 6.13–3.

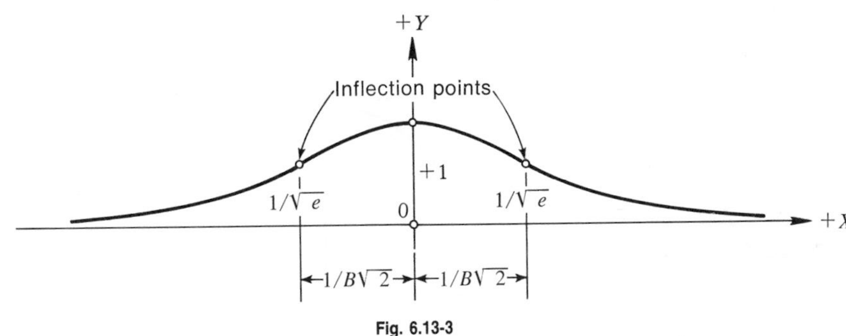

Fig. 6.13-3

(d) Function $y = \log_A x$ defines the general logarithmic curve of base A. The function exists only for $x > 0$ (Fig. 6.13–4).

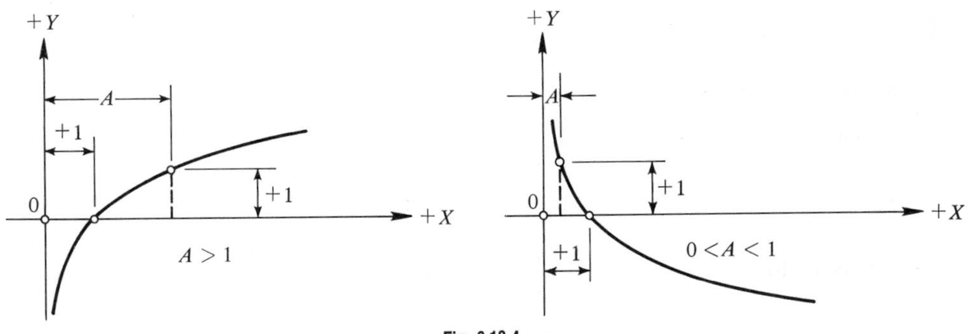

Fig. 6.13-4

(2) Composite Functions

(a) Function $y = Ax^m e^{nx}$ defines a family of curves shown in Fig. 6.13–5.

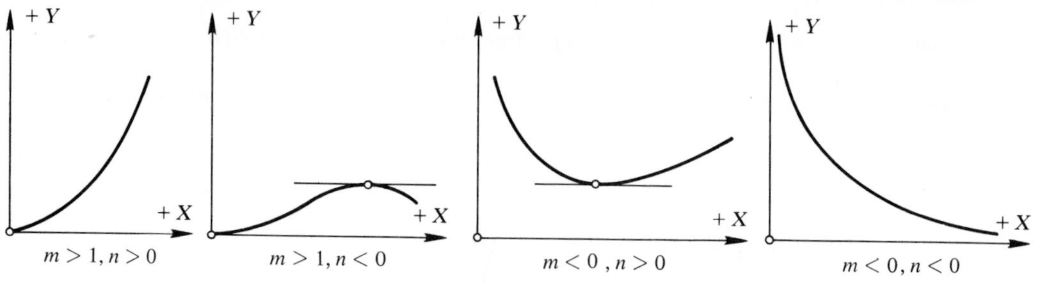

Fig. 6.13-5

(b) Function $y = Ae^{-\alpha x} \cos \beta x$ defines the damped vibration curve of Fig. 6.13–6.

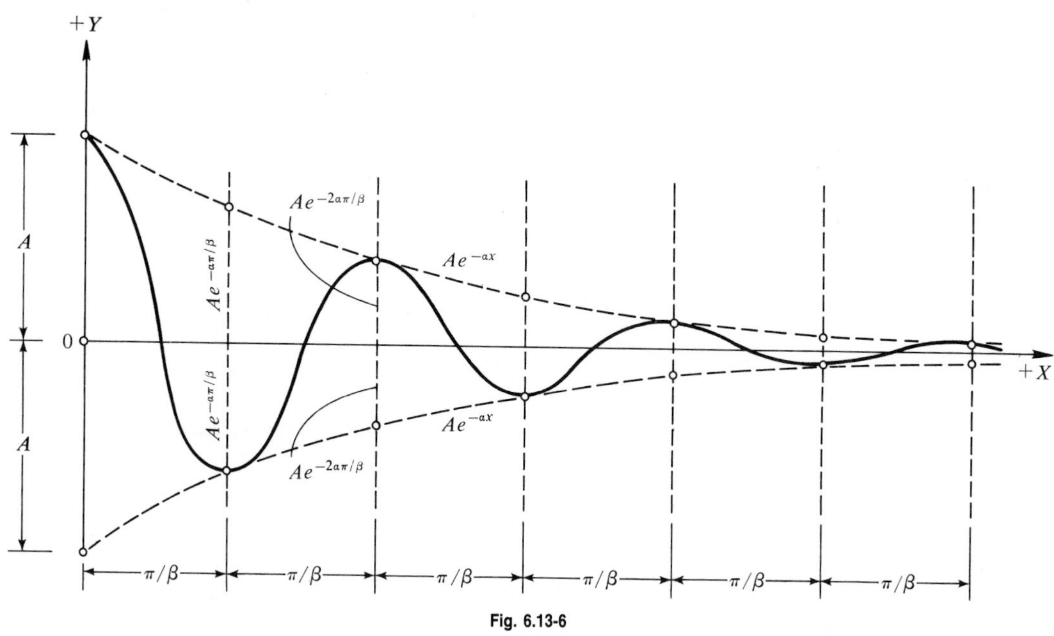

Fig. 6.13-6

6.14 HYPERBOLIC FUNCTIONS

(1) Definitions and Relationships

(a) Hyperbolic functions are defined as follows:

Hyperbolic sine of $x = \sinh x = \dfrac{e^x - e^{-x}}{2}$

Hyperbolic cosine of $x = \cosh x = \dfrac{e^x + e^{-x}}{2}$

Hyperbolic tangent of $x = \tanh x = \dfrac{e^x - e^{-x}}{e^x + e^{-x}}$

Hyperbolic cotangent of $x = \coth x = \dfrac{e^x + e^{-x}}{e^x - e^{-x}}$

Hyperbolic secant of $x = \operatorname{sech} x = \dfrac{2}{e^x + e^{-x}}$

Hyperbolic cosecant of $x = \operatorname{csch} x = \dfrac{2}{e^x - e^{-x}}$

(b) Relationships

$\cosh^2 x - \sinh^2 x = 1$	$\text{sech}\, x \cosh x = 1$
$\tanh^2 x + \text{sech}^2 x = 1$	$\text{csch}\, x \sinh x = 1$
$\coth^2 x - \text{csch}^2 x = 1$	$\tanh x \coth x = 1$
$\tanh x = \dfrac{\sinh x}{\cosh x}$	$\coth x = \dfrac{\cosh x}{\sinh x}$

(2) Analysis of Functions

(a) Graphical representation of hyperbolic functions is shown in Fig. 6.14–1. Contrary to their trigonometric counterparts they are not periodical for the real argument x.

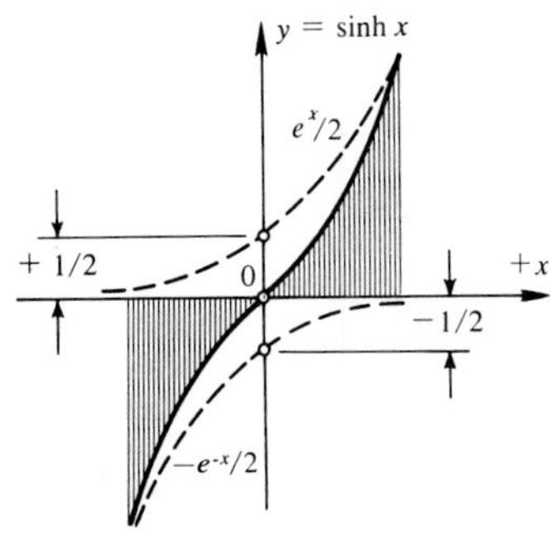

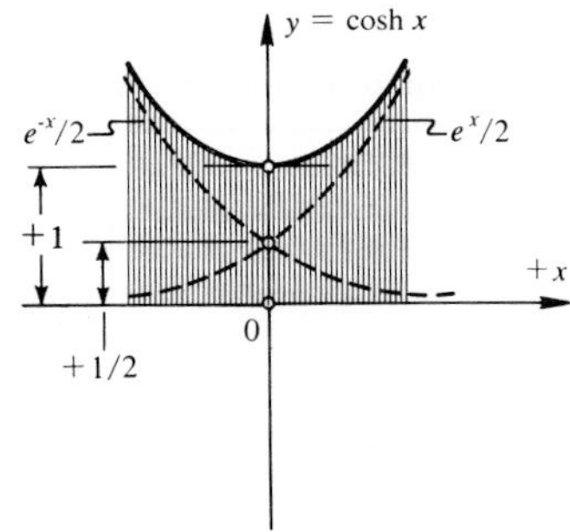

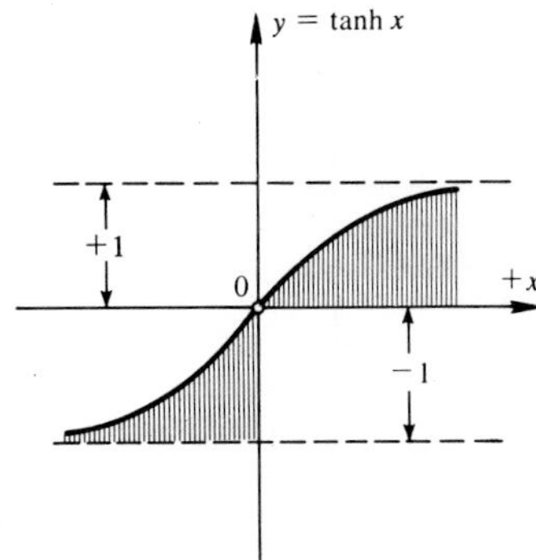

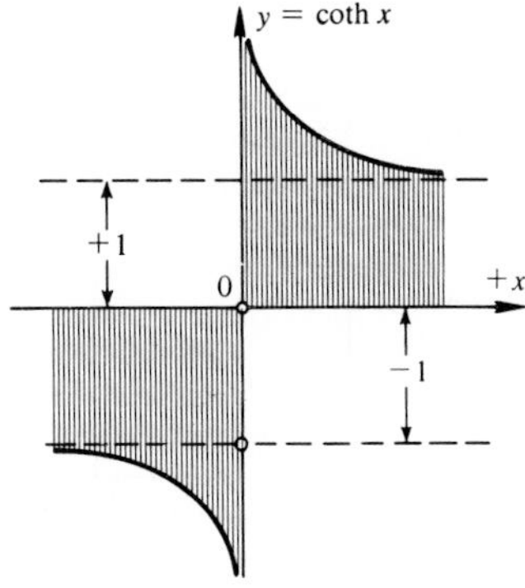

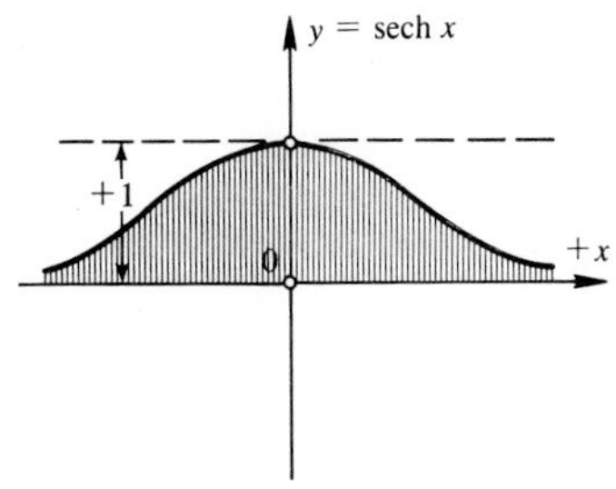

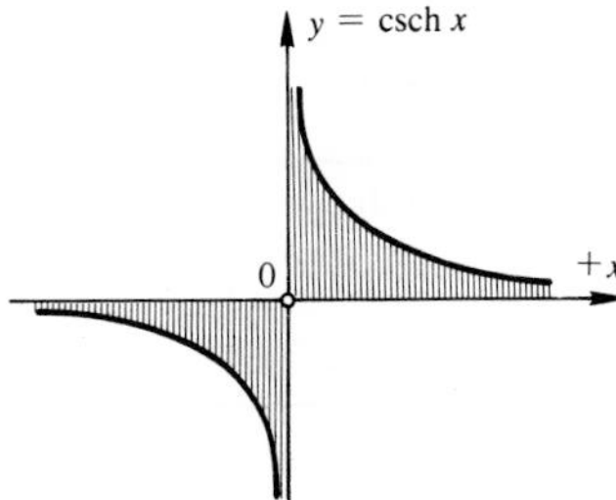

Fig. 6.14-1

(b) Limit values.

x	$\sinh x$	$\cosh x$	$\tanh x$	$\coth x$	$\operatorname{sech} x$	$\operatorname{csch} x$
$-\infty$	$-\infty$	$+\infty$	-1	-1	0	0
-1	-1.1752	$+1.5431$	-0.7616	-1.3130	$+0.6480$	-0.8509
0	0	$+1$	0	$\mp\infty$	$+1$	$\mp\infty$
$+1$	$+1.1752$	$+1.5431$	$+0.7616$	$+1.3130$	$+0.6480$	$+0.8509$
$+\infty$	$+\infty$	$+\infty$	$+1$	$+1$	0	0

(c) Transformation relations between the six hyperbolic functions (based on the formulas of Sec. 6.14–1*b*) are given in Table 6.14–1.

TABLE 6.14–1 Transformations of Hyperbolic Functions*

	$\sinh x$	$\cosh x$	$\tanh x$	$\coth x$	$\operatorname{sech} x$	$\operatorname{csch} x$
$\sinh x$	$\sinh x$	$k\sqrt{\cosh^2 x - 1}$	$\dfrac{\tanh x}{\sqrt{1-\tanh^2 x}}$	$\dfrac{k}{\sqrt{\coth^2 x - 1}}$	$\dfrac{k\sqrt{1-\operatorname{sech}^2 x}}{\operatorname{sech} x}$	$\dfrac{1}{\operatorname{csch} x}$
$\cosh x$	$\sqrt{1+\sinh^2 x}$	$\cosh x$	$\dfrac{1}{\sqrt{1-\tanh^2 x}}$	$\dfrac{k\coth x}{\sqrt{\coth^2 x - 1}}$	$\dfrac{1}{\operatorname{sech} x}$	$\dfrac{k\sqrt{1+\operatorname{csch}^2 x}}{\operatorname{csch} x}$
$\tanh x$	$\dfrac{\sinh x}{\sqrt{1+\sinh^2 x}}$	$\dfrac{k\sqrt{\cosh^2 x - 1}}{\cosh x}$	$\tanh x$	$\dfrac{1}{\coth x}$	$k\sqrt{1-\operatorname{sech}^2 x}$	$\dfrac{1}{\sqrt{1+\operatorname{csch}^2 x}}$
$\coth x$	$\dfrac{\sqrt{1+\sinh^2 x}}{\sinh x}$	$\dfrac{k\cosh x}{\sqrt{\cosh^2 x - 1}}$	$\dfrac{1}{\tanh x}$	$\coth x$	$\dfrac{k}{\sqrt{1-\operatorname{sech}^2 x}}$	$\sqrt{1+\operatorname{csch}^2 x}$
$\operatorname{sech} x$	$\dfrac{1}{\sqrt{1+\sinh^2 x}}$	$\dfrac{1}{\cosh x}$	$\sqrt{1-\tanh^2 x}$	$\dfrac{k\sqrt{\coth^2 x - 1}}{\coth x}$	$\operatorname{sech} x$	$\dfrac{k\operatorname{csch} x}{\sqrt{1+\operatorname{csch}^2 x}}$
$\operatorname{csch} x$	$\dfrac{1}{\sinh x}$	$\dfrac{k}{\sqrt{\cosh^2 - 1}}$	$\dfrac{\sqrt{1-\tanh^2 x}}{\tanh x}$	$k\sqrt{\coth^2 x - 1}$	$\dfrac{k\operatorname{sech} x}{\sqrt{1-\operatorname{sech}^2 x}}$	$\operatorname{csch} x$

*If $x > 0$, $k = +1$; if $x < 0$, $k = -1$.

(d) Negative argument

$\sinh(-x) = -\sinh x$	$\cosh(-x) = +\cosh x$
$\operatorname{sech}(-x) = +\operatorname{sech} x$	$\operatorname{csch}(-x) = -\operatorname{csch} x$
$\tanh(-x) = -\tanh x$	$\coth(-x) = -\coth x$

(3) Sums ($x_1 = a, x_2 = b$)

(a) $F(a \pm b)$.

$$\sinh(a \pm b) = \sinh a \cosh b \pm \cosh a \sinh b$$

$$\cosh(a \pm b) = \cosh a \cosh b \pm \sinh a \sinh b$$

$$\tanh(a \pm b) = \frac{\tanh a \pm \tanh b}{1 \pm \tanh a \tanh b}$$

$$\coth(a \pm b) = \frac{\coth a \coth b \pm 1}{\coth b \pm \coth a}$$

(b) $F(a) \pm F(b)$.

$$\sinh a + \sinh b = 2 \sinh \frac{a+b}{2} \cosh \frac{a-b}{2}$$

$$\sinh a - \sinh b = 2 \cosh \frac{a+b}{2} \sinh \frac{a-b}{2}$$

$$\cosh a + \cosh b = 2 \cosh \frac{a+b}{2} \cosh \frac{a-b}{2}$$

$$\cosh a - \cosh b = 2 \sinh \frac{a+b}{2} \sinh \frac{a-b}{2}$$

$$\tanh a + \tanh b = \frac{\sinh(a+b)}{\cosh a \cosh b}$$

$$\tanh a - \tanh b = \frac{\sinh(a-b)}{\cosh a \cosh b}$$

$$\coth a + \coth b = \frac{\sinh(a+b)}{\sinh a \sinh b}$$

$$\coth a - \coth b = \frac{-\sinh(a-b)}{\sinh a \sinh b}$$

(c) $F(a) \pm G(a)$.

$$\sinh a + \cosh a = e^{a} \qquad \sinh a - \cosh a = -e^{-a}$$

$$\tanh a + \coth a = 2 \coth 2a \qquad \tanh a - \coth a = -2 \operatorname{csch} 2a$$

(d) $F(a+b) \pm F(a-b)$.

$$\sinh(a+b) + \sinh(a-b) = 2 \sinh a \cosh b$$

$$\sinh(a+b) - \sinh(a-b) = 2 \sinh b \cosh a$$

$$\cosh(a+b) + \cosh(a-b) = 2 \cosh a \cosh b$$

$$\cosh(a+b) - \cosh(a-b) = 2 \sinh a \sinh b$$

(4) Multiple Arguments ($x = na$)

(a) $\sinh na \ (n = 2, 3, 4, \ldots)$.

$$\sinh 2a = 2 \sinh a \cosh a$$

$$\sinh 3a = 3 \sinh a \cosh^2 a + \sinh^3 a = \sinh a(3 + 4 \sinh^2 a)$$

$$\sinh 4a = 4 \sinh a \cosh^3 a + 4 \sinh^3 a \cosh a = 4 \sinh a \cosh a(1 + 2 \sinh^2 a)$$

$$\sinh 5a = 5 \sinh a \cosh^4 a + 10 \sinh^3 a \cosh^2 a + \sinh^5 a$$

$$\sinh 6a = 6 \sinh a \cosh^5 a + 20 \sinh^3 a \cosh^3 a + 6 \sinh^5 a \cosh a$$

$$\sinh 7a = 7 \sinh a \cosh^6 a + 35 \sinh^3 a \cosh^4 a + 21 \sinh^5 a \cosh^2 a + \sinh^7 a$$

$$\sinh na = \binom{n}{1} \sinh a \cosh^{n-1} a + \binom{n}{3} \sinh^3 a \cosh^{n-3} a + \binom{n}{5} \sinh^5 a \cosh^{n-5} a + \cdots$$

(b) $\cosh na \ (n = 2, 3, 4, \ldots)$.

$$\cosh 2a = \cosh^2 a + \sinh^2 a$$

$$\cosh 3a = \cosh^3 a + 3 \sinh^2 a \cosh a = \cosh a \,(4 \cosh^2 a - 3)$$

$$\cosh 4a = \cosh^4 a + 6 \sinh^2 a \cosh^2 a + \sinh^4 a = 1 - 8 \cosh^2 a + 8 \cosh^4 a$$

$$\cosh 5a = \cosh^5 a + 10 \sinh^2 a \cosh^3 a + 5 \sinh^4 a \cosh a$$

$$\cosh 6a = \cosh^6 a + 15 \sinh^2 a \cosh^4 a + 15 \sinh^4 a \cosh^2 a + \sinh^6 a$$

$$\cosh 7a = \cosh^7 a + 21 \sinh^2 a \cosh^5 a + 35 \sinh^4 a \cosh^3 a + 7 \sinh^6 a \cosh a$$

$$\cosh na = \cosh^n a + \binom{n}{2} \sinh^2 a \cosh^{n-2} a + \binom{n}{4} \sinh^4 a \cosh^{n-4} a + \cdots$$

(c) $\tanh na \ (n = 2, 3, 4, \ldots)$.

$$\tanh 2a = \frac{2 \tanh a}{1 + \tanh^2 a}$$

$$\tanh 3a = \frac{3 \tanh a + \tanh^3 a}{1 + 3 \tanh^2 a}$$

$$\tanh 4a = \frac{4 \tanh a + 4 \tanh^3 a}{1 + 6 \tanh^2 a + \tanh^4 a}$$

$$\tanh 5a = \frac{5 \tanh a + 10 \tanh^3 a + \tanh^5 a}{1 + 10 \tanh^2 a + 5 \tanh^4 a}$$

$$\tanh 6a = \frac{6 \tanh a + 20 \tanh^3 a + 6 \tanh^5 a}{1 + 15 \tanh^2 a + 15 \tanh^4 a + \tanh^6 a}$$

$$\tanh 7a = \frac{7 \tanh a + 35 \tanh^3 a + 21 \tanh^5 a + \tanh^7 a}{1 + 21 \tanh^2 a + 35 \tanh^4 a + 7 \tanh^6 a}$$

$$\tanh na = \frac{\binom{n}{1} \tanh a + \binom{n}{3} \tanh^3 a + \binom{n}{5} \tanh^5 a + \cdots}{1 + \binom{n}{2} \tanh^2 a + \binom{n}{4} \tanh^4 a + \binom{n}{6} \tanh^6 a + \cdots}$$

(d) coth na ($n = 2, 3, 4, \ldots$).

$$\coth 2a = \frac{\coth^2 a + 1}{2 \coth a}$$

$$\coth 3a = \frac{\coth^3 a + 3 \coth a}{3 \coth^2 a + 1}$$

$$\coth 4a = \frac{\coth^4 a + 6 \coth^2 a + 1}{4 \coth^3 a + 4 \coth a}$$

$$\coth 5a = \frac{\coth^5 a + 10 \coth^3 a + 5 \coth a}{5 \coth^4 a + 10 \coth^2 a + 1}$$

$$\coth 6a = \frac{\coth^6 a + 15 \coth^4 a + 15 \coth^2 a + 1}{6 \coth^5 a + 20 \coth^3 a + 6 \coth a}$$

$$\coth 7a = \frac{\coth^7 a + 21 \coth^5 a + 35 \coth^3 a + 7 \coth a}{7 \coth^6 a + 35 \coth^4 a + 21 \coth^2 a + 1}$$

$$\coth na = \frac{\coth^n a + \binom{n}{2} \coth^{n-2} a + \binom{n}{4} \coth^{n-4} a + \cdots}{\binom{n}{1} \coth^{n-1} a + \binom{n}{3} \coth^{n-3} a + \binom{n}{5} \coth^{n-5} a + \cdots}$$

(5) Half Arguments ($x = a/2$)

(a) $F(a/2) = G(a)$.

$$\sinh \frac{a}{2} = \sqrt{\tfrac{1}{2}(\cosh a - 1)} = \tfrac{1}{2}\sqrt{\cosh a + \sinh a} - \tfrac{1}{2}\sqrt{\cosh a - \sinh a}$$

$$\cosh \frac{a}{2} = \sqrt{\tfrac{1}{2}(\cosh a + 1)} = \tfrac{1}{2}\sqrt{\cosh a + \sinh a} + \tfrac{1}{2}\sqrt{\cosh a - \sinh a}$$

$$\tanh \frac{a}{2} = \frac{\sinh a}{\cosh a + 1} = \frac{\cosh a - 1}{\sinh a} = \sqrt{\frac{\cosh a - 1}{\cosh a + 1}}$$

$$\coth \frac{a}{2} = \frac{\sinh a}{\cosh a - 1} = \frac{\cosh a + 1}{\sinh a} = \sqrt{\frac{\cosh a + 1}{\cosh a - 1}}$$

(b) $F(a) = H(a/2)$.

$$\sinh a = 2 \sinh \frac{a}{2} \cosh \frac{a}{2} \qquad \cosh a = \cosh^2 \frac{a}{2} + \sinh^2 \frac{a}{2}$$

$$\tanh a = \frac{2 \tanh \frac{a}{2}}{1 + \tanh^2 \frac{a}{2}} \qquad \coth a = \frac{1 + \coth^2 a}{2 \coth \frac{a}{2}}$$

(6) Powers $(x = a)$

(a) $\sinh^n a \ (n = 2, 3, 4, \ldots)$.

$\sinh^2 a = \frac{1}{2}(\cosh 2a - 1)$

$\sinh^3 a = \frac{1}{4}(\sinh 3a - 3 \sinh a)$

$\sinh^4 a = \frac{1}{8}(\cosh 4a - 4 \cosh 2a + 3)$

$\sinh^5 a = \frac{1}{16}(\sinh 5a - 5 \sinh 3a + 10 \sinh a)$

(b) $\cosh^n a \ (n = 2, 3, 4, \ldots)$.

$\cosh^2 a = \frac{1}{2}(\cosh 2a + 1)$

$\cosh^3 a = \frac{1}{4}(\cosh 3a + 3 \cosh a)$

$\cosh^4 a = \frac{1}{8}(\cosh 4a + 4 \cosh 2a + 3)$

$\cosh^5 a = \frac{1}{16}(\cosh 5a + 5 \cosh 3a + 10 \cosh a)$

(c) $[F(a) \pm G(a)]^n \ (n = 2, 3, 4, \ldots)$.

$(\sinh a \pm \cosh a)^2 = \sinh 2a \pm \cosh 2a$

$$(\tanh a + \coth a)^2 = 4\frac{\cosh 4a + 1}{\cosh 4a - 1}$$

$$(\tanh a - \coth a)^2 = \frac{8}{\cosh 4a - 1}$$

$(\cosh a \pm \sinh a)^n = \cosh na \pm \sinh na$

6.15 INVERSE HYPERBOLIC FUNCTIONS

(1) Definitions and Relationships

(a) Inverse hyperbolic functions are defined as follows:

If $y = \sinh x$ then $x = \operatorname{arsinh} y = \sinh^{-1} y$

If $y = \cosh x$ then $x = \operatorname{arcosh} y = \cosh^{-1} y$

If $y = \tanh x$ then $x = \operatorname{artanh} y = \tanh^{-1} y$

If $y = \coth x$ then $x = \operatorname{arcoth} y = \coth^{-1} y$

If $y = \operatorname{sech} x$ then $x = \operatorname{arsech} y = \operatorname{sech}^{-1} y$

If $y = \operatorname{csch} x$ then $x = \operatorname{arcsch} y = \operatorname{csch}^{-1} y$

(b) Relationships.

$\sinh^{-1} y = \ln (y + \sqrt{y^2 + 1})$	$\operatorname{csch}^{-1} y = \ln \dfrac{1 \pm \sqrt{1 + y^2}}{y}$
$\cosh^{-1} y = \ln (y \pm \sqrt{y^2 - 1})$	$\operatorname{sech}^{-1} y = \ln \dfrac{1 \pm \sqrt{1 - y^2}}{y}$
$\tanh^{-1} y = \dfrac{1}{2} \ln \dfrac{1 + y}{1 - y}$	$\coth^{-1} y = \dfrac{1}{2} \ln \dfrac{y + 1}{y - 1}$

(2) Analysis of Functions

(a) Graphical representation of inverse hyperbolic functions is shown in Fig. 6.15–1. According to these graphs:

$\sinh^{-1} y$ is an odd, single-branch, single-valued function, in range $-\infty < y < \infty$.

$\cosh^{-1} y$ is an even, single-branch, double-valued function, in range $1 < y$, with principle values $\cosh^{-1} y > 0$.

$\tanh^{-1} y$ is an odd, single-branch, single-valued function, in range $-1 < y < 1$.

$\coth^{-1} y$ is an odd, double-branch, single-valued function, in range $y < -1$ and $y > 1$.

$\text{sech}^{-1} y$ is an even, single-branch, double-valued function, in range $0 < y \leq 1$, with principal values $\text{sech}^{-1} y > 0$.

$\text{csch}^{-1} y$ is an odd, double-branch, single-valued function with range $y > -\infty$, $y \neq 0$, $y < \infty$.

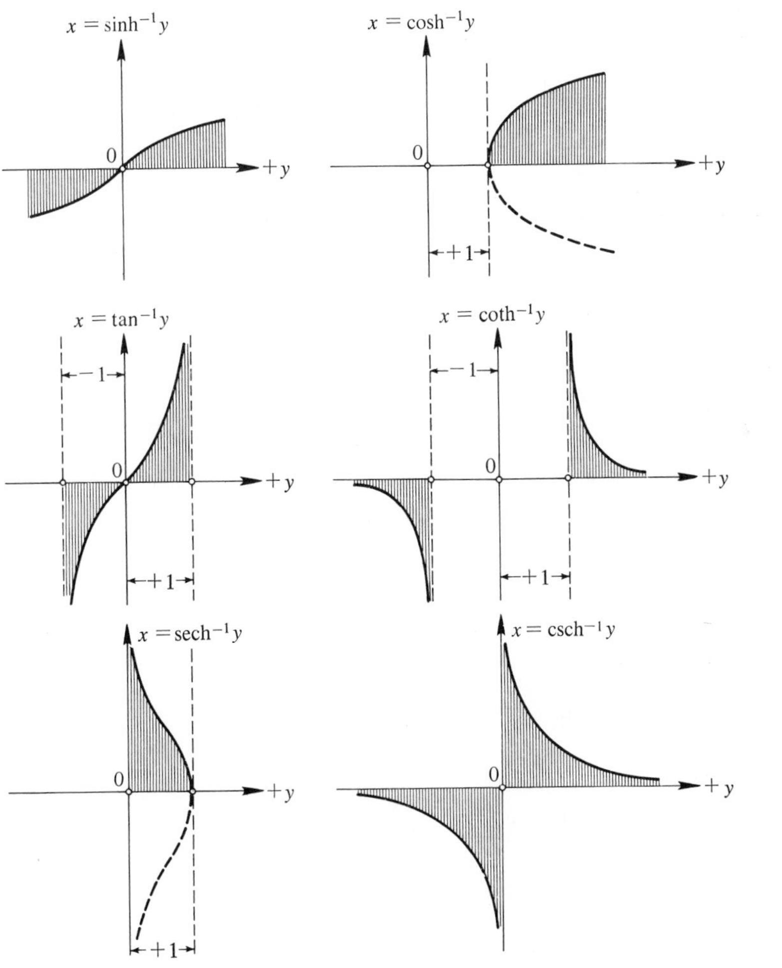

Fig. 6.15-1

(b) Transformation relations between the six inverse hyperbolic functions (based on the formulas of Sec. 6.14–1*b*) are given in Table 6.15–1.

TABLE 6.15–1 Transformations of Inverse Hyperbolic Functions*

	$\sinh^{-1} y$	$\cosh^{-1} y$	$\tanh^{-1} y$	$\coth^{-1} y$	$\operatorname{sech}^{-1} y$	$\operatorname{csch}^{-1} y$
$\sinh^{-1} y$	$\sinh^{-1} y$	$k \cosh^{-1} \sqrt{y^2+1}$	$\tanh^{-1} \frac{y}{\sqrt{y^2+1}}$	$\coth^{-1} \frac{\sqrt{y^2+1}}{y}$	$k \operatorname{sech}^{-1} \frac{1}{\sqrt{y^2+1}}$	$\operatorname{csch}^{-1} \frac{1}{y}$
$\cosh^{-1} y$	$k \sinh^{-1} \sqrt{y^2-1}$	$\cosh^{-1} y$	$k \tanh^{-1} \frac{\sqrt{y^2-1}}{y}$	$k \coth^{-1} \frac{y}{\sqrt{y^2-1}}$	$\operatorname{sech}^{-1} \frac{1}{y}$	$k \operatorname{csch}^{-1} \frac{1}{\sqrt{y^2-}}$
$\tanh^{-1} y$	$\sinh^{-1} \frac{y}{\sqrt{1-y^2}}$	$k \cosh^{-1} \frac{1}{\sqrt{1-y^2}}$	$\tanh^{-1} y$	$\coth^{-1} \frac{1}{y}$	$k \operatorname{sech}^{-1} \sqrt{1-y^2}$	$\operatorname{csch}^{-1} \frac{\sqrt{1-y^2}}{y}$
$\coth^{-1} y$	$\sinh^{-1} \frac{1}{\sqrt{y^2-1}}$	$k \cosh^{-1} \frac{y}{\sqrt{y^2-1}}$	$\tanh^{-1} \frac{1}{y}$	$\coth^{-1} y$	$\operatorname{sech}^{-1} \frac{\sqrt{y^2-1}}{y}$	$\operatorname{csch}^{-1} \sqrt{y^2-1}$
$\operatorname{sech}^{-1} y$	$k \sinh^{-1} \frac{\sqrt{1-y^2}}{y}$	$\cosh^{-1} \frac{1}{y}$	$k \tanh^{-1} \sqrt{1-y^2}$	$k \coth^{-1} \frac{1}{\sqrt{1-y^2}}$	$\operatorname{sech}^{-1} y$	$k \operatorname{csch}^{-1} \frac{y}{\sqrt{y^2-}}$
$\operatorname{csch}^{-1} y$	$\sinh^{-1} \frac{1}{y}$	$k \cosh^{-1} \frac{\sqrt{y^2+1}}{y}$	$\tanh^{-1} \frac{1}{\sqrt{y^2-1}}$	$\coth^{-1} \sqrt{y^2+1}$	$k \operatorname{sech}^{-1} \frac{y}{\sqrt{y^2+1}}$	$\operatorname{csch}^{-1} y$

*If $y > 0$, $k = +1$; or if $y < 0$, $k = -1$.

examples:

Derive $\sinh^{-1} y = k \cosh^{-1} \sqrt{y^2+1}$, for $y > 0$, $k = +1$.

From Table 6.14–1, $\sinh x = \sqrt{\cosh^2 x - 1}$, for $x > 0$, $k = +1$. Placing each side equal to y,

$$y = \sinh x \quad \text{and} \quad y = \sqrt{\cosh^2 x - 1} \quad \text{or} \quad \cosh x = \sqrt{y^2+1}$$

and their inverses are

$$x = \sin^{-1} y \quad \text{and} \quad x = \cosh^{-1} \sqrt{y^2+1}$$

Derive $\tanh^{-1} y = \sinh^{-1} y/\sqrt{1-y^2}$.

From Table 6.14–1, $\tanh x = (\sinh x)/\sqrt{1+\sinh^2 x}$. Placing each side equal to y,

$$y = \tanh x \quad \text{and} \quad y = \frac{\sinh x}{\sqrt{1+\sinh^2 x}} \quad \text{or} \quad \sinh x = \frac{y}{\sqrt{1-y^2}}$$

and their inverses are

$$x = \tanh^{-1} y \quad \text{and} \quad x = \sin^{-1} \frac{y}{\sqrt{1-y^2}}$$

7
DIFFERENTIAL CALCULUS

7.01 DEFINITIONS AND NOTATIONS

(1) Intervals and Functions

(a) Interval. A real number x may be represented by a point on a straight line (Fig. 7.01–1). A set of values of x, such that $a < x < b$, is called a bounded open interval (a,b), which means x is any real number greater than a and less than b, but is not equal to a or b. A set of values of x, such that $a \leq x \leq b$, is called a bounded closed interval $[a,b]$, which means x is any real number equal to or greater than a, or equal to or less than b.

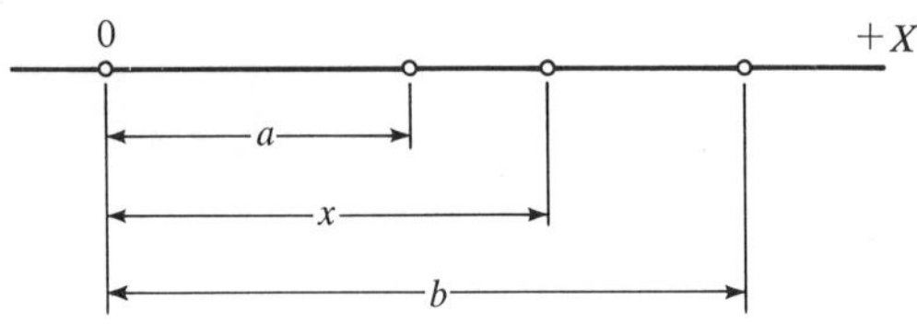

Fig. 7.01-1

(b) Variable and constant. A quality which changes its value is called a variable and its range of variation is known as its interval. A quality which remains unchanged is called a constant and its range is a single number.

(c) Function of one variable. A variable y is a function of another variable x,

$$y = f(x)$$

if for each value in the range of x there are one or more values in the range of y, where x is called the *independent variable* (*argument*) and y is called the *dependent variable.*

examples:

The area A of a circle, $A = \pi R^2$, is a function of the radius R. For $R = 0$, $A = 0$; for $R = 1$, $A = \pi$; for $R = 2$, $A = 4\pi, \ldots$.

The sine of angle α, $y = \sin \alpha$, is a function of the angle α. For $\alpha = 0°$, $y = 0$; for $\alpha = 1°$, $y = 0.01745$; for $\alpha = 2°$, $y = 0.03490, \ldots$.

An algebraic expression $y = x^2 + 3$ is a function of x. For $x = 0$, $y = 3$; for $x = 1$, $y = 4$; for $x = 2$, $y = 7, \ldots$.

(d) Function of several variables. A variable y is a function of several variables $x_1, x_2, \ldots, x_n$,

$$y = f(x_1, x_2, \ldots, x_n)$$

if for each set of values of x's there is one or more values of y.

examples:

The length $c = \sqrt{a^2 + b^2}$ is a function of a and b. For $a = 0$, $b = 1$, $c = 1$; for $a = 1$, $b = 1$, $c = \sqrt{2}$; for $a = 1$, $b = 2$, $c = \sqrt{5}, \ldots$.

The sine of $(\alpha + \beta)$, $y = \sin(\alpha + \beta)$, is a function of α and β. For $\alpha = 0°$, $\beta = 1°$, $y = 0.01745$; for $\alpha = 1°$, $\beta = 1°$, $y = 0.03490$; for $\alpha = 1°$, $\beta = 2°$, $y = 0.05234, \ldots$.

(e) Single and multivalued functions. A function $y = f(x)$ is a single-valued function if it has one value of y for any given value of x. A function $y = f(x)$ is a multivalued function if it has more than one value of y for any given value of x.

examples:

$y = kx$ is a single-valued function of x. For $x = 1$, $y = k$; for $x = 2$, $y = 2k, \ldots$.

$y^2 = kx$ is a double-valued function of x. For $x = 1$, $y_1 = \sqrt{k}$, $y_2 = -\sqrt{k}$; for $x = 2$, $y_1 = \sqrt{2k}$, $y_2 = -\sqrt{2k}, \ldots$.

(f) Even and odd functions. A function is even if $f(-x) = f(x)$ and odd if $f(-x) = -f(x)$.

examples:

$y = \cos\alpha$ is an even function since $\cos(-\alpha) = \cos\alpha$.

$y = \sin\alpha$ is an odd function since $\sin(-\alpha) = -\sin\alpha$ (Sec. 5.02–4*a*).

(g) Analytical representation. Three forms of representation are:

(α) Explicit function: $y = f(x_1, x_2, x_3, \ldots, x_n)$
(β) Implicit function: $0 = g(x_1, x_2, x_3, \ldots, x_n, y)$
(γ) Parametric function: $x_1 = h_1(\tau), x_2 = h_2(\tau), x_3 = h_3(\tau), \ldots, x_n = h_n(\tau), y = h(\tau)$

examples: The circle of radius R can be represented analytically as

$y = \sqrt{R^2 - x^2} \equiv$ explicit function

$0 = x^2 + y^2 - R^2 \equiv$ implicit function

$x = R\sin\tau,\ y = R\cos\tau \equiv$ parametric function

(2) Finite and Infinite Limits

(a) Limit of variable. A variable x has a limit $a(\lim x = a$ or $a \to x)$ as x takes on consecutively the values $x_1, x_2, x_3, \ldots,$ if for any positive value ϵ, however small, the numerical value of

$$|x - a| < \epsilon$$

examples:

If $x = 3.9, 3.99, 3.999, 3.9999, \ldots,$ where the nth term is $a_n = 4 - 0.1^n$, $\epsilon_n = 0.1^n$, then $\lim x = 4$, since $|3.99 - 4| < 0.1$, $|3.999 - 4| < 0.01, \ldots$.

If $x = \frac{4}{3}, \frac{10}{9}, \frac{28}{27}, \ldots,$ where the nth term is $a_n = 1 + \frac{1}{3^n}$, $\epsilon_n = \frac{1}{3^n}$, then $\lim x = 1$, since

$$\left|\frac{10}{9} - 1\right| < \frac{1}{3},\quad \left|\frac{28}{27} - 1\right| < \frac{1}{9}, \ldots$$

(b) Limit of a function. A function $f(x)$ has a limit b as $x \to a$ ($\lim_{x \to a} f(x) = b$) if, as x approaches its limit without assuming the value of a, the numerical value of

$$|f(x) - b| < \epsilon$$

where ϵ is any preassigned number, however small.

example:

If $f(x) = x + 2 = 0.9 + 2, 0.99 + 2, 0.999 + 2, \ldots$, where $f(x_n) = 3 - 0.1^n$, $\epsilon_n = 0.1^n$, then $\lim_{x \to 1} f(x) = 3$, since

$$|2.99 - 3| < 0.1, \quad |2.999 - 3| < 0.01, \quad \ldots$$

(c) Positively infinite variable. A variable x becomes positively infinite ($x \to +\infty$) if for any preassigned positive number N, however large, x becomes larger than N.

example:

$x \to +\infty$ as it takes on consecutively the values of the sequence 3, 6, 9, . . . , (any sequence of positive numbers) where N is any large number in this sequence.

(d) Negatively infinite variable. A variable x becomes negatively infinite ($x \to -\infty$) if for any preassigned number N, however small, x becomes less than N.

example:

$x \to -\infty$ as it takes on consecutively the values of the sequence $-2, -4, -6, \ldots$ (any sequence of negative numbers), where N is any large negative number in this sequence.

(e) Positively infinite function. A function $f(x)$ becomes positively infinite as $x \to a$ [$\lim_{x \to a} f(x) = +\infty$] if, as x approaches its limit a, $f(x)$ is greater than any preassigned positive number N, however large.

examples:

$$\lim_{x \to 0} \frac{1}{x^3} = +\infty \qquad \lim_{x \to +\infty} = \frac{1}{1/x + 1/x^2} = +\infty$$

(f) Negatively infinite function. A function $f(x)$ becomes negatively infinite as $x \to a$ [$\lim_{x \to a} f(x) = -\infty$] if, as x approaches its limit a, $f(x)$ is less than any preassigned negative number N.

examples:

$$\lim_{x \to 0} \frac{-1}{x^3} = -\infty \qquad \lim_{x \to +\infty} = \frac{-1}{1/x + 1/x^2} = -\infty$$

(g) Infinity is not a number and rules (c), (d), (e), (f) are used as terms of convenience denoting that the variable or the function is increasing or decreasing without limit.

example:

Statements such as $1/0 = +\infty$, $1/\infty = 0$ are therefore absurd statements indicating a behavior but not a value.

(3) Continuity

(a) Continuity at a point. A function $f(x)$ is continuous through the neighborhood of $x = a$ if and only if $\lim_{x \to a} f(x) = f(a)$ and $f(a+0) = f(a-0)$.

(b) Continuity in an interval. A function $f(x)$ is continuous in an interval (a,b) or $[a,b]$ if and only if it is continuous at each point of this interval.

(c) Discontinuity at a point. A function $f(x)$ is discontinuous at $x = a$ if $f(a+\epsilon) \neq f(a-\epsilon)$ where $\epsilon \to 0$. The difference of these two values is called the jump of $f(x)$ at that point.

examples:

$\tan \alpha$ is discontinuous at $\alpha = 90°$ where $\tan(90° - \epsilon) = +\infty$ and $\tan(90° + \epsilon) = -\infty$ (see Fig. 5.02–5).
$\coth x$ is discontinuous at $x = 0$, where $\coth(0-\epsilon) = -\infty$ and $\coth(0+\epsilon) = +\infty$ (see Fig. 6.14–1).

7.02 ORDINARY DERIVATIVES AND DIFFERENTIALS

(1) First and Higher Derivatives

(a) First derivative of $y = f(x)$ with respect to x is the limit approached by the ratio of an increment Δy of the function to the increment Δx of its argument, as Δx tends to zero ($\Delta x \to 0$).

$$\lim_{\Delta x \to 0} \frac{\Delta y}{\Delta x} = \lim_{\Delta x \to 0} \frac{f(x+\Delta x) - f(x)}{\Delta x} = \frac{dy}{dx} = \text{first derivative}$$

where dx and dy are the differentials of x and y respectively.

The necessary and sufficient conditions of the differentiability of a function at a certain point are (1) the continuity of this function at this point, and (2) the existence of the said limit.

(b) Alternative notations for the first derivative of $y = f(x)$ are

$$\frac{dy}{dx} = \frac{df(x)}{dx} = f'(x) = y'$$

If $y = f(t)$,

$$\frac{dy}{dt} = \frac{df(t)}{dt} = \dot{f}(t) = \dot{y}$$

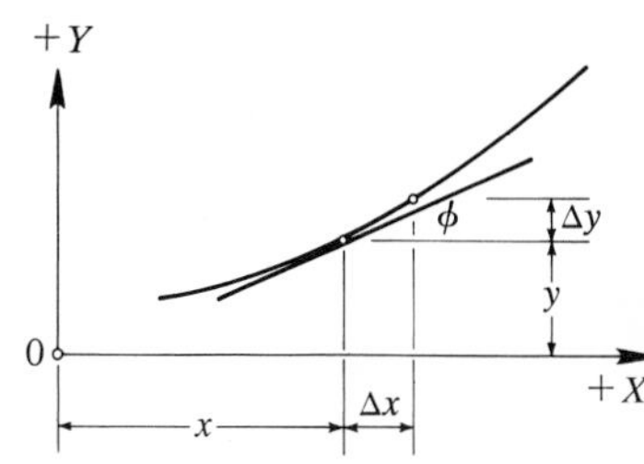

Fig. 7.02-1

(c) Interpretation. The first derivative of a function is a measure of the rate of change of y with respect to x at each point x, where the limit exists. Graphically, $f'(x)$ corresponds to the slope of the tangent to $f(x)$ at x (Fig. 7.02–1).

$$\frac{dy}{dx} = f'(x) = \tan \phi$$

(d) Second and higher derivatives of $f(x)$ with respect to x at the point x are respectively:

$$\frac{d^2y}{dx^2} = \frac{d}{dx}\left(\frac{dy}{dx}\right) = \frac{d}{dx}[f'(x)] = f''(x) = \text{2d derivative}$$

$$\frac{d^3y}{dx^3} = \frac{d}{dx}\left(\frac{d^2y}{dx^2}\right) = \frac{d}{dx}[f''(x)] = f'''(x) = \text{3d derivative}$$

. . . .

$$\frac{d^ny}{dx^n} = \frac{d}{dx}\left(\frac{d^{n-1}y}{dx^{n-1}}\right) = \frac{d}{dx}[f^{(n-1)}(x)] = f^{(n)}(x) = n\text{th derivative}$$

(2) First and Higher Differentials

(a) First differential of $f(x)$ at x is the product of the first derivative of $f(x)$ and of dx.

$$dy = \frac{dy}{dx}\,dy = d[f(x)] = f'(x)\,dx = y'\,dx = \text{first differential}$$

(b) Second and higher differentials of $f(x)$ are obtained by successive differentiation of the first differential.

$$d^2y = \frac{d^2y}{dx^2}\,dx^2 = f''(x)\,dx^2 = \text{2d differential}$$

$$d^3y = \frac{d^3y}{dx^3}\,dx^3 = f'''(x)\,dx^3 = \text{3d differential}$$

. . . .

$$d^ny = \frac{d^ny}{dx^n}\,dx^n = f^{(n)}(x)\,dx^n = n\text{th differential}$$

7.03 FIRST DERIVATIVES OF ELEMENTARY FUNCTIONS

(1) Basic Cases

(a) Algebraic functions. If a is a constant $(a \neq 0)$ and m is an integer, then

$$(ax)' = a \qquad \left(\frac{a}{x}\right)' = -\frac{a}{x^2}$$

$$(ax^m)' = max^{m-1} \qquad \left(\frac{a}{x^m}\right)' = -\frac{ma}{x^{m+1}}$$

$$(a\sqrt{x})' = \frac{a}{2\sqrt{x}} \qquad \left(\frac{a}{\sqrt{x}}\right)' = -\frac{a}{2x\sqrt{x}}$$

$$(a\sqrt[m]{x})' = \frac{a\sqrt[m]{x}}{mx} \qquad \left(\frac{a}{\sqrt[m]{x}}\right)' = -\frac{a}{mx\sqrt[m]{x}}$$

examples:

$$(-2x)' = -2 \qquad (x^3)' = 3x^{3-1} = 3x^2$$

$$\left(\frac{-2}{x}\right)' = (-2x^{-1})' = (-1)(-2)x^{-1-1} = 2x^{-2} = \frac{2}{x^2}$$

$$(3\sqrt{x})' = (3x^{1/2})' = \tfrac{1}{2}(3)x^{1/2-1} = \tfrac{3}{2}x^{-1/2} = \frac{3}{2\sqrt{x}}$$

$$\left(\frac{-2}{\sqrt{x^3}}\right)' = (-2x^{-3/2})' = -\tfrac{3}{2}(-2)x^{-3/2-1} = 3x^{-5/2} = \frac{3}{\sqrt{x^5}} = \frac{3}{x^2\sqrt{x}}$$

(b) Exponential functions. If a is a constant ($a \neq 0$), $e = 2.71828\ldots$, and m is an integer, then

$$(e^x)' = e^x \qquad (e^{-x})' = -e^{-x}$$

$$(e^{mx})' = me^{mx} \qquad (e^{-mx})' = -me^{-mx}$$

$$(a^{mx})' = ma^{mx} \ln a \qquad (a^{-mx})' = -ma^{-mx} \ln a$$

$$(x^x)' = x^x(1 + \ln x) \qquad (x^{-x})' = -x^{-x}(1 + \ln x)$$

$$(x^{mx})' = mx^{mx}(1 + \ln x) \qquad (x^{-mx})' = -mx^{-mx}(1 + \ln x)$$

where ln is the natural logarithm to the base e (Sec. 2.13–1).

examples:

$$(e^{2x})' = 2e^{2x} \qquad (e^{-2x})' = -2e^{-2x} = -\frac{2}{e^{2x}}$$

$$(5^{2x})' = 2(5^{2x}) \ln 5 = 2(25^x) \ln 5 \quad \text{or} \quad (5^{2x})' = (25^x)' = 25^x \ln 25 \quad \text{where} \quad \ln 25 = \ln 5^2 = 2 \ln 5$$

(c) Logarithmic functions. If a is a constant ($a \neq 0$), ln is the natural logarithm to the base e (Sec. 2.13–1), and log is the decadic logarithm to the base 10 (Sec. 2.13–1), then

$$(\ln x)' = \frac{1}{x} \qquad \left(\ln \frac{1}{x}\right)' = -\frac{1}{x}$$

$$(\ln ax)' = \frac{1}{x} \qquad \left(\ln \frac{a}{x}\right)' = -\frac{1}{x}$$

$$(\log x)' = \frac{\log e}{x} \qquad \left(\log \frac{1}{x}\right)' = -\frac{\log e}{x}$$

$$(\log ax)' = \frac{\log e}{x} \qquad \left(\log \frac{a}{x}\right)' = -\frac{\log e}{x}$$

where $\log e = 0.43429\ldots$.

(d) Trigonometric functions. If a is a constant ($a \neq 0$), then

$$(\sin ax)' = a \cos ax \qquad \left(\frac{1}{\sin ax}\right)' = -\frac{a \cos ax}{\sin^2 ax}$$

$$(\cos ax)' = -a \sin ax \qquad \left(\frac{1}{\cos ax}\right)' = \frac{a \sin ax}{\cos^2 ax}$$

$$(\tan ax)' = \frac{a}{\cos^2 ax} \qquad \left(\frac{1}{\tan ax}\right)' = -\frac{a}{\sin^2 ax}$$

$$(\cot ax)' = -\frac{a}{\sin^2 ax} \qquad \left(\frac{1}{\cot ax}\right)' = \frac{a}{\cos^2 ax}$$

$$(\sec ax)' = \frac{a \sin ax}{\cos^2 ax} \qquad \left(\frac{1}{\sec ax}\right)' = -a \sin ax$$

$$(\csc ax)' = -\frac{a \cos ax}{\sin^2 ax} \qquad \left(\frac{1}{\csc ax}\right)' = a \cos ax$$

(e) Inverse trigonometric functions. If a is a constant ($a \neq 0$), then

$$(\sin^{-1} ax)' = \frac{a}{\sqrt{1-(ax)^2}} \qquad -\frac{\pi}{2} < \sin^{-1} ax < \frac{\pi}{2}$$

$$(\cos^{-1} ax)' = \frac{-a}{\sqrt{1-(ax)^2}} \qquad 0 < \cos^{-1} ax < \pi$$

$$(\tan^{-1} ax)' = \frac{a}{1+(ax)^2} \qquad -\frac{\pi}{2} < \tan^{-1} ax < \frac{\pi}{2}$$

$$(\cot^{-1} ax)' = \frac{-a}{1+(ax)^2} \qquad 0 < \cot^{-1} ax < \pi$$

$$(\sec^{-1} ax)' = \frac{\pm 1}{x\sqrt{(ax)^2-1}} \qquad \begin{cases} + \text{ if } 0 < \sec^{-1} ax < \frac{\pi}{2} \\ - \text{ if } \frac{\pi}{2} < \sec^{-1} ax < \pi \end{cases}$$

$$(\csc^{-1} ax)' = \frac{\mp 1}{x\sqrt{(ax)^2-1}} \qquad \begin{cases} - \text{ if } 0 < \csc^{-1} ax < \frac{\pi}{2} \\ + \text{ if } -\frac{\pi}{2} < \csc^{-1} ax < 0 \end{cases}$$

where the second statements define the range of existence of the derivative (see Fig. 5.07–1).

(f) Hyperbolic functions. If a is a constant ($a \neq 0$), then

$$(\sinh ax)' = a \cosh ax \qquad \left(\frac{1}{\sinh ax}\right)' = -\frac{a \cosh ax}{\sinh^2 ax}$$

$$(\cosh ax)' = a \sinh ax \qquad \left(\frac{1}{\cosh ax}\right)' = -\frac{a \sinh ax}{\cosh^2 ax}$$

$$(\tanh ax)' = \frac{a}{\cosh^2 ax} \qquad \left(\frac{1}{\tanh ax}\right)' = -\frac{a}{\sinh^2 ax}$$

$$(\coth ax)' = -\frac{a}{\sinh^2 ax} \qquad \left(\frac{1}{\coth ax}\right)' = \frac{a}{\cosh^2 ax}$$

$$(\operatorname{sech} ax)' = -\frac{a \sinh ax}{\cosh^2 ax} \qquad \left(\frac{1}{\operatorname{sech} ax}\right)' = a \sinh ax$$

$$(\operatorname{csch} ax)' = -\frac{a \cosh ax}{\sinh^2 ax} \qquad \left(\frac{1}{\operatorname{csch} ax}\right)' = a \cosh as$$

(g) Inverse hyperbolic functions. If a is a constant ($a \neq 0$), then

$$(\sinh^{-1} ax)' = \frac{a}{\sqrt{(ax)^2+1}} \qquad \begin{cases} -\infty < \sinh^{-1} ax < \infty \\ -\infty < ax < \infty \end{cases}$$

$$(\cosh^{-1} ax)' = \frac{\pm a}{\sqrt{(ax)^2-1}} \qquad \begin{cases} + \text{ if } \cosh^{-1} ax > 0 & ax > 1 \\ - \text{ if } \cosh^{-1} ax < 0 & ax > 1 \end{cases}$$

$$(\tanh^{-1} ax)' = \frac{a}{1-(ax)^2} \qquad \begin{cases} -\infty < \tanh^{-1} ax < \infty \\ -1 < ax < 1 \end{cases}$$

$$(\coth^{-1} ax)' = \frac{a}{1-(ax)^2} \qquad \begin{cases} -\infty < \coth^{-1} ax < \infty \\ -\infty < ax < -1 \qquad 1 < ax < \infty \end{cases}$$

$$(\operatorname{sech}^{-1} ax)' = \frac{\mp 1}{x\sqrt{1-(ax)^2}} \qquad \begin{cases} - \text{ if } \operatorname{sech}^{-1} ax > 0 & 0 < ax < 1 \\ + \text{ if } \operatorname{sech}^{-1} ax < 0 & 0 < ax < 1 \end{cases}$$

$$(\operatorname{csch}^{-1} ax)' = \frac{\mp 1}{x\sqrt{1+(ax)^2}} \qquad \begin{cases} - \text{ if } \cosh^{-1} ax < \infty & ax > 0 \\ + \text{ if } \cosh^{-1} ax > -\infty & ax < 0 \end{cases}$$

where the statements behind the brace define the existence of the derivative (see Fig. 6.15–1).

(2) Fundamental Rules

(a) Notations. u, v, w = differentiable functions of x, u', v', w' = first derivatives of u, v, w with respect to x, du, dv, dw = differentials of u, v, w.

(b) Derivative and differential of a sum.

$$(u+v+w)' = u'+v'+w' \qquad d(u+v+w) = du+dv+dw$$

examples:

$$y = 5x^3-2x^2+6x-7 \qquad y' = 15x^2-4x+6 \qquad dy = (15x^2-4x+6)\,dx$$

(c) Derivative and differential of a product.

$$(uv)' = u'v + uv'$$

$$d(uv) = v\,du + u\,dv$$

$$(uvw)' = u'vw + uv'w + uvw'$$

$$d(uvw) = vw\,du + uw\,dv + uv\,dw$$

examples:

$$y = ax \sin x \qquad u = ax \qquad v = \sin x \qquad u' = a \qquad v' = \cos x$$

$$y' = a \sin x + ax \cos x \qquad dy = (a \sin x + ax \cos x)\,dx$$

$$y = ax \sin x \cos x \qquad u = ax \qquad v = \sin x \qquad w = \cos x$$

$$u' = a \qquad v' = \cos x \qquad w' = -\sin x$$

$$y' = a \sin x \cos x + ax \cos^2 x - ax \sin^2 x$$

$$dy = (a \sin x \cos x + ax \cos^2 x - ax \sin^2 x)\,dx$$

(d) Derivative and differential of a quotient.

$$\left(\frac{u}{v}\right)' = \frac{u'v - uv'}{v^2}$$

$$d\frac{u}{v} = \frac{v\,du - u\,dv}{v^2}$$

$$\left(\frac{uv}{w}\right)' = \frac{uv}{w}\left(\frac{u'}{u} + \frac{v'}{v} - \frac{w'}{w}\right)$$

$$d\frac{uv}{w} = \frac{uv}{w}\left(\frac{du}{u} + \frac{dv}{v} - \frac{dw}{w}\right)$$

examples:

$$y = \frac{\sin x}{\cos x} \qquad u = \sin x \qquad v = \cos x \qquad u' = \cos x \qquad v' = -\sin x$$

$$y' = \frac{\cos^2 x + \sin^2 x}{\cos^2 x} \qquad dy = \frac{\cos^2 x + \sin^2 x}{\cos^2 x}\,dx$$

$$y = \frac{\sin x \cos x}{e^x} \qquad u = \sin x \qquad v = \cos x \qquad w = e^x \qquad u' = \cos x \qquad v' = -\sin x \qquad w' = e^x$$

$$y' = \frac{\sin x \cos x}{e^x}\left(\frac{\cos x}{\sin x} - \frac{\sin x}{\cos x} - \frac{e^x}{e^x}\right) = \frac{\cos^2 x - \sin^2 x - \cos x \sin x}{e^x}$$

$$dy = \frac{\sin x \cos x}{e^x}\left(\frac{\cos x}{\sin x} - \frac{\sin x}{\cos x} - \frac{e^x}{e^x}\right) dx = \frac{\cos^2 x - \sin^2 x - \cos x \sin x}{e^x}\,dx$$

(e) Derivative and differential of a composite function. If $y = f_0(x_1)$, $x_1 = f_1(x_2)$, $x_2 = f_2(x_3), \ldots, x_n = f_n(x)$, then

$$y' = \frac{dy}{dx_1}\frac{dx_1}{dx_2}\frac{dx_2}{dx_3}\cdots\frac{dx_n}{dx}$$

$$dy = \left(\frac{dy}{dx_1}\frac{dx_1}{dx_2}\frac{dx_2}{dx_3}\cdots\frac{dx_n}{dx}\right) dx$$

examples:

$$y = (ax^2 + b)^2 \qquad y = x_1^2 \qquad x_1 = ax^2 + b$$

$$y' = \frac{dy}{dx_1}\frac{dx_1}{dx} = 2(ax^2 + b)(2ax) = 4ax(ax^2 + b)$$

$$y = \sin^2\sqrt{ax^2 + b} \qquad y = x_1^2 \qquad x_1 = \sin x_2 \qquad x_2 = \sqrt{x_3} \qquad x_3 = ax^2 + b$$

$$y' = \underbrace{(2 \sin \sqrt{ax^2 + b})}_{\frac{dy}{dx_1}}\ \underbrace{(\cos \sqrt{ax^2 + b})}_{\frac{dx_1}{dx_2}}\ \underbrace{[\tfrac{1}{2}(ax^2 + b)^{-1/2}]}_{\frac{dx_2}{dx_3}}\ \underbrace{(2ax)}_{\frac{dx_3}{dx}} = \frac{ax \sin 2\sqrt{ax^2 + b}}{\sqrt{ax^2 + b}}$$

(f) Derivative and differential of an exponential composite function. If $y = u^v$, then $\ln y = v \ln u$ and the first derivative of this logarithmic equation is

$$\frac{y'}{y} = v' \ln u + v\frac{u'}{u}$$

from which

$$y' = u^v\left(v' \ln u + v\frac{u'}{u}\right)$$

$$dy = u^v\left(v' \ln u + v\frac{u'}{u}\right) dx$$

example:

$$y = (ax+b)^{2x} \qquad u = ax+b \qquad v = 2x$$

$$y' = \underbrace{(ax+b)^{2x}}_{u^v}\left[\underbrace{2}_{v'}\ \underbrace{\ln(ax+b)}_{\ln u} + \underbrace{2x}_{v}\ \underbrace{\frac{a}{ax+b}}_{u'/u}\right]$$

7.04 HIGHER DERIVATIVES OF ELEMENTARY FUNCTIONS

(1) Basic Cases

(a) Algebraic functions. If a is a constant ($a \neq 0$) and n is the order of the derivative, then the nth derivatives of the respective algebraic functions are

$$\frac{d^n(ax^m)}{dx^n} = am(m-1)(m-2)\cdots(m-n+1)x^{m-n} \qquad m > n$$

$$\frac{d^n(ax^m)}{dx^n} = an! \qquad m = n$$

$$\frac{d^n(ax^m)}{dx^n} = 0 \qquad m < n$$

$$\frac{d^n(ax^{-m})}{dx^n} = \frac{(-1)^n m(m+1)(m+2)\cdots(m+n-1)a}{x^{m+n}} \qquad m \lesseqgtr n$$

$$\frac{d^n\sqrt[m]{ax}}{dx^n} = \frac{(-1)^{n-1}(m-1)(2m-1)\cdots[(n-1)m-1]\sqrt[m]{a}}{m^n\sqrt[m]{x^{mn-1}}} \qquad m \lesseqgtr n$$

$$\frac{d^n(a_0 + a_1x + a_2x^2 + \cdots + a_nx^n)}{dx^n} = a_n n! \qquad m = n$$

examples:

$$\frac{d^3(6x^4)}{dx^3} = (6)(4)(3)(2)x^{4-3} = 144x$$

$$\frac{d^3\left(\dfrac{1}{2x^2}\right)}{dx^3} = \frac{1}{2}\,\frac{(-1)^3(2)(3)(4)}{x^{2+3}} = -\frac{12}{x^5}$$

(b) Exponential and logarithmic functions. If a, k are constants ($a \neq 0$, $k \neq 0$), ln is the natural logarithm to the base e (Sec. 2.13–1), log is the decadic logarithm to the base 10 (Sec. 2.13–1), and n is the order of the derivative, then the nth derivatives of the respective exponential and logarithmic functions are

$$\frac{d^n(e^x)}{dx^n} = e^x \qquad \frac{d^n(a^x)}{dx^n} = (\ln a)^n a^x$$

$$\frac{d^n(e^{kx})}{dx^n} = k^n e^{kx} \qquad \frac{d^n(a^{kx})}{dx^n} = (k \ln a)^n a^{kx}$$

$$\frac{d^n(e^{-kx})}{dx^n} = (-k)^n e^{-kx} \qquad \frac{d^n(a^{-kx})}{dx^n} = (-k \ln a)^n a^{-kx}$$

$$\frac{d^n(\ln x)}{dx^n} = \frac{(-1)^{n-1}(n-1)!}{x^n} \qquad \frac{d^n(\log x)}{dx^n} = \frac{(-1)^{n-1}(n-1)!\log e}{x^n}$$

$$\frac{d^n(\ln ax)}{dx^n} = \frac{(-1)^{n-1}(n-1)!}{x^n} \qquad \frac{d^n(\log ax)}{dx^n} = \frac{(-1)^{n-1}(n-1)!\log e}{x^n}$$

where $\log e = 0.43429\ldots$.

(c) Trigonometric and hyperbolic functions. If k is a constant ($k \neq 0$) and n is the order of the derivative, then the nth derivatives of the respective trigonometric and hyperbolic functions are

$$\frac{d^n(\sin x)}{dx^n} = \sin\left(x + \frac{n\pi}{2}\right) \qquad \frac{d^n(\cos x)}{dx^n} = \cos\left(x + \frac{n\pi}{2}\right)$$

$$\frac{d^n(\sin kx)}{dx^n} = k^n \sin\left(kx + \frac{n\pi}{2}\right) \qquad \frac{d^n(\cos kx)}{dx^n} = k^n \cos\left(kx + \frac{n\pi}{2}\right)$$

$$\frac{d^n(\sinh x)}{dx^n} = \begin{cases} \sinh x & n \text{ even} \\ \cosh x & n \text{ odd} \end{cases}$$

$$\frac{d^n(\cosh x)}{dx^n} = \begin{cases} \cosh x & n \text{ even} \\ \sinh x & n \text{ odd} \end{cases}$$

$$\frac{d^n(\sinh kx)}{dx^n} = \begin{cases} k^n \sinh kx & n \text{ even} \\ k^n \cosh kx & n \text{ odd} \end{cases}$$

$$\frac{d^n(\cosh kx)}{dx^n} = \begin{cases} k^n \cosh kx & n \text{ even} \\ k^n \sinh kx & n \text{ odd} \end{cases}$$

(2) Fundamental Rules

(a) Notations. u, v, w = differentiable functions of x, $u^{(n)}, v^{(n)}, w^{(n)}$ = nth derivatives of u, v, w, with respect to x, $d^{(n)}u, d^{(n)}v, d^{(n)}w$ = nth differentials of u, v, w.

(b) Derivative and differential of a sum.

$$\frac{d^n(u+v+w)}{dx^n} = \frac{d^n u}{dx^n} + \frac{d^n v}{dx^n} + \frac{d^n w}{dx^n} = u^{(n)} + v^{(n)} + w^{(n)}$$

$$d^n(u+v+w) = d^n u + d^n v + d^n w$$

(c) Derivative and differential of a product and a quotient.

$$\frac{d^n(uv)}{dx^n} = uv^{(n)} + \binom{n}{1} u^{(1)} v^{(n-1)} + \binom{n}{2} u^{(2)} v^{(n-2)} + \cdots + u^{(n)} v$$

$$\frac{d^n\left(\frac{u}{v}\right)}{dx^n} = u\left(\frac{1}{v}\right)^{(n)} + \binom{n}{1} u^{(1)} \left(\frac{1}{v}\right)^{(n-1)} + \binom{n}{2} u^{(2)} \left(\frac{1}{v}\right)^{(n-2)} + \cdots + \frac{u^{(n)}}{v}$$

$$d^n(uv) = ud^n v + \binom{n}{1} du\, d^{n-1} v + \binom{n}{2} d^2 u\, d^{n-2} v + \cdots + d^n uv$$

$$d^n\left(\frac{u}{v}\right) = u\, d^n\left(\frac{1}{v}\right) + \binom{n}{1} du\, d^{n-1}\left(\frac{1}{v}\right) + \binom{n}{2} d^2 u\, d^{n-2}\left(\frac{1}{v}\right) + \cdots + \frac{d^n u}{v}$$

where $\binom{n}{1}, \binom{n}{2}, \ldots$ are binomial coefficients (Sec. 2.19–1*b*) and

$$d^n u = u^{(n)}\, dx^n \qquad d^{n-1} u = u^{(n-1)}\, dx^{n-1}, \quad \ldots$$

$$d^n v = v^{(n)}\, dx^n \qquad d^{n-1} v = v^{(n-1)}\, dx^{n-1}, \quad \ldots$$

$$d^n\left(\frac{1}{v}\right) = \left(\frac{1}{v}\right)^{(n)} dx^n \qquad d^{n-1}\left(\frac{1}{v}\right) = \left(\frac{1}{v}\right)^{(n-1)} dx^{n-1}, \quad \ldots$$

examples:

$$\frac{d^2(uv)}{dx^2} = uv'' + 2u'v' + u''v$$

$$\frac{d^3(uv)}{dx^3} = uv + 3u'v'' + 3u''v' + uv$$

$$d^2(uv) = u\, d^2 v + 2du\, dv + vd^2 u$$

$$d^3(uv) = u\, d^3 v + 3du\, d^2 v + 3d^2 u\, dv + v\, d^3 u$$

7.05 PARTIAL DERIVATIVES AND DIFFERENTIALS

(1) First and Higher Partial Derivatives

(a) First partial derivative of $y = f(x_1, x_2, \ldots, x_m)$ is the limit approached by the ratio of an increment Δy of the function to the increment of one of its arguments $\Delta x_i\ (i = 1, 2, \ldots, m)$, when Δx_i tends to zero and all the other arguments remain constant.

$$\lim_{\Delta x_i \to 0} \frac{\Delta y}{\Delta x_i} = \lim_{\Delta x_i \to 0} \frac{f(x_1, x_2, \ldots, x_i + \Delta x_i, \ldots, x_m) - f(x_1, x_2, \ldots, x_i, \ldots, x_m)}{\Delta x_i} = \frac{\partial y}{\partial x_i}$$

The necessary and sufficient conditions of the differentiability of a function at a certain point are: (1) the continuity of this function at this point; and (2) the existence of the said limit.

(b) Alternative notations for the first partial derivatives of $y = f(x_1, x_2, \ldots, x_m)$ are

$$\frac{\partial y}{\partial x_i} = \frac{\partial f(x_1, x_2, \ldots, x_i, \ldots, x_m)}{\partial x_i} = F_i$$

(c) Number of first partial derivatives. If there are m independent arguments, then the function y may have as many as m first partial derivatives.

$$\frac{\partial y}{\partial x_1} = F_1 \qquad \frac{\partial y}{\partial x_2} = F_2, \ldots, \frac{\partial y}{\partial x_m} = F_m$$

example:

If $y = x_1^2 x_2^2 + x_3^2$, then

$$\frac{\partial y}{\partial x_1} = 2x_1 x_2^2 \qquad \frac{\partial y}{\partial x_2} = 2x_1^2 x_2 \qquad \frac{\partial y}{\partial x_3} = 2x_3$$

(d) Interpretation. The first partial derivative of a function of several variables is the measure of change of y with respect to a particular variable x_i at each point where the limit exists.

(e) Second and higher partial derivatives of y with respect to $x_i (i = 1, 2, \ldots, m)$ and/or $x_j (j = 1, 2, \ldots, m)$ are

$$\frac{\partial^2 y}{\partial x_i^2} = F_{ii}, \quad \frac{\partial^2 y}{\partial x_i\, \partial x_j} = F_{ij}, \quad \ldots$$

$$\frac{\partial^3 y}{\partial x_i^3} = F_{iii}, \quad \frac{\partial^3 y}{\partial x_i^2\, \partial x_j} = F_{iij}, \quad \frac{\partial^3 y}{\partial x_i\, \partial x_j\, \partial x_i} = F_{iji}, \quad \ldots$$

. . . .

if the limits in question exist. The number of differentiations performed is the order of the derivative and if the highest derivative involved is continuous, then the result is independent of the order in which the differentiation is performed.

$$F_{ij} = F_{ji} \qquad F_{iij} = F_{iji} = F_{jii}$$

(2) First and Higher Partial Differentials

(a) First partial differential of $y = f(x_1, x_2, \ldots, x_m)$ with respect to $x_i (i = 1, 2, \ldots, m)$ is

$$d_i y = \frac{\partial y}{\partial x_i} dx_i = d_i[f(x_1, x_2, \ldots, x_m)]$$

(b) Second and higher partial differentials of $y = f(x_1, x_2, \ldots, x_m)$ with respect to $x_i (i = 1, 2, \ldots, m)$ and/or $x_j (j = 1, 2, \ldots, m)$ are

$$d_{ii} y = \frac{\partial^2 y}{\partial x_i^2} dx_i^2 \qquad d_{ij} y = \frac{\partial^2 y}{\partial x_i\, \partial x_j} dx_i\, dx_j$$

$$d_{iii} y = \frac{\partial^3 y}{\partial x_i^3} dx_i^3 \qquad d_{iij} y = \frac{\partial^3 y}{\partial x_i^2\, \partial x_j} dx_i^2\, dx_j, \quad \ldots$$

. . . .

(3) First and Higher Total Differentials

(a) First total differential of $y = f(x_1, x_2, \ldots, x_m)$ is

$$dy = \frac{\partial y}{\partial x_1} dx_1 + \frac{\partial y}{\partial x_2} dx_2 + \cdots + \frac{\partial y}{\partial x_m} dx_m$$

(b) Second and higher total differentials of $y = f(x_1, x_2, \ldots, x_m)$ are

$$d^2y = \left(\frac{\partial}{\partial x_1}dx_1 + \frac{\partial}{\partial x_2}dx_2 + \cdots + \frac{\partial}{\partial x_m}dx_m\right)^2 y$$

$$d^3y = \left(\frac{\partial}{\partial x_1}dx_1 + \frac{\partial}{\partial x_2}dx_2 + \cdots + \frac{\partial}{\partial x_m}dx_m\right)^3 y$$

. . . .

$$d^ny = \left(\frac{\partial}{\partial x_1}dx_1 + \frac{\partial}{\partial x_2}dx_2 + \cdots + \frac{\partial}{\partial x_m}dx_m\right)^n y$$

example:

If $z = x^3 \sin y$, then

$$dz = \frac{\partial z}{\partial x}dx + \frac{\partial z}{\partial y}dy = (3x^2 \sin y)\,dx + (x^3 \cos y)\,dy$$

$$d^2z = \left(\frac{\partial}{\partial x}dx + \frac{\partial}{\partial y}dy\right)^2 z$$

$$= \frac{\partial^2 z}{\partial x^2}dx^2 + 2\frac{\partial^2 z}{\partial x\,\partial y}dx\,dy + \frac{\partial^2 z}{\partial y^2}dy^2$$

$$= (6x \sin y)\,dx^2 + 2(3x^2 \cos y)\,dx\,dy + (-x^3 \sin y)\,dy^2$$

7.06 DERIVATIVES AND DIFFERENTIALS OF SPECIAL FUNCTIONS

(1) Inverse Functions

(a) Differentials. If $y = f(x)$ has the unique inverse function $x = g(y)$ and $dx/dy \neq 0$, then

$$dy = \frac{\partial f(x)}{\partial x}dx \qquad dx = \frac{\partial g(y)}{\partial y}dy$$

$$d^2y = \frac{\partial^2 f(x)}{\partial x^2}dx^2 \qquad d^2x = \frac{\partial^2 g(y)}{\partial y^2}dy^2$$

(b) Derivatives. From these relations (see above),

$$\frac{dy}{dx} = \frac{\partial f(x)}{\partial x} = \frac{1}{\partial g(y)/\partial y} = \frac{1}{dx/dy}$$

$$\frac{d^2y}{dx^2} = \frac{\partial^2 f(x)}{\partial x^2} = \frac{\partial\left[\dfrac{1}{g(y)/\partial y}\right]}{\partial x} = \frac{\partial\left(\dfrac{1}{dx/dy}\right)}{\partial x} = -\frac{d^2x/dy^2}{(dx/dy)^3}$$

. . . .

$$\frac{d^ny}{dx^n} = \frac{\partial^n f(x)}{\partial x^n} = \frac{\partial^{n-1}\left[\dfrac{1}{\partial g(y)/\partial y}\right]}{\partial x^{n-1}} = \frac{\partial^{n-1}\left(\dfrac{1}{dx/dy}\right)}{\partial x^{n-1}}$$

example:

If $y = \sin^{-1} x$ and $x = \sin y$, then

$$y' = \frac{1}{\partial(\sin y)/\partial y} = \frac{1}{\cos y} = \frac{1}{\sqrt{1-\sin^2 y}} = \frac{1}{\sqrt{1-x^2}}$$

(2) Implicit Functions

(a) Differentials. If $y = f(x)$ is given implicitly as $F(x,y) = 0$, then

$$dF(x,y) = \frac{\partial F(x,y)}{\partial x} dx + \frac{\partial F(x,y)}{\partial y} dy = 0$$

$$d^2F(x,y) = \left[\frac{\partial}{\partial x} dx + \frac{\partial}{\partial y} dy\right]^2 F(x,y) = 0$$

. . . .

$$d^nF(x,y) = \left(\frac{\partial}{\partial x} dx + \frac{\partial}{\partial y} dy\right)^n F(x,y) = 0$$

(b) Derivatives. From these relations (see above),

$$F_x dx + F_y\, dy = 0 \qquad \text{or} \qquad F_x + F_y \frac{dy}{dx} = 0$$

where

$$F_x = \frac{\partial F(x,y)}{\partial x} \qquad F_y = \frac{\partial F(x,y)}{\partial y} \neq 0$$

from which

$$\frac{dy}{dx} = -\frac{F_x}{F_y}$$

$$\frac{d^2y}{dx^2} = -\frac{F_{xx}F_y^2 - 2F_xF_yF_{xy} + F_x^2F_{yy}}{F_y^3}$$

where

$$F_{xx} = \frac{\partial^2 F(x,y)}{\partial x^2} \qquad F_{xy} = \frac{\partial^2 F(x,y)}{\partial x\, \partial y} \qquad F_{yy} = \frac{\partial^2 F(x,y)}{\partial y^2}$$

example:

If $F(x,y) = x^2 + y^2 - 1 = 0$,

$$F_x = 2x \qquad F_y = 2y \qquad F_{xx} = 2 \qquad F_{yy} = 2 \qquad F_{xy} = 0$$

and

$$\frac{dy}{dx} = -\frac{x}{y} \qquad \frac{d^2y}{dx^2} = -\frac{8y^2 + 8x^2}{8y^3} = -\frac{1}{y^3}$$

where $x^2 + y^2 = 1$ from the statement of the problem.

(3) Parametric Functions

(a) Differentials. If the dependence of x and y is given parametrically as $x = x(t)$ and $y = y(t)$, then

$$dx = \frac{\partial x(t)}{\partial t} dt = \dot{x}(t)\, dt \qquad dy = \frac{\partial y(t)}{\partial t} dt = \dot{y}(t)\, dt$$

and for $z = f(x,y)$, the total differential

$$dz = \frac{\partial z}{\partial x}\frac{\partial x}{\partial t} dt + \frac{\partial z}{\partial y}\frac{\partial y}{\partial t} dt$$

(b) Derivatives. From these relations (see above),

$$\frac{dy}{dx} = \frac{\partial y/\partial t}{\partial x/\partial t} = \frac{\dot{y}(t)}{\dot{x}(t)} \qquad \dot{x}(t) \neq 0$$

$$\frac{d^2y}{dx^2} = \frac{\partial(dy/dx)}{\partial t}\frac{dt}{dx} = \frac{\ddot{y}(t)\dot{x}(t) - \dot{y}(t)\ddot{x}(t)}{[\dot{x}(t)]^3}$$

. . . .

$$\frac{d^ny}{dx^n} = \frac{\partial^{n-1}(dy/dx)}{\partial t}\frac{dt}{dx}$$

example:

If $x = \sin at$, $y = \cos bt$, then

$$\frac{\partial x}{\partial t} = \dot{x}(t) = a\cos at \qquad \frac{\partial y}{\partial t} = \dot{y}(t) = -b\sin bt$$

$$\frac{\partial^2 x}{\partial t^2} = \ddot{x}(t) = -a^2\sin at \qquad \frac{\partial^2 y}{\partial t^2} = \ddot{y}(t) = -b^2\cos bt$$

$$\frac{dy}{dx} = -\frac{b\sin bt}{a\cos at} \qquad \frac{d^2y}{dx^2} = \frac{-b^2\cos bt\, a\cos at - b\sin bt\, a^2\sin at}{a^3\cos^3 at}$$

7.07 INVESTIGATION OF A FUNCTION

(1) Function of One Variable

(a) Rates of change of $f(x)$ at $x = x_i$ (Fig. 7.07–1).
If $f'(x_i) > 0$, $f(x)$ is rising at $x_i\,(x_1 < x_i < x_2,\ x_4 < x_i < x_5)$.
If $f'(x_i) = 0$, $f(x)$ has a tangent parallel to the X axis at x_i (at x_2 and x_4).
If $f'(x_i) < 0$, $f(x)$ is falling at $x_i\,(x_2 < x_i < x_4)$.

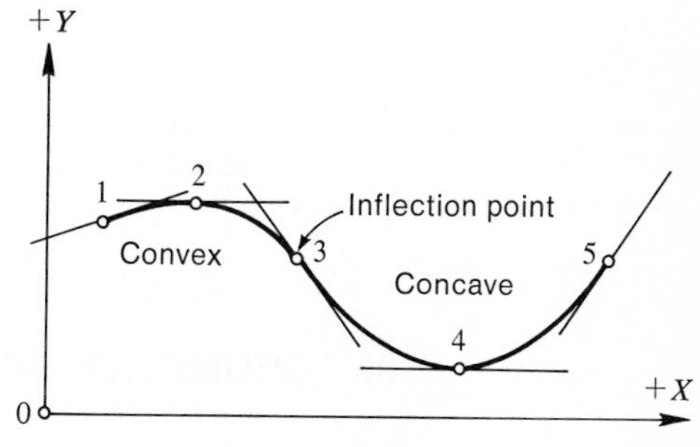

Fig. 7.07-1

(b) Shape of curve of $f(x)$ at $x = x_i$ (Fig. 7.07–1).
If $f''(x_i) < 0$, $f(x)$ is convex at $x_i\,(x_1 < x_i < x_3)$.
If $f''(x_i) = 0$, $f(x)$ has an inflection point at x_i (point 3).
If $f''(x_i) > 0$, $f(x)$ is concave at $x_i\,(x_3 < x_i < x_5)$.

(c) Maximum and minimum of $f(x)$. If $f(x)$ has a continuous second derivative at $x = x_i$, then $f(x_i)$ is:
A maximum, if $f'(x_i) = 0$ and $f''(x_i) < 0$.
A minimum, if $f'(x_i) = 0$ and $f''(x_i) > 0$.

(d) General case. If $f(x)$ has n continuous derivatives $f'(x), f''(x), \ldots, f^{(n)}(x)$ at $x = x_i$, and $f'(x_i) = f''(x_i) = \cdots = f^{(n-1)}(x_i) = 0$ but $f^{(n)}(x_i) \neq 0$, then $f(x_i)$ is
A maximum, if n is even and $f^n(x_i) < 0$.
A minimum, if n is even and $f''(x_i) > 0$.
An inflection point, if n is odd and $f^n(x_i) \gtreqless 0$.

example:

If $y = x^3/3 + x^2/2 - 6x + 1$, then

$$y' = x^2 + x - 6 \qquad \text{and} \qquad y'' = 2x + 1$$

From $y' = 0$, $x_1 = 2$, $x_2 = -3$, and in terms of x_1, x_2, $y_1'' = 5$ and $y_2'' = -5$. Thus y is a maximum at $x_2 = -3$, where $y_2' = 0$, $y_2'' < 0$ and is a minimum at $x_1 = 2$, where $y_1' = 0$, $y_1'' > 0$.

(2) Function of Two Variables

(a) Conditions of extremum. If $f(x,y)$ is continuously differentiable at $x = x_i$ and $y = y_i$, then $f(x_i,y_i)$ is a maximum or minimum if

$$F_x = 0 \qquad F_y = 0 \qquad F_{xx}F_{yy} - F_{xy}^2 = \Delta > 0$$

where

$$F_x = \frac{\partial f(x_i,y_i)}{\partial x} \qquad F_y = \frac{\partial f(x_i,y_i)}{\partial y}$$

$$F_{xx} = \frac{\partial^2 f(x_i,y_i)}{\partial x^2} \qquad F_{yy} = \frac{\partial^2 f(x_i,y_i)}{\partial y^2} \qquad F_{xy} = \frac{\partial^2 f(x_i,y_i)}{\partial x\, \partial y}$$

(b) Maximum and minimum. If $f(x,y)$ has continuous second derivatives at x_i, y_i, then $f(x_i,y_i)$ is:

A maximum, if at least $F_{xx} < 0$ or $F_{yy} < 0$ and $\Delta > 0$.

A minimum, if at least $F_{xx} > 0$ or $F_{yy} > 0$ and $\Delta > 0$.

example:

If $z = x^3 + y^3 - 3x - 3y + 10$, then

$$F_x = 3x^2 - 3 \qquad F_y = 3y^2 - 3 \qquad F_{xx} = 6x \qquad F_{yy} = 6y \qquad F_{xy} = 0$$

From $F_x = 0$, $x_1 = 1$, $x_2 = -1$; from $F_y = 0$, $y_1 = 1$, $y_2 = -1$; and for $x_1 = 1$, $y_1 = 1$, $\Delta_{11} = (6)(6) = 36 > 0$. For $x_2 = -1$, $y_2 = -1$, $\Delta_{22} = (-6)(-6) = 36 > 0$. Thus z is a maximum at $x_2 = -1$, $y_2 = -1$, where $F_{xx} = -6 < 0$, $F_{yy} = -6 < 0$ and is a minimum at $x_1 = 1$, $y_1 = 1$, where $F_{xx} = 6 > 0$, $F_{yy} = 6 > 0$.

7.08 GEOMETRY OF PLANE CURVE

(1) Explicit Equation

(a) Derivatives of the plane curve defined by $y = f(x)$ are

$$y' = \frac{dy}{dx} = \frac{df(x)}{dx} \qquad y'' = \frac{d^2y}{dx^2} = \frac{d^2f(x)}{dx^2}$$

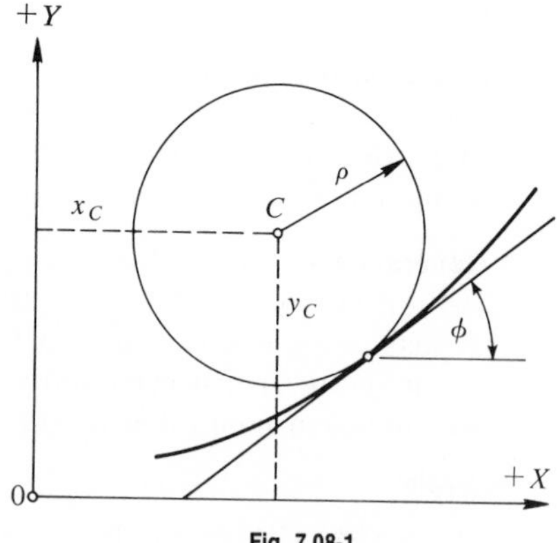

Fig. 7.08-1

(b) Direction functions are

$$\sin\phi = \frac{dy}{ds} \qquad \cos\phi = \frac{dx}{ds} \qquad \tan\phi = \frac{dy}{dx}$$

where

$$ds = \sqrt{dx^2 + dy^2} = \sqrt{1 + \left(\frac{dy}{dx}\right)^2}\, dx = \sqrt{1 + \left(\frac{dx}{dy}\right)^2}\, dy$$

(c) Radius of curvature ρ and the coordinates of the center of curvature x_C, y_C are

$$\rho = A\sqrt{1+\left(\frac{dy}{dx}\right)^2} \qquad x_C = x - A\frac{dy}{dx} \qquad y_C = y + A$$

where

$$A = \frac{1+(dy/dx)^2}{d^2y/dx^2}$$

(2) Implicit Equation

(a) Derivatives of a plane curve defined by $F(x,y) = 0$ are

$$F_x = \frac{\partial F(x,y)}{\partial x} \qquad F_y = \frac{\partial F(x,y)}{\partial y} \qquad F_{xx} = \frac{\partial^2 F(x,y)}{\partial x^2} \qquad \ldots$$

(b) Direction functions are

$$\sin\phi = \frac{-F_x}{\pm\sqrt{F_x^2+F_y^2}} \qquad \cos\phi = \frac{F_y}{\pm\sqrt{F_x^2+F_y^2}} \qquad \tan\phi = -\frac{F_x}{F_y}$$

where

$$ds = \frac{\pm\sqrt{F_x^2+F_y^2}}{F_y}\,dx = \frac{\pm\sqrt{F_x^2+F_y^2}}{-F_x}\,dy$$

and the sign of $\sqrt{F_x^2+F_y^2}$ is governed by the quadrant in which ϕ terminates.

(c) Radius of curvature ρ and the coordinates of the center of curvature x_C, y_C are

$$\rho = B\sqrt{F_x^2+F_y^2} \qquad x_C = x - BF_x \qquad y_C = y - BF_y$$

where

$$B = \frac{F_x^2+F_y^2}{F_{xx}F_y^2 - 2F_xF_yF_{xy} + F_{yy}F_x^2}$$

(3) Parametric Equations

(a) Derivatives of the plane curve defined by $x = x(t)$, $y = y(t)$ are

$$\dot{x} = \frac{dx(t)}{dt} \qquad \dot{y} = \frac{dy(t)}{dt} \qquad \ddot{x} = \frac{d^2x(t)}{dt^2} \qquad \ddot{y} = \frac{d^2y(t)}{dt^2}$$

(b) Direction functions are

$$\sin\phi = \frac{\dot{y}}{\sqrt{\dot{x}^2+\dot{y}^2}} \qquad \cos\phi = \frac{\dot{x}}{\sqrt{\dot{x}^2+\dot{y}^2}} \qquad \tan\phi = \frac{\dot{y}}{\dot{x}}$$

where $ds = \sqrt{\dot{x}^2+\dot{y}^2}\,dt$.

(c) Radius of curvature ρ and the coordinates of the center of curvature x_C, y_C are

$$\rho = C\sqrt{\dot{x}^2+\dot{y}^2} \qquad x_C = x - C\dot{y} \qquad y_C = y + C\dot{x}$$

where

$$C = \frac{\dot{x}^2+\dot{y}^2}{\dot{x}\ddot{y} - \ddot{x}\dot{y}}$$

(4) Examples

(a) Catenary.
For the hyperbolic cosine curve (catenary), of Fig. 7.08–2, $y = \cosh x$.

$$A = \frac{1+\sinh^2 x}{\cosh x} = \cosh x \qquad \rho = \cosh^2 x$$

and $x_C = x - \cosh x \sinh x \qquad y_C = y + \cosh x$

At the vertex V, $x_V = 0 \qquad y_V = +1 \qquad \rho = +1$
$x_C = 0 \qquad y_C = +2$

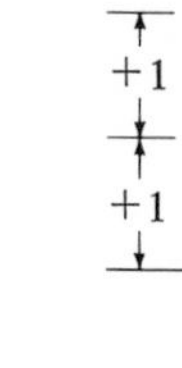

Fig. 7.08-2

(b) Hyperbola.
For the hyperbola of Fig. 7.08–3, $b^2x^2 - a^2y^2 = a^2b^2$

$$B = -\frac{b^4x^2 + a^4y^2}{2a^4b^4} \qquad \rho = \left|\frac{(b^4x^2 + a^4y^2)^{3/2}}{a^4b^4}\right|$$

and $\quad x_C = \dfrac{a^2+b^2}{a^4}x^3 \qquad y_C = -\dfrac{a^2+b^2}{b^4}y^3$

At the vertex 2, $x_2 = a$, $y_2 = 0$, and

$$\rho = \left|\frac{b^2}{a}\right| \qquad x_C = \frac{a^2+b^2}{a} \qquad y_C = 0$$

which leads to the simple geometric construction of Fig. 7.08–3.

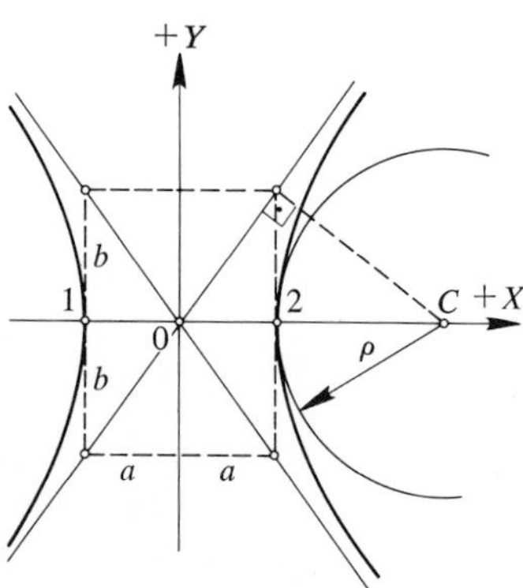

Fig. 7.08-3

(c) Ellipse.
For the ellipse of Fig. 7.08–4, $b^2x^2 + a^2y^2 = a^2b^2$

$$B = \frac{b^4y^2 + a^4x^2}{2a^4b^4} \qquad \rho = \left|\frac{(b^4x^2 + a^4y^2)^{3/2}}{a^4b^4}\right|$$

and $\quad x_C = \dfrac{a^2-b^2}{a^4}x^3 \qquad y_C = -\dfrac{a^2-b^2}{b^4}y^3$

At the vertex 1, $x_1 = -a$, $y_1 = 0$, and

$$\rho_1 = \left|\frac{b^2}{a}\right| \qquad x_{C_1} = -\frac{a^2-b^2}{a} \qquad y_{C_1} = 0$$

At the vertex 4, $x_4 = 0$, $y_4 = b$, and

$$\rho_4 = \left|\frac{a^2}{b}\right| \qquad x_{C_4} = 0 \qquad y_{C_4} = -\frac{a^2-b^2}{b}$$

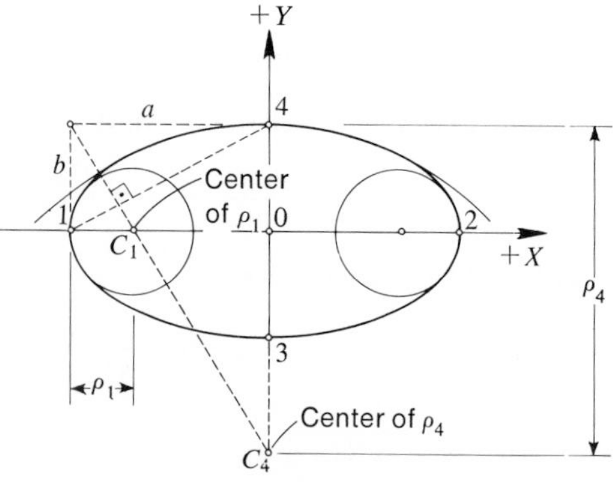

Fig. 7.08-4

which again leads to the simple geometric constructions of Fig. 7.08–4.

(d) Parabola.
For the parabola of Fig. 6.09–1, $x = (p/2)t^2$, $y = pt$, $\dot{x} = pt$, $\ddot{x} = p$, $\dot{y} = p$, $\ddot{y} = 0$, $C = -(1+t^2)$, $\rho = |p(1+t^2)^{3/2}|$, and $x_C = (p/2)(2+3t^2)$, $y_C = pt(2+t^2)$. At the vertex 0, $\rho = |p|$, $x_C = p$, $y_C = 0$, which leads again to a very simple geometric construction.

8
SEQUENCES AND SERIES

8.01 SERIES OF CONSTANT TERMS

(1) Definitions

(a) Sequence is a set of n numbers, $a_1, a_2, a_3, \ldots, a_n$, arranged in a prescribed order and formed according to a definite rule. Each member of the sequence is called a term and the sequence is defined by the number of terms as finite or infinite.

(b) Finite series is the sum of the terms of a finite sequence.

$$\sum_{k=1}^{n} a_k = a_1 + a_2 + a_3 + \cdots + a_{n-1} + a_n = S_n$$

where $n < \infty$ and $-\infty < S_n < \infty$.

example:

$2+4+6+8 = 20$ where $n = 4$, $S_n = 20$

(c) Infinite series is the sum of the terms of an infinite sequence.

$$\sum_{k=1}^{\infty} = a_1 + a_2 + a_3 + \cdots = S_\infty = S$$

where $n = \infty$ and $-\infty \leq S_\infty \leq \infty$.

example:

$\frac{1}{2}+\frac{1}{4}+\frac{1}{8}+\cdots = 1$ where $n = \infty$, $S_\infty = 1$

(d) General term of a sequence defines the *law of formation* (governing law of sequence).

examples ($k = 1, 2, 3, \ldots$):

$a_k = 2k$ $\qquad 2+4+6+8+\cdots$

$a_k = 2^k$ $\qquad 2+2^2+2^3+2^4+\cdots$

$a_k = \dfrac{1}{1+k}$ $\qquad \frac{1}{2}+\frac{1}{3}+\frac{1}{4}+\frac{1}{5}+\cdots$

$a_k = \dfrac{1}{k!}$ $\qquad 1+\frac{1}{2}+\frac{1}{6}+\frac{1}{24}+\cdots$

(e) Convergent series is an infinite series that approaches the limit S as the number of terms approaches infinity ($n \to \infty$); that is, if

$$S_n = \sum_{k=1}^{n} a_k \qquad \text{and} \qquad \lim_{n\to\infty} S_n = S$$

exist, the series is said to be convergent and S is the sum ($-\infty < S < \infty$).

example:

Infinite series $\frac{1}{2}+\frac{1}{4}+\frac{1}{8}+\frac{1}{16}+\cdots$ governed by $a_k = 1/2^k$ is convergent since $S = 1$.

(f) Divergent series is an infinite series that increases without bound as the number of terms increases.

example:

Infinite series $2+4+8+16+\cdots$ governed by $a_k = 2^k$ is divergent, $S = \infty$.

(g) Absolutely convergent series is a series whose absolute terms form a convergent series.

$$\sum_{k=1}^{\infty} |a_k| = |S|$$

example:

Infinite series, $\frac{1}{2}-\frac{1}{4}+\frac{1}{8}-\frac{1}{16}+\cdots$, governed by $a_k = -1/(-2)^k$ is absolutely convergent since $\frac{1}{2}+\frac{1}{4}+\frac{1}{8}+\frac{1}{16}+\cdots$ is a convergent series (Sec. 8.01–1*c*).

(h) Conditionally convergent series is a series which is not absolutely convergent.

example:

Infinite series, $1-\frac{1}{2}+\frac{1}{3}-\frac{1}{4}+\cdots$, governed by $a_k = -(-1)^k/k$ is conditionally convergent since the infinite series of its absolute terms, $1+\frac{1}{2}+\frac{1}{3}+\frac{1}{4}+\cdots$, is divergent (after 1 million terms the sum is less than 15 and still slowly diverging).

(i) Alternating series is a series whose terms alternate in sign.

example:

$$1-\frac{1}{2}+\frac{1}{3}-\frac{1}{4}+\cdots$$

(j) Convergent alternating series is an alternating series satisfying the condition

$$|a_k + a_{k+1}| > |a_{k+2} + a_{k+3}|$$

example:

Alternating series, $1-\frac{1}{2}+\frac{1}{3}-\frac{1}{4}+\cdots$, is convergent, since $|1-\frac{1}{2}| > |\frac{1}{3}-\frac{1}{4}|$.

(2) Tests of Convergence

(a) Comparison test. If

$$|a_k| < c|b_k|$$

where c is a constant independent of k and b_k is the kth term of another series which is known to be absolutely convergent, the series of a terms is also absolutely convergent. Typically absolutely convergent series of cb terms useful in the comparison test are:

$$c + cr + cr^2 + cr^3 + \cdots, \quad b_k = r^{k-1} \qquad -1 < r < 1$$

$$\frac{1}{1(2)}+\frac{1}{2(3)}+\frac{1}{3(4)}+\frac{1}{4(5)}+\cdots, \quad b_k = \frac{1}{k(k+1)}$$

$$\frac{1}{1^n}+\frac{1}{2^n}+\frac{1}{3^n}+\frac{1}{4^n}+\cdots, \quad b_k = \frac{1}{k^n} \qquad n > 1$$

example:

Infinite series $1/1! + 1/2! + 1/3! + 1/4! + \cdots$ governed by $a_k = 1/k!$ is convergent, since $|a_k| = |1/k!| < |b_k| = |1/k(k+1)|$ (see second series above). Note in this case $c = 1$.

(b) Ratio test. If

$$\lim_{k\to\infty} \frac{a_{k+1}}{a_k} = L$$

then the series is absolutely convergent for $L < 1$, divergent for $L > 1$, and the test fails for $L = 1$.

example:

Infinite series $1/2 + 2/2^2 + 3/2^3 + 4/2^4 + \cdots$ has

$$\lim_{k\to\infty} \left|\frac{a_{k+1}}{a_k}\right| = \lim_{k\to\infty} \left|\frac{n+1}{2n}\right| = \lim_{k\to\infty} \left|\frac{1}{2} + \frac{1}{2n}\right| = \frac{1}{2} = L$$

Since $L < 1$, the series is convergent.

(c) Other tests. Several other tests of convergence of infinite series are available: Cauchy's test, Raabe's test, and integral test.

(3) Operations with Absolutely Convergent Series

(a) Change in order. The terms of an absolutely convergent series can be rearranged in any order, and the new series will converge to the same term.

(b) Sum, difference, and product. The sum, difference, and product of two or more absolutely convergent series is an absolutely convergent series.

8.02 FINITE SERIES OF CONSTANT TERMS

(1) Arithmetic Series

(a) Definition. The finite arithmetic series is

$$a_1 + (a_1 + d) + (a_1 + 2d) + \cdots + [a_1 + (k-1)d] + \cdots + [a_1 + (n-1)d]$$

where n = number of terms
a_1 = first term d = difference of two successive terms $(a_{k+1} - a_k)$
a_k = kth term a_n = last term

(b) Governing law. Each term of the arithmetic series is defined as

$$a_k = a_1 + (k-1)d$$

Consequently, the last term is $a_n = a_1 + (n-1)d$

(c) Sum.

$$S_n = \left(\frac{a_1 + a_n}{2}\right)n = \frac{n}{2}[2a_1 + (n-1)d]$$

(d) Number of terms.

$$n = \frac{a_n - a_1}{d} + 1$$

example:

Find the sum of $3 + 6 + 9 + \cdots + 300$.

By (a), $d = 3$, by (d), $n = [(300 - 3)/3] + 1 = 100$, and by (c), $S_n = (3 + 300)100/2 = 15{,}150$.

(2) Geometric Series

(a) Definition. The finite geometric series is

$$a_1 + a_1 r + a_1 r^2 + \cdots + a_1 r^{k-1} + \cdots + a_1 r^{n-1}$$

where n = number of terms
a_1 = first term
a_k = kth term
r = ratio of two successive terms $\dfrac{a_{k+1}}{a_k}$
a_n = last term

(b) Governing law. Each term of the geometric series is defined as

$$a_k = a_1 r^{k-1}$$

Consequently, the last term is $a_n = a_1 r^{n-1}$.

(c) Classification. The geometric series is divergent for $r > 1$, convergent for $0 < r < 1$, alternating for $r < 0$, and is a series of n constant terms $(a_1 + a_1 + a_1 + \cdots = na_1)$ for $r = 1$.

(d) Sum of finite series $(n < \infty)$.

$$S_n = \frac{a_n r - a_1}{r - 1} = \frac{a_1(r^n - 1)}{r - 1} \qquad r > 1$$

$$S_n = \frac{a_1 - a_n r}{1 - r} = \frac{a_1(1 - r^n)}{1 - r} \qquad r < 1$$

$$S_n = na_1 \qquad r = 1$$

example:

Find the sum of $2 + 6 + 18 + 54$.

By (a), $r = 3$, $n = 4$; and by (c) for $r > 1$, $S_n = [54(3) - 2]/(3 - 1) = 80$.

(e) Sum of infinite series $(n = \infty)$.

$$S_\infty = \frac{a_1}{1 - r} \qquad -1 < r < 1$$

example:

Find the sum of $2 + 1 + \frac{1}{2} + \frac{1}{4} + \frac{1}{8} + \cdots$.

By (a), $r = \frac{1}{2}$ and by (e), $S_\infty = 2/(1 - \frac{1}{2}) = 4$.

Note that $2 - 1 + \frac{1}{2} - \frac{1}{4} + \frac{1}{8} - \cdots$ is also an infinite geometric series:

$$r = -\frac{1}{2} \quad \text{and} \quad S_\infty = \frac{2}{1 + \frac{1}{2}} = \frac{4}{3}$$

(3) Finite Series of Powers of Integers $(n < \infty)$

(a) $1^m + 2^m + 3^m + \cdots + n^m = S(n).$

$$\sum_{k=1}^{n} k = 1 + 2 + 3 + \cdots + n = \frac{(n+1)n}{2}$$

$$\sum_{k=1}^{n} k^2 = 1^2 + 2^2 + 3^2 + \cdots + n^2 = \frac{(2n+1)(n+1)n}{6}$$

$$\sum_{k=1}^{n} k^3 = 1^3 + 2^3 + 3^3 + \cdots + n^3 = \frac{(n+1)^2 n^2}{4}$$

$$\sum_{k=1}^{n} k^4 = 1^4 + 2^4 + 3^4 + n^4 = \frac{(3n^2 + 3n - 1)(2n+1)(n+1)n}{30}$$

(b) $1^m + 3^m + 5^m + \cdots + (2n-1)^m = S(n).$

$$\sum_{k=1}^{n} (2k-1) = 1 + 3 + 5 + \cdots + (2n-1) = n^2$$

$$\sum_{k=1}^{n} (2k-1)^2 = 1^2 + 3^2 + 5^2 + \cdots + (2n-1)^2 = \frac{n(4n^2 - 1)}{3}$$

$$\sum_{k=1}^{n} (2k-1)^3 = 1^3 + 3^3 + 5^3 + \cdots + (2n-1)^3 = n^2(2n^2 - 1)$$

$$\sum_{k=1}^{n} (2k-1)^4 = 1^4 + 3^4 + 5^4 + \cdots + (2n-1)^4 = \frac{n(4n^2 - 1)(12n^2 - 7)}{15}$$

(4) Finite Series of Products of Numbers $(n < \infty)$

(a) $(1+a)(1+b) + (2+a)(2+b) + \cdots + (n+a)(n+b) = S(n).$

$$\sum_{k=1}^{n} (k+a)(k+b) = \frac{n(n+1)(3a + 3b + 2n + 1)}{6} + abn$$

where a, b are integers or fractions.

examples:

$a = 0, b = 1$: $\quad 1(2) + 2(3) + 3(4) + \cdots + n(n+1) = \dfrac{n(n+1)(n+2)}{3}$

$a = 0, b = 2$: $\quad 1(3) + 2(4) + 3(5) + \cdots + n(n+2) = \dfrac{n(n+1)(2n+7)}{6}$

$a = 1, b = 3$: $\quad 2(4) + 3(5) + 4(6) + \cdots + (n+1)(n+3) = \dfrac{n(n+1)(2n+13)}{6} + 3n$

$a = \frac{1}{3}, b = \frac{1}{3}$: $\quad \frac{4}{3}(\frac{4}{3}) + \frac{7}{3}(\frac{7}{3}) + \frac{10}{3}(\frac{10}{3}) + \cdots + (n + \frac{1}{3})(n + \frac{1}{3}) = \dfrac{n(n+1)(2n+3)}{6} + \dfrac{n}{9}$

(b) $(1+a)(1+b)(1+c)+(2+a)(2+b)(2+c)+\cdots+(n+a)(n+b)(n+c) = S(n).$

$$\sum_{k=1}^{n}(k+a)(k+b)(k+c) = \frac{n(n+1)[3n^2+(3+4d)n+2d+6e]}{12}+abcn$$

where $d = a+b+c$, $e = ab+bc+ca$, and a, b, c are integers or fractions.

examples:

$a = 0,\ b = 1,\ c = 2$:

$$1(2)(3)+2(3)(4)+\cdots+n(n+1)(n+2) = \frac{n(n+1)(n+2)(n+3)}{4}$$

$a = 0,\ b = -1,\ c = -1$:

$$2(1^2)+3(2^2)+\cdots+n(n-1)^2 = \frac{n(n+1)(3n^2-5n+2)}{12}$$

(5) Finite Trigonometric Series ($n < \infty$)

(a) Sine series.

$$\sin\alpha+\sin 2\alpha+\sin 3\alpha+\cdots+\sin n\alpha = \frac{\sin(n\alpha/2)\sin[(n+1)\alpha/2]}{\sin(\alpha/2)}$$

$$\sin\alpha+\sin 3\alpha+\sin 5\alpha+\cdots+\sin(2n-1)\alpha = \frac{\sin^2 n\alpha}{\sin\alpha}$$

$$\sin\alpha+2\sin 2\alpha+3\sin 3\alpha+\cdots+n\sin n\alpha = \frac{\sin(n+1)\alpha}{4\sin^2(\alpha/2)}-\frac{(n+1)\cos[(2n+1)\alpha/2]}{2\sin(\alpha/2)}$$

$$\sin^2\alpha+\sin^2 2\alpha+\sin^2 3\alpha+\cdots+\sin^2 n\alpha = \frac{n}{2}-\frac{\sin n\alpha\cos(n+1)\alpha}{2\sin\alpha}$$

where α is a positive or negative integer or fraction in radians.

(b) Cosine series.

$$\cos\alpha+\cos 2\alpha+\cos 3\alpha+\cdots+\cos n\alpha = \frac{\sin(n\alpha/2)\cos[(n+1)\alpha/2]}{\sin(\alpha/2)}$$

$$\cos\alpha+\cos 3\alpha+\cos 5\alpha+\cdots+\cos(2n-1)\alpha = \frac{\sin 2n\alpha}{\sin\alpha}$$

$$\cos\alpha+2\cos 2\alpha+3\cos 3\alpha+\cdots+n\cos n\alpha = \frac{\cos(n+1)\alpha-1}{4\sin^2(\alpha/2)}+\frac{(n+1)\sin[(2n+1)\alpha/2]}{2\sin(\alpha/2)}$$

$$\cos^2\alpha+\cos^2 2\alpha+\cos^2 3\alpha+\cdots+\cos^2 n\alpha = \frac{n}{2}+\frac{\sin n\alpha\cos(n+1)\alpha}{2\sin\alpha}$$

where α is a positive or negative integer or fraction in radians.

8.03 INFINITE SERIES OF CONSTANT TERMS

(1) Infinite Series of Fractions ($n = \infty$)

(a) Series converging to $a^2/(a^2+1)$, ($a \neq 0, 1$).

$$1-\frac{1}{2^2}+\frac{1}{2^4}-\frac{1}{2^6}+\cdots=\frac{4}{5} \qquad a=2$$

$$1-\frac{1}{3^2}+\frac{1}{3^4}-\frac{1}{3^6}+\cdots=\frac{9}{10} \qquad a=3$$

$$\cdots\cdots\cdots\cdots\cdots\cdots\cdots\cdots$$

$$1-\frac{1}{a^2}+\frac{1}{a^4}-\frac{1}{a^6}+\cdots=\frac{a^2}{a^2+1} \qquad a=a$$

where $a = 2, 3, 4, \ldots$.

(b) Factorial series.

$$1+\frac{1}{1!}+\frac{1}{2!}+\frac{1}{3!}+\cdots = 2.71828 = e = \text{base of natural logarithm}$$

$$1-\frac{1}{1!}+\frac{1}{2!}-\frac{1}{3!}+\cdots = 0.36788 = \frac{1}{e}$$

$$\frac{1}{2!}+\frac{2}{3!}+\frac{3}{4!}+\frac{4}{5!}+\cdots = 1$$

(c) Series converging to $1, \frac{3}{4}, \frac{1}{2}, \frac{1}{4}$, and $1/(m-1)(m-1)!$

$$\frac{1}{(1)(2)}+\frac{1}{(2)(3)}+\frac{1}{(3)(4)}+\cdots=1 \qquad \frac{1}{(1)(3)}+\frac{1}{(3)(5)}+\frac{1}{(5)(7)}+\cdots=\frac{1}{2}$$

$$\frac{1}{(1)(2)(3)}+\frac{1}{(2)(3)(4)}+\frac{1}{(3)(4)(5)}+\cdots=\frac{1}{4} \qquad \frac{1}{(1)(3)}+\frac{1}{(2)(4)}+\frac{1}{(3)(5)}+\cdots=\frac{3}{4}$$

$$\frac{1}{(1)(2)\cdots m}+\frac{1}{(2)(3)\cdots(m+1)}+\frac{1}{(3)(4)\cdots(m+2)}+\cdots=\frac{1}{(m-1)(m-1)!}$$

(d) Series of powers of fractions.

$$\frac{1}{1^m}+\frac{1}{2^m}+\frac{1}{3^m}+\frac{1}{4^m}+\cdots=\alpha(m) \qquad \frac{1}{1^m}+\frac{1}{3^m}+\frac{1}{5^m}+\frac{1}{7^m}+\cdots=\gamma(m)$$

$$\frac{1}{1^m}-\frac{1}{2^m}+\frac{1}{3^m}-\frac{1}{4^m}+\cdots=\beta(m) \qquad \frac{1}{1^m}-\frac{1}{3^m}+\frac{1}{5^m}-\frac{1}{7^m}+\cdots=\delta(m)$$

The values of $\alpha(m)$, $\beta(m)$, $\gamma(m)$, $\delta(m)$ for $m = 1, 2, 3, 4, 5$ are given below.

m*	$\alpha(m)$	$\beta(m)$	$\gamma(m)$	$\delta(m)$
1	∞	0.69314...	∞	0.78539...
2	1.64493...	0.82246...	1.23370...	0.91596...
3	1.20206...	0.90154...	1.05179...	0.96894...
4	1.08232...	0.94703...	1.01467...	0.98894...
5	1.03693...	0.97211...	1.00144...	0.99615...

*In technical calculations for $m > 5$, $\alpha(m) \simeq \beta(m) \simeq \gamma(m) \simeq \delta(m) \simeq 1$.

(2) Infinite Trigonometric Series ($n = \infty$)

(a) Sine series ($a \neq 0$).

$$a \sin \alpha + a^2 \sin 2\alpha + a^3 \sin 3\alpha + \cdots = \frac{a \sin \alpha}{1 + a^2 - 2a \cos \alpha} \qquad a^2 < 1$$

$$\sin \alpha + \frac{\sin 2\alpha}{2} + \frac{\sin 3\alpha}{3} + \cdots = \frac{\pi - \alpha}{2} \qquad 0 < \alpha < 2\pi$$

where α is in radians.

(b) Cosine series ($a \neq 0$).

$$a \cos \alpha + a^2 \cos 2\alpha + a^3 \cos 3\alpha + \cdots = \frac{1 - a \cos \alpha}{1 + a^2 - 2a \cos \alpha} \qquad a^2 < 1$$

$$\cos + \frac{\cos 2\alpha}{2} + \frac{\cos 3\alpha}{3} + \cdots = \frac{1}{2} \ln \frac{1}{2(1 - \cos \alpha)} \qquad 0 < \alpha < 2\pi$$

where α is in radians.

8.04 BINOMIAL SERIES

(1) General Case

(a) Definition. The nth power of $(1 + x)$ can be expanded by the Newton's formula (Sec. 2.19) in a power series in x, called *binomial series.*

$$(1 \pm x)^n = 1 \pm \binom{n}{1} x + \binom{n}{2} x^2 \pm \binom{n}{3} x^3 + \cdots$$

$$= 1 \pm nx + \frac{n(n-1)}{2!} x^2 \pm \frac{n(n-1)(n-2)}{3!} x^3 + \cdots$$

where $\binom{n}{k} = \dfrac{n!}{(n-k)!k!}$ and n is a positive or negative integer or a positive or negative fraction (see Sec. 8.04–2 and Table A.43).

(b) Classification. If:

(α) $n = 0, 1, 2, 3, \ldots$, then the series consists of $(n + 1)$ terms (finite series) (Sec. 2.10–3*a*).

(β) $n \neq 0, 1, 2, 3, \ldots$ and $x^2 < 1$, then the series is convergent.

(γ) $n \neq 0, 1, 2, 3, \ldots$ and $x^2 > 1$, then the series is divergent.

(c) Transformation.

$$[a \pm f(x)]^n = a^n (1 \pm u)^n = a^n \left[1 \pm \binom{n}{1} u + \binom{n}{2} u^2 \pm \binom{n}{3} u^3 + \cdots \right]$$

where $u = f(x)/a$ is a function in x.

(2) Special Cases $[u = u(x),\ u^2 < 1]$.

(a) Exponent $n = 1/m\ (m = 2, 3, 4, \ldots)$.

$$\sqrt[m]{1 \pm u} = 1 \pm \frac{u}{m} - \frac{m-1}{2!}\left(\frac{u}{m}\right)^2 \pm \frac{(m-1)(2m-1)}{3!}\left(\frac{u}{m}\right)^3 - \frac{(m-1)(2m-1)(3m-1)}{4!}\left(\frac{u}{m}\right)^4 \pm \cdots$$

$$\sqrt{1 \pm u} = 1 \pm \frac{u}{2} - \frac{1}{2}\left(\frac{u}{2}\right)^2 \pm \frac{(1)(3)}{(2)(3)}\left(\frac{u}{2}\right)^3 - \frac{(1)(3)(5)}{(2)(3)(4)}\left(\frac{u}{2}\right)^4 \pm \cdots$$

$$\sqrt[3]{1 \pm u} = 1 \pm \frac{u}{3} - \frac{2}{2}\left(\frac{u}{3}\right)^2 \pm \frac{(2)(5)}{(2)(3)}\left(\frac{u}{3}\right)^3 - \frac{(2)(5)(8)}{(2)(3)(4)}\left(\frac{u}{3}\right)^4 \pm \cdots$$

$$\sqrt[4]{1 \pm u} = 1 \pm \frac{u}{4} - \frac{3}{2}\left(\frac{u}{4}\right)^2 \pm \frac{(3)(7)}{(2)(3)}\left(\frac{u}{4}\right)^3 - \frac{(3)(7)(11)}{(2)(3)(4)}\left(\frac{u}{4}\right)^4 \pm \cdots$$

$$\sqrt[5]{1 \pm u} = 1 \pm \frac{u}{5} - \frac{4}{2}\left(\frac{u}{5}\right)^2 \pm \frac{(4)(9)}{(2)(3)}\left(\frac{u}{5}\right)^3 - \frac{(4)(9)(14)}{(2)(3)(4)}\left(\frac{u}{5}\right)^4 \pm \cdots$$

(b) Exponent $n = -m\ (m = 1, 2, 3, \ldots)$.

$$\frac{1}{(1 \pm u)^m} = 1 \mp mu + \frac{m(m+1)}{2!}u^2 \mp \frac{m(m+1)(m+2)}{3!}u^3 + \frac{m(m+1)(m+2)(m+3)}{4!}u^4 \mp \cdots$$

$$\frac{1}{1 \pm u} = 1 \mp u + u^2 \mp u^3 + u^4 \mp \cdots$$

$$\frac{1}{(1 \pm u)^2} = 1 \mp 2u + 3u^2 \mp 4u^3 + 5u^4 \mp \cdots$$

$$\frac{1}{(1 \pm u)^3} = 1 \mp 3u + 6u^2 \mp 10u^3 + 15u^4 \mp \cdots$$

$$\frac{1}{(1 \pm u)^4} = 1 \mp 4u + 10u^2 \mp 20u^3 + 35u^4 \mp \cdots$$

$$\frac{1}{(1 + u)^5} = 1 \mp 5u + 15u^2 \mp 35u^3 + 70u^4 \mp \cdots$$

(c) Exponent $n = -1/m\ (m = 2, 3, 4, \ldots)$.

$$\frac{1}{\sqrt[m]{1 \pm u}} = 1 \mp \frac{u}{m} + \frac{m+1}{2!}\left(\frac{u}{m}\right)^2 \mp \frac{(m+1)(2m+1)}{3!}\left(\frac{u}{m}\right)^3 + \frac{(m+1)(2m+1)(3m+1)}{4!}\left(\frac{u}{m}\right)^4 \mp \cdots$$

$$\frac{1}{\sqrt{1 \pm u}} = 1 \mp \frac{u}{2} + \frac{3}{2}\left(\frac{u}{2}\right)^2 \mp \frac{(3)(5)}{(2)(3)}\left(\frac{u}{2}\right)^3 + \frac{(3)(5)(7)}{(2)(3)(4)}\left(\frac{u}{2}\right)^4 \mp \cdots$$

$$\frac{1}{\sqrt[3]{1 \pm u}} = 1 \mp \frac{u}{3} + \frac{4}{2}\left(\frac{u}{3}\right)^2 \mp \frac{(4)(7)}{(2)(3)}\left(\frac{u}{3}\right)^3 + \frac{(4)(7)(10)}{(2)(3)(4)}\left(\frac{u}{3}\right)^4 \mp \cdots$$

$$\frac{1}{\sqrt[4]{1 \pm u}} = 1 \mp \frac{u}{4} + \frac{5}{2}\left(\frac{u}{4}\right)^2 \mp \frac{(5)(9)}{(2)(3)}\left(\frac{u}{4}\right)^3 + \frac{(5)(9)(13)}{(2)(3)(4)}\left(\frac{u}{4}\right)^4 \mp \cdots$$

$$\frac{1}{\sqrt[5]{1 \pm u}} = 1 \mp \frac{u}{5} + \frac{6}{2}\left(\frac{u}{5}\right)^2 \mp \frac{(6)(11)}{(2)(3)}\left(\frac{u}{5}\right)^3 + \frac{(6)(11)(16)}{(2)(3)(4)}\left(\frac{u}{5}\right)^4 \mp \cdots$$

(d) Exponent $n = \alpha/\beta$ ($\alpha = 1, 2, 3, \ldots; \beta = 2, 3, 4, \ldots$).

$$\sqrt[\beta]{(1 \pm u)^{\alpha}} = 1 \pm \alpha\left(\frac{u}{\beta}\right) - \frac{\alpha(\beta-\alpha)}{2!}\left(\frac{u}{\beta}\right)^2 \pm \frac{\alpha(\beta-\alpha)(2\beta-\alpha)}{3!}\left(\frac{u}{\beta}\right)^3 - \frac{\alpha(\beta-\alpha)(2\beta-\alpha)(3\beta-\alpha)}{4!}\left(\frac{u}{\beta}\right)^4 \pm \cdots$$

$$\sqrt[3]{(1 \pm u)^2} = 1 \pm 2\left(\frac{u}{3}\right) - \frac{2}{2}\left(\frac{u}{3}\right)^2 \pm \frac{(2)(4)}{(2)(3)}\left(\frac{u}{3}\right)^3 - \frac{(2)(4)(7)}{(2)(3)(4)}\left(\frac{u}{3}\right)^4 \pm \cdots$$

$$\sqrt[4]{(1 \pm u)^3} = 1 \pm 3\left(\frac{u}{4}\right) - \frac{3}{2}\left(\frac{u}{4}\right)^2 \pm \frac{(3)(5)}{(2)(3)}\left(\frac{u}{4}\right)^3 - \frac{(3)(5)(9)}{(2)(3)(4)}\left(\frac{u}{4}\right)^4 \pm \cdots$$

(e) Exponent $n = -\alpha/\beta$ ($\alpha = 1, 2, 3, \ldots; \beta = 2, 3, 4, \ldots$).

$$\frac{1}{\sqrt[\beta]{(1 \pm u)^{\alpha}}} = 1 \mp \alpha\left(\frac{u}{\beta}\right) + \frac{\alpha(\alpha+\beta)}{2!}\left(\frac{u}{\beta}\right)^2 \mp \frac{\alpha(\alpha+\beta)(\alpha+2\beta)}{3!}\left(\frac{u}{\beta}\right)^3 + \frac{\alpha(\alpha+\beta)(\alpha+2\beta)(\alpha+3\beta)}{4!}\left(\frac{u}{\beta}\right)^4 \mp \cdots$$

$$\frac{1}{\sqrt[3]{(1 \pm u)^2}} = 1 \mp 2\left(\frac{u}{3}\right) + \frac{2(5)}{2}\left(\frac{u}{3}\right)^2 \mp \frac{(2)(5)(8)}{(2)(3)}\left(\frac{u}{3}\right)^3 + \frac{(2)(5)(8)(11)}{(2)(3)(4)}\left(\frac{u}{3}\right)^4 \mp \cdots$$

$$\frac{1}{\sqrt[4]{(1 \pm u)^3}} = 1 \mp 3\left(\frac{u}{4}\right) + \frac{(3)(7)}{2}\left(\frac{u}{4}\right)^2 \mp \frac{(3)(7)(11)}{(2)(3)}\left(\frac{u}{4}\right)^3 + \frac{(3)(7)(11)(15)}{(2)(3)(4)}\left(\frac{u}{4}\right)^4 \mp \cdots$$

8.05 SERIES OF FUNCTIONS

(1) Definitions

(a) Sequence of functions in x is a set of n functions, $f_1(x), f_2(x), \ldots, f_n(x)$, arranged in a prescribed order and formed according to a definite rule. The sequence is finite if $n < \infty$ or infinite if $n = \infty$.

(b) Finite series of functions is the sum of a finite sequence.

$$\sum_{k=1}^{n} f_k(x) = f_1(x) + f_2(x) + \cdots + f_n(x) = S_n(x)$$

where $f_k(x)$ is the kth function, $n < \infty$, and $S_n(x)$ is the sum.

example:

$\sin x + \sin^2 3x + \sin 5x = (\sin^2 3x)/(\sin x)$ (Sec. 8.02–5*a*), where $n = 3$ and $S_3(x) = (\sin^2 3x)/(\sin x)$.

(c) Infinite series of functions is the sum of an infinite sequence.

$$\sum_{k=1}^{\infty} f_k(x) = f_1(x) + f_2(x) + \cdots = S(x)$$

where $n = \infty$ and $S(x)$ is the sum.

example:

$\sin x + (\sin 2x)/2 + (\sin 3x)/3 + \cdots = (\pi - x)/2$ if $0 < x < 2\pi$ (Sec. 8.03–2*a*) where $n = \infty$ and $S(x) = (\pi - x)/2$.

(d) Generating function of a series of functions defines the law of formation of the series.

examples ($k = 1, 2, 3, \ldots$):

$f_k(x) = kx^k$ $\qquad x + 2x^2 + 3x^3 + \cdots$

$f_k(x) = \dfrac{x^k}{(-2)^k}$ $\qquad -\dfrac{x}{2} + \dfrac{x^2}{4} - \dfrac{x^3}{8} + \cdots$

$f_k(x) = 3^k \sin^k x$ $\qquad 3 \sin x + 9 \sin^2 x + 27 \sin^3 x + \cdots$

(e) Region of convergence of a series of functions is the set of all values of the argument x for which this series converges; i.e., if

$$S_n(x) = \sum_{k=1}^{n} f_k(x) \qquad \text{and} \qquad \lim_{n\to\infty} S_n(x) = S(x)$$

exists for $a < x < b$, then the series is said to be convergent at all points of the interval $a < x < b$.

example:

Interval of convergence of the series of example (c) is $0 < x < 2\pi$. At the ends of this interval ($x = 0$, $x = 2\pi$) the series is zero and for $x < 0$ and $x > 2\pi$ the series diverges, $S(x < 0) = \infty$, $S(x > 2\pi) = \infty$.

(f) Infinite power series in $(x - a)$ is a special and most important case of series of functions defined as

$$\sum_{k=0}^{\infty} c_k (x - a)^k = c_0 + c_1(x - a) + c_2(x - a)^2 + \cdots = S(x)$$

where $a, c_0, c_1, c_2, \ldots$ are constants, $n = \infty$, and $S(x)$ is the sum. If $a = 0$, the series is called a power series in x.

example:

If $f_k(x) = -(-x)^k/k$ and $k = 1, 2, 3, \ldots$, then

$$x - \frac{x^2}{2} + \frac{x^3}{3} - \frac{x^4}{4} + \cdots$$

is a power series in x.

(2) Operations with Power Series

(a) Uniqueness theorem. If two power series

$$\sum_{k=0}^{\infty} a_k x^k = S(x) \qquad \text{and} \qquad \sum_{k=0}^{\infty} b_k x^k = S(x)$$

converge to the same sum $S(x)$ for all real values of x, then

$$a_0 = b_0, \quad a_1 = b_1, \quad a_2 = b_2, \quad \ldots$$

(b) Summation theorem. Two power series in x can be added or subtracted term by term for each value of x common to their interval of convergence.

$$\sum_{k=0}^{\infty} a_k x^k + \sum_{k=0}^{\infty} b_k x^k = \sum_{k=0}^{\infty} (a_k + b_k) x^k$$

Their sum is a new power series which converges at least in the common interval of convergence of the two series.

example:

Since

$$1 + \frac{x}{1!} + \frac{x^2}{2!} + \frac{x^3}{3!} + \cdots = e^x \qquad \text{and} \qquad 1 + \frac{x^2}{2!} + \frac{x^4}{4!} + \frac{x^6}{6!} + \cdots = \cosh x$$

are convergent for all $|x| < \infty$, then their sum

$$2 + \frac{x}{1!} + \frac{2x^2}{2!} + \frac{x^3}{3!} + \frac{2x^4}{4!} + \cdots = e^x + \cosh x$$

is also convergent for all $|x| < \infty$.

(c) Multiplication theorem. Two power series in x can be multiplied term by term for each value of x common to their interval of convergence.

$$\left(\sum_{k=0}^{\infty} a_k x^k\right)\left(\sum_{k=0}^{\infty} b_k x^k\right) = a_0 b_0 (1 + A_1 x + A_2 x^2 + \cdots)(1 + B_1 x + B_2 x^2 + \cdots)$$

$$= a_0 b_0 (1 + C_1 x + C_2 x^2 + C_3 x^3 + C_4 x^4 + \cdots)$$

where $C_1 = A_1 + B_1 \qquad A_1 = \dfrac{a_1}{a_0} \qquad B_1 = \dfrac{b_1}{b_0}$

$C_2 = A_2 + A_1 B_1 + B_2 \qquad A_2 = \dfrac{a_2}{a_0} \qquad B_2 = \dfrac{b_2}{b_0}$

$C_3 = A_3 + A_2 B_1 + A_1 B_2 + B_3 \qquad A_3 = \dfrac{a_3}{a_0} \qquad B_3 = \dfrac{b_3}{b_0}$

$C_4 = A_4 + A_3 B_1 + A_2 B_2 + A_1 B_3 + B_4 \qquad A_4 = \dfrac{a_4}{a_0} \qquad B_4 = \dfrac{b_4}{b_0}$

. . . .

The product is a new series which converges at least in the common interval of convergence of the two series.

example:

Since

$$\frac{x}{1!} - \frac{x^3}{3!} + \frac{x^5}{5!} - \frac{x^7}{7!} + \cdots = \sin x \qquad \text{and} \qquad \frac{x}{1!} + \frac{x^3}{3!} + \frac{x^5}{5!} + \frac{x^7}{7!} + \cdots = \sinh x$$

are convergent for all $|x| < \infty$, then their product

$$\frac{x^2}{2!}2 - \frac{x^6}{6!}2^3 + \frac{x^{10}}{10!}2^5 - \frac{x^{14}}{14!}2^7 + \cdots = \sin x \sinh x$$

is also convergent for all $|x| < \infty$.

(d) Division theorem. The quotient of two power series in x is

$$\frac{\sum_{k=0}^{\infty} a_k x^k}{\sum_{k=0}^{\infty} b_k x^k} = \frac{a_0\,(1 + A_1 x + A_2 x^2 + \cdots)}{b_0\,(1 + B_1 x + B_2 x^2 + \cdots)} = \frac{a_0}{b_0}(1 + C_1 x + C_2 x^2 + C_3 x^3 + C_4 x^4 + \cdots)$$

where $C_1 = A_1 - B_1$

$$C_2 = A_2 - (B_2 + B_1 C_1)$$

$$C_3 = A_3 - (B_3 + B_2 C_1 + B_1 C_2)$$

$$C_4 = A_4 - (B_4 + B_3 C_1 + B_2 C_2 + B_1 C_3)$$

....

The interval of convergence of the quotient series cannot be (in general) determined from the intervals of convergence of the two series.

example:

$$\frac{x}{1!} - \frac{x^3}{3!} + \frac{x^5}{5!} - \frac{x^7}{7!} + \cdots = \sin x \qquad \text{and} \qquad 1 - \frac{x^2}{2!} + \frac{x^4}{4!} - \frac{x^6}{6!} + \cdots = \cos x$$

are convergent for all $|x| < \infty$. Their quotient

$$\frac{x}{1} + \frac{x^3}{3} + \frac{2x^5}{15} + \frac{17x^7}{315} + \frac{62x^9}{2{,}835} + \cdots = \frac{\sin x}{\cos x} = \tan x$$

is not convergent for all $|x| < \infty$ but only for all $|x| < \pi/2$.

(e) Power theorem. The mth power of a power series in x [obtained by theorem (c)] is

$$\left(\sum_{k=0}^{\infty} a_k x^k\right)^m = a_0^m (1 + A_1 x + A_2 x^2 + \cdots)^m = a_0^m (1 + C_1 x + C_2 x^2 + C_3 x^3 + C_4 x^4 + \cdots)$$

where $C_1 = \binom{n}{1} A_1$

$$C_2 = \binom{n}{1} A_2 + \binom{n}{2} A_1^2$$

$$C_3 = \binom{n}{1} A_3 + 2\binom{n}{2} A_2 A_1 + \binom{n}{3} A^3$$

$$C_4 = \binom{n}{1} A_4 + \binom{n}{2}(2A_3 A_1 + A_2^2) + 3\binom{n}{3} A_2 A_1^2 + \binom{n}{4} A_1^4$$

....

The resulting power series is convergent in the interval of convergence of the original series and $n = m$.

example:

For $m = 2$

$C_1 = 2A_1 \qquad C_2 = (A_1^2 + 2A_2) \qquad C_3 = 2(A_1 A_2 + A_3) \qquad C_4 = (A_2^2 + 2A_1 A_3 + 2A_4)$

(f) Reversion theorem. *If* $y = x + A_2x^2 + A_3x^3 + A_4x^4 + \cdots$, then the reversed series is

$$x = y + C_2y^2 + C_3y^3 + C_4y^4 + \cdots$$

where $C_2 = -A_2 \qquad C_3 = -(A_3 - 2A_2^2)$

$$C_4 = -(A_4 - 5A_3A_2 + 5A_2^3) \qquad C_5 = -(A_5 - 6A_4A_2 - 3A_3^2 + 21A_3A_2^2 - 14A_2^4)$$

$$C_6 = -(A_6 - 7A_5A_2 + 28A_4A_2^2 + 28A_3^2A_2 - 84A_3A_2^2 + 42A_2^5)$$

$$\cdots\cdots\cdots\cdots\cdots\cdots\cdots\cdots\cdots\cdots\cdots\cdots$$

The interval of convergence of the reversed series cannot be determined from the interval of convergence of the given series.

example:

If $\qquad y = x - \dfrac{x^3}{3!} + \dfrac{x^5}{5!} - \dfrac{x^7}{7!} + \dfrac{x^9}{9!} - \cdots = \sin x$

then $\qquad x = y + \dfrac{1}{2}\dfrac{y^3}{3} + \dfrac{1(3)}{2(4)}\dfrac{y^5}{5} + \dfrac{1(3)(5)}{2(4)(6)}\dfrac{y^7}{7} - \dfrac{1(3)(5)(7)}{2(4)(6)(8)}\dfrac{y^9}{9} + \cdots = \sin^{-1} y$

where $\qquad A_2 = A_4 = A_6 = \cdots = 0$

and $\qquad C_2 = 0, \quad C_3 = -A_3, \quad C_4 = 0, \quad C_5 = -A_5 + 3A_3^2, \quad \ldots$

8.06 REPRESENTATION OF FUNCTIONS BY SERIES

(1) General Expansions

(a) Taylor's series about $x = a$. If $f(x)$ and its derivatives $f'(x), f''(x), \ldots, f^{(n)}(x)$ exist and are continuous in the closed interval $a \le x \le b$ and $f^{(n+1)}(x)$ exists in the open interval $a < x < b$, then

$$f(x) = f(a) + \frac{f'(a)}{1!}(x-a) + \frac{f''(a)}{2!}(x-a)^2 + \cdots + \frac{f^{(n)}(a)}{n!}(x-a)^n + R_n(x-a)$$

where $R_n(x-a) = \dfrac{f^{(n+1)}(\theta x)}{(n+1)!}(x-a)^n \qquad a < \theta x < x$

is the remainder of the Taylor's series expansion of $f(x)$.

(b) MacLaurin's series about $a = 0$. A special case of Taylor's series expansion of $f(x)$ for $a = 0$ is

$$f(x) = f(0) + \frac{f'(0)}{1!}x + \frac{f''(0)}{2!}x^2 + \cdots + \frac{f^{(n)}(0)}{n!}x^n + R_n(x)$$

where $R_n(x) = \dfrac{f^{(n+1)}(\theta x)}{(n+1)!}x^{n+1} \qquad 0 < \theta < 1$

is the remainder of the MacLaurin's series expansion of $f(x)$.

(c) Approximate values of functions (algebraic and transcendent) are calculated by means of series (a) or (b). The error involved in replacing the given function by the partial sum of the respective series is the remainder R_n. The series represents $f(x)$ for those values of x for which $R_n \to 0$ as $n \to \infty$.

(2) Particular Cases

(a) Exponential functions $(a > 0)$.

$$e = \sum_{k=0}^{\infty} \frac{1}{k!} = 1 + \frac{1}{1!} + \frac{1}{2!} + \frac{1}{3!} + \cdots = 2.71828\ldots$$

$$\frac{1}{e} = \sum_{k=0}^{\infty} \frac{(-1)^k}{k!} = 1 - \frac{1}{1!} + \frac{1}{2!} - \frac{1}{3!} + \cdots = 0.36787\ldots$$

$$e^x = \sum_{k=0}^{\infty} \frac{x^k}{k!} = 1 + \frac{x}{1!} + \frac{x^2}{2!} + \frac{x^3}{3!} + \cdots \qquad -\infty < x < \infty$$

$$e^{-x} = \sum_{k=0}^{\infty} \frac{(-x)^k}{k!} = 1 - \frac{x}{1!} + \frac{x^2}{2!} - \frac{x^3}{3!} + \cdots \qquad -\infty < x < \infty$$

$$a = \sum_{k=0}^{\infty} \frac{(\ln a)^k}{k!} = 1 + \frac{\ln a}{1!} + \frac{(\ln a)^2}{2!} + \frac{(\ln a)^3}{3!} + \cdots$$

$$\frac{1}{a} = \sum_{k=0}^{\infty} \frac{(-\ln a)^k}{k!} = 1 - \frac{\ln a}{1!} + \frac{(\ln a)^2}{2!} - \frac{(\ln a)^3}{3!} + \cdots$$

$$a^x = \sum_{k=0}^{\infty} \frac{(x \ln a)^k}{k!} = 1 + \frac{x \ln a}{1!} + \frac{(x \ln a)^2}{2!} + \frac{(x \ln a)^3}{3!} + \cdots \qquad -\infty < x < \infty$$

$$a^{-x} = \sum_{k=0}^{\infty} \frac{(-x \ln a)^k}{k!} = 1 - \frac{x \ln a}{1!} + \frac{(x \ln a)^2}{2!} - \frac{(x \ln a)^3}{3!} + \cdots \qquad -\infty < x < \infty$$

(b) Logarithmic functions $(a > 0)$.

$$\ln 2 = \sum_{k=1}^{\infty} \frac{1}{k2^k} = \frac{1}{2} + \frac{1}{2(2^2)} + \frac{1}{3(2^3)} + \frac{1}{4(2^4)} + \cdots = 0.69314\ldots$$

$$\ln x = 2 \sum_{k=1}^{\infty} \frac{1}{2k-1} \left(\frac{x-1}{x+1}\right)^{2k-1} = 2\left[\frac{x-1}{x+1} + \frac{1}{3}\left(\frac{x-1}{x+1}\right)^3 + \frac{1}{5}\left(\frac{x-1}{x-1}\right)^5 + \cdots\right] \qquad x > 0$$

$$\ln x = \sum_{k=1}^{\infty} (-1)^{k+1} \frac{(x-1)^k}{k} = (x-1) - \frac{(x-1)^2}{2} + \frac{(x-1)^3}{3} - \cdots \qquad 0 < x \le 2$$

$$\ln x = \sum_{k=1}^{\infty} \frac{1}{k}\left(\frac{x-1}{x}\right)^k = \frac{x-1}{x} + \frac{1}{2}\left(\frac{x-1}{x}\right)^2 + \frac{1}{3}\left(\frac{x-1}{x}\right)^3 + \cdots \qquad x \ge \tfrac{1}{2}$$

$$\ln(1+x) = \sum_{k=1}^{\infty} \frac{(-x)^k}{-k} = \frac{x}{1} - \frac{x^2}{2} + \frac{x^3}{3} - \frac{x^4}{4} + \cdots \qquad -1 < x \le 1$$

$$\ln(1-x) = \sum_{k=1}^{\infty} \left(-\frac{x^k}{k}\right) = -\frac{x}{1} - \frac{x^2}{2} - \frac{x^3}{3} - \frac{x^4}{4} - \cdots \qquad -1 \le x < 1$$

$$\ln \frac{x+1}{x-1} = \sum_{k=1}^{\infty} \frac{2}{(2k-1)x^{2k-1}} = 2\left(\frac{1}{x} + \frac{1}{3x^3} + \frac{1}{5x^5} + \cdots\right) = 2 \coth^{-1} x \qquad |x| > 1$$

$$\ln \frac{1+x}{1-x} = \sum_{k=1}^{\infty} \frac{2x^{2k-1}}{2k-1} = 2\left(\frac{x}{1} + \frac{x^3}{3} + \frac{x^5}{5} + \cdots\right) = 2 \tanh^{-1} x \qquad |x| < 1$$

(c) Trigonometric functions.

$$\sin x = \sum_{k=0}^{\infty} (-1)^k \frac{x^{2k+1}}{(2k+1)!} = x - \frac{x^3}{3!} + \frac{x^5}{5!} - \frac{x^7}{7!} + \cdots \qquad -\infty < x < \infty$$

$$\cos x = \sum_{k=0}^{\infty} (-1)^k \frac{x^{2k}}{(2k)!} = 1 - \frac{x^2}{2!} + \frac{x^4}{4!} - \frac{x^6}{6!} + \cdots \qquad -\infty < x < \infty$$

$$\tan x = \sum_{k=1}^{\infty} \frac{2^{2k}(2^{2k}-1)B_k x^{2k-1}}{(2k)!} = x + \frac{x^3}{3} + \frac{2x^5}{15} + \frac{17x^7}{315} + \frac{62x^9}{2{,}835} + \cdots \qquad |x| < \frac{\pi}{2}$$

$$\cot x = \frac{1}{x} - \sum_{k=1}^{\infty} \frac{2^{2k} B_k x^{2k-1}}{(2k)!} = \frac{1}{x} - \left(\frac{x}{3} + \frac{x^3}{45} + \frac{2x^5}{945} + \frac{x^7}{4{,}725} + \cdots\right) \qquad 0 < |x| < \pi$$

$$\sec x = \sum_{k=0}^{\infty} \frac{E_k x^{2k}}{(2k)!} = 1 + \frac{x^2}{2} + \frac{5x^4}{24} + \frac{61x^6}{720} + \frac{277x^8}{8{,}064} + \cdots \qquad |x| < \frac{\pi}{2}$$

$$\csc x = \frac{1}{x} + \sum_{k=1}^{\infty} \frac{2(2^{2k-1}-1)B_k x^{2k-1}}{(2k)!}$$

$$= \frac{1}{x} + \frac{x}{6} + \frac{7x^3}{360} + \frac{31x^5}{15{,}120} + \frac{127x^7}{604{,}800} + \cdots \qquad 0 < |x| < \frac{\pi}{2}$$

$$\sin(\alpha + x) = \sin\alpha + \frac{x}{1!}\cos\alpha - \frac{x^2}{2!}\sin\alpha - \frac{x^3}{3!}\cos\alpha + \frac{x^4}{4!}\sin\alpha + \cdots \qquad -\infty < \alpha + x < \infty$$

$$\cos(\alpha + x) = \cos\alpha - \frac{x}{1!}\sin\alpha - \frac{x^2}{2!}\cos\alpha + \frac{x^3}{3!}\sin\alpha + \frac{x^4}{4!}\cos\alpha - \cdots \qquad -\infty < \alpha + x < \infty$$

where B_k, E_k are constants given in Sec. 8.06–2*l* and α is an arbitrary constant.

(d) Powers of trigonometric functions.

$$\sin^2 x = \sum_{k=1}^{\infty} (-1)^{k+1} \frac{2^{2k-1} x^{2k}}{(2k)!} \qquad \cos^2 x = 1 - \sum_{k=1}^{\infty} (-1)^{k+1} \frac{2^{2k-1} x^{2k}}{(2k)!}$$

$$\sin^3 x = \frac{1}{4}\sum_{k=1}^{\infty} (-1)^{k+1} \frac{(3^{2k+1}-3)x^{2k+1}}{(2k+1)!} \qquad \cos^3 x = \frac{1}{4}\sum_{k=0}^{\infty} (-1)^k \frac{(3^{2k}+3)x^{2k}}{(2k)!}$$

(e) Products of exponential and trigonometric functions $(-\infty < x < \infty)$.

$$e^x \sin x = F_1 x + F_2 x^2 + F_3 x^3 - F_5 x^5 - F_6 x^6 - F_7 x^7 + F_9 x^9 + F_{10} x^{10} + F_{11} x^{11} - F_{13} x^{13} - F_{14} x^{14} - F_{15} x^{15} + F_{17} x^{17} + F_{18} x^{18} + F_{19} x^{19} + \cdots$$

$$e^{-x} \sin x = F_1 x - F_2 x^2 + F_3 x^3 - F_5 x^5 + F_6 x^6 - F_7 x^7 + F_9 x^9 - F_{10} x^{10} + F_{11} x^{11} - F_{13} x^{13} + F_{14} x^{14} - F_{15} x^{15} + F_{17} x^{17} - F_{18} x^{18} + F_{19} x^{19} + \cdots$$

$$e^x \cos x = 1 + F_1 x - F_3 x^3 - F_4 x^4 - F_5 x^5 + F_7 x^7 + F_8 x^8 + F_9 x^9 - F_{11} x^{11} - F_{12} x^{12} - F_{13} x^{13} + F_{15} x^{15} + F_{16} x^{16} + F_{17} x^{17} + \cdots$$

$$e^{-x} \cos x = 1 - F_1 x + F_3 x^3 - F_4 x^4 + F_5 x^5 - F_7 x^7 + F_8 x^8 - F_9 x^9 + F_{11} x^{11} - F_{12} x^{12} + F_{13} x^{13} - F_{15} x^{15} + F_{16} x^{16} - F_{17} x^{17} + \cdots$$

where $F_1, F_2, F_3, \ldots$ are constants given in Sec. 8.06–2*l*.

(f) Hyperbolic functions.

$$\sinh x = \sum_{k=0}^{\infty} \frac{x^{2k+1}}{(2k+1)!} = x + \frac{x^3}{3!} + \frac{x^5}{5!} + \frac{x^7}{7!} + \cdots \qquad |x| < \infty$$

$$\cosh x = \sum_{k=1}^{\infty} \frac{x^{2k}}{(2k)!} = 1 + \frac{x^2}{2!} + \frac{x^4}{4!} + \frac{x^6}{6!} + \cdots \qquad |x| < \infty$$

$$\tanh x = \sum_{k=1}^{\infty} \frac{2^{2k}(2^{2k}-1)B_k x^{2k-1}}{(-1)^{k+1}(2k)!} = x - \frac{x^3}{3} + \frac{2x^5}{15} - \frac{17x^7}{315} + \frac{62x^9}{3{,}835} - \cdots \qquad |x| < \frac{\pi}{2}$$

$$\coth x = \frac{1}{x} + \sum_{k=1}^{\infty} \frac{2^{2k} B_k x^{2k-1}}{(-1)^{k-1}(2k)!} = \frac{1}{x} + \frac{x}{3} - \frac{x^3}{45} + \frac{2x^5}{945} - \frac{x^7}{4{,}725} + \cdots \qquad 0 < |x| < \pi$$

$$\operatorname{sech} x = 1 + \sum_{k=1}^{\infty} \frac{E_k x^{2k}}{(-1)^k (2k)!} = 1 - \frac{x^2}{2} + \frac{5x^4}{24} - \frac{61x^6}{720} + \frac{1{,}385x^8}{40{,}320} - \cdots \qquad |x| < \frac{\pi}{2}$$

$$\operatorname{csch} x = \frac{1}{x} + \sum_{k=1}^{\infty} \frac{2(2^{2k-1}-1)B_k x^{2k-1}}{(-1)^k (2k)!} = \frac{1}{x} - \frac{x}{6} + \frac{7x^3}{360} - \frac{31x^5}{15{,}120} + \cdots \qquad 0 < |x| < \pi$$

$$\sinh(a+x) = \sinh a + \frac{x}{1!}\cosh a + \frac{x^2}{2!}\sinh a + \frac{x^3}{3!}\cosh x + \cdots \qquad |x+a| < \infty$$

$$\cosh(a+x) = \cosh a + \frac{x}{1!}\sinh a + \frac{x^2}{2!}\cosh a + \frac{x^3}{3!}\sinh x + \cdots \qquad |x+a| < \infty$$

where B_k, E_k are constants given in Sec. 8.06–2*l*.

(g) Sums of hyperbolic and trigonometric functions.

$$\cosh x + \cos x = 2\left(1 + \frac{x^4}{4!} + \frac{x^8}{8!} + \frac{x^{12}}{12!} + \cdots\right) \qquad |x| < \infty$$

$$\sinh x + \sin x = 2\left(\frac{x}{1!} + \frac{x^5}{5!} + \frac{x^9}{9!} + \frac{x^{13}}{13!} + \cdots\right) \qquad |x| < \infty$$

$$\cosh x - \cos x = 2\left(\frac{x^2}{2!} + \frac{x^6}{6!} + \frac{x^{10}}{10!} + \frac{x^{14}}{14!} + \cdots\right) \qquad |x| < \infty$$

$$\sinh x - \sin x = 2\left(\frac{x^3}{3!} + \frac{x^7}{7!} + \frac{x^{11}}{11!} + \frac{x^{15}}{15!} + \cdots\right) \qquad |x| < \infty$$

(h) Products of hyperbolic and trigonometric functions.

$$\cosh x \cos x = 1 - \frac{x^4}{4!}2^2 + \frac{x^8}{8!}2^4 - \frac{x^{12}}{12!}2^6 + \cdots \qquad |x| < \infty$$

$$\sinh x \sin x = \frac{x^2}{2!}2 - \frac{x^6}{6!}2^3 + \frac{x^{10}}{10!}2^5 - \frac{x^{14}}{14!}2^7 + \cdots \qquad |x| < \infty$$

$$\cosh x \sin x = \frac{x}{1!} + \frac{x^3}{3!}2 - \frac{x^5}{5!}2^2 - \frac{x^7}{7!}2^3 + \cdots \qquad |x| < \infty$$

$$\sinh x \cos x = \frac{x}{1!} - \frac{x^3}{3!}2 - \frac{x^5}{5!}2^2 + \frac{x^7}{7!}2^3 + \cdots \qquad |x| < \infty$$

(i) Sums of products of hyperbolic and trigonometric functions.

$$\cosh x \sin x + \sinh x \cos x = 2\left(\frac{x}{1!} - \frac{x^5}{5!}2^2 + \frac{x^9}{9!}2^4 - \frac{x^{13}}{13!}2^6 + \cdots\right) \qquad |x| < \infty$$

$$\cosh x \sin x - \sinh x \cos x = 2\left(\frac{x^3}{3!}2 - \frac{x^7}{7!}2^3 + \frac{x^{11}}{11!}2^5 - \frac{x^{15}}{15!}2^7 + \cdots\right) \qquad |x| < \infty$$

(j) Inverse trigonometric functions.

$$\sin^{-1} x = \sum_{k=0}^{\infty} \frac{(2k)!x^{2k+1}}{2^{2k}(2k+1)(k!)^2} = x + \frac{(1)(x^3)}{(2)(3)} + \frac{(1)(3)(x^5)}{(2)(4)(5)} + \frac{(1)(3)(5)(x^7)}{(2)(4)(6)(7)} + \cdots \qquad |x| < 1$$

$$\cos^{-1} x = \frac{\pi}{2} - \sin^{-1} x \quad \text{(for } \sin^{-1} x \text{ use the series above)} \qquad |x| < 1$$

$$\tan^{-1} x = \begin{cases} \displaystyle\sum_{k=0}^{\infty} \frac{(-1)^k x^{2k+1}}{2k+1} = x - \frac{x^3}{3} + \frac{x^5}{5} - \frac{x^7}{7} + \cdots & |x| < 1 \\ \displaystyle\pm\frac{\pi}{2} - \sum_{k=0}^{\infty} \frac{(-1)^k}{(2k+1)x^{2k+1}} = \pm\frac{\pi}{2} - \frac{1}{x} + \frac{1}{3x^3} - \frac{1}{5x^5} + \cdots & \begin{cases} +\text{if} & x \geq 1 \\ -\text{if} & x \leq -1 \end{cases} \end{cases}$$

$$\cot^{-1} x = \frac{\pi}{2} - \tan^{-1} x \quad \text{(for } \tan^{-1} x \text{ use the respective series above corresponding to the interval of } x\text{)}$$

$$\sec^{-1} x = \cos^{-1}\frac{1}{x} = \frac{\pi}{2} - \sin^{-1}\frac{1}{x} = \frac{\pi}{2} - \left(\frac{1}{x} + \frac{1}{(2)(3)(x^3)} + \frac{1(3)}{(2)(4)(5)(x^5)} + \cdots\right) \qquad |x| > 1$$

$$\csc^{-1} x = \sin^{-1}\frac{1}{x} = \frac{1}{x} + \frac{1}{(2)(3)(x^3)} + \frac{(1)(3)}{(2)(4)(5)(x^5)} + \frac{(1)(3)(5)}{(2)(4)(5)(7)(x^7)} + \cdots \qquad |x| > 1$$

(k) Inverse hyperbolic functions.

$$\sinh^{-1} x = \begin{cases} \displaystyle\sum_{k=0}^{\infty} \frac{(-1)^k (2k)!x^{2k+1}}{2^{2k}(2k+1)(k!)^2} = x - \frac{1(x^3)}{2(3)} + \frac{(1)(3)(x^5)}{(2)(4)(5)} - \frac{(1)(3)(5)(x^7)}{(2)(4)(6)(7)} + \cdots & |x| < 1 \\ \displaystyle\ln 2x + \sum_{k=1}^{\infty} \frac{(-1)^{k+1}(2k)!x^{-2k}}{2^{2k}2k(k!)^2} = \ln 2x + \frac{1}{2}\frac{1}{2x^2} - \frac{(1)(3)}{2(4)}\frac{1}{4x^4} + \cdots & |x| \geq 1 \end{cases}$$

$$\cosh^{-1} x = \pm \ln 2x - \sum_{k=1}^{\infty} \frac{(2k)!x^{-2k}}{2^{2k}2k(k!)}$$

$$= \pm\left[\ln 2x - \left(\frac{1}{(2)(2x^2)} + \frac{(1)(3)}{(2)(4)(4)(x^4)} + \frac{(1)(3)(5)}{(2)(4)(6)(6)(x^6)} + \cdots\right]\right. \qquad |x| \geq 1$$

$$\tanh^{-1} x = \sum_{k=0}^{\infty} \frac{x^{2k+1}}{2k+1} = x + \frac{x^3}{3} + \frac{x^5}{5} + \frac{x^7}{7} + \cdots \qquad |x| < 1$$

$$\coth^{-1} x = \sum_{k=0}^{\infty} \frac{1}{(2k+1)x^{2k+1}} = \frac{1}{x} + \frac{1}{3x^3} + \frac{1}{5x^5} + \frac{1}{7x^7} + \cdots \qquad |x| > 1$$

$$\operatorname{sech}^{-1} x = \ln\frac{2}{x} - \sum_{k=1}^{\infty} \frac{(2k-1)!x^{2k}}{2^{2k}(k!)^2} = \ln\frac{2}{x} - \frac{x^2}{2(2)} - \frac{(1)(3x^4)}{(2)(4)(4)} - \frac{(1)(3)(5x^6)}{(2)(4)(6)(6)} - \cdots \qquad 0 < x < 1$$

$$\operatorname{csch}^{-1} x = \ln\frac{2}{x} + \sum_{k=1}^{\infty} (-1)^{k+1} \frac{(2k-1)!x^{2k}}{2^{2k}(k!)^2}$$

$$= \ln\frac{2}{x} + \frac{x^2}{(2)(2)} - \frac{(1)(3x^4)}{(2)(4)(4)} + \frac{(1)(3)(5x^6)}{(2)(4)(6)(6)} - \cdots \qquad 0 < x < 1$$

(I) Constants B_k, E_k, F_k and $n!$ (Sec. 8.06–2c, e, f).

B_k	E_k	F_k	$n!$
$B_1 = 1/6$	$E_1 = 1$	$F_1 = 1$	$1! = 1$
$B_2 = 1/30$	$E_2 = 5$	$F_2 = 2/2!$	$2! = 2$
$B_3 = 1/42$	$E_3 = 61$	$F_3 = 2/3!$	$3! = 6$
$B_4 = 1/30$	$E_4 = 1385$	$F_4 = 2^2/4!$	$4! = 24$
$B_5 = 5/66$	$E_5 = 50{,}521$	$F_5 = 2^2/5!$	$5! = 120$
$B_6 = 691/2{,}730$	$E_6 = 2{,}702{,}765$	$F_6 = 2^3/6!$	$6! = 720$
$B_7 = 7/6$	$E_7 = 199{,}360{,}981$	$F_7 = 2^3/7!$	$7! = 5{,}040$
$B_8 = 3{,}617/510$	$E_8 = 19{,}391{,}512{,}145$	$F_8 = 2^4/8!$	$8! = 40{,}320$
$B_9 = 43{,}867/798$	$E_9 = 2{,}404{,}879{,}675{,}441$	$F_9 = 2^4/9!$	$9! = 362{,}880$
$B_{10} = 174{,}611/330$	$E_{10} = 370{,}371{,}188{,}237{,}525$	$F_{10} = 2^5/10!$	$10! = 3{,}628{,}800$
$B_{11} = 854{,}513/138$	$E_{11} = 69{,}348{,}874{,}393{,}137{,}901$	$F_{11} = 2^5/11!$	$11! = 39{,}916{,}800$
$B_{12} = 236{,}364{,}091/2730$	$E_{12} = 15{,}514{,}534{,}163{,}557{,}086{,}905$	$F_{12} = 2^6/12!$	$12! = 479{,}001{,}600$

8.07 SERIES OF FUNCTIONS OF COMPLEX VARIABLE

(1) Pure Imaginary Argument, $ix = x\sqrt{-1}$

(a) Exponential functions. By the series-expansion procedure introduced in Sec. 8.06 it can be shown that for the pure imaginary argument,

$$e^{ix} = 1 + \frac{ix}{1!} + \frac{(ix)^2}{2!} + \frac{(ix)^3}{3!} + \cdots$$

$$= \left(1 - \frac{x^2}{2!} + \frac{x^4}{4!} - \cdots\right) + i\left(\frac{x}{1!} - \frac{x^3}{3!} + \frac{x^5}{4!} - \cdots\right)$$

$$= \cos x + i \sin x$$

$$e^{-ix} = 1 - \frac{ix}{1!} + \frac{(ix)^2}{2!} - \frac{(ix)^3}{3!} + \cdots$$

$$= \left(1 - \frac{x^2}{2!} + \frac{x^4}{4!} - \cdots\right) - i\left(\frac{x}{1!} - \frac{x^3}{3!} + \frac{x^5}{4!} - \cdots\right)$$

$$= \cos x - i \sin x$$

from which

$$\sin x = \frac{e^{ix} - e^{-ix}}{2i} \qquad \cos x = \frac{e^{ix} + e^{-ix}}{2} \qquad \sec x = \frac{2}{e^{ix} + e^{-ix}}$$

$$\tan x = -\frac{i(e^{ix} - e^{-ix})}{e^{ix} + e^{-ix}} \qquad \cot x = \frac{i(e^{ix} + e^{-ix})}{e^{ix} - e^{-ix}} \qquad \csc x = \frac{2i}{e^{ix} - e^{-ix}}$$

and also for $n = 1, 2, 3, \ldots,$

$$(\cos x \pm i \sin x)^n = \cos nx \pm i \sin nx$$

(b) Trigonometric functions.

$$\sin ix = \frac{ix}{1!} - \frac{(ix)^3}{3!} + \frac{(ix)^5}{4!} - \cdots = i \sinh x$$

$$\cos ix = 1 - \frac{(ix)^2}{2!} + \frac{(ix)^4}{4!} - \cdots = \cosh x$$

and similarly,

$$\tan ix = i \tanh x \qquad \cot ix = -i \coth x$$

$$\sec ix = \operatorname{sech} x \qquad \csc ix = -i \operatorname{csch} x$$

(c) Hyperbolic functions.

$$\sinh ix = \frac{ix}{1!} + \frac{(ix)^3}{3!} + \frac{(ix)^5}{5!} - \cdots = i \sin x$$

$$\cosh ix = 1 + \frac{(ix)^2}{2!} + \frac{(ix)^4}{4!} + \cdots = \cos x$$

and similarly

$$\tanh ix = i \tan x \qquad \coth ix = -i \cot x$$

$$\operatorname{sech} ix = \sec x \qquad \operatorname{csch} ix = -i \operatorname{csh} x$$

(d) Inverse functions. By similar reasoning,

$$\sin^{-1} ix = i \sinh^{-1} x \qquad \sinh^{-1} ix = i \sin^{-1} x$$

$$\cos^{-1} ix = ki \cosh^{-1} x \qquad \cosh^{-1} ix = ki \cos^{-1} x$$

$$\tan^{-1} ix = i \tanh^{-1} x \qquad \tanh^{-1} ix = i \tan^{-1} x$$

$$\cot^{-1} ix = -i \coth^{-1} x \qquad \coth^{-1} ix = -i \cot^{-1} x$$

$$\sec^{-1} ix = ki \operatorname{sech}^{-1} x \qquad \operatorname{sech}^{-1} ix = ki \sec^{-1} x$$

$$\csc^{-1} ix = -i \operatorname{csch}^{-1} x \qquad \operatorname{csch}^{-1} ix = -i \csc^{-1} x$$

where $k = +1$ if $x > 0$, and $k = -1$ if $x < 0$.

(2) Complex Argument, $z = x \pm iy$, $i = \sqrt{-1}$

(a) Exponential functions. By the series-expansion procedure introduced in Sec. 8.07–1*a* it can be shown that for the complex argument,

$$e^z = e^x(\cos y \pm i \sin y) \qquad e^{-z} = e^{-x}(\cos y \mp i \sin y)$$

from which

$$\sin y = \pm \frac{e^{z-x} - e^{x-z}}{2i} \qquad \cos y = \frac{e^{z-x} + e^{x-z}}{2}$$

$$\tan y = \mp \frac{i(e^{z-x} - e^{x-z})}{e^{z-x} + e^{x-z}} \qquad \cot y = \pm \frac{i(e^{z-x} + e^{x-z})}{e^{z-x} - e^{x-z}}$$

(b) Trigonometric functions

$$\sin z = \sin x \cosh y \pm i \cos x \sinh y$$

$$\cos z = \cos x \cosh y \mp i \sin x \sinh y$$

$$\tan z = \frac{\sin 2x \pm i \sinh 2y}{\cos 2x + \cosh 2y}$$

$$\cot z = \frac{\sin 2x \mp i \sinh 2y}{\cos 2x - \cosh 2y}$$

(c) Hyperbolic functions

$$\sinh z = \sinh x \cos y \pm i \cosh x \sin y$$

$$\cosh z = \cosh x \cos y \pm i \sinh x \sin y$$

$$\tanh z = \frac{\sinh 2x \pm i \sin 2y}{\cosh 2x + \cos 2y}$$

$$\coth z = \frac{\sinh 2x \mp i \sin 2y}{\cosh 2x - \cos 2y}$$

(d) Periodicity, $k = 1, 2, \ldots$

$$\sin(x + 2k\pi) = \sin x \qquad \sinh(x + 2k\pi i) = \sinh x$$

$$\cos(x + 2k\pi) = \cos x \qquad \cosh(x + 2k\pi i) = \cosh x$$

$$\tan(x + 2k\pi) = \tan x \qquad \tanh(x + 2k\pi i) = \tanh x$$

$$\cot(x + 2k\pi) = \cot x \qquad \coth(x + 2k\pi i) = \coth x$$

9

INTEGRAL CALCULUS

9.01 DEFINITIONS AND CONCEPTS

(1) Definitions

(a) Primitive function. The function $F(x)$ is said to be a primitive function (indefinite integral) of the function $f(x)$ in the interval (a,b) if

$$\frac{dF(x)}{dx} = f(x)$$

for all x in (a,b).

(b) Indefinite integral. Since the derivative of $F(x)+C$ where C is a constant (or zero) is also equal to $f(x)$, all indefinite integrals of $f(x)$ are included in the expression

$$\int f(x)\,dx = F(x)+C$$

where $f(x)$ is called the *integrand* and C is the *constant of integration.*

(c) Constant of integration. Because of the indeterminancy of C, there is an infinite number of $F(x)+C$ differing only in their position relative to the axis.

example:

$$y = \int \cos x\,dx = \sin x + C$$

Figure 9.01–1 represents y for $C = -2, -1, 0, +1, +2$.

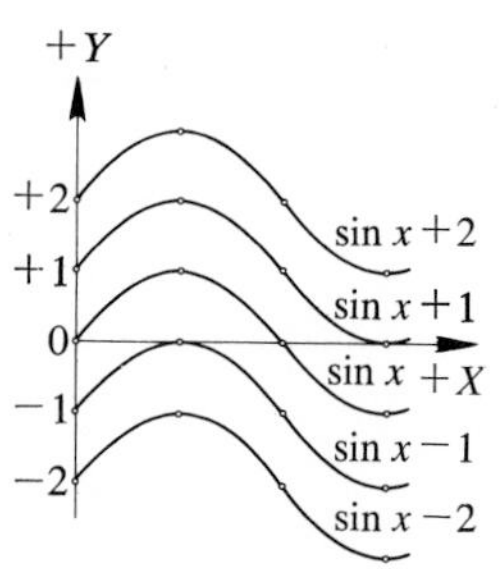

Fig. 9.01-1

(2) Basic Cases

(a) Antiderivatives. If $f(x)$ is a derivative of a known function $F(x)$ such as those given in Sec. 7.03–1, then the indefinite integral of $f(x)$ is this function $F(x)$ plus the constant of integration.

$$\int \frac{dF(x)}{dx}\,dx = F(x)+C$$

The application of this relationship leads to the so-called basic integrals listed below.

(b) Integrals of constants and of simple algebraic functions.

$$\int (0)\,dx = C \qquad \int x^m\,dx = \frac{x^{m+1}}{m+1}+C \qquad m \neq -1$$

$$\int (1)\,dx = x+C \qquad \int \frac{dx}{x} = \ln x + C \qquad x \neq 0$$

$$\int (a)\,dx = ax+C \qquad a \neq 0$$

examples:

$$\int 5x^2\,dx = 5\frac{x^{2+1}}{2+1}+C = \frac{5x^3}{3}+C \qquad \int 5\sqrt{x}\,dx = 5\frac{x^{(1/2)+1}}{\frac{1}{2}+1}+C = \frac{10x\sqrt{x}}{3}+C$$

$$\int \frac{5}{x^2}\,dx = 5\frac{x^{-2+1}}{-2+1}+C = -\frac{5}{x}+C \qquad \int \frac{5}{\sqrt{x}}\,dx = 5\frac{x^{-(1/2)+1}}{-\frac{1}{2}+1}+C = 10\sqrt{x}+C$$

(c) Integrals of simple transcendental functions.

$$\int e^x\,dx = e^x + C \qquad \int a^x\,dx = \frac{a^x}{\ln a} + C$$

$$\int e^{-x}\,dx = -e^{-x} + C \qquad \int a^{-x}\,dx = -\frac{a^{-x}}{\ln a} + C$$

$$\int \cos x\,dx = \sin x + C \qquad \int \sin x\,dx = -\cos x + C$$

$$\int \frac{dx}{\cos^2 x} = \tan x + C \qquad \int \frac{dx}{\sin^2 x} = -\cot x + C$$

$$\int \cosh x\,dx = \sinh x + C \qquad \int \sinh x\,dx = \cosh x + C$$

$$\int \frac{dx}{\cosh^2 x} = \tanh x + C \qquad \int \frac{dx}{\sinh^2 x} = -\coth x + C$$

9.02 TECHNIQUES OF INTEGRATION

(1) Algebra of Integration

(a) Integral of a sum of functions equals the sum of their integrals plus C.

$$\int [f_1(x) \pm f_2(x) \pm \cdots \pm f_m(x)]\,dx = F_1(x) \pm F_2(x) \pm \cdots \pm F_m(x) + C$$

example:

According to Sec. 9.01–1*b*, *c*

$$\int (3 - x^2 + \sin x)\,dx = 3x - \frac{x^3}{3} - \cos x + C$$

(b) Integral of a product $af(x)$, where $a \neq 0$ is a constant, equals the product of a and of the integral of $f(x)$ plus C.

$$\int [(a)f(x)]\,dx = (a)\int f(x)\,dx = (a)F(x) + C$$

example:

According to Sec. 9.01–1*c*

$$\int 7\cos x\,dx = 7\int \cos x\,dx = 7\sin x + C$$

(c) Integral of a quotient $f(x)/a$, where $a \neq 0$ is a constant, equals the quotient of the integral of $f(x)$ and of a plus C.

$$\int \frac{f(x)}{a}\,dx = \frac{1}{a}\int f(x)\,dx = \frac{F(x)}{a} + C$$

example:

According to Sec. 9.01–1*c*

$$\int \frac{\cos x}{5}\,dx = \frac{1}{5}\int \cos x\,dx = \frac{\sin x}{5} + C$$

(d) Integral of a product of two functions $f_1(x)f_2(x)$ does not equal the product of their integrals plus C.

$$\int [f_1(x)f_2(x)]\,dx \neq F_1(x)F_2(x)+C$$

(e) Integral of a quotient of two functions $f_1(x)f_2(x)$ does not equal the quotient of their integrals plus C.

$$\int \frac{f_1(x)}{f_2(x)}\,dx \neq \frac{F_1(x)}{F_2(x)}+C$$

(2) Substitution Method

(a) Change in variable. If the argument of the integrand is a function in x such as $f(x) = f[g(x)]$, it serves frequently to an advantage to introduce the substitution $t = g(x)$ and $dt = g'(x)\,dx$ so that

$$\int f[g(x)]\,dx = \int \frac{f(t)}{g'(x)}\,dt = \int \frac{f(t)}{h(t)}\,dt$$

where $g'(x)$ is either a constant or must be expressed in terms of t as $g'(x) = h(t)$.

examples:

$\int \sin 2x\,dx$ can be expressed in terms of $t = 2x$, $dt = 2\,dx$ or $dx = dt/2$ as

$$\int \frac{\sin t}{2}\,dt = -\frac{\cos t}{2}+C = -\frac{\cos 2x}{2}+C$$

$\int x\cos x^2\,dx$ can be expressed in terms of $t = x^2$, $dt = 2x\,dx$ or $dx = dt/2x$ as

$$\int \frac{x\cos t\,dt}{2x} = \int \frac{\cos t}{2}\,dt = \frac{\sin t}{2}+C = \frac{\sin x^2}{2}+C$$

(b) Integral of a product $f(x)f'(x)$ can be expressed as

$$\int [\underbrace{f(x)}_{u}\cdot \underbrace{f'(x)]\,dx}_{du} = \frac{u^2}{2}+C = \frac{[f(x)]^2}{2}+C$$

example:

$$\int \underbrace{\sin x}_{u}\cdot \underbrace{\cos x\,dx}_{du} = \frac{\sin^2 x}{2}+C$$

(c) Integral of a quotient $f'(x)/f(x)$ can be expressed as

$$\int \frac{f'(x)\,dx}{f(x)} = \int \frac{du}{u} = \ln u + C = \ln\,[f(x)]+C$$

example:

$$\int \frac{\cos x}{\sin x}\,dx = \int \frac{d(\sin x)/dx}{\sin x}\,dx = \ln\,(\sin x)+C$$

(3) Integration by Parts

(a) Integral of a product of two functions. If the integrand $f(x)$ can be expressed as a product of two functions in x such as

$$f(x) = u(x)v'(x)$$

or

$$f(x) = u'(x)v(x)$$

where

$$u'(x) = \frac{du(x)}{dx} \quad \text{and} \quad v'(x) = \frac{dv(x)}{dx}$$

then respectively

$$\int f(x)\,dx = u(x)v(x) - \int [u'(x)v(x)]\,dx$$

or

$$\int f(x)\,dx = u(x)v(x) - \int [u(x)v'(x)]\,dx$$

where the integral of the right side may be one of the basic cases (Sec. 9.01–2) or it may be one of the tabulated values (Sec. 9.03).

example:

$$\int \underbrace{x^2}_{u'} \cdot \underbrace{\ln x}_{v}\,dx = \underbrace{\frac{x^3}{3}}_{u}\underbrace{\ln x}_{v} - \int \underbrace{\frac{x^3}{3}}_{u} \cdot \underbrace{\frac{1}{x}}_{v'}\,dx = \frac{x^3 \ln x}{3} - \frac{1}{3}\int x^2\,dx = \frac{x^3}{3}\ln x - \frac{x^3}{9} + C$$

(b) Integral of the derivative of a product of two functions. If the integrand

$$f(x) = u'v + uv'$$

where $u = f_1(x)$, $v = f_2(x)$, and u', v' are their respective derivatives, then

$$\int f(x)\,dx = \int (u'v + uv')\,dx = uv + C$$

examples:

$$\int e^x(\sin x + \cos x)\,dx = \int (\underbrace{e^x}_{u'}\underbrace{\sin x}_{v} + \underbrace{e^x}_{u}\underbrace{\cos x}_{v'})\,dx = e^x \sin x + C$$

$$\int (\cos^2 x - \sin^2 x)\,dx = \int \left(\underbrace{\frac{d \sin x}{dx}}_{u'}\underbrace{\cos x}_{v} + \underbrace{\sin x}_{u}\underbrace{\frac{d \cos x}{dx}}_{v'}\right)dx = \sin x \cos x + C = \frac{\sin 2x}{2} + C$$

9.03 TABLES OF INDEFINITE INTEGRALS

(1) Notations

(a) Tables of the most common integrals are given in this section.

(b) The constant C is omitted but implied in all cases.

(c) Logarithmic expressions ln( ) (if involved) must be evaluated for the absolute value of the argument.

(d) All angles are in radians and *all arguments* of inverse functions represent the principal values only.

(2) Rational Algebraic Functions

(a) $f(x) = x^m$.

$$\int x\,dx = \frac{x^2}{2} \qquad \int \frac{dx}{x} = \ln x$$

$$\int x^m\,dx = \frac{x^{m+1}}{m+1} \qquad \int \frac{dx}{x^m} = \frac{x^{1-m}}{1-m} \qquad m \neq 1$$

(b) $f(x) = (a+bx)^n,\ R = a+bx,\ b \neq 0.$

$$\int R\,dx = \frac{R^2}{2b} \qquad \int \frac{dx}{R} = \frac{\ln R}{b}$$

$$\int R^n\,dx = \frac{R^{n+1}}{b(n+1)} \quad n \neq -1 \qquad \int \frac{dx}{R^n} = -\frac{1}{b(n-1)R^{n-1}} \quad n \neq 1$$

(c) $f(x) = x^m(a+bx)^n,\ R = a+bx,\ b \neq 0.$

$$\int xR\,dx = \frac{R^3}{3b^2} - \frac{aR^2}{2b^2}$$

$$\int xR^n\,dx = \frac{R^{m+2}}{(n+2)b^2} - \frac{aR^{m+1}}{(n+1)b^2} \qquad n \neq -1, -2$$

$$\int x^mR^n\,dx = \frac{1}{b(m+n+1)}\left(x^mR^{n+1} - ma\int x^{m-1}R^n\,dx\right) = \frac{1}{m+n+1}\left(x^{m+1}R^n + na\int x^mR^{n-1}\,dx\right)$$

(d) $f(x) = \dfrac{x^m}{(a+bx)^n},\ R = a+bx,\ b \neq 0.$

$$\int \frac{dx}{R} = \frac{\ln R}{b} \qquad \int \frac{x\,dx}{R} = \frac{1}{b^2}(R - a\ln R)$$

$$\int \frac{x\,dx}{R^n} = \frac{-1}{(n-2)b^2R^{n-1}}\left(R - \frac{n-2}{n-1}a\right) \qquad n \neq 0, 1, 2$$

$$\int \frac{x^m\,dx}{R^n} = \frac{1}{b^{n+1}}\int \frac{(R-a)^m}{R^n}\,dR$$

$$\int \frac{dx}{xR} = -\frac{1}{a}\ln\frac{R}{x}$$

$$\int \frac{dx}{x^2R} = \frac{b}{a}\ln\frac{R}{x} - \frac{1}{ax}$$

$$\int \frac{dx}{x^2R^2} = \frac{2b}{a^3}\ln\frac{R}{x} - \frac{1}{a^2x} - \frac{b}{a^2R}$$

(e) $f(x) = x^m/(a^2+x^2)^n\ R = a^2+x^2,\ a \neq 0.$

$$\int \frac{dx}{R} = \frac{1}{a}\tan^{-1}\frac{x}{a}$$

$$\int \frac{x\,dx}{R} = \frac{1}{2}\ln R \qquad \int \frac{dx}{xR} = \frac{1}{2a^2}\ln\frac{x^2}{R}$$

(f) $f(x) = x^m/(a^2-x^2)^n,\ R = a^2-x^2,\ a \neq 0.$

$$\int \frac{dx}{R} = \frac{1}{2a}\ln\frac{a+x}{a-x}$$

$$\int \frac{x\,dx}{R} = -\frac{1}{2}\ln R \qquad \int \frac{dx}{xR} = \frac{1}{2a^2}\ln\frac{x^2}{R}$$

(g) $f(x) = [x^m/(a+bx+cx^2)^n]\ x,\ R = a+bx+cx^2,\ c \neq 0,\ \gamma = 4ac-b^2.$

$$\int \frac{dx}{R} = \begin{cases} \dfrac{2}{\sqrt{\gamma}}\tan^{-1}\dfrac{2cx+b}{\sqrt{\gamma}} & \gamma > 0 \\[2ex] -\dfrac{2}{2cx+b} & \gamma = 0 \\[2ex] \dfrac{1}{\sqrt{-\gamma}}\ln\dfrac{2cx+b-\sqrt{-\gamma}}{2cx+b+\sqrt{-\gamma}} & \gamma < 0 \end{cases}$$

$$\int \frac{x\,dx}{R} = \frac{1}{2c}\ln R - \frac{b}{2c}\int\frac{dx}{R}$$

$$\int \frac{dx}{R^n} = \frac{2cx+b}{(n-1)\gamma R^{n-1}} + \frac{2(2n-3)c}{(n-1)\gamma}\int\frac{dx}{R^{n-1}} \qquad n \neq 1$$

(3) Irrational Algebraic Functions

(a) $f(x) = \sqrt[m]{x},\ m \neq 1.$

$$\int \sqrt{x}\,dx = \frac{2x\sqrt{x}}{3} \qquad \int \frac{dx}{\sqrt{x}} = 2\sqrt{x}$$

$$\int \sqrt[m]{x}\,dx = \frac{mx\sqrt[m]{x}}{m+1} \qquad \int \frac{dx}{\sqrt[m]{x}} = \frac{mx}{(m-1)\sqrt[m]{x}}$$

(b) $f(x) = \sqrt[m]{a+bx},\ R = a+bx,\ b \neq 0.$

$$\int \sqrt{R}\,dx = \frac{2R\sqrt{R}}{3b} \qquad \int \frac{dx}{\sqrt{R}} = \frac{2\sqrt{R}}{b}$$

$$\int \sqrt[m]{R}\,dx = \frac{m}{(m+1)b}\sqrt[m]{R^{m+1}} \qquad \int \frac{dx}{\sqrt[m]{R}} = \frac{m}{(m-1)b}\sqrt[m]{R^{m-1}}$$

(c) $f(x) = x^m\sqrt[n]{a+bx},\ R = a+bx,\ b \neq 0,\ n \neq 1.$

$$\int x\sqrt{R}\,dx = \frac{2R\sqrt{R}}{b^2}\left(\frac{R}{5}-\frac{a}{3}\right)$$

$$\int x\sqrt[n]{R}\,dx = \frac{nx}{(n+1)b}\sqrt[n]{R^{n+1}} - \frac{n^2}{(n+1)(2n+1)b^2}\sqrt[n]{R^{n+2}}$$

(d) $f(x) = x^m/\sqrt[n]{a+bx},\ R = a + bx,\ b \neq 0.$

$$\int \frac{x\,dx}{\sqrt{R}} = \frac{2\sqrt{R}}{3b^2}(R-3a)$$

$$\int \frac{x^2\,dx}{\sqrt{R}} = \frac{2\sqrt{R}}{b^3}\left(\frac{R^2}{5} - \frac{2aR}{3} + a^2\right)$$

$$\int \frac{dx}{x\sqrt{R}} = \begin{cases} -\dfrac{2}{\sqrt{a}}\tanh^{-1}\sqrt{\dfrac{R}{a}} & a > R > 0 \\ -\dfrac{2}{\sqrt{a}}\coth^{-1}\sqrt{\dfrac{R}{a}} & R > a > 0 \\ \dfrac{2}{\sqrt{-a}}\tanh^{-1}\sqrt{\dfrac{R}{-a}} & a < 0,\ R > 0 \\ \dfrac{1}{\sqrt{a}}\ln\dfrac{\sqrt{R}-\sqrt{a}}{\sqrt{R}+\sqrt{a}} & a > 0,\ R > 0 \end{cases}$$

(e) $f(x) = x^m\sqrt{a^2+x^2},\ R = a^2+x^2,\ a \neq 0.$

$$\int \sqrt{R}\,dx = \tfrac{1}{2}[x\sqrt{R} + a^2\ln(x+\sqrt{R})]$$

$$\int x\sqrt{R}\,dx = \tfrac{1}{3}R\sqrt{R} \qquad \int \frac{\sqrt{R}\,dx}{x} = \sqrt{R} - a\ln\frac{a+\sqrt{R}}{x}$$

(f) $f(x) = x^m/\sqrt{a^2+x^2},\ R = a^2+x^2,\ a \neq 0.$

$$\int \frac{dx}{\sqrt{R}} = \sinh^{-1}\frac{x}{a} = \ln(x+\sqrt{R})$$

$$\int \frac{x\,dx}{\sqrt{R}} = \sqrt{R} \qquad \int \frac{dx}{x\sqrt{R}} = -\frac{1}{a}\ln\frac{a+\sqrt{R}}{x}$$

(g) $f(x) = x^m\sqrt{a^2-x^2},\ R = a^2-x^2,\ a \neq 0.$

$$\int \sqrt{R}\,dx = \frac{1}{2}\left(x\sqrt{R} + a^2\sin^{-1}\frac{x}{a}\right)$$

$$\int x\sqrt{R}\,dx = -\tfrac{1}{3}R\sqrt{R} \qquad \int \frac{\sqrt{R}\,dx}{x} = \sqrt{R} - a\ln\frac{a+\sqrt{R}}{x}$$

(h) $f(x) = x^m/\sqrt{a^2-x^2},\ R = a^2-x^2,\ a \neq 0.$

$$\int \frac{dx}{\sqrt{R}} = \sinh^{-1}\frac{x}{a} = \ln(x+\sqrt{R})$$

$$\int \frac{x\,dx}{\sqrt{R}} = -\sqrt{R} \qquad \int \frac{dx}{x\sqrt{R}} = -\frac{1}{a}\ln\frac{a+\sqrt{R}}{x}$$

(i) $f(x) = x^m\sqrt{x^2 - a^2}$, $R = a^2 - x^2$, $a \neq 0$.

$$\int \sqrt{R}\,dx = \frac{x}{2}\sqrt{R} - \frac{a^2}{2}\ln(x + \sqrt{R})$$

$$\int x\sqrt{R}\,dx = \tfrac{1}{3}R\sqrt{R} \qquad \int \frac{\sqrt{R}\,dx}{x} = \sqrt{R} - a\cosh^{-1}\frac{a}{x}$$

(j) $f(x) = x^m/\sqrt{x^2 - a^2}$, $R = a^2 - x^2$, $a \neq 0$.

$$\int \frac{dx}{\sqrt{R}} = \cosh^{-1}\frac{x}{a} = \ln(x + \sqrt{R})$$

$$\int \frac{x\,dx}{\sqrt{R}} = \sqrt{R} \qquad \int \frac{dx}{x\sqrt{R}} = \frac{1}{a}\cosh^{-1}\frac{a}{x}$$

(k) $f(x) = x^m\sqrt{a + bx + cx^2}$, $R = a + bx + cx^2$, $a \neq 0$, $\gamma = 4ac - b^2$.

$$\int \sqrt{R}\,dx = \frac{b + 2cx}{4c}\sqrt{R} + \frac{\gamma}{8c}\int \frac{dx}{\sqrt{R}} \qquad \text{see (l)}$$

$$\int x\sqrt{R}\,dx = \frac{R\sqrt{R}}{3c} - \frac{b}{2c}\int \sqrt{R}\,dx \qquad \text{see above}$$

(l) $f(x) = x^m/\sqrt{a + bx + cx^2}$, $R = a + bx + cx^2$, $a \neq 0$, $\gamma = 4ac - b^2$.

$$\int \frac{dx}{\sqrt{R}} = \begin{cases} \dfrac{1}{\sqrt{c}}\ln(b + 2cx + 2\sqrt{cR}) & c > 0 \\ \dfrac{1}{\sqrt{c}}\sinh^{-1}\dfrac{b + 2cx}{\sqrt{\gamma}} & c > 0,\ \gamma > 0 \\ \dfrac{1}{\sqrt{c}}\cosh^{-1}\dfrac{b + 2cx}{\sqrt{-\gamma}} & c > 0, \gamma < 0 \\ \dfrac{-1}{\sqrt{-c}}\sin^{-1}\dfrac{b + 2cx}{\sqrt{-\gamma}} & c < 0, \gamma < 0 \end{cases}$$

$$\int \frac{x\,dx}{\sqrt{R}} = \frac{\sqrt{R}}{c} - \frac{b}{2c}\int \frac{dx}{\sqrt{R}} \qquad \text{see above}$$

(4) Trigonometric Functions

(a) $f(x) = \sin^n A$, $A = bx$, $b \neq 0$.

$$\int \sin A\,dx = -\frac{\cos A}{b}$$

$$\int \sin^2 A\,dx = -\frac{\sin 2A}{4b} + \frac{A}{2b}$$

$$\int \sin^n A\,dx = -\frac{\cos A\sin^{n-1} A}{nb} + \frac{n-1}{n}\int \sin^{n-2} A\,dx \qquad n > 0$$

(b) $f(x) = \cos^n A,\ A = bx,\ b \neq 0.$

$$\int \cos A\, dx = \frac{\sin A}{b}$$

$$\int \cos^2 A\, dx = \frac{\sin 2A}{4b} + \frac{A}{2b}$$

$$\int \cos^n A\, dx = \frac{\sin A \cos^{n-1} A}{nb} + \frac{n-1}{n}\int \cos^{n-2} A\, dx \qquad n > 0$$

(c) $f(x) = 1/(\sin^n A),\ A = bx,\ b \neq 0.$

$$\int \frac{dx}{\sin A} = \frac{1}{b}\ln\left(\tan\frac{A}{2}\right)$$

$$\int \frac{dx}{\sin^2 A} = -\frac{\cot A}{b}$$

$$\int \frac{dx}{\sin^n A} = \frac{-\cos A}{b(n-1)\sin^{n-1} A} + \frac{n-2}{n-1}\int \frac{dx}{\sin^{n-2} A} \qquad n > 1$$

(d) $f(x) = 1/(\cos^n A),\ A = bx,\ b \neq 0.$

$$\int \frac{dx}{\cos A} = \frac{1}{b}\ln\left[\tan\left(\frac{\pi}{4} + \frac{A}{2}\right)\right]$$

$$\int \frac{dx}{\cos^2 A} = \frac{\tan A}{b}$$

$$\int \frac{dx}{\cos^n A} = \frac{\sin A}{b(n-1)\cos^{n-1} A} + \frac{n-2}{n-1}\int \frac{dx}{\cos^{n-2} A} \qquad n > 1$$

(e) $f(x) = x^m \sin^n A,\ A = bx,\ b \neq 0.$

$$\int x \sin A\, dx = -\frac{x\cos A}{b} + \frac{\sin A}{b^2}$$

$$\int x \sin^2 A\, dx = \frac{x^2}{4} - \frac{x \sin 2A}{4b} - \frac{\cos 2A}{8b^2}$$

$$\int x^2 \sin A\, dx = -\frac{x^2 \cos A}{b} + \frac{2x \sin A}{b^2} + \frac{2\cos A}{b^3}$$

(f) $f(x) = x^m \cos^n A,\ A = bx,\ b \neq 0.$

$$\int x \cos A\, dx = \frac{x\sin A}{b} + \frac{\cos A}{b^2}$$

$$\int x \cos^2 A\, dx = \frac{x^2}{4} + \frac{x \sin 2A}{4b} + \frac{\cos 2A}{8b^2}$$

$$\int x^2 \cos A\, dx = \frac{x^2 \sin A}{b} + \frac{2x \cos A}{b^2} - \frac{2\sin A}{b^3}$$

(g) $f(x) = f(1 \pm \sin A),\ A = bx,\ b \neq 0.$

$$\int \frac{dx}{1+\sin A} = -\frac{1}{b}\tan\left(\frac{\pi}{4}-\frac{A}{2}\right)$$

$$\int \frac{x\,dx}{1+\sin A} = -\frac{x}{b}\tan\left(\frac{\pi}{4}-\frac{A}{2}\right)+\frac{2}{b^2}\ln\left[\cos\left(\frac{\pi}{4}-\frac{A}{2}\right)\right]$$

$$\int \frac{dx}{1-\sin A} = \frac{1}{b}\tan\left(\frac{\pi}{4}+\frac{A}{2}\right)$$

$$\int \frac{x\,dx}{1-\sin A} = \frac{x}{b}\cot\left(\frac{\pi}{4}+\frac{A}{2}\right)+\frac{2}{b^2}\ln\left[\sin\left(\frac{\pi}{4}-\frac{A}{2}\right)\right]$$

(h) $f(x) = f(1 \pm \cos A),\ A = bx,\ b \neq 0.$

$$\int \frac{dx}{1+\cos A} = \frac{1}{b}\tan\frac{A}{2}$$

$$\int \frac{x\,dx}{1+\cos A} = \frac{x}{b}\tan\frac{A}{2}+\frac{2}{b^2}\ln\left(\cos\frac{A}{2}\right)$$

$$\int \frac{dx}{1-\cos A} = -\frac{1}{b}\cot\frac{A}{2}$$

$$\int \frac{x\,dx}{1-\cos A} = -\frac{x}{b}\cot\frac{A}{2}+\frac{2}{b^2}\ln\left(\sin\frac{A}{2}\right)$$

(i) $f(x) = \sin^m A \cos^n A,\ A = bx,\ b \neq 0.$

$$\int \sin A \cos A\,dx = -\frac{\cos^2 A}{2b}$$

$$\int \sin^m A \cos A\,dx = \frac{\sin^{m+1} A}{(m+1)b} \qquad m \neq -1$$

$$\int \sin A \cos^n A\,dx = -\frac{\cos^{n+1} A}{(n+1)b} \qquad n \neq -1$$

(j) $f(x) = 1/(\sin^m A \cos^n A),\ A = bx,\ b \neq 0.$

$$\int \frac{dx}{\sin A \cos A} = \frac{1}{b}\ln(\tan A)$$

$$\int \frac{dx}{\sin^m A \cos A} = \frac{-1}{b(m-1)\sin^{m-1} A}+\int \frac{dx}{\sin^{m-2} A \cos A} \qquad m \neq 1$$

$$\int \frac{dx}{\sin A \cos^n A} = \frac{1}{b(n-1)\cos^{n-1} A}+\int \frac{dx}{\sin A \cos^{n-2} A} \qquad n \neq 1$$

(k) $f(x) = (\sin^m A)/(\cos^n A),\ A = bx,\ \mathrm{b} \neq 0.$

$$\int \frac{\sin A\,dx}{\cos A} = -\frac{\ln(\cos A)}{b}$$

$$\int \frac{\sin^m A\,dx}{\cos A} = -\frac{\sin^{m-1} A}{b(m-1)}+\int \frac{\sin^{m-2} A\,dx}{\cos A} \qquad m \neq 1$$

$$\int \frac{\sin A\,dx}{\cos^n A\,dx} = \frac{1}{b(n-1)\cos^{n-1} A} \qquad n \neq 1$$

(l) $f(x) = (\cos^n A)/(\sin^m A),\ A = bx,\ b \neq 0.$

$$\int \frac{\cos A\,dx}{\sin A} = \frac{\ln(\sin A)}{b}$$

$$\int \frac{\cos^n A\,dx}{\sin A} = \frac{\cos^{n-1} A}{b(n-1)} + \int \frac{\cos^{n-2} A\,dx}{\sin A} \qquad n \neq 1$$

$$\int \frac{\cos A\,dx}{\sin^m A} = \frac{-1}{b(m-1)\sin^{m-1} A} \qquad m \neq 1$$

(m) $f(x) = f(\sin \alpha x, \sin \beta x, \cos \alpha x, \cos \beta x),\ \alpha \neq 0,\ \beta \neq 0.$

$$\left.\begin{aligned}
\int \sin \alpha x \sin \beta x\,dx &= \frac{\sin(\alpha-\beta)x}{2(\alpha-\beta)} - \frac{\sin(\alpha+\beta)x}{2(\alpha+\beta)}\\
\int \sin \alpha x \cos \beta x\,dx &= -\frac{\cos(\alpha-\beta)x}{2(\alpha-\beta)} - \frac{\cos(\alpha+\beta)x}{2(\alpha+\beta)}\\
\int \cos \alpha x \cos \beta x\,dx &= \frac{\sin(\alpha-\beta)x}{2(\alpha-\beta)} + \frac{\sin(\alpha+\beta)x}{2(\alpha+\beta)}
\end{aligned}\right\} \qquad \alpha^2 \neq \beta^2$$

(n) $f(x) = f(\tan^m A, \cot^n A),\ A = bx,\ b \neq 0.$

$$\int \tan A\,dx = -\frac{\ln(\cos A)}{b}$$

$$\int \tan^m A\,dx = \frac{\tan^{m-1} A}{b(m-1)} - \int \tan^{m-2} A\,dx \qquad m > 1$$

$$\int \cot A\,dx = \frac{\ln(\sin A)}{b}$$

$$\int \cot^n A\,dx = -\frac{\cot^{n-1} A}{b(n-1)} - \int \cot^{n-2} A\,dx \qquad n > 1$$

(5) Hyperbolic Functions

(a) $f(x) = f(x^m, \sinh^n A,\ A = bx,\ b \neq 0.$

$$\int \sinh A\,dx = \frac{\cosh A}{b}$$

$$\int \sinh^n A\,dx = \frac{\sinh^{n-1} A \cosh A}{bn} - \frac{n-1}{n}\int \sinh^{n-2} A\,dx$$

$$\int x \sinh A\,dx = \frac{x \cosh A}{b} - \frac{\sinh A}{b^2}$$

$$\int x^m \sinh A\,dx = \frac{x^m \cosh A}{b} - \frac{m}{b}\int x^{m-1} \cosh A\,dx$$

(b) $f(x) = f(x^m, \cosh A),\ A = bx,\ b \neq 0.$

$$\int \cosh A\, dx = \frac{\sinh A}{b}$$

$$\int \cosh^n A\, dx = \frac{\cosh^{n-1} A \sinh A}{bn} + \frac{n-1}{n}\int \cosh^{n-2} A\, dx$$

$$\int x \cosh A\, dx = \frac{x \sinh A}{b} - \frac{\cosh A}{b^2}$$

$$\int x^m \cosh A\, dx = \frac{x^m \sinh A}{b} - \frac{m}{b}\int x^{m-1} \sinh A\, dx$$

(c) $f(x) = f(\sinh^m A, \cosh^n A),\ A = bx,\ b \neq 0$

$$\int \sinh A \cosh A\, dx = \frac{\sinh^2 A}{2b}$$

$$\int \sinh^m A \cosh A\, dx = \frac{\sinh^{m+1} A}{(m+1)b} \qquad m \neq -1$$

$$\int \sinh A \cosh^n A\, dx = \frac{\cosh^{n+1} A}{(n+1)b} \qquad n \neq -1$$

$$\int \frac{dx}{\sinh A \cosh A} = \frac{\ln(\tanh A)}{b}$$

$$\int \frac{\cosh A\, dx}{\sinh^m A} = -\frac{1}{(m-1)b \sinh^{m-1} A} \qquad m \neq 1$$

$$\int \frac{\sinh A\, dx}{\cosh^n A} = -\frac{1}{(n-1)b \cosh^{n-1} A} \qquad n \neq 1$$

(d) $f(x) = f(\sinh \alpha x, \sinh \beta x, \cosh \alpha x, \cosh \beta x),\ \alpha \neq 0,\ \beta \neq 0.$

$$\left.\begin{aligned}
\int \sinh \alpha x \sinh \beta x\, dx &= \frac{\sinh(\alpha+\beta)x}{2(\alpha+\beta)} - \frac{\sinh(\alpha-\beta)x}{2(\alpha-\beta)}\\
\int \sinh \alpha x \cosh \beta x\, dx &= \frac{\cosh(\alpha+\beta)x}{2(\alpha+\beta)} + \frac{\cosh(\alpha-\beta)x}{2(\alpha-\beta)}\\
\int \cosh \alpha x \cosh \beta x\, dx &= \frac{\sinh(\alpha+\beta)x}{2(\alpha+\beta)} + \frac{\sinh(\alpha-\beta)x}{2(\alpha-\beta)}
\end{aligned}\right\} \qquad \alpha^2 \neq \beta^2$$

(e) $f(x) = f(\tanh A, \coth A),\ A = bx,\ b \neq 0.$

$$\int \tanh A\, dx = \frac{\ln(\cosh A)}{b}$$

$$\int \tanh^m A\, dx = -\frac{\tanh^{m-1} A}{b(m-1)} + \int \tanh^{m-2} A\, dx \qquad m \neq 1$$

$$\int \coth A\, dx = \frac{\ln(\sinh A)}{b}$$

$$\int \coth^n A\, dx = -\frac{\coth^{n-1} A}{b(n-1)} + \int \coth^{n-2} A\, dx \qquad n \neq 1$$

(6) Exponential and Logarithmic Functions

(a) $f(x) = f(x^m, e^A),\ A = bx,\ b \neq 0.$

$$\int e^A\,dx = \frac{e^A}{b}$$

$$\int xe^A\,dx = \frac{(A-1)e^A}{b^2}$$

$$\int x^m e^A\,dx = \frac{x^m e^A}{b} - \frac{m}{b}\int x^{m-1}e^A\,dx$$

(b) $f(x) = f(x^m, a^A),\ A = bx,\ a \neq 0,\ b \neq 0.$

$$\int a^A\,dx = \frac{a^A}{b\ln a}$$

$$\int xa^A\,dx = \frac{(A\ln a - 1)a^A}{(b\ln a)^2}$$

$$\int x^m a^A\,dx = \frac{x^m a^A}{b\ln a} - \frac{m}{b\ln a}\int x^{m-1}a^A\,dx$$

(c) $f(x) = f(e^{\alpha x})\sin\beta x,\ \cos\beta x),\ \alpha \neq 0,\ \beta \neq 0.$

$$\int e^{\alpha x}\sin\beta x\,dx = \frac{e^{\alpha x}(\alpha\sin\beta x - \beta\cos\beta x)}{\alpha^2+\beta^2}$$

$$\int e^{\alpha x}\cos\beta x\,dx = \frac{e^{\alpha x}(\alpha\cos\beta x + \beta\sin\beta x)}{\alpha^2+\beta^2}$$

$$\int \frac{\sin\beta x\,dx}{e^{\alpha x}} = -\frac{(\alpha\sin\beta x + \beta\cos\beta x)}{(\alpha^2+\beta^2)e^{\alpha x}}$$

$$\int \frac{\cos\beta x\,dx}{e^{\alpha x}} = -\frac{(\alpha\cos\beta x - \beta\sin\beta x}{(\alpha^2+\beta^2)e^{\alpha x}}$$

(d) $f(x) = f(x^m, \ln A),\ A = bx,\ b \neq 0.$

$$\int \ln A\,dx = x(\ln A - 1)$$

$$\int x^m\ln A\,dx = \frac{x^{m+1}}{(m+1)^2}[(m+1)\ln A - 1] \qquad m \neq -1$$

(7) Inverse Trigonometric Functions

(a) $f(x) = f(x^m, \sin^{-1}B, \cos^{-1}B),\ B = x/b,\ b \neq 0.$

$$\int \sin^{-1}B\,dx = x\sin^{-1}B + \sqrt{b^2-x^2}$$

$$\int x^m\sin^{-1}B\,dx = \frac{x^{m+1}}{m+1}\sin^{-1}B - \frac{1}{m+1}\int\frac{x^{m+1}}{\sqrt{b^2-x^2}}\,dx \qquad m \neq -1$$

$$\int \cos^{-1}B\,dx = x\cos^{-1}B - \sqrt{b^2-x^2}$$

$$\int x^m\cos^{-1}B\,dx = \frac{x^{m+1}}{m+1}\cos^{-1}B + \frac{1}{m+1}\int\frac{x^{m+1}}{\sqrt{b^2-x^2}}\,dx \qquad m \neq -1$$

(b) $f(x) = f(x^m, \tan^{-1} B, \cot^{-1} B),\ B = x/b,\ b \neq 0.$

$$\int \tan^{-1} B\,dx = x \tan^{-1} B - b \ln \sqrt{b^2 + x^2}$$

$$\int x^m \tan^{-1} B\,dx = \frac{x^{m+1}}{m+1} \tan^{-1} B - \frac{b}{m+1} \int \frac{x^{m+1}}{b^2 + x^2}\,dx \qquad m \neq -1$$

$$\int \cot^{-1} B\,dx = x \cot^{-1} B + b \ln \sqrt{b^2 + x^2}$$

$$\int x^m \cot^{-1} B\,dx = \frac{x^{m+1}}{m+1} \cot^{-1} B + \frac{b}{m+1} \int \frac{x^{m+1}}{b^2 + x^2}\,dx \qquad m \neq -1$$

(8) Inverse Hyperbolic Functions

(a) $f(x) = f(x^m, \sinh^{-1} B, \cosh^{-1} B),\ B = x/b,\ b \neq 0.$

$$\int \sinh^{-1} B\,dx = x \sinh^{-1} B - \sqrt{x^2 - b^2}$$

$$\int x^m \sinh^{-1} B\,dx = \frac{x^{m+1}}{m+1} \sinh^{-1} B - \frac{1}{m+1} \int \frac{x^{m+1}}{\sqrt{x^2 + b^2}}\,dx \qquad m \neq -1$$

$$\int \cosh^{-1} B\,dx = x \cosh^{-1} B \mp \sqrt{x^2 - b^2}$$

$$\int x^m \cosh^{-1} B\,dx = \frac{x^{m+1}}{m+1} \cosh^{-1} B \mp \frac{1}{m+1} \int \frac{x^{m+1}}{\sqrt{x^2 - b^2}}\,dx \qquad m \neq -1$$

($-$ if $\cosh^{-1} B > 0$, $+$ if $\cosh^{-1} B < 0$)

(b) $f(x) = f(x^m, \tanh^{-1} B, \coth^{-1} B),\ B = x/b,\ b \neq 0.$

$$\int \tanh^{-1} B\,dx = x \tanh^{-1} B + b \ln \sqrt{b^2 - x^2}$$

$$\int x^m \tanh^{-1} B\,dx = \frac{x^{m+1}}{m+1} \tanh^{-1} B - \frac{b}{m+1} \int \frac{x^{m+1}}{b^2 - x^2}\,dx \qquad m \neq -1$$

$$\int \coth^{-1} B\,dx = x \coth^{-1} B + b \ln \sqrt{x^2 - b^2}$$

$$\int \coth^{-1} B\,dx = \frac{x^{m+1}}{m+1} \coth^{-1} B - \frac{b}{m+1} \int \frac{x^{m+1}}{x^2 - b^2}\,dx \qquad m \neq -1$$

(9) Tables of Integrals

For more extensive tables of indefinite integrals, refer to:

Tuma, Jan J.: "Engineering Mathematics Handbook," McGraw-Hill Book Company, New York, 1970.

Dwight, H. B.: "Tables of Integrals and Other Mathematical Data," 4th ed., Macmillan, New York, 1961.

Gradshteyn, I.C., and I. M. Ryzhik: "Tables of Integrals, Series and Products," Academic Press, New York, 1965.

9.04 DEFINITE INTEGRAL

(1) Definitions and Notations

(a) Definition. If $F(x)$ and its derivative $f(x)$ are single-valued and continuous in the closed interval $[a,b]$ and if this interval is subdivided into n equal parts by the end points of coordinates $x_1, x_2, \ldots, x_{n-1}$ (Fig. 9.04–1) so that

$$\Delta x = \frac{b-a}{n}$$

then the definite integral of $f(x)$ with respect to x, between limits $x = a$ and $x = b$, is

$$\int_a^b f(x)\,dx = \lim_{\substack{n\to\infty \\ \Delta x\to 0}} \sum_{0,1}^{n-1,n} f(\xi_{i,j})\,\Delta x = \left[\int f(x)\,dx\right]_a^b = [F(x)]_a^b = F(b) - F(a)$$

where $F(x)$ is the indefinite integral of $f(x)$.

examples:

$$\int_2^5 x^2\,dx = \left[\frac{x^3}{3}\right]_2^5 = \frac{5^3}{3} - \frac{2^3}{3} = 39$$

$$\int_0^5 x^2\,dx = \left[\frac{x^3}{3}\right]_0^5 = \frac{5^3}{3} - 0 = \frac{125}{3}$$

$$\int_2^0 x^2\,dx = \left[\frac{x^3}{3}\right]_2^0 = 0 - \frac{2^3}{3} = -\frac{8}{3}$$

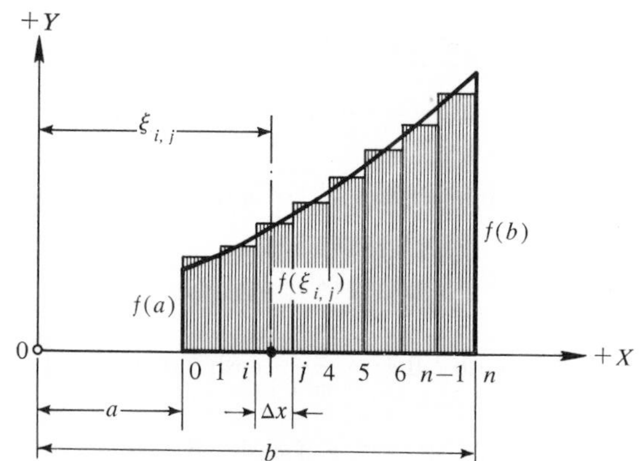

Fig. 9.04-1

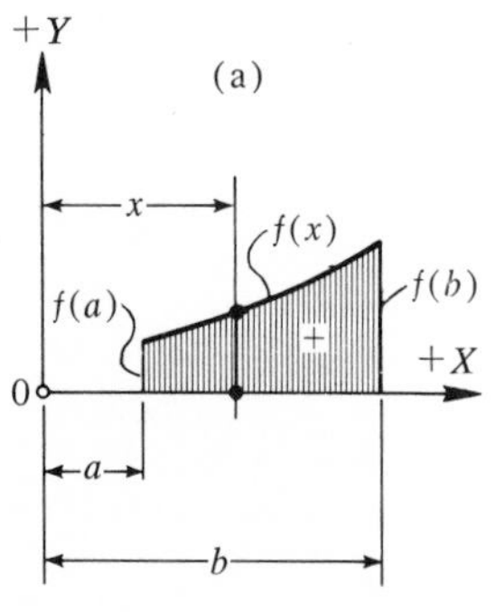

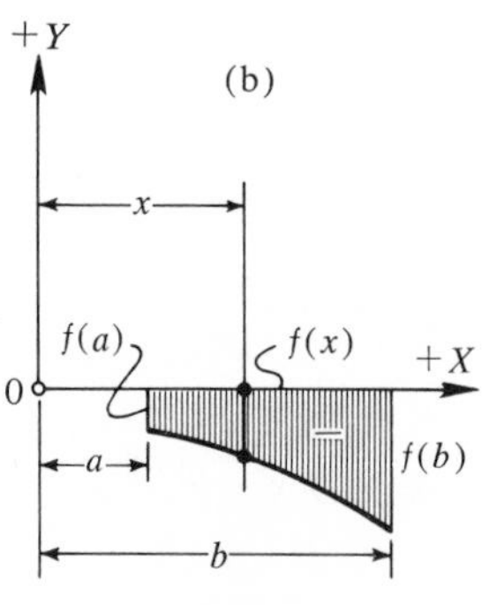

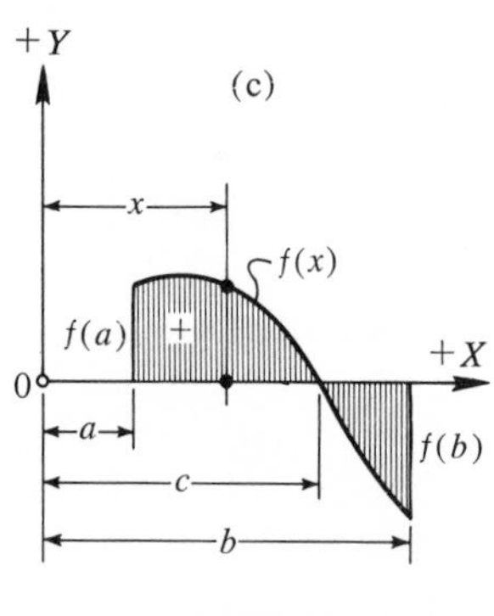

Fig. 9.04-2

(b) Limits. The numbers a and b are called respectively the lower and upper limits of integration and the closed interval $[a,b]$ is called the range of integration.

(c) Geometric interpretation. The definite integral equals numerically the area bounded by the graph of $f(x)$, the X axis, and ordinates $f(a)$, $f(b)$ as shown in Fig. 9.04–2*a, b, c*. The sign of this area is governed by the signs of $f(x)$. If the graph intersects the X axis one or several times inside the interval $[a,b]$, then the integral equals the algebraic sum of all areas above and below the X axis. If the sum of all the areas above the axis (+ areas) and the sum of all the areas below the axis (− areas) are equal, then the definite integral equals zero.

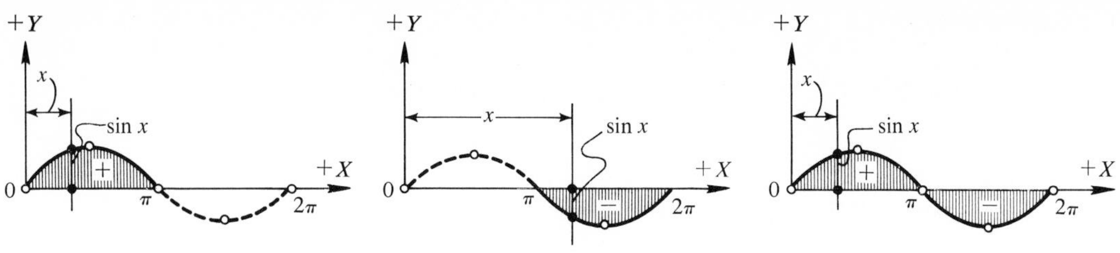

Fig. 9.04-3

examples:

In Fig. 9.04–3*a*: $\int_0^{\pi} \sin x\, dx = [-\cos x]_0^{\pi} = (-\cos \pi) - (-\cos 0) = 2$

In Fig. 9.04–3*b*: $\int_{\pi}^{2\pi} \sin x\, dx = [-\cos x]_{\pi}^{2\pi} = [(-\cos 2\pi) - (-\cos \pi)] = -2$

In Fig. 9.04–3*c*: $\int_0^{2\pi} \sin x\, dx = [-\cos x]_0^{2\pi} = [(-\cos 2\pi) - (-\cos 0)] = 0$

(2) Rules and Theorems

(a) Rules of limits.

$$\int_a^b = -\int_b^a \qquad \int_a^c + \int_c^b = \int_a^b \qquad \int_a^c - \int_b^c = \int_a^b$$

Fig. 9.04-4

(b) Special theorems. In addition to the theorems of Sec. 9.02 the following special theorems govern the integration of $f(x)$ in $[a,b]$.

$$\int_a^a f(x)\, dx = 0$$

$$\int_a^b \lambda f(x)\, dx = \lambda \int_a^b f(x)\, dx \qquad \lambda = \text{const.}$$

$$\int_{-a}^{a} f(x)\, dx = \begin{cases} 2\int_0^a f(x)\, dx & \text{if } f(-x) = f(x) \quad \text{(Fig. 9.04–4}a\text{)} \\ 0 & \text{if } f(-x) = -f(x) \quad \text{(Fig. 9.04–4}b\text{)} \end{cases}$$

examples:

$$\int_0^{\pi} 2\sin x\,dx = 2\int_0^{\pi}\sin x\,dx = 2[-\cos x]_0^{\pi} = 2[-\cos\pi + \cos 0] = 4$$

$$\int_{-\pi/2}^{\pi/2}\sin x\,dx = 0 \qquad \text{since } \sin(-x) = -\sin x$$

$$\int_{-\pi/2}^{\pi/2}\cos x\,dx = 2\int_0^{\pi/2}\cos x\,dx = 2[\sin x]_0^{\pi/2} = 2\left[\sin\frac{\pi}{2} - \sin 0\right] = 2 \qquad \text{since } \cos(-x) = \cos x$$

(c) Change in variable. If $x = \phi(t)$, $a = \phi(\alpha)$, $b = \phi(\beta)$, then

$$\int_a^b f(x)\,dx = \int_\alpha^\beta f[\phi(t)]\phi'(t)\,dt$$

where α and β are the new limits and t is the new variable.

example:

$\int_0^1 \sqrt{1-x^2}\,dx$ in terms $x = \sin t$, $dx = \cos t\,dt$, and $t = \sin^{-1}x$ has the lower limit $\sin^{-1}0 = 0$ and $\sin^{-1}1 = \pi/2$. Thus

$$\int_0^1 \sqrt{1-x^2}\,dx = \int_0^{\pi/2}\sqrt{(1-\sin^2 t)}\cos t\,dt = \int_0^{\pi/2}\cos^2 t\,dt = \left[\frac{\sin 2t}{4} + \frac{t}{2}\right]_0^{\pi/2} = \frac{\pi}{4} \qquad \text{(Sec. 9.03–4b)}$$

(3) Tables of Definite Integrals

(a) Algebraic functions in $[0,a]$.

$$\int_0^a \sqrt{a^2+x^2}\,dx = \frac{a^2}{2}[\sqrt{2} + \ln(1+\sqrt{2})] = (1.14774\ldots)a^2$$

$$\int_0^a \sqrt{a^2-x^2}\,dx = \frac{\pi a^2}{4} = (0.78539\ldots)a^2$$

$$\int_0^a \frac{dx}{\sqrt{a^2+x^2}}dx = \ln(1+\sqrt{2}) = 0.88137\ldots$$

$$\int_0^a \frac{dx}{\sqrt{a^2-x^2}}dx = \frac{\pi}{2} = 1.57079\ldots$$

(b) Trigonometric functions in $[-a,a]$ $(m, n = 1, 2, 3, \ldots)$.

$$\int_{-a}^a \sin\frac{m\pi x}{a}\sin\frac{n\pi x}{a}\,dx = \begin{cases} 0 & \text{if } m \neq n \\ a & \text{if } m = n \end{cases}$$

$$\int_{-a}^a \sin\frac{m\pi x}{a}\cos\frac{n\pi x}{a}\,dx = 0$$

$$\int_{-a}^a \cos\frac{m\pi x}{a}\cos\frac{n\pi x}{a}\,dx = \begin{cases} 0 & \text{if } m \neq n \\ a & \text{if } m = n \end{cases}$$

(c) Trigonometric functions in $[0,\pi/2]$ $(m,n = 1,2,3,\ldots)$.

$$\int_0^{\pi/2} \sin x\,dx = \int_0^{\pi/2} \cos x\,dx = 1$$

$$\int_0^{\pi/2} \sin^2 x\,dx = \int_0^{\pi/2} \cos^2 x\,dx = \frac{\pi}{4}$$

$$\int_0^{\pi/2} \sin^3 x\,dx = \int_0^{\pi/2} \cos^3 x\,dx = \frac{2}{3}$$

$$\int_0^{\pi/2} \sin^{2n} x\,dx = \int_0^{\pi/2} \cos^{2n} x\,dx = \frac{(1)(3)(5)\cdots(2n-1)}{(2)(4)(6)\cdots(2n)}\,\frac{\pi}{2}$$

$$\int_0^{\pi/2} \sin^{2n+1} x\,dx = \int_0^{\pi/2} \cos^{2n+1} x\,dx = \frac{(2)(4)(6)\cdots(2n)}{(3)(5)(7)\cdots(2n+1)}$$

$$\int_0^{\pi/2} \sin^{2m-1} x \cos^{2n-1} x\,dx = \frac{(m-1)!(n-1)!}{2(m+n-1)!}$$

(d) Trigonometric functions in $[0,\pi]$ $(m,\, n = 1,2,3,\ldots)$.

$$\int_0^{\pi} \sin x\,dx = 2 \qquad \int_0^{\pi} \cos x\,dx = 0$$

$$\int_0^{\pi} \sin^2 x\,dx = \frac{\pi}{2} \qquad \int_0^{\pi} \cos^2 x\,dx = \frac{\pi}{2}$$

$$\int_0^{\pi} \sin^3 x\,dx = \frac{4}{3} \qquad \int_0^{\pi} \cos^3 x\,dx = 0$$

$$\int_0^{\pi} \cos mx \cos nx\,dx = \begin{cases} 0 & \text{if } m \neq n \\ \dfrac{\pi}{2} & \text{if } m = n \end{cases}$$

$$\int_0^{\pi} \cos mx \sin nx\,dx = \begin{cases} 0 & \text{if } m \neq n \text{ and } m+n \text{ is even} \\ \dfrac{2n}{n^2 - m^2} & \text{if } m \neq n \text{ and } m+n \text{ is odd} \\ 0 & \text{if } m = n \end{cases}$$

$$\int_0^{\pi} \sin mx \sin mx\,dx = \begin{cases} 0 & \text{if } m \neq n \\ \dfrac{\pi}{2} & \text{if } m = n \end{cases}$$

example:

$$\int_0^{\pi} \cos x \sin 2x\,dx = \frac{4}{4-1} = \frac{4}{3} \qquad \text{since } 1 \neq 2 \text{ and } 1+2 \text{ is odd}$$

9.05 PROPERTIES OF PLANE CURVES

(1) Length of a Plane Arc

(a) Cartesian coordinates. The length L of the arc $\overline{12}$ of the plane curve of Fig. 9.05–1a given explicitly by $y = f(x)$ or $x = g(y)$ is, respectively,

$$L = \int_{x_1}^{x_2} \sqrt{1+[f'(x)]^2}\, dx \qquad L = \int_{y_1}^{y_2} \sqrt{1+[g'(y)]^2}\, dy$$

where $f'(x) = \dfrac{df(x)}{dx}$

$g'(y) = \dfrac{dg(y)}{dy}$

x_1, y_1, x_2, y_2 are the cartesian coordinates of the end points of the arc

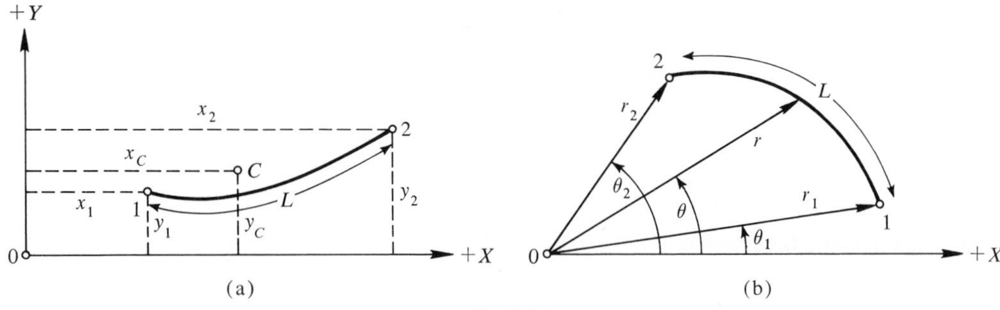

Fig. 9.05-1

(b) Polar coordinates. If the curve is defined in polar coordinates (Fig. 9.05–1b) by $r = r(\theta)$ or $\theta = \theta(r)$, then

$$L = \int_{r_1}^{r_2} \sqrt{1+r^2[\theta'(r)]^2}\, dr \qquad L = \int_{\theta_1}^{\theta_2} \sqrt{r^2+[r'(\theta)]^2}\, d\theta$$

where $\theta'(r) = \dfrac{d\theta(r)}{dr}$

$r'(\theta) = \dfrac{dr(\theta)}{d\theta}$

$r_1, r_2, \theta_1, \theta_2$ are the polar coordinates of the end points of the arc

(c) Parametric form. If the same curve (Fig. 9.05–1a) is defined parametrically by $x = x(t)$, $y = y(t)$, then

$$L = \int_{t_1}^{t_2} \sqrt{(\dot{x})^2 + (\dot{y})^2}\, dt$$

where $\dot{x} = \dfrac{dx(t)}{dt}$

$\dot{y} = \dfrac{dy(t)}{dt}$

t_1, t_2 are the parameters of the end points of the arc

(2) Static Functions of a Plane Arc

(a) Static moments of L in Fig. 9.05–1a are

$$Q_x = \int_{y_1}^{y_2} y\sqrt{1+[g'(y)]^2}\,dy = \int_{x_1}^{x_2} f(x)\sqrt{1+[f'(x)]^2}\,dx \qquad \text{about the } X \text{ axis}$$

$$Q_y = \int_{x_1}^{x_2} x\sqrt{1+[f'(x)]^2}\,dx = \int_{y_1}^{y_2} g(y)\sqrt{1+[g'(y)]^2}\,dy \qquad \text{about the } Y \text{ axis}$$

(b) Coordinates of the centroid of L in Fig. 9.05–1a are

$$x_C = \frac{Q_y}{L} \qquad y_C = \frac{Q_x}{L}$$

(3) Inertia Functions of a Plane Arc

(a) Moments of inertia of L in Fig. 9.05–1a are

$$I_{xx} = \int_{y_1}^{y_2} y^2\sqrt{1+[g'(y)]^2}\,dy = \int_{x_1}^{x_2} [f(x)]^2\sqrt{1+[f'(x)]^2}\,dx \qquad \text{about the } X \text{ axis}$$

$$I_{yy} = \int_{x_1}^{x_2} x^2\sqrt{1+[f'(x)]^2}\,dx = \int_{y_1}^{y_2} [g(y)]^2\sqrt{1+[g'(y)]^2}\,dy \qquad \text{about the } Y \text{ axis}$$

(b) Products of inertia of L in Fig. 9.05–1a are

$$I_{xy} = I_{yx} = \int_{y_1}^{y_2} g(y)y\sqrt{1+[g'(y)]^2}\,dy \qquad \text{in terms of } y$$

$$= \int_{x_1}^{x_2} xf(x)\sqrt{1+[f'(x)]^2}\,dx \qquad \text{in terms of } x$$

(c) Polar moment of inertia of L in Fig. 9.05–1a is

$$J_z = I_{xx} + I_{yy} \qquad \text{about the } Z \text{ axis}$$

where I_{xx} and I_{yy} are the moments of inertia given in (a).

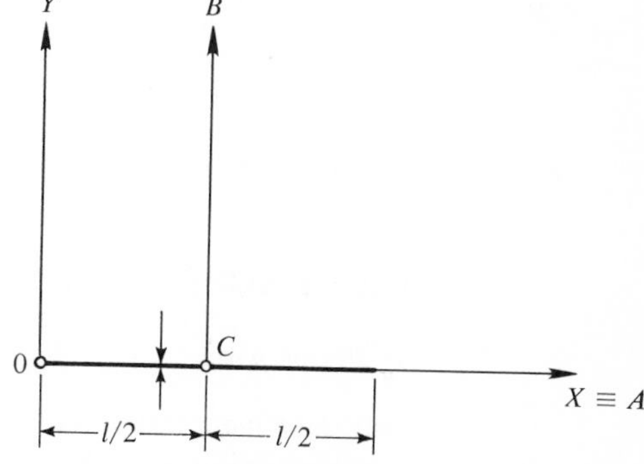

Fig. 9.05-4a

(4) Particular Cases

(a) Horizontal straight segment (Fig. 9.05–4a).*

$$L = l \qquad x_C = l/2 \qquad y_C = 0$$

$$I_{xx} = 0 \qquad I_{yy} = l^3/3 \qquad I_{xy} = 0$$

$$I_{AA} = 0 \qquad I_{BB} = l^3/12 \qquad I_{AB} = 0$$

(b) Inclined straight segment (Fig. 9.05–4b).

$$L = l \qquad x_C = \frac{l\cos\alpha}{2} \qquad y_C = \frac{l\sin\alpha}{2}$$

$$I_{xx} = \frac{l^3\sin^2\alpha}{3} \qquad I_{yy} = \frac{l^3\cos^2\alpha}{3} \qquad I_{xy} = \frac{l^3\sin\alpha\cos\alpha}{3}$$

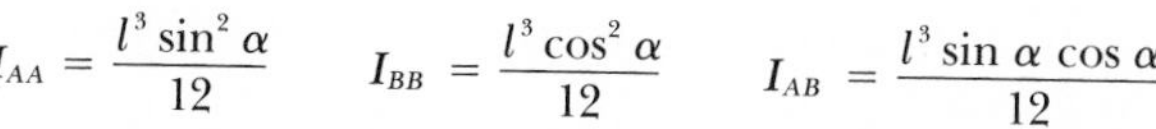

$$I_{AA} = \frac{l^3\sin^2\alpha}{12} \qquad I_{BB} = \frac{l^3\cos^2\alpha}{12} \qquad I_{AB} = \frac{l^3\sin\alpha\cos\alpha}{12}$$

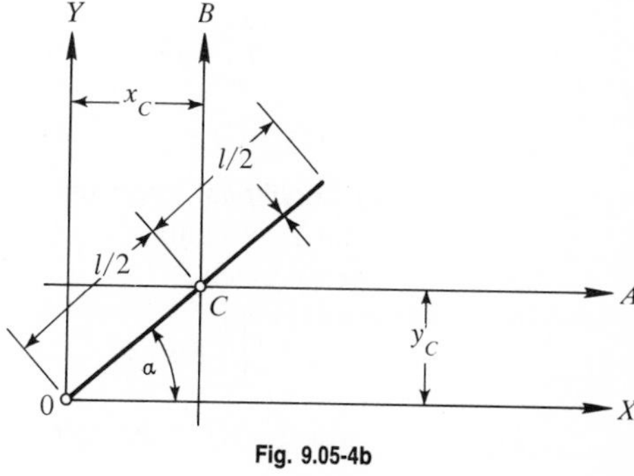

Fig. 9.05-4b

*There are no figures missing. Numbers of figures have been adjusted to match their case numbers.

(c) Circle (Fig. 9.05–4*c*).

$$L = 2\pi R \qquad x_C = y_C = R$$

$$I_{xx} = I_{yy} = 3\pi R^3 \qquad I_{xy} = 2\pi R^3$$

$$I_{AA} = I_{BB} = \pi R^3 \qquad I_{AB} = 0$$

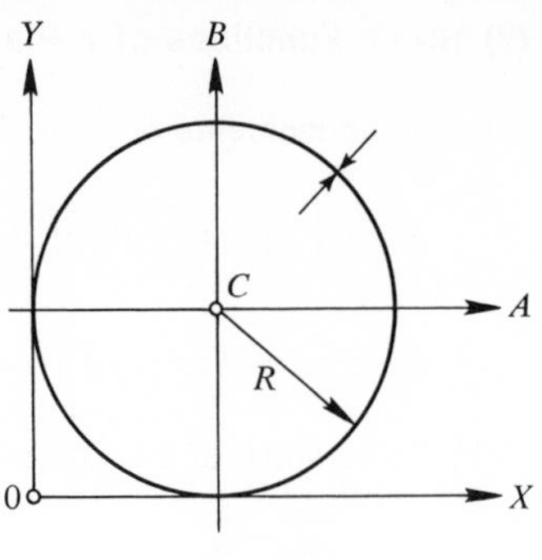

Fig. 9.05-4c

(d) Half circle (Fig. 9.05–4*d*).

$$L = \pi R \qquad x_C = R \qquad y_C = 2R/\pi$$

$$I_{xx} = \frac{\pi R^3}{2} \qquad I_{yy} = \frac{3\pi R^3}{2} \qquad I_{xy} = 2R^3$$

$$I_{AA} = \frac{(\pi^2 - 8)R^3}{2\pi} \qquad I_{BB} = \frac{\pi R^3}{2} \qquad I_{AB} = 0$$

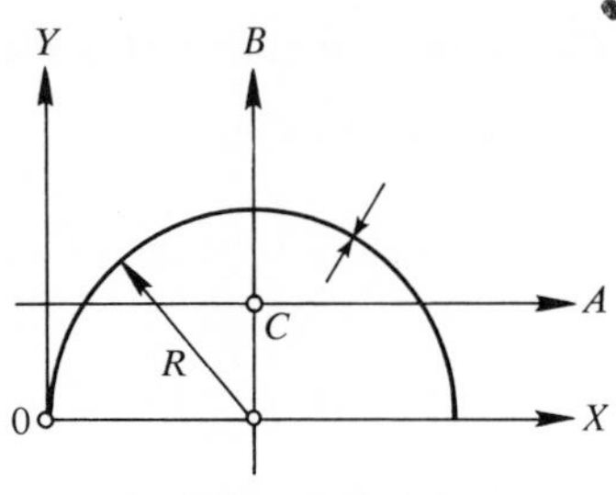

Fig. 9.05-4d

(e) Circular segment (Fig. 9.05–4*e*).

$$L = 2R\alpha \qquad x_C = R \sin\alpha \qquad y_C = R\left(\frac{\sin\alpha}{\alpha} - \cos\alpha\right)$$

$$I_{xx} = R^3(\alpha - 3\sin\alpha\cos\alpha + 2\alpha\cos^2\alpha)$$

$$I_{yy} = R^3(\alpha - \sin\alpha\cos\alpha + 2\alpha\sin^2\alpha)$$

$$I_{xy} = 2R^3\alpha\sin\alpha\left(\frac{\sin\alpha}{\alpha} - \cos\alpha\right)$$

$$I_{AA} = R^3\left(\alpha + \sin\alpha\cos\alpha - \frac{2\sin^2\alpha}{\alpha}\right)$$

$$I_{BB} = R^3(\alpha - \sin\alpha\cos\alpha) \qquad I_{AB} = 0$$

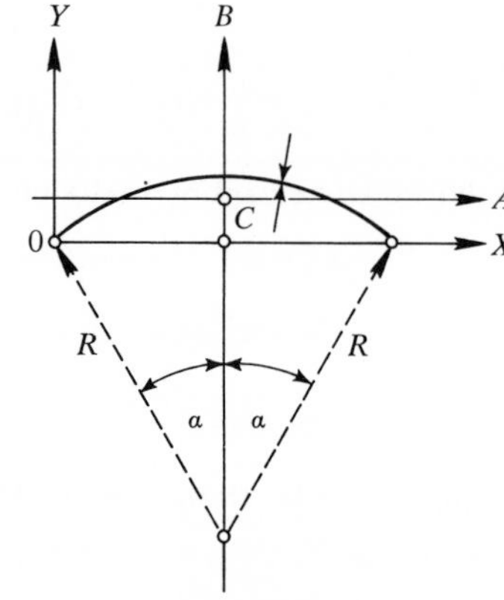

Fig. 9.05-4e

9.06 PROPERTIES OF PLANE FIGURES

(1) Plane Area

(a) Curvilinear trapezoid of Fig. 9.06–1*a*. The area of the curvilinear trapezoid bounded by the *X* axis and $y = f(x)$ in limits $a_1 \le x \le a_2$ or by $x = x(t)$, $y = y(t)$ in limits $t_1 \le t \le t_2$ is, respectively,

$$A = \int_{a_1}^{a_2} f(x)\,dx \qquad \text{or} \qquad A = \int_{t_1}^{t_2} y(t)\dot{x}(t)\,dt$$

where $\dot{x}(t) = dx(t)/dt$.

(b) Curvilinear trapezoid of Fig. 9.06–1*b*. The area of the curvilinear trapezoid bounded by the *Y* axis and $x = g(y)$ in limits $b_1 \le y \le b_2$ or by $x = x(t)$, $y = y(t)$ in limits $t_1 \le t \le t_2$ is, respectively,

$$A = \int_{b_1}^{b_2} g(y)\,dy \qquad \text{or} \qquad A = \int_{t_1}^{t_2} x(t)\dot{y}(t)\,dt$$

where $\dot{y}(t) = dy(t)/dt$.

(c) Curvilinear sector of Fig. 9.06–1*c*. The area of the curvilinear sector bounded by $r = r(\theta)$ and radii $r(\theta_1)$, $r(\theta_2)$ in limits $\theta_1 \le \theta \le \theta_2$ is

$$A = \tfrac{1}{2}\int_{\theta_1}^{\theta_2} [r(\theta)]^2\,d\theta$$

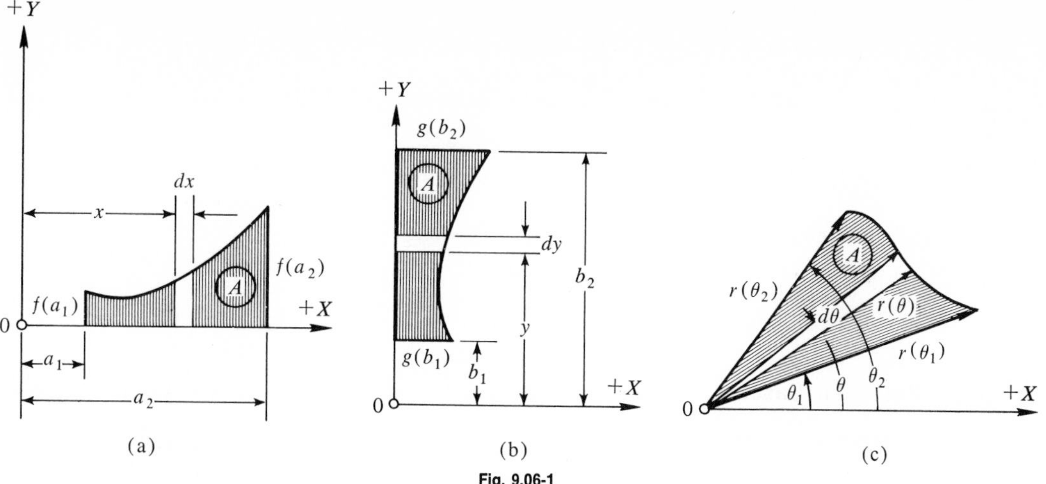

Fig. 9.06-1

(2) Static Functions of a Plane Area

(a) Static moments of A in Fig. 9.06–1a, b are, respectively,

$$Q_x = \tfrac{1}{2}\int_{a_1}^{a_2} [f(x)]^2\,dx \qquad \text{about the } X \text{ axis in Fig. 9.06–1}a$$

$$Q_y = \int_{a_1}^{a_2} xf(x)\,dx \qquad \text{about the } Y \text{ axis in Fig. 9.06–1}a$$

$$Q_x = \int_{b_1}^{b_2} yg(y)\,dy \qquad \text{about the } X \text{ axis in Fig. 9.06–1}b$$

$$Q_y = \tfrac{1}{2}\int_{b_1}^{b_2} [g(y)]^2\,dy \qquad \text{about the } Y \text{ axis in Fig. 9.06–1}b$$

(c) Coordinates of the centroid of A in Fig. 9.06–1a,b are

$$x_C = \frac{Q_y}{A} \qquad y_C = \frac{Q_x}{A}$$

where Q_x, Q_y are the respective static moments and A is the area of the figure.

(3) Inertia Functions of a Plane Area

(a) Moments of inertia of A in Fig. 9.06–1a,b are, respectively,

$$I_{xx} = \tfrac{1}{3}\int_{a_1}^{a_2} [f(x)]^3\,dx \qquad \text{about the } X \text{ axis in Fig. 9.06–1}a$$

$$I_{yy} = \int_{a_1}^{a_2} x^2 f(x)\,dx \qquad \text{about the } Y \text{ axis in Fig. 9.06–1}a$$

$$I_{xx} = \int_{b_1}^{b_2} y^2 g(y)\,dy \qquad \text{about the } X \text{ axis in Fig. 9.06–1}b$$

$$I_{yy} = \tfrac{1}{3}\int_{b_1}^{b_2} [g(y)]^3\,dy \qquad \text{about the } Y \text{ axis in Fig. 9.06–1}b$$

(b) Products of inertia of A in Fig. 9.06–1a,b are, respectively,

$$I_{xy} = I_{yx} = \tfrac{1}{2}\int_{a_1}^{a_2} x[f(x)]^2\,dx \qquad \text{in terms of } x \text{ in Fig. 9.06–1}a$$

$$I_{xy} = I_{yx} = \tfrac{1}{2}\int_{b_1}^{b_2} y[g(y)]^2\,dy \qquad \text{in terms of } y \text{ in Fig. 9.06–1}b$$

(c) Polar moments of inertia of A in Fig. 9.06–1a,b are

$$J_z = I_{xx} + I_{yy} \qquad \text{about the } Z \text{ axis}$$

where I_{xx} and I_{yy} are the moments of inertia of the respective area given in (a).

(4) Particular Cases

(a) Square (Fig. 9.06–4a).*

$$A = a^2 \qquad x_C = y_C = \frac{a}{2}$$

$$I_{xx} = I_{yy} = \frac{a^4}{3} \qquad I_{xy} = \frac{a^4}{4}$$

$$I_{AA} = I_{BB} = \frac{a^4}{12} \qquad I_{AB} = 0 \qquad I_{DD} = \frac{a^4}{12}$$

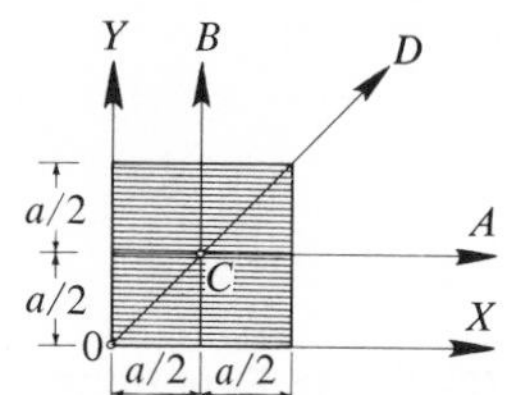

Fig. 9.06-4a

(b) Rectangle (Fig. 9.06–4b).

$$A = ab \qquad x_C = \frac{a}{2} \qquad y_C = \frac{b}{c}$$

$$I_{xx} = \frac{ab^3}{3} \qquad I_{yy} = \frac{a^3b}{3} \qquad I_{xy} = \frac{a^2b^2}{4}$$

$$I_{AA} = \frac{ab^3}{12} \qquad I_{BB} = \frac{a^3b}{12} \qquad I_{AB} = 0 \qquad I_{DD} = \frac{a^3b^3}{6(a^2+b^2)}$$

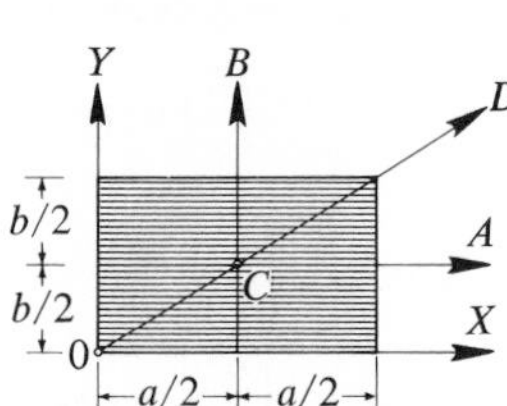

Fig. 9.06-4b

(c) Triangle (Fig. 9.06–4c).

$$A = \frac{bh}{2} \qquad x_C = \frac{b_2 - b_1}{3} \qquad y_C = \frac{h}{3}$$

$$I_{xx} = \frac{bh^3}{12} \qquad I_{yy} = \frac{(b_1^3 + b_2^3)h}{12} \qquad I_{xy} = \frac{(b_2^2 - b_1^2)h^2}{24}$$

$$I_{AA} = \frac{bh^3}{36} \qquad I_{BB} = \frac{(b_1^3 + 2bb_1b_2 + b_2^3)h}{36} \qquad I_{TT} = \frac{bh^3}{4} \qquad I_{AB} = \frac{(b_1^2 - b_2^2)h^2}{72}$$

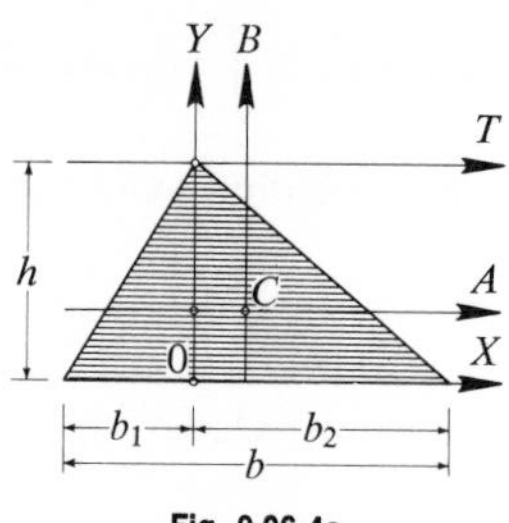

Fig. 9.06-4c

(d) Trapezoid (Fig. 9.06–4d).

$$A = \frac{(a+b)h}{2} \qquad x_C = 0 \qquad y_C = \frac{h(a+2b)}{3(a+b)}$$

$$I_{xx} = \frac{(a+3b)h^3}{12} \qquad I_{yy} = \frac{(a^4 - b^4)h}{48(a-b)} = I_{BB}$$

$$I_{AA} = \frac{(a^2 + 4ab + b^2)h^3}{36(a+b)} \qquad I_{TT} = \frac{(3a+b)h^3}{12}$$

$$I_{xy} = I_{AB} = I_{TB} = 0 \qquad I_{DD} = \frac{(e_1^3 + e_2^3)f}{12}$$

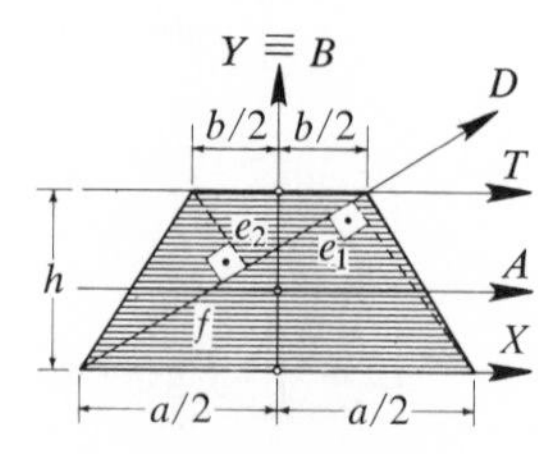

Fig. 9.06-4d

*There are no figures missing. See footnote on page 193.

(e) Regular polygon (Fig. 9.06–4*e*).

$$A = \frac{na^2}{4}\cot\frac{\alpha}{2} \qquad r = \frac{a}{2}\cot\frac{\alpha}{2} \qquad R = \frac{a}{2\sin(\alpha/2)}$$

$$I_{AA} = I_{BB} = \frac{nar(12r^2 + a^2)}{96} \qquad \alpha = \frac{360°}{n}$$

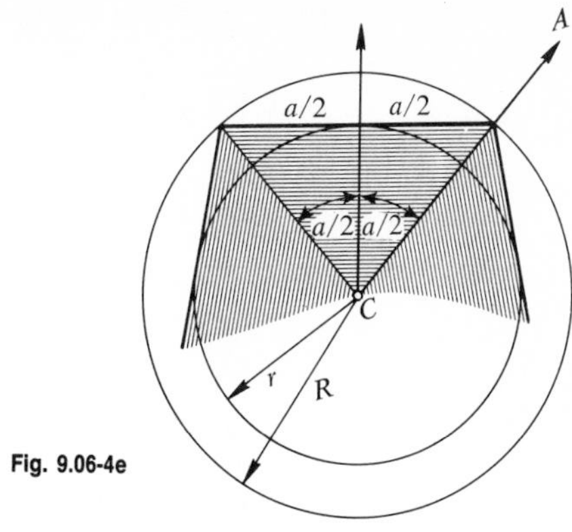

Fig. 9.06-4e

(f) Circle (Fig. 9.06–4*f*).

$$A = \pi R^2 \qquad x_C = R \qquad y_C = R$$

$$I_{xx} = I_{yy} = \frac{5\pi R^4}{4} \qquad I_{xy} = \pi R^4$$

$$I_{AA} = I_{BB} = \frac{\pi R^4}{4} \qquad I_{AB} = 0$$

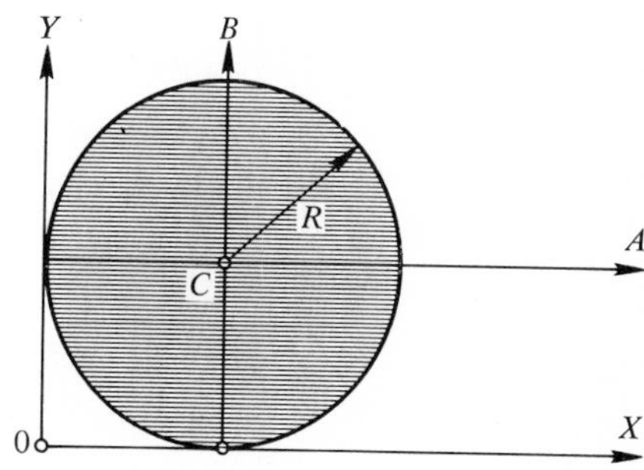

Fig. 9.06-4f

(g) Half circle (Fig. 9.06–4*g*).

$$A = \frac{\pi R^2}{2} \qquad x_C = R \qquad y_C = \frac{4R}{3\pi}$$

$$I_{xx} = \frac{\pi R^4}{8} \qquad I_{yy} = \frac{5\pi R^4}{8} \qquad I_{xy} = \frac{2R^4}{3}$$

$$I_{AA} = 0.1098R^4 \qquad I_{BB} = \frac{\pi R^4}{8} \qquad I_{AB} = 0$$

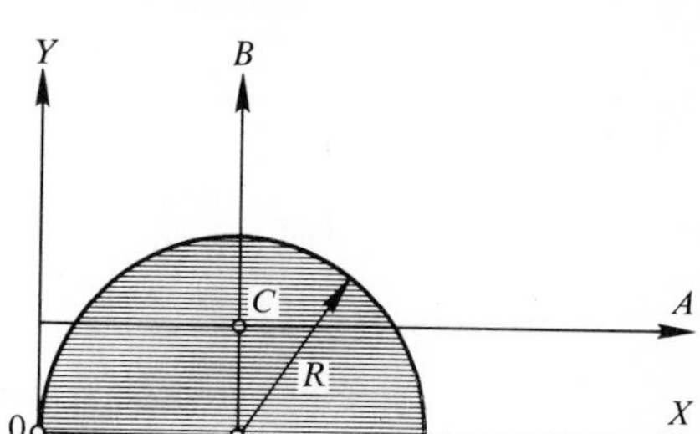

Fig. 9.06-4g

(h) Quarter circle (Fig. 9.06–4*h*).

$$A = \frac{\pi R^2}{4} \qquad x_C = \frac{4R}{3\pi} \qquad y_C = \frac{4R}{3\pi}$$

$$I_{xx} = I_{yy} = \frac{\pi R^4}{16} \qquad I_{xy} = \frac{R^4}{8}$$

$$I_{AA} = I_{BB} = 0.0549R^4 \qquad I_{AB} = -0.0165R^4$$

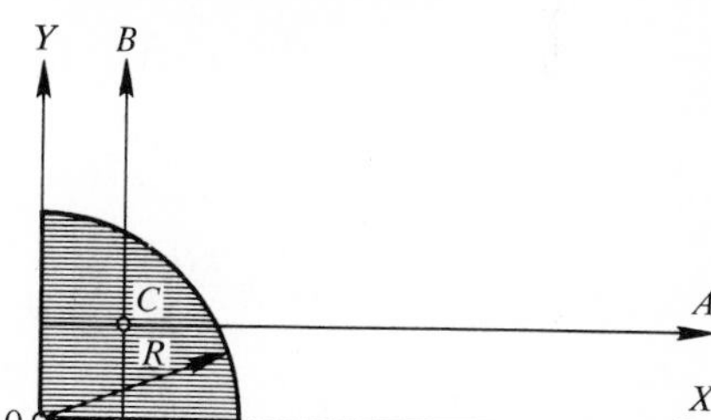

Fig. 9.06-4h

(i) 2° **parabolic complement** (Fig. 9.06–4*i*).

$$A = \frac{ab}{3} \qquad x_C = \frac{3a}{4} \qquad y_C = \frac{3b}{10}$$

$$I_{xx} = \frac{ab^3}{21} \qquad I_{yy} = \frac{a^3b}{5} \qquad I_{xy} = \frac{a^2b^2}{12}$$

$$I_{AA} = \frac{37ab^3}{2100} \qquad I_{BB} = \frac{a^3b}{80} \qquad I_{AB} = \frac{a^2b^2}{120}$$

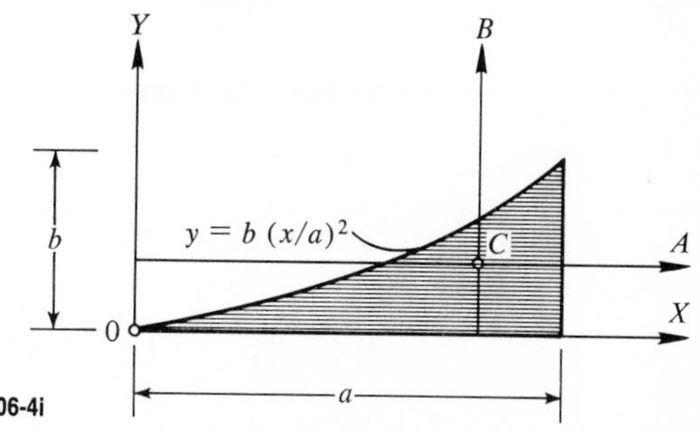

Fig. 9.06-4i

(j) $k°$ **parabolic complement** (Fig. 9.06–4j).

$$A = \frac{ab}{k+1} \qquad x_C = \frac{k+1}{k+2}a \qquad y_C = \frac{(k+1)b}{2(2k+1)}$$

$$I_{xx} = \frac{ab^3}{3(3k+1)} \qquad I_{yy} = \frac{a^3b}{k+3} \qquad I_{xy} = \frac{a^2b^2}{4(k+1)}$$

$$I_{AA} = \frac{(7k^2+4k+1)ab^3}{12(3k+1)(2k+1)^2} \qquad I_{BB} = \frac{a^3b}{(k+2)^2(k+3)} \qquad I_{AB} = \frac{ka^2b^2}{4(k+1)(k+2)(2k+1)}$$

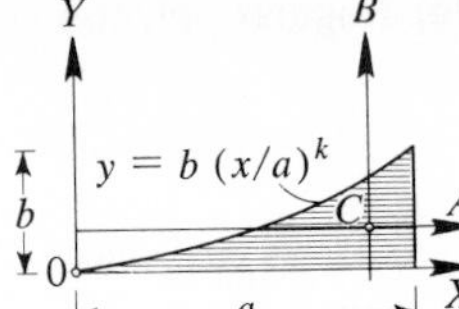

Fig. 9.06-4j

(k) 2° **parabolic segment** (Fig. 9.06–4k).

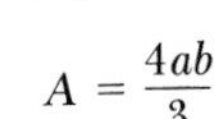

$$A = \frac{4ab}{3} \qquad x_C = \frac{3a}{5} \qquad y_C = 0$$

$$I_{xx} = \frac{4ab^3}{15} \qquad I_{yy} = \frac{4a^3b}{7} \qquad I_{xy} = 0$$

$$I_{BB} = \frac{16a^3b}{175} \qquad I_{CC} = \frac{32a^3b}{105} \qquad I_{AB} = 0$$

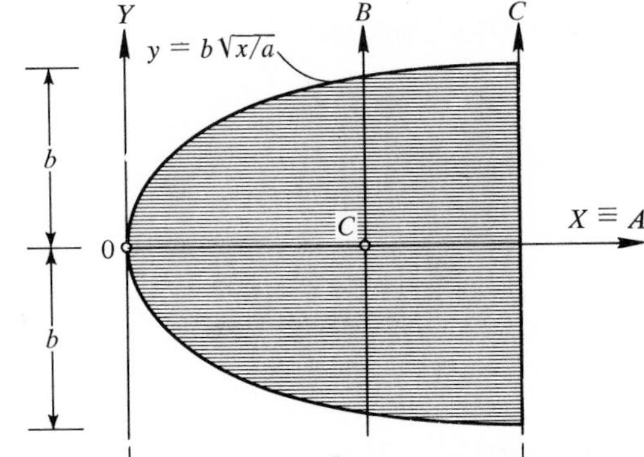

Fig. 9.06-4k

(l) 2° **parabolic sector** (Fig. 9.06–4l).

$$A = \frac{2ab}{3} \qquad x_C = \frac{3a}{5} \qquad y_C = \frac{3b}{8}$$

$$I_{xx} = \frac{2ab^3}{15} \qquad I_{yy} = \frac{2a^3b}{7} \qquad I_{xy} = \frac{a^2b^2}{6}$$

$$I_{AA} = \frac{19ab^3}{480} \qquad I_{BB} = \frac{8a^3b}{175} \qquad I_{AB} = \frac{a^2b^2}{60}$$

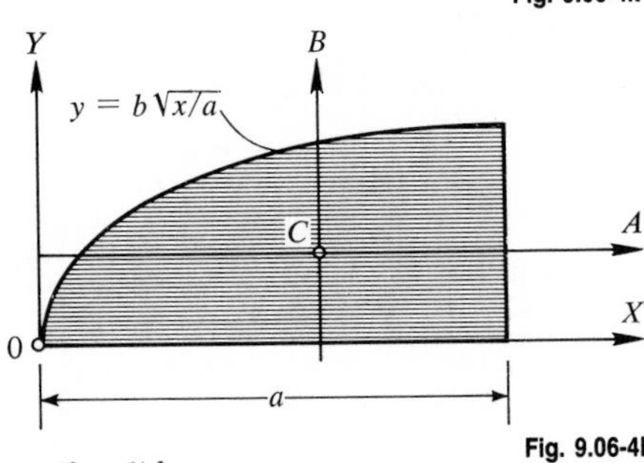

Fig. 9.06-4l

(m) $k°$ **parabolic sector** (Fig. 9.06–4m).

$$A = \frac{kab}{k+1} \qquad x_C = \frac{(k+1)a}{2k+1} \qquad y_C = \frac{(k+1)b}{2(k+2)}$$

$$I_{xx} = \frac{kab^3}{3(k+3)} \qquad I_{yy} = \frac{ka^3b}{3k+1} \qquad I_{xy} = \frac{ka^2b^2}{4(k+1)}$$

$$I_{AA} = \frac{k(k^2+4k+7)ab^3}{12(k+2)^2(k+3)} \qquad I_{BB} = \frac{k^3a^3b}{(3k+1)(2k+1)^2}$$

$$I_{AB} = \frac{k^2a^2b^2}{4(k+1)(k+2)(2k+1)}$$

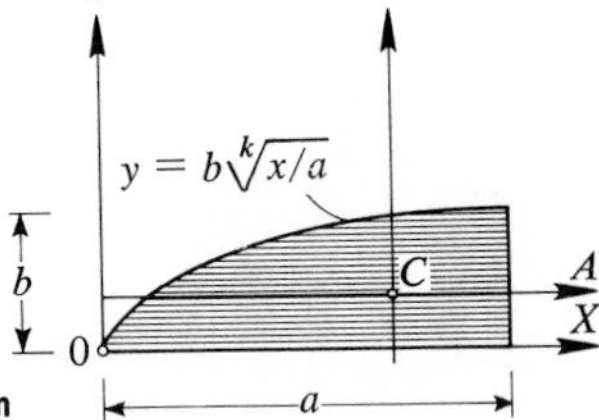

Fig. 9.06-4m

(n) Ellipse (Fig. 9.06–4n).

$$A = \pi ab \qquad x_C = a \qquad y_C = b$$

$$I_{xx} = \frac{5\pi ab^3}{4} \qquad I_{yy} = \frac{5\pi a^3b}{4} \qquad I_{xy} = \pi a^2b^2$$

$$I_{AA} = \frac{\pi ab^3}{4} \qquad I_{BB} = \frac{\pi a^3b}{4} \qquad I_{AB} = 0$$

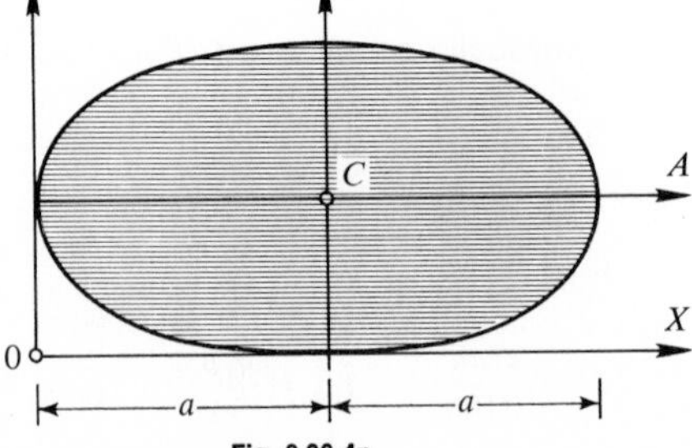

Fig. 9.06-4n

(o) Half ellipse (Fig. 9.06–4*o*).

$$A = \frac{\pi ab}{2} \qquad x_C = a \qquad y_C = \frac{4b}{3\pi}$$

$$I_{xx} = \frac{\pi ab^3}{8} \qquad I_{yy} = \frac{\pi a^3 b}{8} \qquad I_{xy} = \frac{2a^2b^2}{3}$$

$$I_{AA} = \frac{\pi ab^3}{8}\left(1 - \frac{64}{9\pi^2}\right) \qquad I_{BB} = \frac{\pi a^3 b}{8} \qquad I_{AB} = 0$$

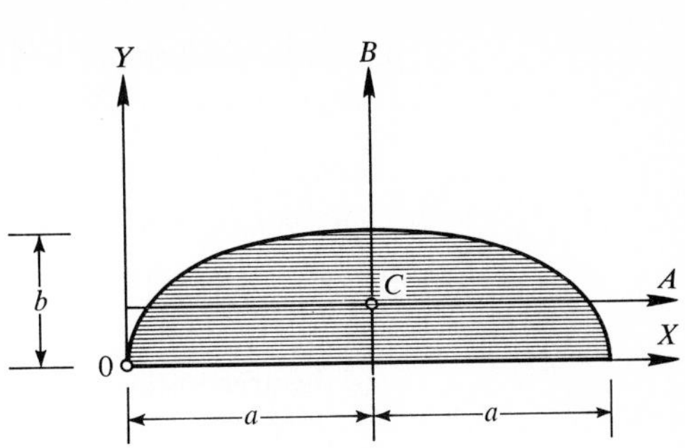

Fig. 9.06-4o

(p) Circular sector (Fig. 9.06–4*p*).

$$A = \alpha R^2 \qquad x_C = 0 \qquad y_C = \frac{2R\sin\alpha}{3\alpha}$$

$$I_{xx} = R^4 M \qquad I_{yy} = R^4 N \qquad I_{xy} = 0$$

$$I_{AA} = R^4(M - P) \qquad I_{BB} = R^4 N \qquad I_{AB} = 0$$

where $M = \dfrac{\alpha + \sin\alpha\cos\alpha}{4}$

$N = \dfrac{\alpha - \sin\alpha\cos\alpha}{4}$

$P = \dfrac{4\sin^2\alpha}{9\alpha}$

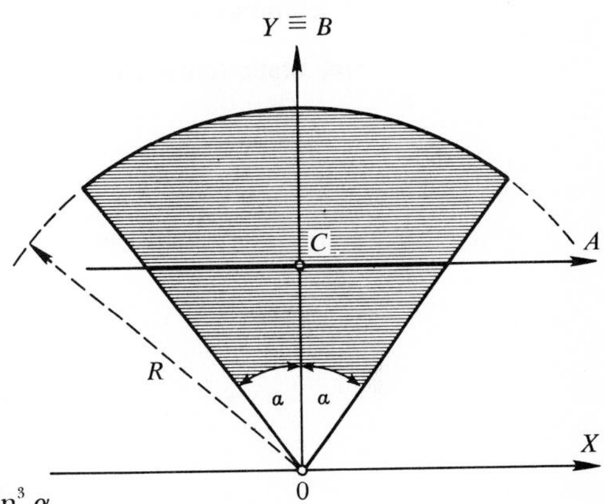

Fig. 9.06-4p

(q) Circular segment (Fig. 9.06–4*q*).

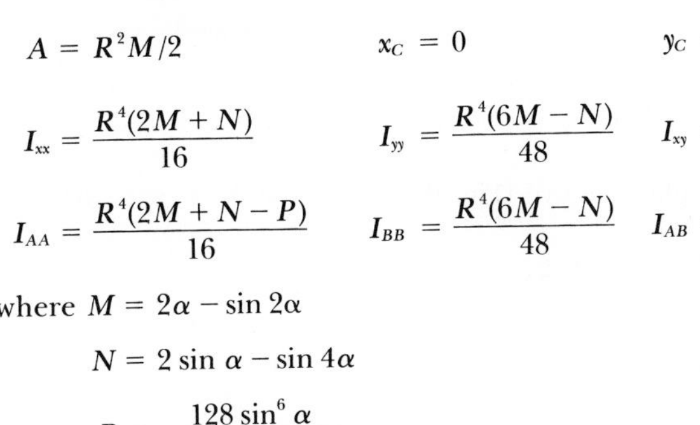

$$A = R^2M/2 \qquad x_C = 0 \qquad y_C = \frac{4R\sin^3\alpha}{3M}$$

$$I_{xx} = \frac{R^4(2M+N)}{16} \qquad I_{yy} = \frac{R^4(6M-N)}{48} \qquad I_{xy} = 0$$

$$I_{AA} = \frac{R^4(2M+N-P)}{16} \qquad I_{BB} = \frac{R^4(6M-N)}{48} \qquad I_{AB} = 0$$

where $M = 2\alpha - \sin 2\alpha$

$N = 2\sin\alpha - \sin 4\alpha$

$P = \dfrac{128\sin^6\alpha}{9(2\alpha - \sin 2\alpha)}$

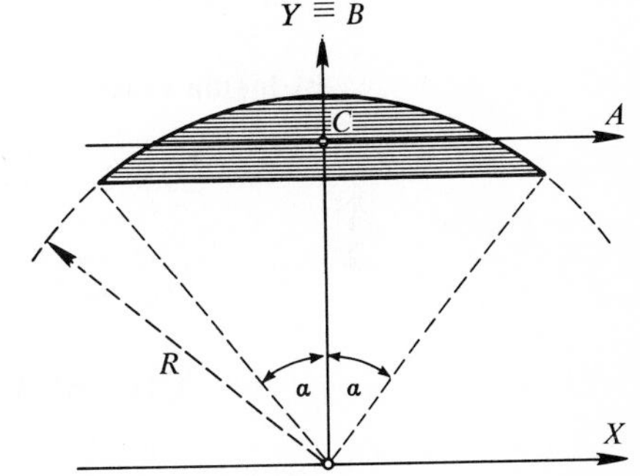

Fig. 9.06-4q

9.07 PROPERTIES OF COMPOSITE PLANE FIGURES

(1) Parallel-Axes Theorem

(a) Static moments of a single plane area. Since by definition the static moments of a single plane area with respect to its centroidal axes *A*, *B* are zero, the static moments of the same area with respect to any other set of parallel axes *X*, *Y* (Fig. 9.07–1) can be expressed as

$$Q_x = y_C A + \cancelto{0}{Q_A} \qquad Q_y = x_X A + \cancelto{0}{Q_B}$$

where A is the area, $Q_A = 0$, $Q_B = 0$ are the static moments of the area with respect to the centroidal axes *A*, *B*, and x_c, y_c are the coordinates of the centroid *C* in the *X*, *Y* axes.

(b) Inertia functions of a single plane area. Analogically, the inertia functions of the same plane area (Fig. 9.07–1) with respect to the same set of axes X, Y are

$$I_{xx} = y_C^2 A + I_{AA} \qquad I_{yy} = x_C^2 A + I_{BB}$$

$$I_{xy} = I_{yx} = x_C y_C A + I_{AB}$$

$$J_z = (x_C^2 + y_C^2)A + I_{AA} + I_{BB} = r_C^2 A + J_C$$

where I_{AA}, I_{BB}, $I_{AB} = I_{BA}$, J_C are the inertia functions of the area in the A, B axes, parallel to the X, Y axes.

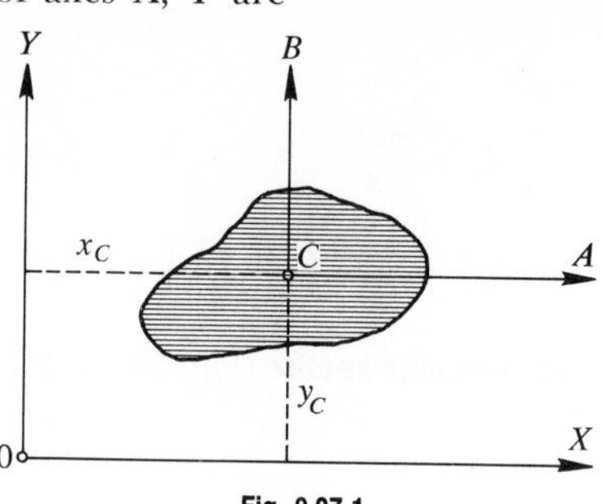

Fig. 9.07-1

(2) Summation Formulas

(a) Static functions of a composite plane figure in the X, Y axes are

$$A = A_1 + A_2 + \cdots + A_m = \sum_m A_i$$

$$Q_x = y_1 A_1 + y_2 A_2 + \cdots + y_m A_m = \sum_m y_i A_i$$

$$Q_y = x_1 A_1 + x_2 A_2 + \cdots + x_m A_m = \sum_m x_i A_i$$

$$x_C = \frac{\sum_m x_i A_i}{\sum_m A_i} \qquad y_C = \frac{\sum_m y_i A_i}{\sum_m A_i}$$

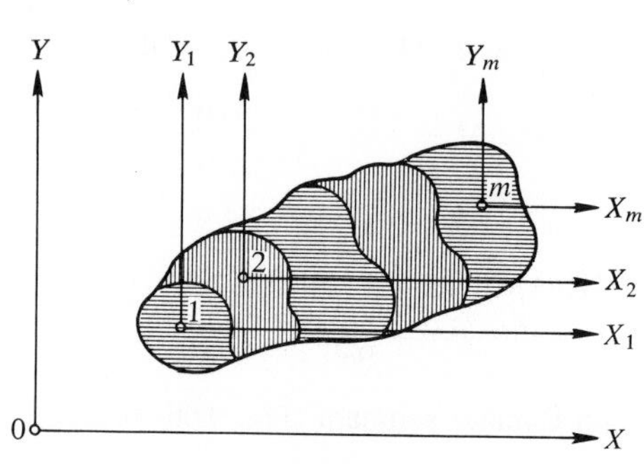

Fig. 9.07-2

where $A_1, A_2, \ldots, A_m$ are the areas of the respective parts, $x_1, x_2, \ldots, x_m, y_1, y_2, \ldots, y_m$ are the coordinates of the respective centroids, and $i = 1, 2, \ldots, m$ (Fig. 9.07–2).

(b) Inertia functions of a composite plane figure (Fig. 9.07–2) in the X, Y axes are

$$I_{xx} = \sum_m y_i^2 A + \sum_m I_{ixx} \qquad I_{yy}' = \sum_m x_i^2 A + \sum_m I_{iyy}$$

$$I_{xy} = I_{xy} = \sum_m x_i y_i A_i + \sum_m I_{ixy} = \sum_m x_i y_i A_i + \sum_m I_{iyx}$$

$$J_z = \sum_m (x_i^2 + y_i^2) A_i + \sum_m (I_{ixx} + I_{iyy}) = \sum_m r_i^2 A_i + \sum_m J_{iz}$$

where I_{ixx}, I_{iyy}, $I_{ixy} = I_{iyx}$, J_{iz} are the inertia functions of A_i in its centroidal axes X_i, Y_i parallel to the X, Y axes.

(3) Particular Cases

(a) Hollow circle (Fig. 9.07–3a).

$$A = \pi(a^2 - b^2) \qquad x_C = a \qquad y_C = a$$

$$I_{xx} = I_{yy} = \frac{A}{4}(5a^2 + b^2) \qquad I_{xy} = a^2 A$$

$$I_{AA} = I_{BB} = \frac{A}{4}(a^2 + b^2) \qquad I_{AB} = 0$$

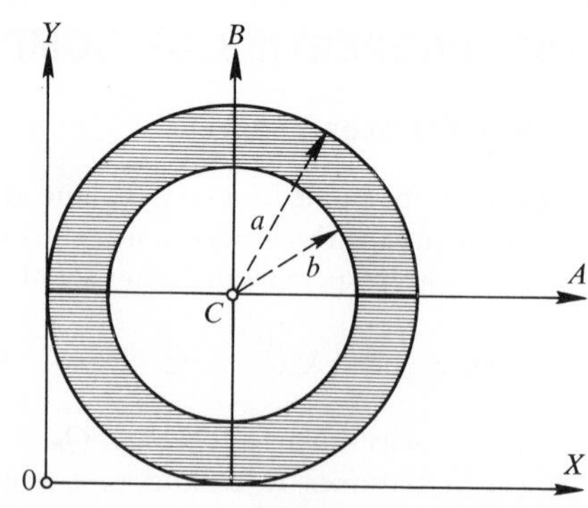

Fig. 9.07-3a

(b) Half hollow circle (Fig. 9.07–3*b*).

$$A = \frac{\pi(a^2 - b^2)}{2} \qquad x_C = a \qquad y_C = \frac{2(a^3 - b^3)}{3A}$$

$$I_{xx} = \frac{A}{4}(a^2 + b^2) \qquad I_{yy} = \frac{A}{4}(5a^2 + b^2) \qquad I_{xy} = \frac{2(a^3 - b^3)a}{3\pi}$$

$$I_{AA} = I_{xx} - y_C^2 A \qquad I_{BB} = \frac{A}{4}(a^2 + b^2) \qquad I_{AB} = 0$$

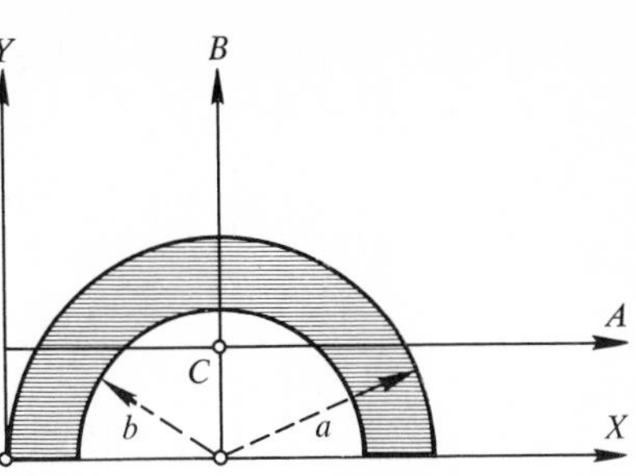

Fig. 9.07-3b

(c) Hollow rectangle (Fig. 9.07–3*c*).

$$A = ab - cd \qquad x_C = \frac{a}{2} \qquad y_C = \frac{b}{2}$$

$$I_{xx} = \frac{b^2}{4}A + I_{AA} \qquad I_{yy} = \frac{a^2}{2}A + I_{BB} \qquad I_{xy} = \frac{ab}{4}A$$

$$I_{AA} = \frac{ab^3 - cd^3}{12} \qquad I_{BB} = \frac{a^3b - c^3d}{12} \qquad I_{AB} = 0$$

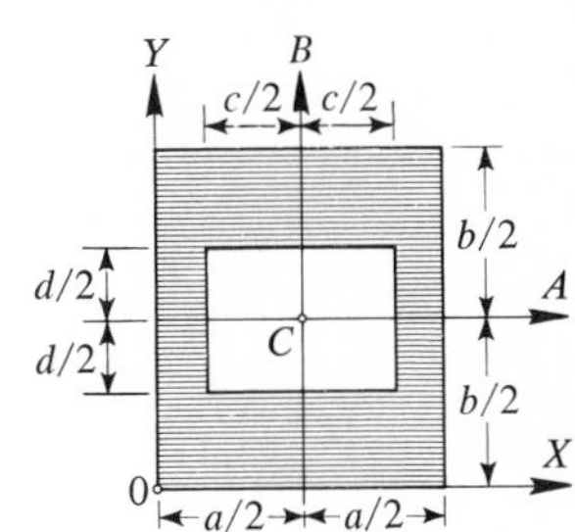

Fig. 9.07-3c

(d)  **-Section** (Fig. 9.07–3*d*).

$$A = ab - cd \qquad x_C = \frac{a}{2} \qquad y_C = \frac{b}{2}$$

$$I_{AA} = \frac{ab^3 - cd^3}{12} \qquad I_{BB} = \frac{a^3b - c^3d}{12} \qquad I_{AB} = 0$$

$$I_{xx} = \frac{b^2}{4}A + I_{AA} \qquad I_{yy} = \frac{a^2}{4}A + I_{BB} \qquad I_{xy} = \frac{ab}{4}A$$

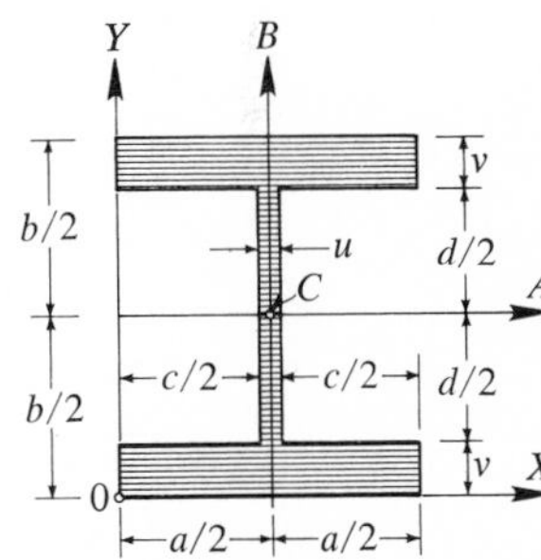

Fig. 9.07-3d

(e) ⊏ -Section (Fig. 9.07–3*e*).

$$A = ab - cd \qquad x_C = e = \frac{1}{2}\frac{2a^2v + u^2d}{A} \qquad y_C = \frac{b}{2}$$

$$I_{AA} = \frac{ab^3 - cd^3}{12} \qquad I_{BB} = \frac{e^3b - f^3d + 2g^3v}{3} \qquad I_{AB} = 0$$

$$I_{xx} = \frac{b^2}{4}A + I_{AA} \qquad I_{yy} = e^2A + I_{BB} \qquad I_{xy} = \frac{be}{2}A$$

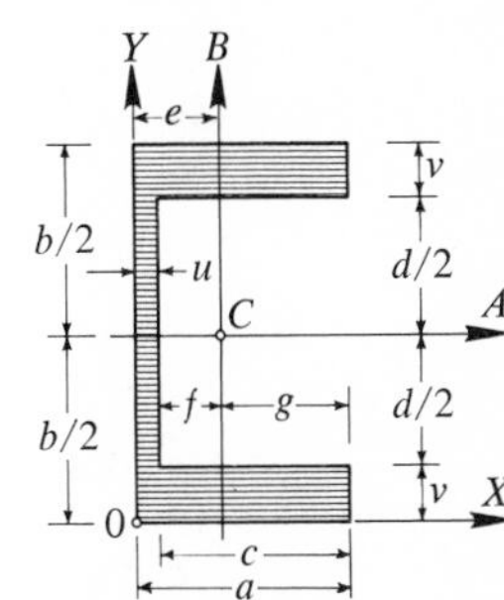

Fig. 9.07-3e

(f) ⊥ -Section (Fig. 9.07–3*f*).

$$A = av + du \qquad x_C = 0$$

$$y_C = \frac{1}{2}\frac{(a-u)v^2 + ub^2}{A} = e \qquad f = b - e$$

$$I_{AA} = \frac{ae^3 - (a-u)(e-v)^3 + uf^3}{3} \qquad I_{BB} = \frac{a^3v + u^3(b-v)}{12} \qquad I_{AB} = 0$$

$$I_{xx} = e^2A + I_{AA} \qquad I_{yy} = I_{BB} \qquad I_{xy} = 0$$

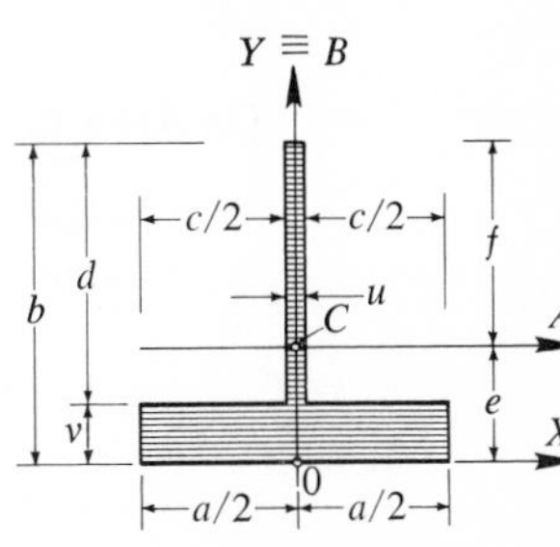

Fig. 9.07-3f

(g) I-Section (Fig. 9.07–3*g*).

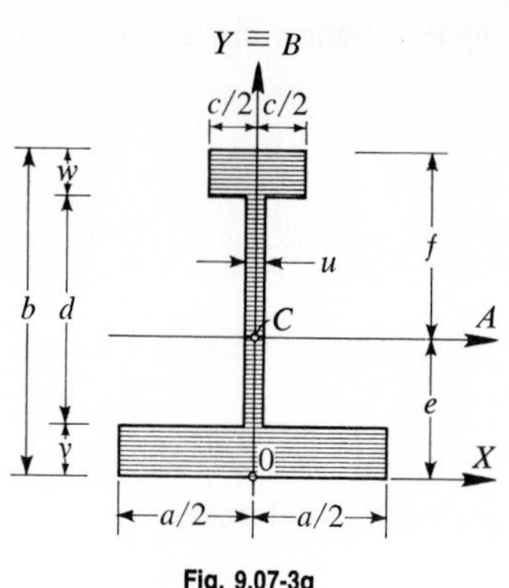

Fig. 9.07-3g

$$A = av + cw + du \qquad f = b - e$$

$$y_C = \frac{1}{2}\frac{(a-u)v^2 + (c-u)(2b-w)w + ud^2}{A} = e$$

$$I_{AA} = \frac{ae^3 - (a-u)(e-v)^3 + cf^3 - (c-u)(f-w)^3}{3}$$

$$I_{BB} = \frac{a^3v + c^3w + u^3d}{12} \qquad I_{AB} = 0$$

$$I_{xx} = e^2A + I_{AA} \qquad I_{yy} = I_{BB} \qquad I_{xy} = 0$$

(h) L-Section (Fig. 9.07–3*h*).

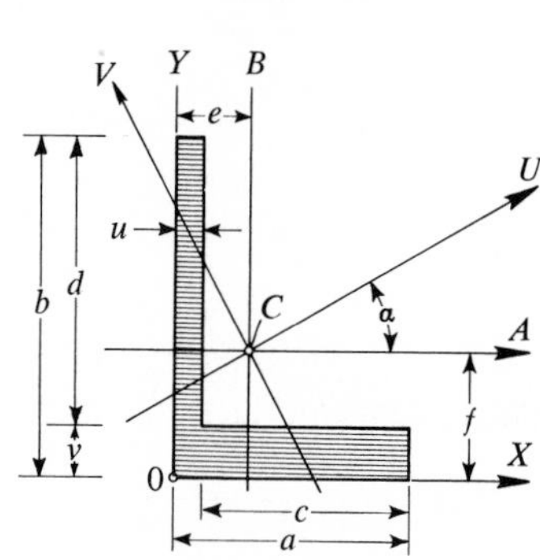

Fig. 9.07-3h

$$A = av + du$$

$$x_C = \frac{1}{2}\frac{u(b-v) + a^2}{a+b-u} = e \qquad y_C = \frac{1}{2}\frac{u(a-u) + b^2}{a+b-v} = f$$

$$I_{AA} = \frac{af^3 - (a-u)(f-v)^3 + u(b-f)^3}{3}$$

$$I_{BB} = \frac{be^3 - d(e-u)^3 + v(a-u)^3}{3} \qquad \tan 2\alpha = -\frac{2I_{AB}}{I_{AA} - I_{BB}}$$

$$I_{AB} = \frac{ab(a-2e)(b-2f) - cd(2u+c-2e)(2v+d-2f)}{4}$$

$$I_{xx} = f^2A + I_{AA} \qquad I_{yy} = e^2A + I_{BB} \qquad I_{xy} = efA + I_{AB}$$

where α is the position angle of the principal axes U, V.

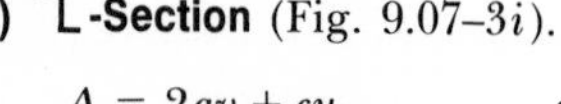

(i) ⅂-Section (Fig. 9.07–3*i*).

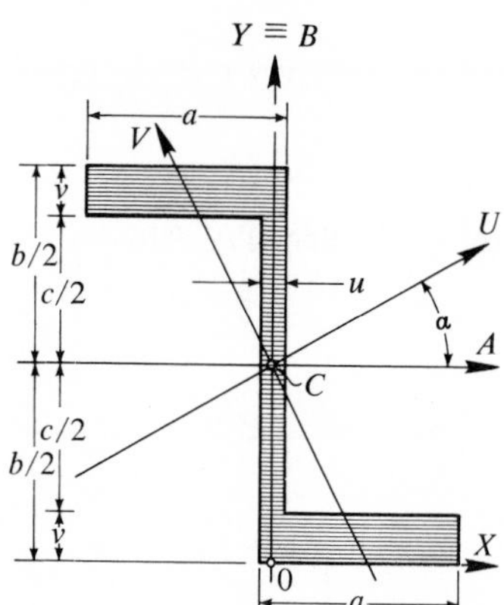

Fig. 9.07-3i

$$A = 2av + cu \qquad x_C = 0 \qquad y_C = \frac{b}{2}$$

$$I_{AA} = \frac{ab^3 - (a-u)c^3}{12} \qquad I_{xx} = \frac{b^2}{4}A + I_{AA}$$

$$I_{BB} = \frac{bu^3 + [(2a-u)^3 - u^3]v}{12} = I_{yy}$$

$$I_{AB} = -\frac{a(a-u)(b-v)v}{2} = I_{xy} \qquad \tan 2\alpha = -\frac{2I_{AB}}{I_{AA} - I_{BB}}$$

where α is the position angle of the principal axes U, V.

9.08 PROPERTIES OF SURFACES OF REVOLUTION

(1) Area of a Surface

(a) Rotation of $f(x)$ about the X axis. The area S of a surface of revolution generated by the rotation of the plane curve defined by $y = f(x)$ about the X axis (Fig. 9.08–1*a*) is

$$S = 2\pi \int_{a_1}^{a_2} f(x)\sqrt{1 + [f'(x)]^2}\, dx$$

where $y' = df(x)/dx = f'(x)$ and $a_1 \le x \le a_2$.

(b) Rotation of $g(y)$ about the Y axis. The area S of a surface of revolution generated by the rotation of a plane curve defined by $x = g(y)$ about the Y axis is

$$S = 2\int_{b_1}^{b_2} g(y)\sqrt{1+[g'(y)]^2}\,dy$$

where $x' = dg(y)/dy = g'(y)$ and $b_1 \le y \le b_2$.

Fig. 9.08-1a

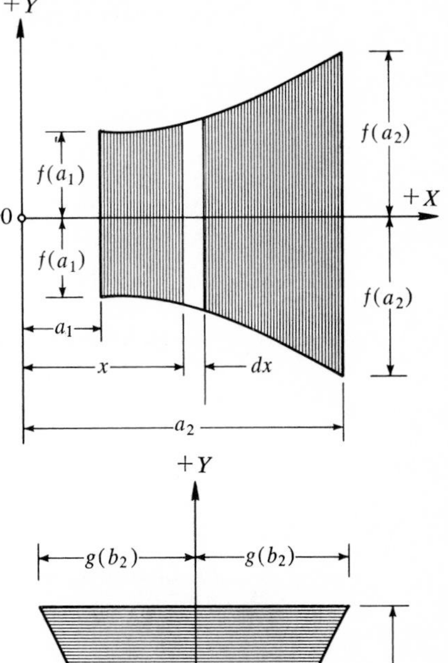

(2) Static Functions of a Surface

(a) Static moments of S are, in Fig. 9.08–1a,

$$Q_x = 0 \qquad \text{about the } X \text{ axis}$$

$$Q_y = Q_z = 2\pi\int_{a_1}^{a_2} xf(x)\sqrt{1+[f'(x)]^2}\,dx \qquad \text{about the } Y \text{ or } Z \text{ axis}$$

and in Fig. 9.08–1b,

$$Q_x = Q_z = 2\pi\int_{b_1}^{b_2} yg(y)\sqrt{1+[g'(y)]^2}\,dy \qquad \text{about the } X \text{ or } Z \text{ axis}$$

$$Q_y = 0 \qquad \text{about the } Y \text{ axis}$$

(b) Coordinates of the centroid of S in Fig. 9.08–1a are

$$x_C = \frac{Q_y}{S} \qquad y_C = 0 \qquad z_C = 0$$

and in Fig. 9.08–1b,

$$x_C = 0 \qquad y_C = \frac{Q_x}{S} \qquad z_C = 0$$

(3) Inertia Functions of a Surface

(a) Moments of inertia of S in Fig. 9.08–1a are

$$I_{xx} = 2\pi\int_{a_1}^{a_2} [f(x)]^3\sqrt{1+[f'(x)]^2}\,dx \qquad \text{about the } X \text{ axis}$$

$$I_{yy} = I_{zz} = 2\pi\int_{a_1}^{a_2} \left\{x^2 + \frac{[f(x)]^2}{2}\right\} f(x)\sqrt{1+[f'(x)]^2}\,dx$$

about the Y or Z axis

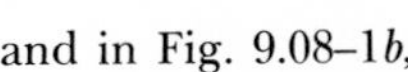

and in Fig. 9.08–1b,

$$I_{xx} = I_{zz} = 2\pi\int_{b_1}^{b_2} \left\{y^2 + \frac{[g(y)]^2}{2}\right\} g(y)\sqrt{1+[g'(y)]^2}\,dy \qquad \text{about the } X \text{ or } Z \text{ axis}$$

$$I_{yy} = 2\pi\int_{b_1}^{b_2} [g(y)]^3\sqrt{1+[g'(y)]^2}\,dy \qquad \text{about the } Y \text{ axis}$$

Fig. 9.08-1b

(b) Products of inertia of S in Figs. 9.08–1a, b are

$$I_{xy} = I_{yz} = I_{zx} = 0$$

since the surfaces are axial symmetrical with respect to the X and Y axis respectively.

(c) Polar moment of inertia of S in Figs. 9.08–1a, b is

$$J = \frac{I_{xx} + I_{yy} + I_{zz}}{2}$$

where I_{xx}, I_{yy}, I_{zz} are the respective moments of inertia given in (a).

(4) Particular Cases (shell thickness $t = 1$)

(a) Spherical shell (Fig. 9.08–4*a*).*

$$S = 4\pi R^2 \qquad x_C = y_C = 0 \qquad z_C = R$$

$$I_{xx} = I_{yy} = \frac{5SR^2}{3} \qquad I_{zz} = \frac{2SR^2}{3}$$

$$I_{AA} = I_{BB} = I_{CC} = \frac{2SR^2}{3}$$

Fig. 9.08-4a

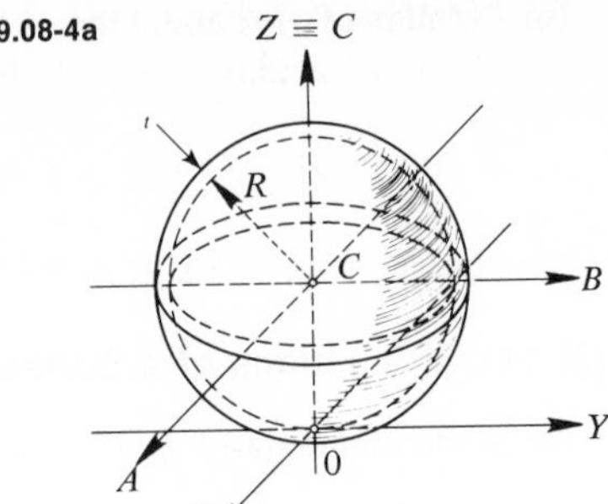

(b) Hemispherical shell (Fig. 9.08–4*b*).

$$S = 2\pi R^2 \qquad x_C = y_C = 0 \qquad z_C = -\frac{R}{2}$$

$$I_{xx} = I_{yy} = I_{zz} = \frac{2SR^2}{3}$$

$$I_{AA} = I_{BB} = \frac{5SR^2}{12} \qquad I_{CC} = \frac{2SR^2}{3}$$

Fig. 9.08-4b

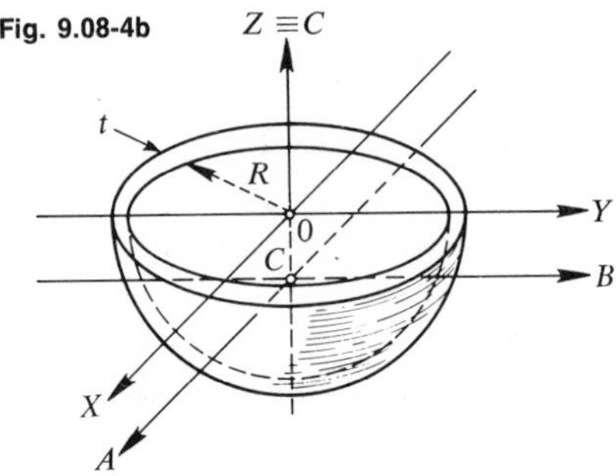

(c) Circular cylindrical shell (Fig. 9.08–4*c*).

$$S = 2\pi Rh \qquad x_C = y_C = 0 \qquad z_C = \frac{h}{2}$$

$$I_{xx} = I_{yy} = \frac{S}{6}(3R^2 + 2h^2) \qquad I_{zz} = SR^2$$

$$I_{AA} = I_{BB} = \frac{S}{12}(6R^2 + h^2) \qquad I_{CC} = SR^2$$

Fig. 9.08-4c

Z ≡ C, t, R, h/2, B, h/2, C, 0, Y, A, X

(d) Half circular cylindrical shell (Fig. 9.08–4*d*).

$$S = \pi Rh \qquad x_C = \frac{h}{2} \qquad y_C = 0 \qquad z_C = -\frac{2R}{\pi}$$

$$I_{xx} = SR^2 \qquad I_{yy} = I_{zz} = \frac{S}{6}(3R^2 + 2h^2)$$

$$I_{AA} = SR^2\left(1 - \frac{4}{\pi^2}\right) \qquad I_{BB} = \frac{S}{12}\left(6R^2 - \frac{48R^2}{\pi^2} + h^2\right)$$

$$I_{CC} = I_{NN} = \frac{S}{12}(6R^2 + h^2)$$

Fig. 9.08-4d

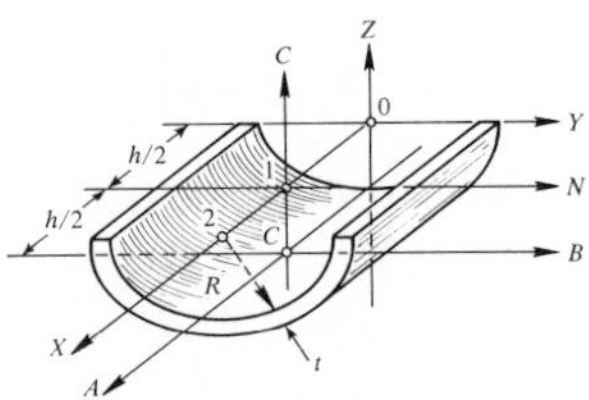

(e) Circular conical shell (Fig. 9.08–4*e*).

$$S = \pi R\sqrt{R^2 + h^2} \qquad x_C = 0 \qquad y_C = 0 \qquad z_C = \frac{2h}{3}$$

$$I_{xx} = I_{yy} = \frac{S}{2}(R^2 + 2h^2) \qquad I_{zz} = \frac{SR^2}{2}$$

$$I_{AA} = I_{BB} = \frac{S}{18}(9R^2 + 10h^2) \qquad I_{CC} = \frac{SR^2}{2}$$

*There are no figures missing. See footnote, p. 193.

Fig. 9.08-4e

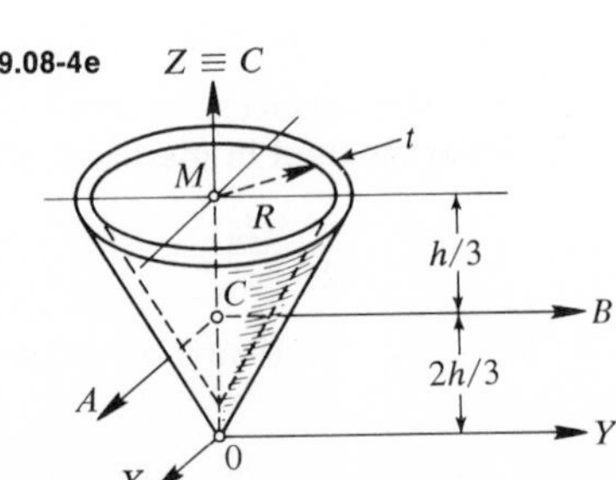

(f) Half circular conical shell (Fig. 9.08–4f).

$$S = \frac{\pi R}{2}\sqrt{R^2+h^2} \qquad x_C = \frac{2h}{3} \qquad y_C = 0 \qquad z_C = -\frac{4R}{3\pi}$$

$$I_{xx} = \frac{SR^2}{2} \qquad I_{yy} = I_{zz} = \frac{S}{2}(R^2+2h^2)$$

$$I_{AA} = \frac{SR^2}{18}\left(9-\frac{32}{\pi^2}\right) \qquad I_{BB} = \frac{S}{18}\left(9R^2-\frac{32}{\pi^2}R^2+10h^2\right)$$

$$I_{CC} = \frac{S}{18}(9R^2+10h^2) \qquad I_{NN} = \frac{S}{6}(3R^2+4h^2)$$

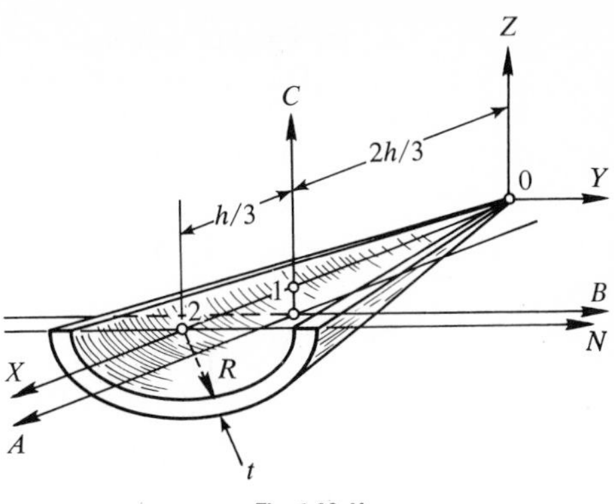

Fig. 9.08-4f

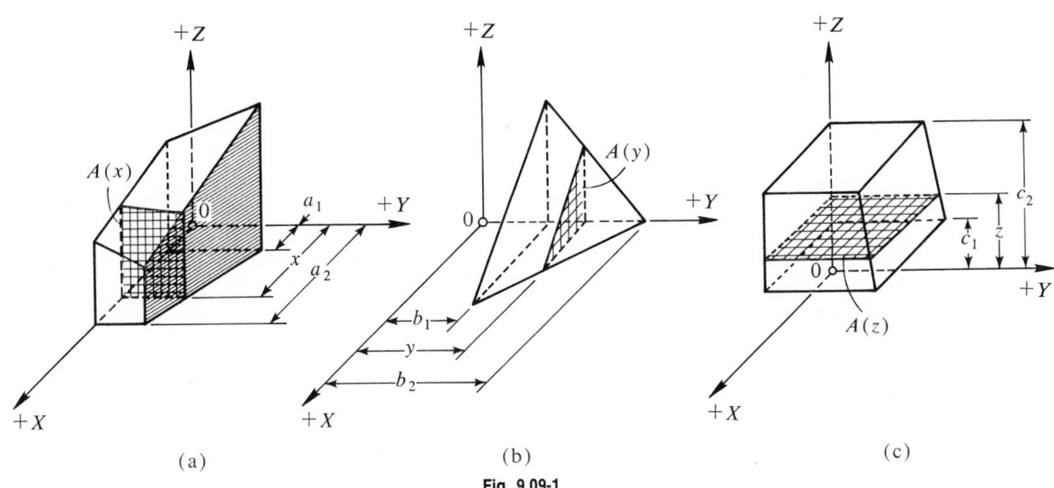

Fig. 9.09-1

9.09 PROPERTIES OF HOMOGENEOUS BODIES

(1) Volume of a Body

(a) Circular sections. The volume V of a body of revolution generated by the rotation of the plane curve $y = f(x)$ about the X axis (Fig. 9.08–1a) is

$$V = \int_{a_1}^{a_2} A(x)\,dx = \pi\int_{a_1}^{a_2} [f(x)]^2\,dx$$

where $A(x)$ is the area of the circular section normal to the X axis. Similarly, the volumes of bodies of revolution generated by the rotation of the plane curved about the Y and Z axes are respectively

$$V = \int_{b_1}^{b_2} A(y)\,dy = \pi\int_{b_1}^{b} [g(y)]^2\,dy$$

$$V = \int_{c_1}^{c_2} A(z)\,dz = \pi\int_{c_1}^{c_2} [h(z)]^2\,dz$$

where $A(y)$ and $A(z)$ are the areas of the circular sections normal to the Y and Z.

(b) Parallel sections. The volumes V of bodies formed by sections parallel to a fixed place (Fig. 9.09–1) are respectively

$$V = \int_{a_1}^{a_2} A(x)\,dx \qquad A(x) \text{ parallel to the } Y, Z \text{ plane, Fig. 9.09–1}a$$

$$V = \int_{b_1}^{b_2} A(y)\,dy \qquad A(y) \text{ parallel to the } Z, X \text{ plane, Fig. 9.09–1}b$$

$$V = \int_{c_1}^{c_2} A(z)\,dz \qquad A(z) \text{ parallel to the } X, Y \text{ plane, Fig. 9.09–1}c$$

(2) Static Functions of a Body

(a) Static moments of V are respectively

$$Q_x = \int_{a_1}^{a_2} \sqrt{(b^2+c^2)}A(x)\,dx \qquad \text{about the } X \text{ axis}$$

$$Q_y = \int_{a_1}^{a_2} \sqrt{(c^2+a^2)}A(x)\,dx \qquad \text{about the } Y \text{ axis}$$

$$Q_z = \int_{a_1}^{a_2} \sqrt{a^2+b^2}A(x)\,dx \qquad \text{about the } Z \text{ axis}$$

where $a = a(x)$, $b = b(x)$, $c = c(x)$ are the coordinates of the centroid of $A(x)$. Q_x, Q_y, Q_z may be also expressed in terms of $A(y)$ or $A(z)$ as the definite integrals in $b_1 \le y \le b_2$ or in $c_1 \le z \le c_2$, respectively.

examples:

$$Q_x = \int_{b_1}^{b_2} \sqrt{(b^2+c^2)}A(y)\,dy \qquad \text{where } b = b(y),\ c = c(y)$$

$$Q_x = \int_{c_1}^{c_2} \sqrt{(b^2+c^2)}A(z)\,dz \qquad \text{where } b = b(z),\ c = c(z)$$

(b) Coordinates of the centroid of V are

$$x_C = \frac{\int_{a_1}^{a_2} aA(x)\,dx}{\int_{a_1}^{a_2} A(x)\,dx} \qquad y_C = \frac{\int_{a_1}^{a_2} bA(x)\,dx}{\int_{a_1}^{a_2} A(x)\,dx} \qquad z_C = \frac{\int_{a_1}^{a_2} cA(x)\,dx}{\int_{a_1}^{a_2} A(x)\,dx}$$

where a, b, c and $A(x)$ have the same meaning as in Sec. 9.09–1a. x_C, y_C, z_C may be also expressed in terms of $A(y)$ or $A(z)$.

$$x_C = \frac{\int_{b_1}^{b_2} aA(y)\,dy}{\int_{b_1}^{b_2} A(y)\,dy} \cdots \qquad \text{where } a = a(y), \ldots$$

or

$$x_C = \frac{\int_{c_1}^{c_2} aA(z)\,dz}{\int_{c_1}^{c_2} A(z)\,dz} \cdots \qquad \text{where } a = a(z), \ldots$$

(3) Inertia Functions of a Body

(a) Moments of inertia of V are respectively,

$$I_{xx} = \int_{a_1}^{a_2} [(b^2 + c^2)A(x) + J_x(x)]\, dx \qquad \text{about the } X \text{ axis}$$

$$I_{yy} = \int_{a_1}^{a_2} [(c^2 + a^2)A(x) + I_{yy}(x)]\, dx \qquad \text{about the } Y \text{ axis}$$

$$I_{zz} = \int_{a_1}^{a_2} [(a^2 + b^2)A(x) + I_{zz}(x)]\, dx \qquad \text{about the } Z \text{ axis}$$

where a, b, c and $A(x)$ have the same meaning as in Sec. 9.09–2a and the $J_x(x)$, $I_{yy}(x)$, $I_{zz}(x)$ are the inertia functions of $A(x)$. I_{xx}, I_{yy}, I_{zz} may be also expressed in terms of $A(y)$ or $A(z)$.

(b) Products of inertia of V are respectively,

$$I_{xy} = I_{yx} = \int_{a_1}^{a_2} abA(x)\, dx \qquad I_{yz} = I_{zy} = \int_{a_1}^{a_2} bcA(x)\, dx \qquad I_{zx} = I_{xz} = \int_{a_1}^{a_2} caA(x)\, dx$$

where a, b, c and $A(x)$ have the same meaning as in Sec. 9.08–2a. I_{xy}, I_{yz}, I_{zx} may be also expressed in terms of $A(y)$ or $A(z)$.

(c) Polar moment of inertia of V is

$$J = \frac{I_{xx} + I_{yy} + I_{zz}}{z}$$

where I_{xx}, I_{yy}, I_{zz} are the moments of inertia of V defined in Sec. 9.09–3a.

(4) Particular Cases (A = area of cross section, t = thickness)

(a) Thin straight bar (Fig. 9.09–4a).*

$$V = lA \qquad x_C = \frac{l}{2} \qquad y_C = 0 \qquad z_C = 0$$

$$I_{xx} = 0 \qquad I_{yy} = I_{zz} = \frac{Vl^2}{3} \qquad I_{xy} = I_{yz} = I_{zx} = 0$$

$$I_{AA} = 0 \qquad I_{BB} = I_{CC} = \frac{Vl^2}{12} \qquad I_{AB} = I_{BC} = I_{CA} = 0$$

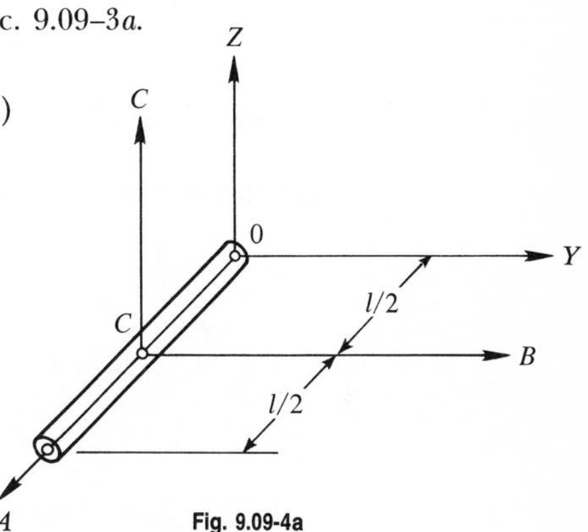

Fig. 9.09-4a

(b) Thin straight bar (Fig. 9.09–4b).

$$V = lA \qquad x_C = \frac{a}{2} \qquad y_C = \frac{b}{2} \qquad z_C = \frac{c}{2}$$

$$I_{xx} = \frac{(b^2 + c^2)V}{3} \qquad I_{yy} = \frac{(c^2 + a^2)V}{3} \qquad I_{zz} = \frac{(a^2 + b^2)V}{3}$$

$$I_{xy} = \frac{abV}{3} \qquad I_{yz} = \frac{bcV}{3} \qquad I_{zx} = \frac{caV}{3}$$

$$I_{AA} = \frac{(b^2 + c^2)V}{12} \qquad I_{BB} = \frac{(c^2 + a^2)V}{12} \qquad I_{CC} = \frac{(a^2 + b^2)V}{12}$$

$$I_{AB} = \frac{abV}{12} \qquad I_{BC} = \frac{bcV}{12} \qquad I_{CA} = \frac{caV}{12}$$

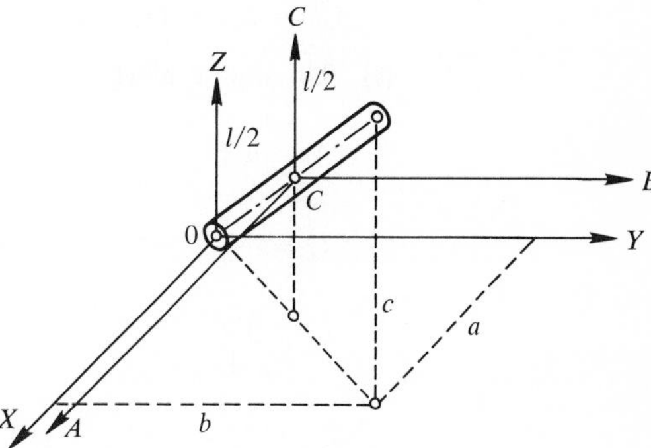

Fig. 9.09-4b

*No figures are missing. Please see footnote, p. 193.

(c) Thin circular ring (Fig. 9.09–4c).

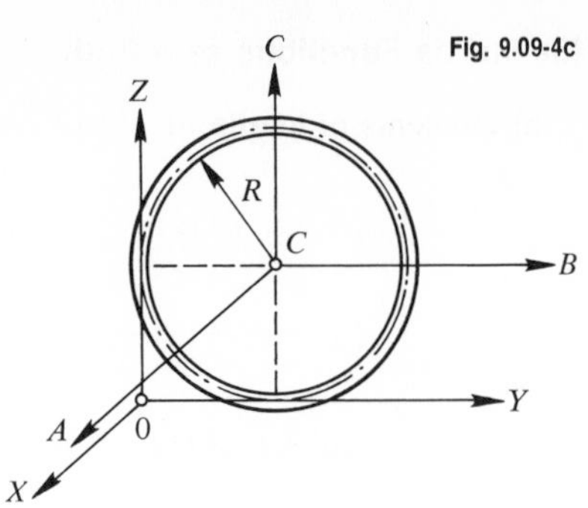

Fig. 9.09-4c

$$V = 2\pi RA \qquad x_C = 0 \qquad y_C = z_C = R$$

$$I_{xx} = 3VR^2 \qquad I_{yy} = I_{zz} = \frac{3VR^2}{2}$$

$$I_{xy} = 0 \qquad I_{yz} = VR^2 \qquad I_{zx} = 0$$

$$I_{AA} = VR^2 \qquad I_{BB} = I_{CC} = \frac{VR^2}{2}$$

$$I_{AB} = I_{BC} = I_{CA} = 0$$

(d) Thin circular bar (Fig. 9.09–4d).

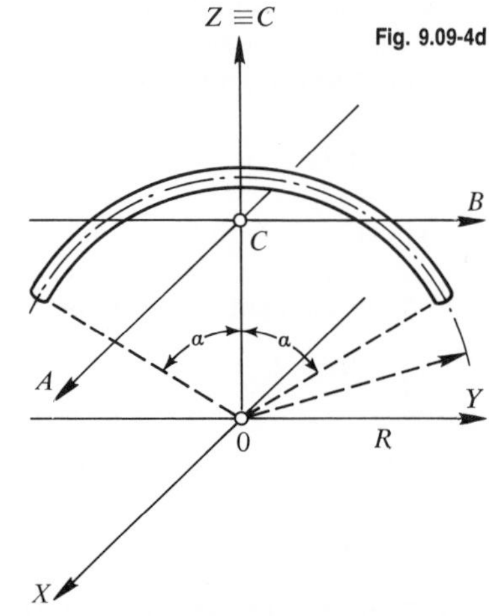

Fig. 9.09-4d

$$V = 2R\alpha A \qquad x_C = 0 \qquad y_C = 0 \qquad z_C = \frac{R\sin\alpha}{\alpha}$$

$$I_{xx} = VR^2 \qquad I_{xy} = 0 \qquad I_{yy} = \frac{VR^2}{2}\left(1+\frac{\sin 2\alpha}{2\alpha}\right)$$

$$I_{zz} = \frac{VR^2}{2}\left(1-\frac{\sin 2\alpha}{2\alpha}\right) \qquad I_{zx} = 0 \qquad I_{yz} = 0$$

$$I_{AA} = VR^2\left(1-\frac{\sin^2\alpha}{\alpha^2}\right) \qquad I_{AB} = 0$$

$$I_{BB} = \frac{VR^2}{2}\left(1+\frac{\sin 2\alpha}{2\alpha}-\frac{2\sin^2\alpha}{\alpha^2}\right) \qquad I_{BC} = 0$$

$$I_{CC} = \frac{VR^2}{2}\left(1-\frac{\sin 2\alpha}{2\alpha}\right) \qquad I_{CA} = 0$$

(e) Thin isosceles triangular plate (Fig. 9.09–4e).

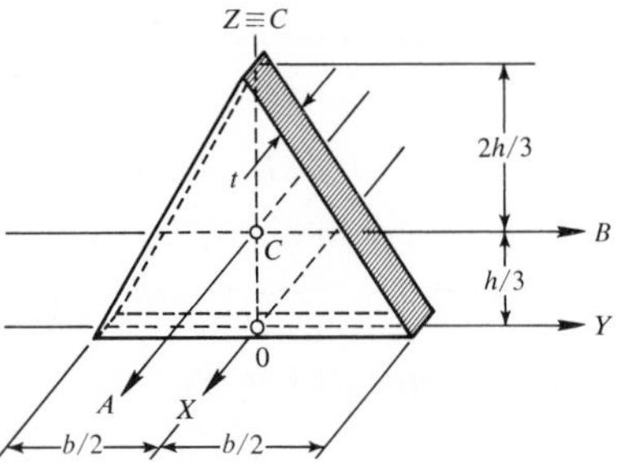

Fig. 9.09-4e

$$V = \frac{bht}{2} \qquad x_C = 0 \qquad y_C = 0 \qquad z_C = \frac{h}{3}$$

$$I_{xx} = \frac{V}{24}(4h^2+b^2) \qquad I_{yy} = \frac{Vh^2}{6} \qquad I_{zz} = \frac{Vb^2}{24}$$

$$I_{xy} = I_{yz} = I_{zx} = 0$$

$$I_{AA} = \frac{V}{72}(4h^2+3b^2) \qquad I_{BB} = \frac{Vh^2}{18} \qquad I_{CC} = \frac{Vb^2}{24}$$

$$I_{AB} = I_{BC} = I_{CA} = 0$$

(f) Thin square plate (Fig. 9.09–4f).

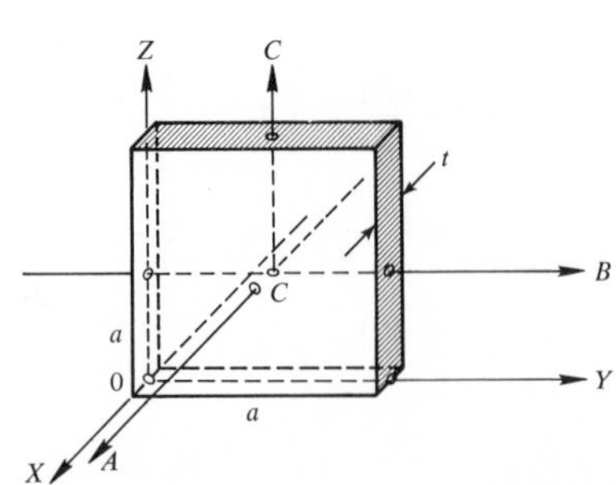

Fig. 9.09-4f

$$V = a^2 t \qquad x_C = 0 \qquad y_C = \frac{a}{2} \qquad z_C = \frac{a}{2}$$

$$I_{xx} = \frac{2Va^2}{3} \qquad I_{yy} = I_{zz} = \frac{Va^2}{3} \qquad I_{xy} = 0$$

$$I_{yz} = \frac{Va^2}{4} \qquad I_{zx} = 0$$

$$I_{AA} = \frac{Va^2}{6} \qquad I_{BB} = I_{CC} = \frac{Va^2}{12}$$

$$I_{AB} = I_{BC} = I_{CA} = 0$$

(g) Thin rectangular plate (Fig. 9.09–4*g*).

$$V = abt \qquad x_C = 0 \qquad y_C = \frac{a}{2} \qquad z_C = \frac{b}{2}$$

$$I_{xx} = \frac{V(a^2 + b^2)}{3} \qquad I_{yy} = \frac{Vb^2}{3} \qquad I_{zz} = \frac{Va^2}{3}$$

$$I_{xy} = 0 \qquad I_{yz} = \frac{Vab}{4} \qquad I_{zx} = 0$$

$$I_{AA} = \frac{V(a^2 + b^2)}{12} \qquad I_{BB} = \frac{Vb^2}{12} \qquad I_{CC} = \frac{Va^2}{12}$$

$$I_{AB} = I_{BC} = I_{CA} = 0$$

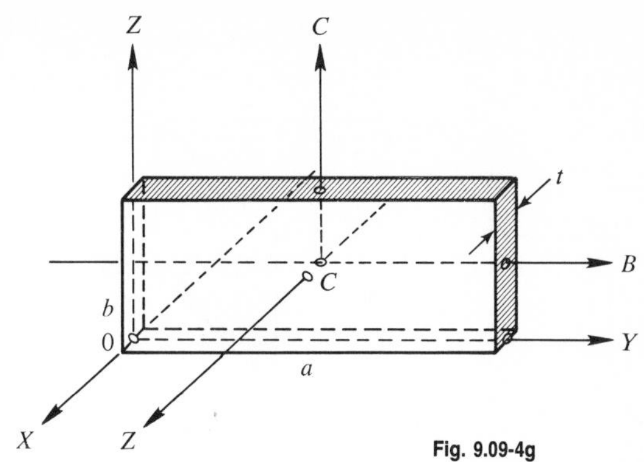

Fig. 9.09-4g

(h) Thin circular plate (Fig. 9.09–4*h*).

$$V = \pi R^2 t \qquad x_C = 0 \qquad y_C = R \qquad z_C = R$$

$$I_{xx} = \frac{5VR^2}{2} \qquad I_{yy} = I_{zz} = \frac{5VR^2}{4}$$

$$I_{xy} = 0 \qquad I_{yz} = VR^2 \qquad I_{zx} = 0$$

$$I_{AA} = \frac{VR^2}{2} \qquad I_{BB} = I_{CC} = \frac{VR^2}{4}$$

$$I_{AB} = I_{BC} = I_{CA} = 0$$

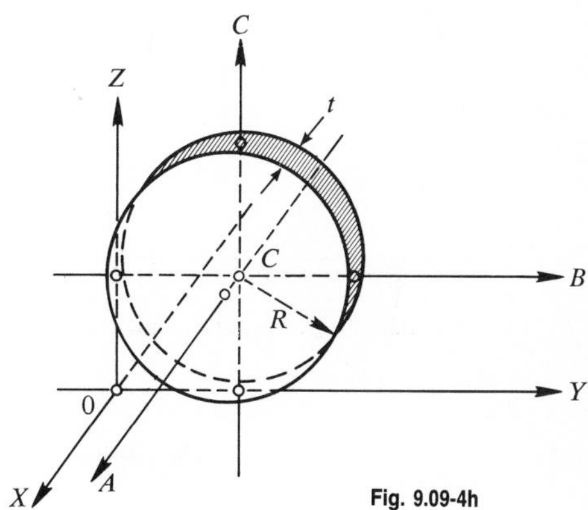

Fig. 9.09-4h

(i) Thin half circular plate (Fig. 9.09–4*i*).

$$V = \frac{\pi R^2 t}{2} \qquad x_C = 0 \qquad y_C = R \qquad z_C = \frac{4R}{3\pi}$$

$$I_{xx} = \frac{3VR^2}{2} \qquad I_{yy} = \frac{VR^2}{4} \qquad I_{zz} = \frac{5VR^2}{4}$$

$$I_{xy} = 0 \qquad I_{yz} = \frac{4VR^2}{3\pi} \qquad I_{zx} = 0$$

$$I_{AA} = 0.3199\,VR^2 \qquad I_{BB} = 0.0699\,VR^2$$

$$I_{CC} = 0.25\,VR^2 \qquad I_{AB} = I_{BC} = I_{CA} = 0$$

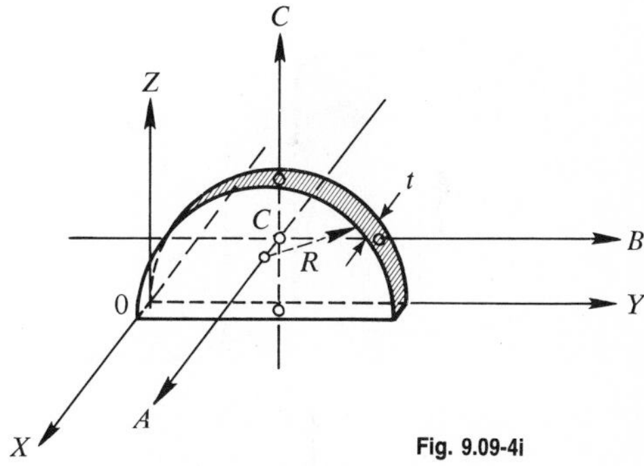

Fig. 9.09-4i

(j) Cube (Fig. 9.09–4*j*).

$$V = a^3 \qquad x_C = \frac{a}{2} \qquad y_C = \frac{a}{2} \qquad z_C = \frac{a}{2}$$

$$I_{xx} = I_{yy} = I_{zz} = \frac{2Va^2}{3}$$

$$I_{xy} = I_{yz} = I_{zx} = \frac{Va^2}{4}$$

$$I_{AA} = I_{BB} = I_{CC} = \frac{Va^2}{6}$$

$$I_{AB} = I_{BC} = I_{CA} = 0$$

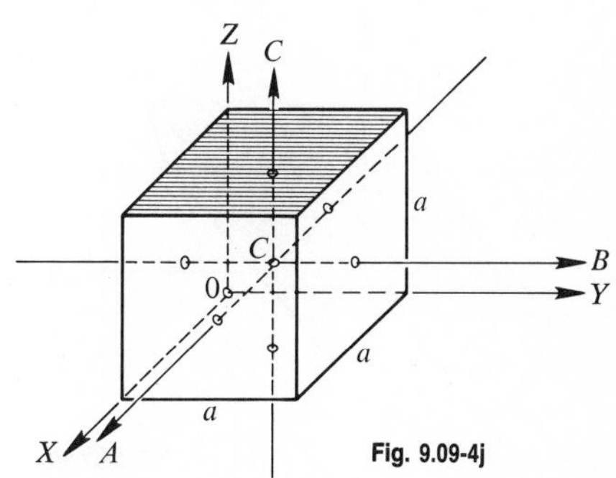

Fig. 9.09-4j

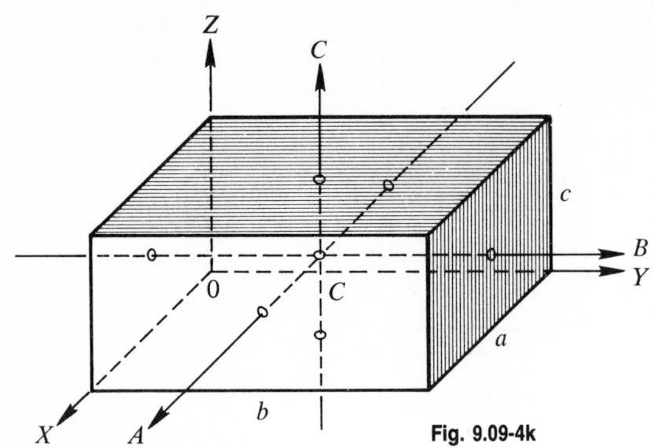

Fig. 9.09-4k

(k) Rectangular prism (Fig. 9.09–4k).

$$V = abc \qquad x_C = \frac{a}{2} \qquad y_C = \frac{b}{2} \qquad z_C = \frac{c}{2}$$

$$I_{xx} = \frac{(b^2+c^2)V}{3} \qquad I_{yy} = \frac{(c^2+a^2)V}{3} \qquad I_{zz} = \frac{(a^2+b^2)V}{3}$$

$$I_{xy} = \frac{abV}{4} \qquad I_{yz} = \frac{bcV}{4} \qquad I_{zx} = \frac{caV}{4}$$

$$I_{AA} = \frac{(b^2+c^2)V}{12} \qquad I_{BB} = \frac{(c^2+a^2)V}{12} \qquad I_{CC} = \frac{(a^2+b^2)V}{12}$$

$$I_{AB} = I_{BC} = I_{CA} = 0$$

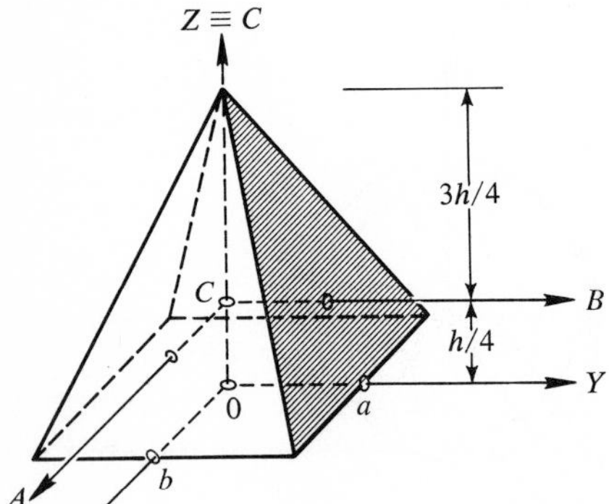

Fig. 9.09-4l

(l) Right rectangular pyramid (Fig. 9.09–4l).

$$V = \frac{abh}{3} \qquad x_C = 0 \qquad y_C = 0 \qquad z_C = \frac{h}{4}$$

$$I_{xx} = \frac{(b^2+2h^2)V}{20} \qquad I_{yy} = \frac{(a^2+2h^2)V}{20} \qquad I_{zz} = \frac{(a^2+b^2)V}{20}$$

$$I_{xy} = I_{yz} = I_{zx} = 0$$

$$I_{AA} = \frac{(4b^2+3h^2)V}{80} \qquad I_{BB} = \frac{(4a^2+3h^2)V}{80} \qquad I_{CC} = I_{zz}$$

$$I_{AB} = I_{BC} = I_{CA} = 0$$

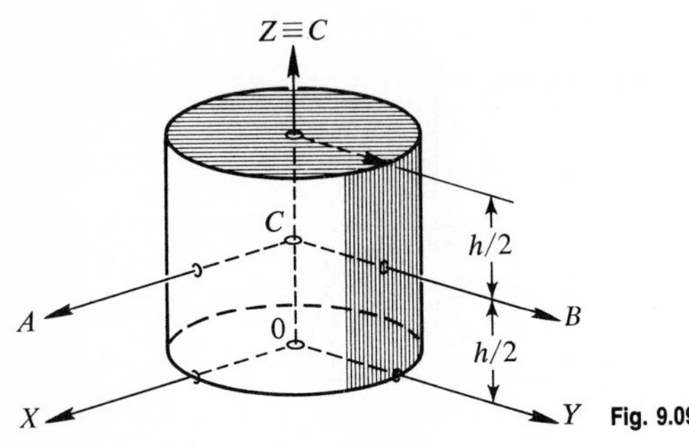

Fig. 9.09-4m

(m) Right circular cylinder (Fig. 9.09–4m).

$$V = \pi R^2 h \qquad x_C = 0 \qquad y_C = 0 \qquad z_C = \frac{h}{2}$$

$$I_{xx} = I_{yy} = \frac{(3R^2+4h^2)V}{12} \qquad I_{zz} = \frac{VR^2}{2}$$

$$I_{xy} = I_{yz} = I_{zx} = 0$$

$$I_{AA} = I_{BB} = \frac{(3R^2+h^2)V}{12} \qquad I_{CC} = I_{zz}$$

$$I_{AB} = I_{BC} = I_{CA} = 0$$

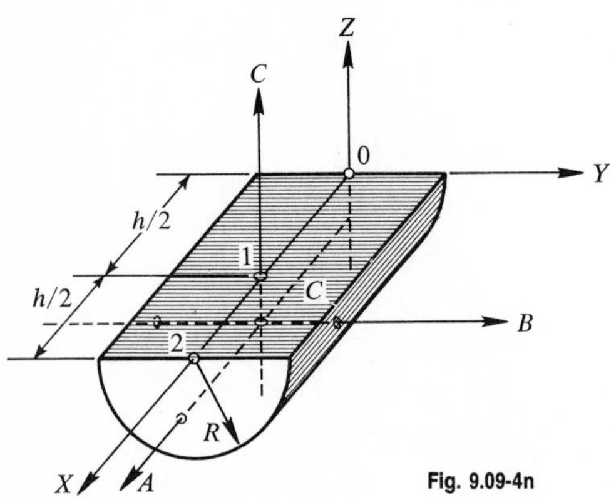

Fig. 9.09-4n

(n) Right circular semicylinder (Fig. 9.09–4n).

$$V = \frac{\pi R^2 h}{2} \qquad x_C = \frac{h}{2} \qquad y_C = 0 \qquad z_C = \frac{4R}{3\pi}$$

$$I_{xx} = \frac{VR^2}{2} \qquad I_{yy} = I_{zz} = \frac{(3R^2+h^2)V}{12}$$

$$I_{xy} = I_{yz} = I_{zx} = 0$$

$$I_{AA} = 0.3199\,VR^2 \qquad I_{BB} = 0.0699\,VR^2 \qquad I_{CC} = 0.25\,VR^2$$

(o) Right circular cone (Fig. 9.09–4*o*).

$$V = \frac{\pi R^2 h}{3} \qquad x_C = 0 \qquad y_C = 0 \qquad z_C = \frac{h}{4}$$

$$I_{xx} = I_{yy} = \frac{(3R^2 + 2h^2)V}{20} \qquad I_{zz} = \frac{3VR^2}{10}$$

$$I_{xy} = I_{yz} = I_{zx} = 0$$

$$I_{AA} = I_{BB} = \frac{3(4R^2 + h^2)V}{80} \qquad I_{CC} = I_{zz}$$

$$I_{AB} = I_{BC} = I_{CA} = 0$$

Fig. 9.09-4o

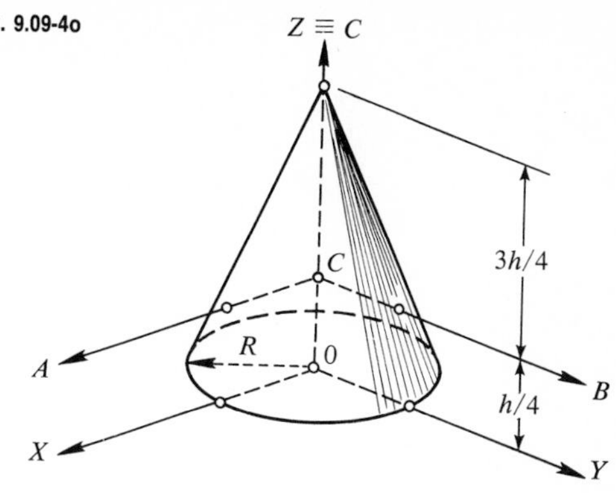

(p) Right circular semicone (Fig. 9.09–4*p*).

$$V = \frac{\pi R^2 h}{6} \qquad x_C = \frac{h}{4} \qquad y_C = 0 \qquad z_C = \frac{R}{\pi}$$

$$I_{xx} = \frac{3VR^2}{10} \qquad I_{yy} = I_{zz} = \frac{(3R^2 + 2h^2)V}{20}$$

$$I_{xy} = I_{yz} = I_{zx} = 0$$

$$I_{AA} \approx \frac{VR^2}{5} \qquad I_{BB} \approx \frac{(4R^2 + 3h^2)V}{80} \qquad I_{CC} = \frac{3(4R^2 + h^2)V}{80}$$

Fig. 9.09-4p

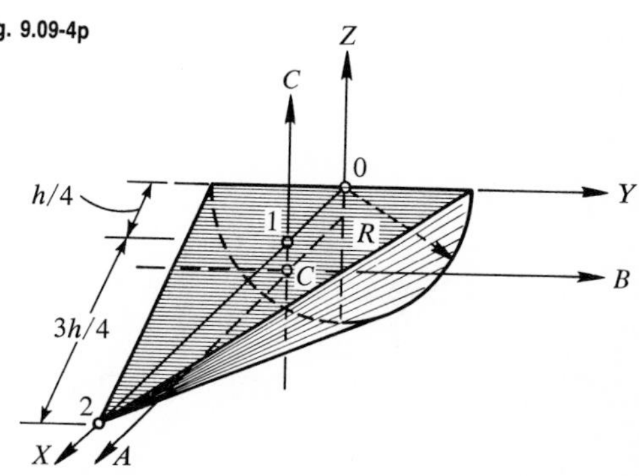

(q) Sphere (Fig. 9.09–4*q*).

$$V = \frac{4\pi R^3}{3} \qquad x_C = 0 \qquad y_C = 0 \qquad z_C = 0$$

$$I_{xx} = I_{yy} = I_{zz} = \frac{2VR^2}{5}$$

$$I_{xy} = I_{yz} = I_{zx} = 0 \qquad I_{TT} = T_{NN} = \frac{7VR^2}{5}$$

Fig. 9.09-4q

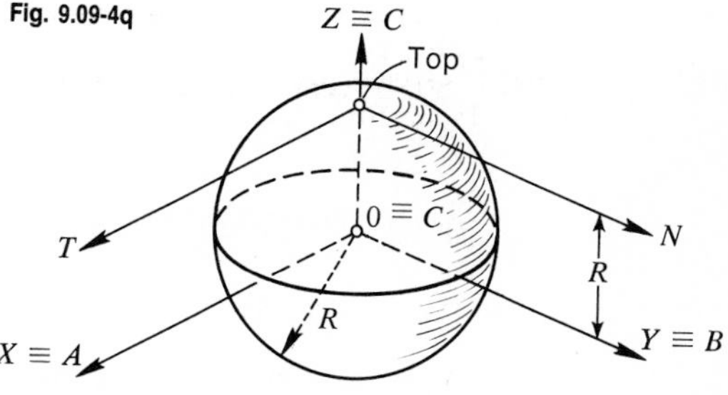

(r) Hemisphere (Fig. 9.09–4*r*).

$$V = \frac{2\pi R^3}{3} \qquad x_C = 0 \qquad y_C = 0 \qquad z_C = \frac{3R}{8}$$

$$I_{xx} = I_{yy} = I_{zz} = \frac{2VR^2}{5}$$

$$I_{xy} = I_{yz} = I_{zx} = 0$$

$$I_{AA} = I_{BB} = \frac{83VR^2}{320} \qquad I_{CC} = I_{zz} \qquad I_{TT} = \frac{208VR^2}{320}$$

$$I_{AB} = I_{BC} = I_{CA} = 0$$

Fig. 9.09-4r

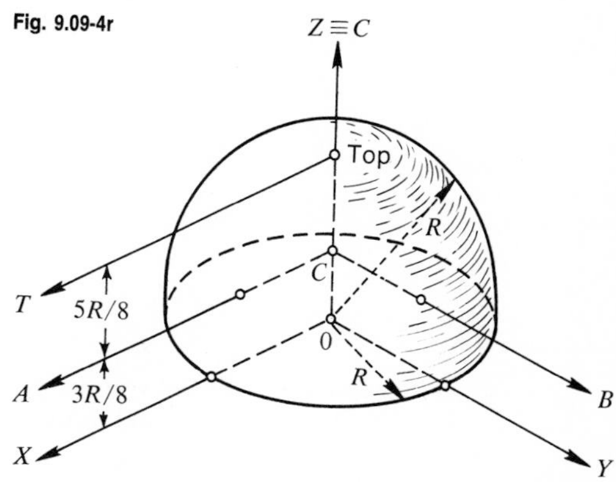

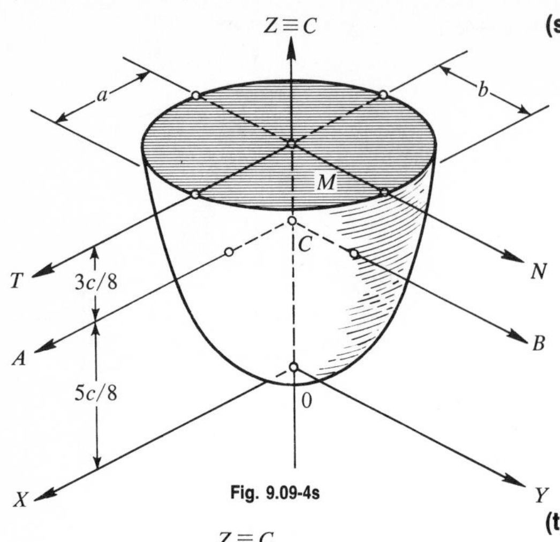

Fig. 9.09-4s

(s) Semiellipsoid (Fig. 9.09–4*s*).

$$V = \frac{2\pi abc}{3} \qquad x_C = 0 \qquad y_C = 0 \qquad z_C = \frac{5c}{8}$$

$$I_{xx} = \frac{(4b^2 + 9c^2)V}{20} \qquad I_{yy} = \frac{(4a^2 + 9c^2)V}{20} \qquad I_{zz} = \frac{(a^2 + b^2)V}{5}$$

$$I_{xy} = I_{yz} = I_{zx} = 0$$

$$I_{AA} = \frac{(64b^2 + 19c^2)V}{320} \qquad I_{BB} = \frac{(64a^2 + 19c^2)V}{320}$$

$$I_{CC} = I_{zz} \qquad I_{AB} = I_{BC} = I_{CA} = 0$$

$$I_{TT} = \frac{(b^2 + c^2)V}{5} \qquad I_{NN} = \frac{(a^2 + c^2)V}{5}$$

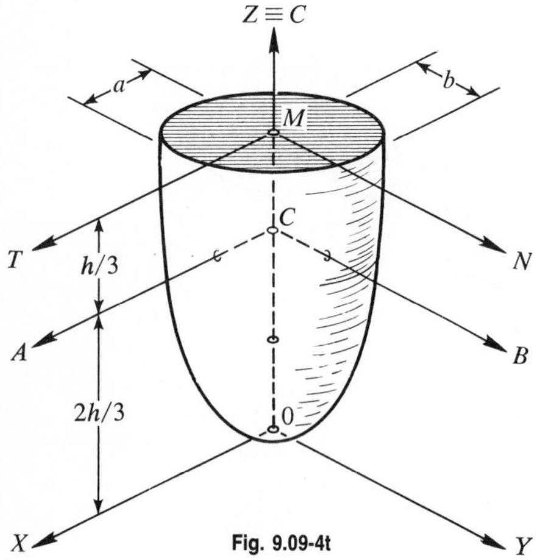

Fig. 9.09-4t

(t) Elliptic paraboloid (Fig. 9.09–4*t*).

$$V = \frac{\pi abh}{2} \qquad x_C = 0 \qquad y_C = 0 \qquad z_C = \frac{2h}{3}$$

$$I_{xx} = \frac{(b^2 + 3h^2)V}{6} \qquad I_{yy} = \frac{(a^2 + 3h^2)V}{6} \qquad I_{zz} = \frac{(a^2 + b^2)V}{6}$$

$$I_{xy} = I_{yz} = I_{zx} = 0$$

$$I_{AA} = \frac{(3b^2 + h^2)V}{18} \qquad I_{BB} = \frac{(3a^2 + h^2)V}{18} \qquad I_{CC} = I_{zz}$$

$$I_{AB} = I_{BC} = I_{CA} = 0$$

$$I_{TT} = \frac{(b^2 + h^2)V}{6} \qquad I_{NN} = \frac{(a^2 + h^2)V}{6}$$

9.10 PROPERTIES OF COMPOSITE HOMOGENEOUS BODIES

(1) Parallel-Axes Theorems

(a) Static moments of a single body. Since by definition the static moments of a single body with respect to its centroidal axes A, B, C are zero, the static moments of the same body with respect to any other set of parallel axes X, Y, Z (Fig. 9.10–1) can be expressed as

$$Q_x = \sqrt{(y_C^2 + z_C^2)}\,V + \cancelto{0}{Q_A}$$

$$Q_y = \sqrt{(z_C^2 + x_C^2)}\,V + \cancelto{0}{Q_B}$$

$$Q_z = \sqrt{(x_C^2 + y_C^2)}\,V + \cancelto{0}{Q_C}$$

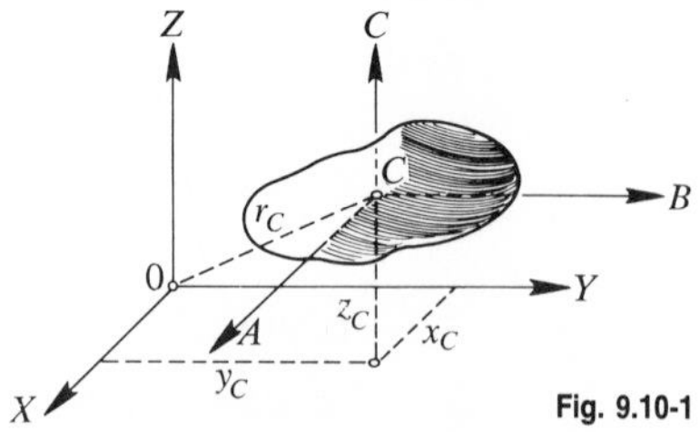

Fig. 9.10-1

where V is the volume of the body, $Q_A = 0$, $Q_B = 0$, $Q_C = 0$ are the static moments of V with respect to the centroidal axes A, B, C, and x_C, y_C, z_C are the coordinates of the centroid in the X, Y, Z axes.

(b) Inertia functions of a single body. Analogically, the inertia functions of the same body (Fig. 9.10–1) with respect to the same axes X, Y, Z are

$$I_{xx} = (y_C^2 + z_C^2)V + I_{AA}$$

$$I_{yy} = (z_C^2 + x_C^2)V + I_{BB}$$

$$I_{zz} = (x_C^2 + y_C^2)V + I_{CC}$$

$$I_{xy} = I_{yx} = x_C y_C V + I_{AB}$$

$$I_{yz} = I_{zy} = y_C z_C V + I_{BC}$$

$$I_{zx} = I_{xz} = z_C x_C V + I_{CA}$$

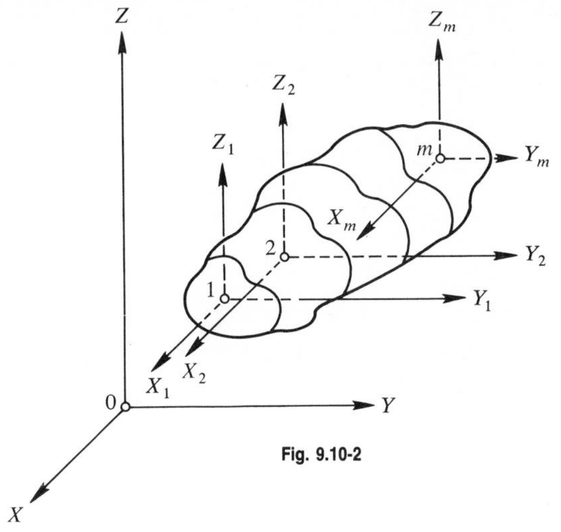

Fig. 9.10-2

$$J_0 = (x_C^2 + y_C^2 + z_C^2)V + \frac{I_{AA} + I_{BB} + I_{CC}}{2} = r_C^2 V + J_C$$

where I_{AA}, I_{BB}, I_{CC}, $I_{AB} = I_{BA}$, $I_{BC} = I_{CB}$, $I_{CA} = I_{AC}$, J_C are the inertia functions of V in the centroidal axes A, B, C, parallel to the X, Y, Z axes respectively.

(2) Summation Formulas

(a) Static functions of a composite body (Fig. 9.10–2) in the X, Y, Z axes are

$$V = V_1 + V_2 + \cdots + V_m = \sum_m V_i$$

$$Q_x = \sqrt{(y_1^2 + z_1^2)}\,V_1 + \sqrt{(y_2^2 + z_2^2)}\,V_2 + \cdots + \sqrt{(y_m^2 + z_m^2)}\,V_m = \sum_m \sqrt{(y_i^2 + z_i^2)}\,V_i$$

$$Q_y = \sqrt{(z_1^2 + x_1^2)}\,V_1 + \sqrt{(z_2^2 + x_2^2)}\,V_2 + \cdots + \sqrt{(z_m^2 + x_m^2)}\,V_m = \sum_m \sqrt{(z_i^2 + x_i^2)}\,V_i$$

$$Q_z = \sqrt{(x_1^2 + y_1^2)}\,V_1 + \sqrt{(x_2^2 + y_2^2)}\,V_2 + \cdots + \sqrt{(x_m^2 + y_m^2)}\,V_m = \sum_m \sqrt{(x_i^2 + y_i^{1})}\,V_i$$

$$x_C = \frac{\sum_m x_u V_i}{\sum_m V_i} \qquad y_C = \frac{\sum_m y_i V_i}{\sum_m V_i} \qquad z_C = \frac{\sum_m z_i V_i}{\sum_m V_i}$$

where $V_1, V_2, \ldots, V_m$ are the volumes of the respective parts, $x_1, x_2, \ldots, x_m$, $y_1, y_2, \ldots, y_m$, $z_1, z_2, \ldots, z_m$ are the coordinates of the respective centroids, $i = 1, 2, \ldots, m$.

(b) Inertia functions of a composite body (Fig. 9.10–2) in the X, Y, Z axes are

$$I_{xx} = \sum_m (y_i^2 + z_i^2) V_i + \sum_m I_{ixx} \qquad I_{xy} = I_{yx} = \sum_m x_i y_i V_i + \sum_m I_{ixy}$$

$$I_{yy} = \sum_m (z_i^2 + x_i^2) V_i + \sum_m I_{iyy} \qquad I_{yz} = I_{zy} = \sum_m y_i z_i V_i + \sum_m I_{iyz}$$

$$I_{zz} = \sum_m (x_i^2 + y_i^2) V_i + \sum_m I_{izz} \qquad I_{zx} = I_{xz} = \sum_m z_i x_i V_i + \sum_m I_{izx}$$

$$J_0 = \sum_m (x_i^2 + y_i^2 + z_i^2) V_i + \frac{\sum_m (I_{ixx} + I_{iyy} + I_{izz})}{2}$$

where I_{ixx}, I_{iyy}, I_{izz}, $I_{ixy} = I_{iyx}$, $I_{iyz} = I_{izy}$, $I_{izx} = I_{ixz}$ are the inertia functions of V_i in its centroidal axes x_i, y_i, z_i parallel to the X, Y, Z axes.

10
MATRICES AND DETERMINANTS

10.01 BASIC CONCEPTS OF MATRICES

(1) Definitions and Notations

(a) Matrix $[A]$ is a rectangular array of $m \times n$ elements arranged in m rows, n columns, and enclosed in brackets.

$$[A] = \begin{bmatrix} a_{11} & a_{12} & a_{13} & \cdots & a_{1n} \\ a_{21} & a_{22} & a_{23} & \cdots & a_{2n} \\ \cdots & \cdots & \cdots & \cdots & \cdots \\ a_{m1} & a_{m2} & a_{m3} & \cdots & a_{nn} \end{bmatrix}$$

(b) Element a_{jk} (any element) of the matrix represents a number, an algebraic expression or a function in the jth row and in the kth column ($j = 1, 2, 3, \ldots, m$ and $k = 1, 2, 3, \ldots, n$).

(c) Order (size, dimension) of a matrix is designated as $(m \times n)$, stating the number of rows m, the number of columns n, and the number of elements $m \times n$.

examples:

$$\begin{bmatrix} 1 & 0 & -7 \\ 3 & 8 & 0 \\ 4 & 11 & -5 \\ 5 & -3 & 17 \end{bmatrix}$$ = (4×3) matrix of signed numbers which consists of 4 rows and 3 columns

$$\begin{bmatrix} a^2 & (a+b)^2 & b^2 \\ (c+d)^2 & e^2 & f^2 \end{bmatrix}$$ = (2×3) matrix of algebraic expressions which consists of 2 rows and 3 columns

$$\begin{bmatrix} \cos x & -\sin x \\ \sin x & \cos x \end{bmatrix}$$ = (2×2) matrix of trigonometric functions which consists of 2 rows and 2 columns

(2) General Types of Matrices

(a) Rectangular matrix, $m \neq n$.

$$\begin{bmatrix} a_{11} & a_{12} & a_{13} \\ a_{21} & a_{22} & a_{23} \end{bmatrix} \quad (2 \times 3)$$

(b) Square matrix, $m = n$.

$$\begin{bmatrix} a_{11} & a_{12} \\ a_{21} & a_{22} \end{bmatrix} \quad (2 \times 2)$$

(c) Row matrix, $m = 1$.

$$[a_1 \quad a_2 \quad a_3] \quad (1 \times 3)$$

(d) Column matrix, $n = 1$.

$$\begin{bmatrix} b_1 \\ b_2 \\ b_3 \end{bmatrix} \quad (3 \times 1)$$

(3) Special Types of Matrices

(a) Zero (null) matrix. A matrix whose elements are all equal to zero is called the zero matrix, $[0]$.

(b) Unit matrix. A square matrix whose principal diagonal elements equal 1 and whose off-diagonal elements are zero is called the unit matrix, $[I]$.

(c) Diagonal matrix. A square matrix whose some or all principal diagonal elements are nonzero and all off-diagonal elements are zero is called the diagonal matrix.

(d) Symmetrical matrix. A square matrix which is symmetrical about the principal diagonal ($a_{jk} = a_{kj}$) is called the symmetrical matrix.

(e) Antisymmetrical (skew) matrix. A square matrix which is antisymmetrical about the principal diagonal ($a_{jk} = -a_{kj}$) and whose principal diagonal elements equal zero is called the antisymmetrical matrix.

examples:

Zero matrix (3 × 3):

$$\begin{bmatrix} 0 & 0 & 0 \\ 0 & 0 & 0 \\ 0 & 0 & 0 \end{bmatrix}$$

Unit matrix (3 × 3):

$$\begin{bmatrix} 1 & 0 & 0 \\ 0 & 1 & 0 \\ 0 & 0 & 1 \end{bmatrix}$$

Diagonal matrix (3 × 3):

$$\begin{bmatrix} a & 0 & 0 \\ 0 & b & 0 \\ 0 & 0 & c \end{bmatrix}$$

Symmetrical matrix (4 × 4):

$$\begin{bmatrix} 1 & 8 & 12 & 0 \\ 8 & 3 & -3 & 5 \\ 12 & -3 & 5 & -6 \\ 0 & 5 & -6 & 9 \end{bmatrix}$$

Antisymmetrical matrix (4 × 4):

$$\begin{bmatrix} 0 & -8 & 12 & 0 \\ 8 & 0 & -3 & -5 \\ -12 & 3 & 0 & 6 \\ 0 & 5 & -6 & 0 \end{bmatrix}$$

(4) Relationships of Two Matrices

(a) Equality of matrices. Two matrices $[A]$ and $[B]$ are equal if they have the same order and if their corresponding elements are equal ($a_{jk} = b_{jk}$).

(b) Transpose of a matrix. The matrix $[A]^T$ is called the transpose of the matrix $[A]$ if each row of $[A]^T$ is identical with the column of $[A]$ and vice versa.

$$[a \quad b \quad c]^T = \begin{bmatrix} a \\ b \\ c \end{bmatrix} \qquad \begin{bmatrix} d \\ e \\ f \end{bmatrix}^T = [d \quad e \quad f]$$

$$\begin{bmatrix} a & d \\ b & e \\ c & f \end{bmatrix}^T = \begin{bmatrix} a & b & c \\ d & e & f \end{bmatrix} \qquad \begin{bmatrix} h & i & j \\ k & l & m \end{bmatrix}^T = \begin{bmatrix} h & k \\ i & l \\ j & m \end{bmatrix}$$

(c) Transpose of a symmetrical matrix. If $[B]$ is a symmetrical matrix, then

$$[B]^T = [B]$$

consequently,

$$[0]^T = [0] \qquad \text{and} \qquad [I]^T = [I]$$

examples:

$$\begin{bmatrix} 1 & 0 & 0 \\ 0 & 2 & 0 \\ 0 & 0 & -3 \end{bmatrix}^T = \begin{bmatrix} 1 & 0 & 0 \\ 0 & 2 & 0 \\ 0 & 0 & -3 \end{bmatrix}$$

$$\begin{bmatrix} 1 & 4 & -5 \\ 4 & 2 & 6 \\ -5 & 6 & -3 \end{bmatrix}^T = \begin{bmatrix} 1 & 4 & -5 \\ 4 & 2 & 6 \\ -5 & 6 & -3 \end{bmatrix}$$

(d) Transpose of an antisymmetrical matrix. If $[C]$ is an antisymmetrical matrix, then

$$[C]^T = -[C]$$

example:

$$\begin{bmatrix} 0 & -1 & 2 \\ 1 & 0 & 3 \\ -2 & -3 & 0 \end{bmatrix}^T = -\begin{bmatrix} 0 & -1 & 2 \\ 1 & 0 & 3 \\ -2 & -3 & 0 \end{bmatrix}$$

(e) Transpose of a transpose. The transpose of $[A]^T$ is the matrix $[A]$.

$$|[A]^T|^T = [A]$$

example:

$$\begin{bmatrix} 1 & -4 \\ -2 & 5 \\ 3 & 6 \end{bmatrix}^T = \begin{bmatrix} 1 & -2 & 3 \\ -4 & 5 & 6 \end{bmatrix} \qquad \begin{bmatrix} 1 & -2 & 3 \\ -4 & 5 & 6 \end{bmatrix}^T = \begin{bmatrix} 1 & -4 \\ -2 & 5 \\ 3 & 6 \end{bmatrix}$$

10.02 ALGEBRAIC OPERATIONS WITH MATRICES

(1) Matrix Addition

(a) Definition. The sum of two or more matrices of order $(m \times n)$ is a matrix of the same order each of whose elements equals the sum of the respective elements of the given matrices.

example:

$$\underbrace{\begin{bmatrix} 1 & 4 \\ 7 & 12 \end{bmatrix}}_{[A]} + \underbrace{\begin{bmatrix} 6 & 9 \\ 13 & 5 \end{bmatrix}}_{[B]} = \underbrace{\begin{bmatrix} 7 & 13 \\ 20 & 17 \end{bmatrix}}_{[A+B]}$$

(b) Order of terms. The matrix addition is commutative.

$$[A]+[B]+[C] = [B]+[C]+[A] = [C]+[A]+[B]$$

(c) Grouping of terms. The matrix addition is associative.

$$[A]+[B]+[C] = [A+B]+[C] = [A]+[B+C]$$

(d) Transpose of a sum. The transpose of a sum of two or more matrices is equal to the sum of their transposes and vice versa.

$$[A+B+C]^T = [A]^T + [B]^T + [C]^T$$

(2) Matrix Subtraction

(a) Definition. The difference of two or more matrices of the same order $(m \times n)$ is a matrix of the same order each of whose elements equals the difference of the respective elements of the given matrices.

example:

$$\underbrace{\begin{bmatrix} 1 & 4 \\ 7 & -12 \end{bmatrix}}_{[A]} - \underbrace{\begin{bmatrix} -6 & 9 \\ 13 & -5 \end{bmatrix}}_{[B]} = \underbrace{\begin{bmatrix} 7 & -5 \\ -6 & -7 \end{bmatrix}}_{[A-B]}$$

(b) Order of terms. The matrix subtraction is commutative.

$$[A]-[B]-[C] = -[B]-[C]+[A] = -[C]+[A]-[B]$$

(c) Grouping of terms. The matrix subtraction is associative.

$$[A]-[B]-[C] = [A-B]-[C] = [A]-[B+C]$$

(d) Transpose of a difference. The transpose of a difference of two or more matrices is equal to the difference of their transposes and vice versa.

$$[A-B-C]^T = [A]^T - [B]^T - [C]^T$$

(3) Matrix and its Transpose

(a) Sum of the square matrix $[A]$ and its transpose $[A]^T$ is a symmetrical matrix.

$$[A]+[A]^T = [\text{symmetrical matrix}]$$

example:

$$\underbrace{\begin{bmatrix} 2 & 6 \\ 8 & 12 \end{bmatrix}}_{[A]} + \underbrace{\begin{bmatrix} 2 & 8 \\ 6 & 12 \end{bmatrix}}_{[A]^T} = \begin{bmatrix} 4 & 14 \\ 14 & 24 \end{bmatrix}$$

(b) Difference of the square matrix $[A]$ and its transpose $[A]^T$ is an antisymmetrical matrix.

$$[A]-[A]^T = [\text{antisymmetrical matrix}]$$

example:

$$\underbrace{\begin{bmatrix} 2 & 6 \\ 8 & 12 \end{bmatrix}}_{[A]} - \underbrace{\begin{bmatrix} 2 & 8 \\ 6 & 12 \end{bmatrix}}_{[A]^T} = \begin{bmatrix} 0 & -2 \\ 2 & 0 \end{bmatrix}$$

(c) Resolution. Every square unsymmetrical matrix $[A]$ can be resolved into the sum of a symmetrical and antisymmetrical matrix.

$$[A] = \tfrac{1}{2}[[A]+[A]^T] + \tfrac{1}{2}[[A]-[A]^T]$$

example:

$$\begin{bmatrix} 1 & -3 \\ 9 & 12 \end{bmatrix} = \underbrace{\tfrac{1}{2}\left[\begin{bmatrix} 1 & -3 \\ 9 & 12 \end{bmatrix} + \begin{bmatrix} 1 & 9 \\ -3 & 12 \end{bmatrix}\right]}_{\begin{bmatrix} 1 & 3 \\ 3 & 12 \end{bmatrix}} + \underbrace{\tfrac{1}{2}\left[\begin{bmatrix} 1 & -3 \\ 9 & 12 \end{bmatrix} - \begin{bmatrix} 1 & 9 \\ -3 & 12 \end{bmatrix}\right]}_{\begin{bmatrix} 0 & -6 \\ 6 & 0 \end{bmatrix}}$$

(4) Product of a Scalar and a Matrix

(a) Definition. The product of a scalar α and a matrix $[A]$ of order $(m \times n)$ is an $(m \times n)$ matrix, each of whose elements equals the product of α and the corresponding element of the matrix.

$$\alpha\begin{bmatrix} a_{11} & a_{12} \\ a_{21} & a_{22} \end{bmatrix} = \begin{bmatrix} \alpha a_{11} & \alpha a_{12} \\ \alpha a_{21} & \alpha a_{22} \end{bmatrix}$$

(b) Order of terms. The product of a matrix and one or more scalars is commutative.

$$\alpha[A] = [A]\alpha \qquad \alpha_1\alpha_2[A] = \alpha_1[A]\alpha_2 = [A]\alpha_1\alpha_2$$

(c) Grouping of terms. The product of a matrix and of one or more scalars is associative.

$$\alpha_1\alpha_2[A] = [\alpha_1 A]\alpha_2 = [\alpha_2 A]\alpha_1 = [\alpha_1\alpha_2 A]$$

(5) Matrix Multiplication

(a) Definition. The product of two conformable matrices of order $(m_1 \times n_1)$, $(m_2 \times n_2)$ is a rectangular matrix of order $(m_1 \times n_2)$ whose elements are equal to the sum of products of the inner elements (see below). Two matrices are conformable if $n_1 = m_2$.

examples:

$$\begin{bmatrix} a_{11} & a_{12} & a_{13} \\ a_{21} & a_{22} & a_{23} \end{bmatrix}\begin{bmatrix} b_{11} & b_{12} \\ b_{21} & b_{22} \\ b_{31} & b_{32} \end{bmatrix} = \begin{bmatrix} (a_{11}b_{11} + a_{12}b_{21} + a_{13}b_{31}) & (a_{11}b_{12} + a_{12}b_{22} + a_{13}b_{32}) \\ (a_{21}b_{11} + a_{22}b_{21} + a_{23}b_{31}) & (a_{21}b_{12} + a_{22}b_{22} + a_{23}b_{32}) \end{bmatrix}$$

$$\begin{bmatrix} +1 & +2 & -3 \\ +4 & -5 & +6 \end{bmatrix}\begin{bmatrix} -7 & +10 \\ +8 & -11 \\ +9 & +12 \end{bmatrix} = \begin{bmatrix} (1)(-7)+(2)(8)+(-3)(9) & (1)(10)+(2)(-11)+(-3)(12) \\ (4)(-7)+(-5)(8)+(6)(9) & (4)(10)+(-5)(-11)+(6)(12) \end{bmatrix}$$

$$= \begin{bmatrix} -18 & -48 \\ -14 & 167 \end{bmatrix}$$

(b) Sequence of terms. The multiplication of conformable matrices is not commutative.

$$[A][B][C] \neq [B][C][A] \neq [C][A][B]$$

(c) Grouping of terms. The multiplication of conformable matrices is associative.

$$[A][B][C] = [AB][C] = [A][BC]$$

(d) Distribution of terms. The multiplication of conformable matrices is distributive.

$$[A][B + C] = [A][B]+[A][C] \qquad [E + F][G] = [E][G]+[F][G]$$

(e) Transpose of a product. The transpose of a product of two or more matrices is the product of their transposes in reverse order.

$$[[A][B][C]]^T = [C]^T[B]^T[A]^T$$

(6) Special Products

(a) Row and column matrix products.

$$[a \quad b \quad c]\begin{bmatrix} x \\ y \\ z \end{bmatrix} = [x \quad y \quad z]\begin{bmatrix} a \\ b \\ c \end{bmatrix} = [ax + by + cz]$$

(b) Column and row matrix products.

$$\begin{bmatrix} x \\ y \\ z \end{bmatrix}[a \quad b \quad c] = \begin{bmatrix} ax & bx & cx \\ ay & by & cy \\ az & bz & cz \end{bmatrix} \neq \begin{bmatrix} a \\ b \\ c \end{bmatrix}[x \quad y \quad z] = \begin{bmatrix} ax & ay & az \\ bx & by & bz \\ cx & cy & cz \end{bmatrix}$$

(c) Diagonal and square matrix products.

$$\begin{bmatrix} x & 0 & 0 \\ 0 & y & 0 \\ 0 & 0 & z \end{bmatrix}\begin{bmatrix} a_{11} & a_{12} & a_{13} \\ a_{21} & a_{22} & a_{23} \\ a_{31} & a_{32} & a_{33} \end{bmatrix} = \begin{bmatrix} a_{11}x & a_{12}x & a_{13}x \\ a_{21}y & a_{22}y & a_{23}y \\ a_{31}z & a_{32}z & a_{33}z \end{bmatrix}$$

$$\begin{bmatrix} a_{11} & a_{12} & a_{13} \\ a_{21} & a_{22} & a_{23} \\ a_{31} & a_{32} & a_{33} \end{bmatrix}\begin{bmatrix} x & 0 & 0 \\ 0 & y & 0 \\ 0 & 0 & z \end{bmatrix} = \begin{bmatrix} a_{11}x & a_{12}y & a_{13}z \\ a_{21}x & a_{22}y & a_{23}z \\ a_{31}x & a_{32}y & a_{33}z \end{bmatrix}$$

(7) Operations with Zero Matrix

(a) Sum.

$$[A]+[0] = [A]$$

(b) Difference.

$$[A]-[0] = [A]$$

(c) Difference of two equal matrices.

$$[A]-[A] = [0]$$

(d) Products.

$$[A][0] = [0] \qquad [0][A] = [0] \qquad [0][0] = [0]$$

(8) Operations with Unit Matrix

(a) Sum and difference.

$$[A]\pm[I] = [A \pm I] = \begin{bmatrix} a_{11} \pm 1 & a_{12} & a_{13} & \cdots & a_{1n} \\ a_{21} & a_{22} \pm 1 & a_{23} & \cdots & a_{2n} \\ \cdots & \cdots & \cdots & \cdots & \cdots \\ a_{m1} & a_{m2} & a_{m3} & \cdots & a_{mn} \pm 1 \end{bmatrix}$$

(b) Products.

$$[A][I] = [A] \qquad [I][A] = [A] \qquad [I][I] = [I]$$

10.03 BASIC CONCEPTS OF DETERMINANTS

(1) Definitions and Notations

(a) Determinant $|A|$ is a square array of $n \times n$ elements arranged in n rows and n columns, and enclosed in two vertical clues.

$$|A| = \begin{vmatrix} a_{11} & a_{12} & a_{13} & \cdots & a_{1n} \\ a_{21} & a_{22} & a_{23} & \cdots & a_{2n} \\ \cdots & \cdots & \cdots & \cdots & \cdots \\ a_{n1} & a_{n2} & a_{n3} & \cdots & a_{nn} \end{vmatrix} = \det A$$

where a_{jk} is again (as in the matrix) the element of the jth row and kth column.

(b) Similarity. The arrangement of elements in the matrix $[A]$ of $(n \times n)$ order and in the determinant $|A|$ of the same order is identical, but the symbols [ ] and | | distinctly indicate that each array is governed by specific laws which are not the same for $[A]$ and $|A|$.

(c) Minor. Any element a_{jk} of $|A|$ is associated with another determinant $|A_{jk}|$ of $(n-1) \times (n-1)$ order obtained by deleting the jth column and kth row in $|A|$.

example:

$$|A_{23}| = \begin{vmatrix} a_{11} & a_{12} & a_{13} & a_{14} & \cdots & a_{1n} \\ a_{21} & a_{22} & a_{23} & a_{24} & \cdots & a_{2n} \\ a_{31} & a_{32} & a_{33} & a_{34} & \cdots & a_{3n} \\ \cdots & \cdots & \cdots & \cdots & \cdots & \cdots \\ a_{n1} & a_{n2} & a_{n3} & a_{n4} & \cdots & a_{nn} \end{vmatrix} = \begin{vmatrix} a_{11} & a_{12} & a_{14} & \cdots & a_{1n} \\ a_{31} & a_{32} & a_{34} & \cdots & a_{3n} \\ \cdots & \cdots & \cdots & \cdots & \cdots \\ a_{n1} & a_{n2} & a_{n4} & \cdots & a_{nn} \end{vmatrix}$$

(d) Cofactor. The product of the minor $|A_{jk}|$ and $(-1)^{j+k}$ is called the cofactor A_{jk}.

$$A_{jk} = (-1)^{j+k}|A_{jk}|$$

example:

In the determinant

$$|A| = \begin{vmatrix} 1 & -3 & 0 \\ 5 & -7 & 8 \\ 9 & 2 & 3 \end{vmatrix}$$

the cofactors are

$$A_{11} = (-1)^{1+1}\begin{vmatrix} -7 & 8 \\ 2 & 3 \end{vmatrix} \qquad A_{12} = (-1)^{1+2}\begin{vmatrix} 5 & 8 \\ 9 & 3 \end{vmatrix} \qquad A_{13} = (-1)^{1+3}\begin{vmatrix} 5 & -7 \\ 9 & 2 \end{vmatrix}$$

$$A_{21} = (-1)^{2+1}\begin{vmatrix} -3 & 0 \\ 2 & 3 \end{vmatrix} \qquad A_{22} = (-1)^{2+2}\begin{vmatrix} 1 & 0 \\ 9 & 3 \end{vmatrix} \qquad A_{23} = (-1)^{2+3}\begin{vmatrix} 1 & -3 \\ 9 & 2 \end{vmatrix}$$

$$A_{31} = (-1)^{3+1}\begin{vmatrix} -3 & 0 \\ -7 & 8 \end{vmatrix} \qquad A_{32} = (-1)^{3+2}\begin{vmatrix} 1 & 0 \\ 5 & 8 \end{vmatrix} \qquad A_{33} = (-1)^{3+3}\begin{vmatrix} 1 & -3 \\ 5 & -7 \end{vmatrix}$$

For the evaluation of these cofactors refer to Sec. 10.03–2*a*.

(e) Evaluation. The value of the determinant $|A|$ is the sum of the products of the elements of any row or any column into their respective cofactors.

$$|A| = \sum_{j=1}^{n} a_{jk}A_{jk} = \sum_{k=1}^{n} a_{jk}A_{jk}$$

(2) Two Basic Cases of Evaluation

(a) Second-order determinant.

$$A = \begin{vmatrix} a_{11} & a_{12} \\ a_{21} & a_{22} \end{vmatrix} = a_{11}a_{22} - a_{12}a_{21}$$

examples:

$$\begin{vmatrix} 1 & -3 \\ 7 & 5 \end{vmatrix} = (1)(5) - (-3)(7) = 5 + 21 = 26$$

$$\begin{vmatrix} 1 & 0 \\ 2 & 3 \end{vmatrix} = (1)(3) - (0)(2) = 3 - 0 = 3$$

(b) Third-order determinant.

$$|A| = \begin{vmatrix} a_{11} & a_{12} & a_{13} \\ a_{21} & a_{22} & a_{23} \\ a_{31} & a_{32} & a_{33} \end{vmatrix} = a_{11}A_{11} + a_{12}A_{12} + a_{13}A_{13}$$

$$= a_{11}(a_{22}a_{33} - a_{23}a_{32}) - a_{12}(a_{21}a_{33} - a_{23}a_{31}) + a_{13}(a_{21}a_{32} - a_{22}a_{31})$$

where A_{11}, A_{12}, A_{13} are the cofactors defined in Sec. 10.03–1*d*, and the minus sign in front of a_{12} is the result of $(-1)^{j+k} = (-1)^{1+2} = (-1)^3 = -1$.

Alternative forms of evaluation are: $a_{21}A_{21} + a_{22}A_{22} + a_{23}A_{23}$, or $a_{31}A_{31} + a_{32}A_{32} + a_{33}A_{33}$, or $a_{11}A_{11} + a_{21}A_{21} + a_{31}A_{31}$, or $a_{12}A_{12} + a_{22}A_{22} + a_{32}A_{32}$, or $a_{13}A_{13} + a_{23}A_{23} + a_{33}A_{33}$, all yielding the same value of the determinant.

example:

$$\begin{vmatrix} 1 & -3 & 0 \\ 5 & -7 & 8 \\ 9 & 2 & 3 \end{vmatrix} = (1)\begin{vmatrix} -7 & 8 \\ 2 & 3 \end{vmatrix} - (-3)\begin{vmatrix} 5 & 8 \\ 9 & 3 \end{vmatrix} + (0)\begin{vmatrix} 5 & -7 \\ 9 & 2 \end{vmatrix} = (1)(-37) + (3)(-57) + (0)(+73) = -208$$

or

$$\begin{vmatrix} 1 & -3 & 0 \\ 5 & -7 & 8 \\ 9 & 2 & 3 \end{vmatrix} = (1)\begin{vmatrix} -7 & 8 \\ 2 & 3 \end{vmatrix} - (5)\begin{vmatrix} -3 & 0 \\ 2 & 3 \end{vmatrix} + (9)\begin{vmatrix} -3 & 0 \\ -7 & 8 \end{vmatrix}$$

$$= (1)(-37) - (5)(-9) + (9)(-24) = -208$$

and so on.

(3) Theorems of Determinants

(a) Same value. The value of the determinant remains unchanged if:

(α) Rows and columns are interchanged, i.e., the determinant is equal to its transpose.

example:

$$\begin{vmatrix} a_{11} & a_{12} & a_{13} \\ a_{21} & a_{22} & a_{23} \\ a_{31} & a_{32} & a_{33} \end{vmatrix} = \begin{vmatrix} a_{11} & a_{21} & a_{31} \\ a_{12} & a_{22} & a_{32} \\ a_{13} & a_{23} & a_{33} \end{vmatrix}$$

(β) Even number of rows or columns is interchanged.

example:

$$\begin{vmatrix} a_{11} & a_{12} & a_{13} \\ a_{21} & a_{22} & a_{23} \\ a_{31} & a_{32} & a_{33} \end{vmatrix} = \begin{vmatrix} a_{21} & a_{22} & a_{23} \\ a_{31} & a_{32} & a_{33} \\ a_{11} & a_{12} & a_{13} \end{vmatrix}$$

(γ) To each element of a row (or column) is added m times the respective element of any other row (or column).

example:

$$\begin{vmatrix} a_{11} & a_{12} & a_{13} \\ a_{21} & a_{22} & a_{23} \\ a_{31} & a_{32} & a_{33} \end{vmatrix} = \begin{vmatrix} a_{11} & a_{12} & a_{13} \\ a_{21}+ma_{11} & a_{22}+ma_{12} & a_{23}+ma_{13} \\ a_{31} & a_{32} & a_{33} \end{vmatrix}$$

(b) Zero value. The value of the determinant is zero if:

(α) Two or more rows (or columns) are identical.

examples:

$$\begin{vmatrix} a_{11} & a_{12} & a_{13} \\ a_{11} & a_{12} & a_{13} \\ a_{31} & a_{32} & a_{33} \end{vmatrix} = 0 \qquad \begin{vmatrix} a_{11} & a_{11} & a_{13} \\ a_{21} & a_{21} & a_{23} \\ a_{31} & a_{31} & a_{33} \end{vmatrix} = 0$$

(β) All elements of at least one row (or column) are zero.

examples:

$$\begin{vmatrix} a_{11} & a_{12} & a_{13} \\ 0 & 0 & 0 \\ a_{31} & a_{32} & a_{33} \end{vmatrix} = 0 \qquad \begin{vmatrix} a_{11} & 0 & a_{13} \\ a_{21} & 0 & a_{23} \\ a_{31} & 0 & a_{33} \end{vmatrix} = 0$$

(γ) Elements of one row (or column) are linear combinations of the elements of another row (or column).

examples:

$$\begin{vmatrix} a_{11} & a_{12} & a_{13} \\ ma_{11} & ma_{12} & ma_{13} \\ a_{31} & a_{32} & a_{33} \end{vmatrix} = 0 \qquad \begin{vmatrix} a_{11} & ma_{11} & a_{13} \\ a_{21} & ma_{21} & a_{23} \\ a_{31} & ma_{31} & a_{33} \end{vmatrix} = 0$$

where m is an integral or fractional factor.

(c) Change in sign. The sign of the determinant is changed by the odd number of interchanges of rows (or columns).

example:

$$\begin{vmatrix} a_{11} & a_{12} & a_{13} \\ a_{21} & a_{22} & a_{23} \\ a_{31} & a_{32} & a_{33} \end{vmatrix} = -\begin{vmatrix} a_{21} & a_{22} & a_{23} \\ a_{11} & a_{12} & a_{13} \\ a_{31} & a_{32} & a_{33} \end{vmatrix} = -\begin{vmatrix} a_{12} & a_{11} & a_{13} \\ a_{22} & a_{21} & a_{23} \\ a_{32} & a_{31} & a_{33} \end{vmatrix}$$

10.04 INVERSE AND RELATED MATRICES

(1) Cofactor and Adjoint Matrices

(a) Cofactor matrix of a given matrix $[A]$ of order $(n \times n)$ denoted as $[A]^C$ is the matrix of order $(n \times n)$ in which each element a_{jk} of $[A]$ is replaced by its respective cofactor A_{jk} (Sec. 10.03–1*d*).

$$[A]^C = \begin{bmatrix} A_{11} & A_{12} & A_{13} & \cdots & A_{1n} \\ A_{21} & A_{22} & A_{23} & \cdots & A_{2n} \\ \cdots & \cdots & \cdots & \cdots & \cdots \\ A_{n1} & A_{n2} & A_{n3} & \cdots & A_{nn} \end{bmatrix}$$

(b) Adjoint matrix adj$[A]$ of a given matrix $[A]$ of order $(n \times n)$ is the transpose of its cofactor matrix $[A]^C$.

$$\text{adj}\,[A] = \begin{bmatrix} A_{11} & A_{21} & A_{31} & \cdots & A_{n1} \\ A_{12} & A_{22} & A_{32} & \cdots & A_{n2} \\ \cdots & \cdots & \cdots & \cdots & \cdots \\ A_{1n} & A_{2n} & A_{3n} & \cdots & A_{nn} \end{bmatrix} = \{[A]^C\}^T$$

(c) Product of a matrix and its adjoint of order $(n \times n)$ is a diagonal matrix of order $(n \times n)$ whose principal diagonal elements are $|A|$.

$$[A]\,\text{adj}\,[A] = \begin{bmatrix} |A| & 0 & 0 & \cdots & 0 \\ 0 & |A| & 0 & \cdots & 0 \\ \cdots & \cdots & \cdots & \cdots & \cdots \\ 0 & 0 & 0 & \cdots & |A| \end{bmatrix} = |A|[I]$$

example:

The cofactor matrix of the determinant

$$|A| = \begin{vmatrix} 1 & -3 & 0 \\ 5 & -7 & 8 \\ 9 & 2 & 3 \end{vmatrix} = -208 \qquad \text{(Sec. 10.03–2}b\text{)}$$

is

$$[A]^C = \begin{bmatrix} (-1)^2\begin{vmatrix} -7 & 8 \\ 2 & 3 \end{vmatrix} & (-1)^3\begin{vmatrix} 5 & 8 \\ 9 & 3 \end{vmatrix} & (-1)^4\begin{vmatrix} 5 & -7 \\ 9 & 2 \end{vmatrix} \\ (-1)^3\begin{vmatrix} -3 & 0 \\ 2 & 3 \end{vmatrix} & (-1)^4\begin{vmatrix} 1 & 0 \\ 9 & 3 \end{vmatrix} & (-1)^5\begin{vmatrix} 1 & -3 \\ 9 & 2 \end{vmatrix} \\ (-1)^4\begin{vmatrix} -3 & 0 \\ -7 & 8 \end{vmatrix} & (-1)^5\begin{vmatrix} 1 & 0 \\ 5 & 8 \end{vmatrix} & (-1)^6\begin{vmatrix} 1 & -3 \\ 5 & -7 \end{vmatrix} \end{bmatrix} = \begin{bmatrix} -37 & 57 & 73 \\ 9 & 3 & -29 \\ -24 & -8 & 8 \end{bmatrix}$$

and the adjoint of $[A]$ is

$$Adj\,[A] = \{[A]^C\}^T = \begin{bmatrix} -37 & 9 & -24 \\ 57 & 3 & -8 \\ 73 & -29 & 8 \end{bmatrix}$$

The product of the matrix $[A]$ of the determinant $|A|$ and its adjoint must satisfy the relation of Sec. 10.04–1*c*; that is,

$$[A]\,\text{adj}\,[A] = \begin{bmatrix} 1 & -3 & 0 \\ 5 & -7 & 8 \\ 9 & 2 & 3 \end{bmatrix}\begin{bmatrix} -37 & 9 & -24 \\ 57 & 3 & -8 \\ 73 & -29 & 8 \end{bmatrix} = \begin{bmatrix} -208 & 0 & 0 \\ 0 & -208 & 0 \\ 0 & 0 & -208 \end{bmatrix} = -208\begin{bmatrix} 1 & 0 & 0 \\ 0 & 1 & 0 \\ 0 & 0 & 1 \end{bmatrix}$$

(2) Inverse of a Matrix

(a) Definition. The inverse of the matrix $[A]$ of order $(n \times n)$ designated as $[A]^{-1}$ is an $(n \times n)$ matrix uniquely defined by the conditions

$$[A][A]^{-1} = [A]^{-1}[A] = [I] \qquad \text{and} \qquad |A| \neq 0$$

where $[I]$ is the unit matrix of order $(n \times n)$ and $|A|$ is the determinant of $[A]$.

(b) Nonsingular matrix. The matrix $[A]$ which satisfies these conditions is said to be a nonsingular matrix.

(c) Formula. Analytically, the inverse of $[A]$ is

$$[A]^{-1} = \frac{\operatorname{adj}[A]}{|A|}$$

where adj $[A]$ is defined in Sec. 10.04–1*b* and its construction is shown in Sec. 10.04–1*c*.

example:

The inverse of

$$[A] = \begin{bmatrix} 1 & -3 & 0 \\ 5 & -7 & 8 \\ 9 & 2 & 3 \end{bmatrix}$$

is

$$[A]^{-1} = \frac{\operatorname{adj}[A]}{|A|} = \frac{1}{-208}\begin{bmatrix} -37 & 9 & -24 \\ 57 & 3 & -8 \\ 73 & -29 & 8 \end{bmatrix} = \begin{bmatrix} 0.17788 & -0.04326 & 0.11538 \\ -0.27403 & -0.01442 & 0.03846 \\ -0.35096 & 0.13942 & -0.03846 \end{bmatrix}$$

which in turn must satisfy

$$[A][A]^{-1} = \begin{bmatrix} 1 & -3 & 0 \\ 5 & -7 & 8 \\ 9 & 2 & 3 \end{bmatrix}\begin{bmatrix} 0.17788 & -0.04326 & 0.11538 \\ -0.27403 & -0.01442 & 0.03846 \\ -0.35096 & 0.13942 & -0.03846 \end{bmatrix} = \begin{bmatrix} 1 & 0 & 0 \\ 0 & 1 & 0 \\ 0 & 0 & 1 \end{bmatrix}$$

(d) Singular matrix. If $|A| = 0$, the inverse of $[A]$ does not exist and $[A]$ is called singular matrix.

(e) Inverse of an inverse of order $(n \times n)$ is the initial matrix of the same order.

$$[[A]^{-1}]^{-1} = [A]$$

(f) Inverse of a transpose of order $(n \times n)$ is the transpose of the inverse of the same order.

$$[[A]^T]^{-1} = [[A]^{-1}]^T$$

Thus the operations of transposing and inversing are commutative.

(g) Inverse of a product. The inverse of a product of two or more square matrices equals the product of their inverses in reversed order.

$$\{[A][B][C]\}^{-1} = [C]^{-1}[B]^{-1}[A]^{-1}$$

(3) Special Inverses

(a) Inverse of a unit matrix of order $(n \times n)$ is the unit matrix of the same order.

$$[I]^{-1} = [I] \qquad [I]^{-1}[I] = [I]$$

(b) Inverse of a diagonal matrix of order $(n \times n)$ is a diagonal matrix of the same order, each of whose principal diagonal elements is the reciprocal of the respective element of the given matrix.

example:

$$\begin{bmatrix} a & 0 & 0 & 0 \\ 0 & b & 0 & 0 \\ 0 & 0 & c & 0 \\ 0 & 0 & 0 & d \end{bmatrix}^{-1} = \begin{bmatrix} 1/a & 0 & 0 & 0 \\ 0 & 1/b & 0 & 0 \\ 0 & 0 & 1/c & 0 \\ 0 & 0 & 0 & 1/d \end{bmatrix}$$

(c) Inverse of a symmetrical matrix of order $(n \times n)$ is another symmetrical matrix of the same order.

(d) Inverse of an antisymmetrical matrix of order $(2n+1) \times (2n+1)$ does not exist (singular matrix).

(e) Inverse of an antisymmetrical matrix of order $(2n) \times (2n)$ is an antisymmetrical matrix of the same order.

(4) Special Matrices

(a) Normal matrix. A square matrix $[A]$ is said to be normal if

$$[A] = [A]^T$$

All zero, unit, diagonal, and symmetrical matrices are normal matrices.

(b) Orthogonal matrix. A square matrix $[A]$ is said to be orthogonal if

$$[A]^T = [A]^{-1}$$

All unit and angular transformation matrices are orthogonal (Sec. 10.07–1*b*).

(c) Orthonormal matrix. A square matrix $[A]$ is said to be orthonormal (involutory) if

$$[A] = [A]^T = [A]^{-1}$$

All unit matrices are orthonormal matrices.

10.05 MATRIX DIVISION

(1) Scaling

(a) Uniform scaling. A matrix is divided by the scalar α if each element is multiplied by the reciprocal of this scalar.

example:

$$\begin{bmatrix} a_{11} & a_{12} & a_{13} \\ a_{21} & a_{22} & a_{23} \\ a_{31} & a_{32} & a_{33} \end{bmatrix} : \alpha = \begin{bmatrix} a_{11}/\alpha & a_{12}/\alpha & a_{13}/\alpha \\ a_{21}/\alpha & a_{22}/\alpha & a_{23}/\alpha \\ a_{31}/\alpha & a_{32}/\alpha & a_{33}/\alpha \end{bmatrix}$$

(b) Specific scaling of each row. Each row of a matrix is divided by the selected scalar if it is premultiplied by the inverse of the diagonal matrix of these scalars.

example:

$$\begin{bmatrix} 1/\alpha & 0 & 0 \\ 0 & 1/\beta & 0 \\ 0 & 0 & 1/\gamma \end{bmatrix} \begin{bmatrix} a_{11} & a_{12} & a_{13} \\ a_{21} & a_{22} & a_{23} \\ a_{31} & a_{32} & a_{33} \end{bmatrix} = \begin{bmatrix} a_{11}/\alpha & a_{12}/\alpha & a_{13}/\alpha \\ a_{21}/\beta & a_{22}/\beta & a_{23}/\beta \\ a_{31}/\gamma & a_{32}/\gamma & a_{33}/\gamma \end{bmatrix}$$

(c) Specific scaling of each column. Each column of a matrix is divided by the selected scalar if the matrix is postmultiplied by the inverse of the diagonal matrix of these scalars.

example:

$$\begin{bmatrix} a_{11} & a_{12} & a_{13} \\ a_{21} & a_{22} & a_{23} \\ a_{31} & a_{32} & a_{33} \end{bmatrix} \begin{bmatrix} 1/\alpha & 0 & 0 \\ 0 & 1/\beta & 0 \\ 0 & 0 & 1/\gamma \end{bmatrix} = \begin{bmatrix} a_{11}/\alpha & a_{12}/\beta & a_{13}/\gamma \\ a_{21}/\alpha & a_{22}/\beta & a_{23}/\gamma \\ a_{31}/\alpha & a_{32}/\beta & a_{33}/\gamma \end{bmatrix}$$

(2) Division of Square Matrices

(a) Simple division. The square matrix $[A]$ of order $(n \times n)$ is divided by the square matrix $[B]$ of the same order if it is premultiplied by the inverse of the second matrix.

$$[A]:[B] = [B]^{-1}[A]$$

(b) Chain division. The square matrix $[A]$ of order $(n \times n)$ is divided by a chain product of square matrices $[B][C]\cdots[N]$ each of which is of order $(n \times n)$ if it is premultiplied by their inverses in reversed order.

example:

$$[A]:\{[B][C]\cdots[N]\} = [N]^{-1}\cdots[C]^{-1}[B]^{-1}[A]$$

(c) Factoring of square matrices of the same order is governed by the following rules:

$$[A]+[B][C] = [B][[B]^{-1}[A]+[C]]$$

$$[A][B]+[C][D] = [A][B]\{[I]+[B]^{-1}[A]^{-1}[C][D]\}$$

(d) Fraction. A simple fraction of two square matrices $[A]$ and $[B]$ of the same order is the indicated division of these two matrices.

$$\frac{[A]}{[B]} = [A]:[B] = [B]^{-1}[A]$$

(e) Formulas. The following relations hold for square matrices of the same order.

$$[A]+\frac{[C]}{[D]} = [A]+[D]^{-1}[C]$$

$$[A]\frac{[C]}{[D]} = [A][D]^{-1}[C]$$

$$[A]:\frac{[C]}{[D]} = [A]:\{[D]^{-1}[C]\} = [C]^{-1}[D][A]$$

10.06 SIMULTANEOUS LINEAR EQUATIONS

(1) Homogeneous and Nonhomogeneous Equations

(a) System of m equations of the form

$$a_{11}x_1 + a_{12}x_2 + \cdots + a_{1n}x_n = b_1$$

$$a_{21}x_1 + a_{22}x_2 + \cdots + a_{2n}x_n = b_2$$

$$\cdots\cdots\cdots\cdots\cdots\cdots$$

$$a_{m1}x_1 + a_{m2}x_2 + \cdots + a_{mn}x_n = b_m$$

is called a system of m linear equations in the n unknowns $x_1, x_2, \ldots, x_n$.

(b) Homogeneous equations. If $b_1, b_2, \ldots, b_m$ are all zero, the system is called homogeneous.

(c) Nonhomogeneous equations. If at least one of the coefficients $b_1, b_2, \ldots, b_m$ is not zero, the system is called nonhomogeneous.

(d) Admissible solution. Any set of numbers $x_1, x_2, \ldots, x_n$ which satisfies the given system of equations is called an admissible solution of this system.

(e) Unique solution. If $m = n$, and the determinant of the factors,

$$|A| = \begin{vmatrix} a_{11} & a_{12} & \cdots & a_{1n} \\ a_{21} & a_{22} & \cdots & a_{2n} \\ \cdots & \cdots & \cdots & \cdots \\ a_{n1} & a_{n2} & \cdots & a_{nn} \end{vmatrix} \neq 0$$

and at least one of the coefficients $b_1, b_2, \ldots, b_n$ is not zero, then there is only one set of values $x_1, x_2, \ldots, x_n$ which satisfies this system and this set is called the unique solution.

(2) Determinant Solution

(a) General solution. If the conditions of Sec. 10.06–1*e* are satisfied, then the unique solution of the given system in the determinant form is

$$x_1 = \frac{|A_1|}{|A|}, \quad x_2 = \frac{|A_2|}{|A|}, \quad \ldots, \quad x_n = \frac{|A_n|}{|A|}$$

where $|A|$ is the determinant of a_{jk} and $|A_1|, |A_2|, \ldots, |A_n|$ are the augmented determinants defined below.

(b) Augmented determinant $|A_k|$ is obtained from $|A|$ by removing the kth column of a elements and replacing it by the column of b elements; that is,

$$|A_1| = \begin{vmatrix} b_1 & a_{12} & \cdots & a_{1n} \\ b_2 & a_{22} & \cdots & a_{2n} \\ \cdots & \cdots & \cdots & \cdots \\ b_n & a_{n1} & \cdots & a_{nn} \end{vmatrix}, \quad |A_2| = \begin{vmatrix} a_{11} & b_1 & \cdots & a_{1n} \\ a_{21} & b_2 & \cdots & a_{2n} \\ \cdots & \cdots & \cdots & \cdots \\ a_{n1} & b_n & \cdots & a_{nn} \end{vmatrix}, \quad \ldots, \quad |A_n| = \begin{vmatrix} a_{11} & a_{12} & \cdots & b_1 \\ a_{21} & a_{22} & \cdots & b_2 \\ \cdots & \cdots & \cdots & \cdots \\ a_{n1} & a_{n2} & \cdots & b_n \end{vmatrix}$$

example:

The unique determinant solution of

$$\begin{aligned} x_1 - 3x_2 \qquad\quad &= -5 \\ 5x_1 - 7x_2 + 8x_3 &= 15 \\ 9x_1 + 2x_2 + 3x_3 &= 22 \end{aligned}$$

is

$$x_1 = \frac{|A_1|}{|A|} = 1 \qquad x_2 = \frac{|A_2|}{|A|} = 2 \qquad x_3 = \frac{|A_3|}{|A|} = 3$$

where $|A| = -208$ (Sec. 10.03–2*b*), and

$$|A_1| = \begin{vmatrix} -5 & -3 & 0 \\ 15 & -7 & 8 \\ 22 & 2 & 3 \end{vmatrix} = -208 \qquad |A_2| = \begin{vmatrix} 1 & -5 & 0 \\ 5 & 15 & 8 \\ 9 & 22 & 3 \end{vmatrix} = -416 \qquad |A_3| = \begin{vmatrix} 1 & -3 & -5 \\ 5 & -7 & 15 \\ 9 & 2 & 22 \end{vmatrix} = -624$$

The back substitution of $x_1 = 1$, $x_2 = 2$, $x_3 = 3$ in the given system of equations must satisfy uniquely these equations, that is, $(1)(1)-(3)(2) = -5$, $(5)(1)-(7)(2)+(8)(3) = 15$, $(9)(1)+(2)(2)+(3)(3) = 22$.

(3) Matrix Solutions

(a) Inverse solution. If the conditions of Sec. 10.6–1e are satisfied, then the unique solution of the given system in the matrix form is

$$\begin{bmatrix} x_1 \\ x_2 \\ \cdots \\ \\ x_n \end{bmatrix} = \frac{1}{|A|} \begin{bmatrix} A_{11} & A_{21} & \cdots & A_{n1} \\ A_{12} & A_{22} & \cdots & A_{n2} \\ \cdots & \cdots & \cdots & \cdots \\ \\ A_{1n} & A_{2n} & \cdots & A_{nn} \end{bmatrix} \begin{bmatrix} b_1 \\ b_2 \\ \cdots \\ \\ b_n \end{bmatrix}$$

or, symbolically,

$$[x] = [A]^{-1}[b]$$

where $[A]^{-1}$ is the inverse of

$$[A] = \begin{bmatrix} a_{11} & a_{12} & \cdots & a_{1n} \\ a_{21} & a_{22} & \cdots & a_{2n} \\ \cdots & \cdots & \cdots & \cdots \\ a_{n1} & a_{n2} & \cdots & a_{nn} \end{bmatrix}$$

example:

The unique inverse solution of

$$x_1 - 3x_2 \qquad = -5$$

$$5x_1 - 7x_2 + 8x_3 = 15$$

$$9x_1 + 2x_2 + 3x_3 = 22$$

is

$$\underbrace{\begin{bmatrix} x_1 \\ x_2 \\ x_3 \end{bmatrix}}_{[x]} = \underbrace{\begin{bmatrix} 0.17788 & -0.04326 & 0.11538 \\ -0.27403 & -0.01442 & 0.03846 \\ -0.35096 & 0.13942 & -0.03846 \end{bmatrix}}_{[A]^{-1}} \underbrace{\begin{bmatrix} -5 \\ 15 \\ 22 \end{bmatrix}}_{[b]} = \begin{bmatrix} 1 \\ 2 \\ 3 \end{bmatrix}$$

where $[A]^{-1}$ was computed in Sec. 10.04–2c.

(b) Transformation. The given system of equations (Sec. 10.06–1a) can be transformed to

$$x_1 = -\frac{a_{12}}{a_{11}} x_2 - \cdots - \frac{a_{1n}}{a_{11}} x_n + \frac{b_1}{a_{11}}$$

$$x_2 = -\frac{a_{21}}{a_{22}} x_1 - \cdots - \frac{a_{2n}}{a_{22}} x_n + \frac{b_2}{a_{22}}$$

$$\cdots\cdots\cdots\cdots\cdots\cdots\cdots\cdots$$

$$x_n = -\frac{a_{n1}}{a_{nn}} x_1 - \cdots + \frac{b_n}{a_{nn}}$$

which is the basis for the iterative solution shown in (c).

(c) Iterative solution. The matrix form of these transformed equations is

$$\underbrace{\begin{bmatrix} x_1 \\ x_2 \\ \vdots \\ x_n \end{bmatrix}}_{[x]} = \underbrace{\begin{bmatrix} 0 & r_{12} & \cdots & r_{1n} \\ r_{21} & 0 & \cdots & r_{2n} \\ \cdots & \cdots & \cdots & \cdots \\ r_{n1} & r_{n2} & \cdots & 0 \end{bmatrix}}_{[r]} \underbrace{\begin{bmatrix} x_1 \\ x_2 \\ \vdots \\ x_n \end{bmatrix}}_{[x]} + \underbrace{\begin{bmatrix} m_1 \\ m_2 \\ \vdots \\ m_n \end{bmatrix}}_{[m]}$$

where $r_{jk} = -a_{jk}/a_{jj}$ and $m_j = b_j/a_{jj}$.
Then the unique iterative solution of this set is

$$[x] = \{[I]+[r]+[r]^2+[r]^3+\cdots\}[m]$$

where $[I]$ is an $(n \times n)$ unit matrix, $[r]$ is the carryover matrix of the same order shown above, $[r]^2 = [r][r]$, $[r]^3 = [r][r][r], \ldots$, and the matrix series is equivalent to $[A]^{-1}$. If the principal diagonal terms $a_{11}, a_{22}, \ldots, a_{nn}$ are large compared to the other a terms in their respective rows, the iteration series converges rapidly to the unique solution; if they are small, it may converge slowly or not at all.

example:

The unique iterative solution of

$$\begin{aligned} 10x_1 - x_2 \qquad &= 80 \\ -x_1 + 10x_2 - x_3 &= 160 \\ -x_2 + 10x_3 &= 280 \end{aligned}$$

is obtained from the transformed set

$$x_1 = (0.1)x_2 + 8$$

$$x_2 = (0.1)x_1 + (0.1)x_3 + 16$$

$$x_3 = (0.1)x_2 + 28$$

as $\qquad [x] = \{[I]+[r]+[r]^2+[r]^3+\cdots\}[m]$

where $\qquad [I] = \begin{bmatrix} 1 & 0 & 0 \\ 0 & 1 & 0 \\ 0 & 0 & 1 \end{bmatrix} \qquad (r) = \begin{bmatrix} 0 & 0.1 & 0 \\ 0.1 & 0 & 0.1 \\ 0 & 0.1 & 0 \end{bmatrix} \qquad [m] = \begin{bmatrix} 8 \\ 16 \\ 28 \end{bmatrix}$

$$[r]^2 = [r][r]\begin{bmatrix} 0.01 & 0 & 0.01 \\ 0 & 0.02 & 0 \\ 0.01 & 0 & 0.01 \end{bmatrix} \qquad [r]^3 = [r][r][r] = \begin{bmatrix} 0 & 0.002 & 0 \\ 0.002 & 0 & 0.002 \\ 0 & 0.002 & 0 \end{bmatrix}$$

and for $\{[I]+[r]+[r]^2+[r]^3\}$ gives

$$\begin{bmatrix} x_1 \\ x_2 \\ x_3 \end{bmatrix} = \begin{bmatrix} 1.0100 & 0.1020 & 0.0100 \\ 0.1020 & 1.0200 & 0.1020 \\ 0.0100 & 0.1020 & 1.0100 \end{bmatrix} \begin{bmatrix} 8 \\ 16 \\ 28 \end{bmatrix} = \begin{bmatrix} 9.9080 \\ 19.9920 \\ 29.9920 \end{bmatrix}$$

which shows a good convergence to the exact solution $x_1 = 10$, $x_2 = 20$, $x_3 = 30$, made possible by the large diagonal terms $a_{11} = a_{22} = a_{33} = 10$, and small off-diagonal terms $a_{12} = a_{21} = a_{23} = a_{32} = -1$, $a_{13} = a_{31} = 0$.

10.07 TRANSFORMATION OF COORDINATES

(1) Direct Transformations

(a) Translation. The transformation equations of translation introduced in the algebraic form in Sec. 6.05–2*a* and referred to in Fig. 6.05–1 can be expressed in matrix form as

$$\underbrace{\begin{bmatrix} x^0 \\ y^0 \end{bmatrix}}_{[s^0]} = \underbrace{\begin{bmatrix} a \\ b \end{bmatrix}}_{[d]} + \underbrace{\begin{bmatrix} x^1 \\ y^1 \end{bmatrix}}_{[s^1]} \qquad \underbrace{\begin{bmatrix} x^1 \\ y^1 \end{bmatrix}}_{[s^1]} = \underbrace{\begin{bmatrix} x^0 \\ y^0 \end{bmatrix}}_{[s^0]} - \underbrace{\begin{bmatrix} a \\ b \end{bmatrix}}_{[d]}$$

where $[s^0]$ is the column matrix of the coordinates (x^0,y^0) of P in the 0 system of axes, $[s^1]$ is the column of coordinates (x^1,y^1) in the 1 system of axes, and $[d]$ is the column matrix of the linear displacements (a,b) of the origin from 0 to 1.

(b) Rotation. The transformation equations of rotation introduced in algebraic form in Sec. 6.05–2*b* and referred to in Fig. 6.05–2 can be also expressed in matrix form as

$$\underbrace{\begin{bmatrix} x^0 \\ y^0 \end{bmatrix}}_{[s^0]} = \underbrace{\begin{bmatrix} \cos\omega & -\sin\omega \\ \sin\omega & \cos\omega \end{bmatrix}}_{[\pi^{0l}]} \underbrace{\begin{bmatrix} x^l \\ y^l \end{bmatrix}}_{[s^l]} \qquad \underbrace{\begin{bmatrix} x^l \\ y^l \end{bmatrix}}_{[s^l]} = \underbrace{\begin{bmatrix} \cos\omega & \sin\omega \\ -\sin\omega & \cos\omega \end{bmatrix}}_{[\pi^{l0}]} \underbrace{\begin{bmatrix} x^0 \\ y^0 \end{bmatrix}}_{[s^0]}$$

where x^0, y^0 are the coordinates of P in the X^0, Y^0 axes, x^l, y^l are the coordinates of the same point in the X^l, Y^l axes, and $[\pi^{0l}]$, $[\pi^{l0}]$ are the angular transformation matrices which satisfy the relations

$$[\pi^{0l}] = [\pi^{l0}]^T = [\pi^{l0}]^{-1} \qquad [\pi^{l0}] = [\pi^{0l}]^T = [\pi^{0l}]^{-1}$$

and consequently are orthogonal.

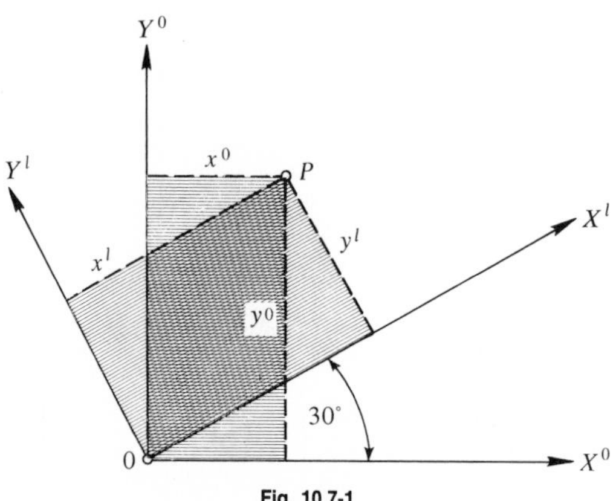

Fig. 10.7-1

example:

If the point P is given in the X^0, Y^0 axes by $x^0 = 10$, $y^0 = 20$, then its coordinates in the X^l, Y^l axes of Fig. 10.07–1 are

$$\begin{bmatrix} x^l \\ y^l \end{bmatrix} = \begin{bmatrix} \cos 30^\circ & \sin 30^\circ \\ -\sin 30^\circ & \cos 30^\circ \end{bmatrix} \begin{bmatrix} 10 \\ 20 \end{bmatrix} = \begin{bmatrix} 0.86603 & 0.50000 \\ -0.50000 & 0.86603 \end{bmatrix} \begin{bmatrix} 10 \\ 20 \end{bmatrix} = \begin{bmatrix} 18.66 \\ 12.32 \end{bmatrix}$$

and inversely

$$\begin{bmatrix} x^0 \\ y^0 \end{bmatrix} = \begin{bmatrix} \cos 30^\circ & -\sin 30^\circ \\ \sin 30^\circ & \cos 30^\circ \end{bmatrix} \begin{bmatrix} 18.66 \\ 12.32 \end{bmatrix} = \begin{bmatrix} 0.86603 & -0.50000 \\ 0.50000 & 0.86603 \end{bmatrix} \begin{bmatrix} 18.66 \\ 12.32 \end{bmatrix} = \begin{bmatrix} 10 \\ 20 \end{bmatrix}$$

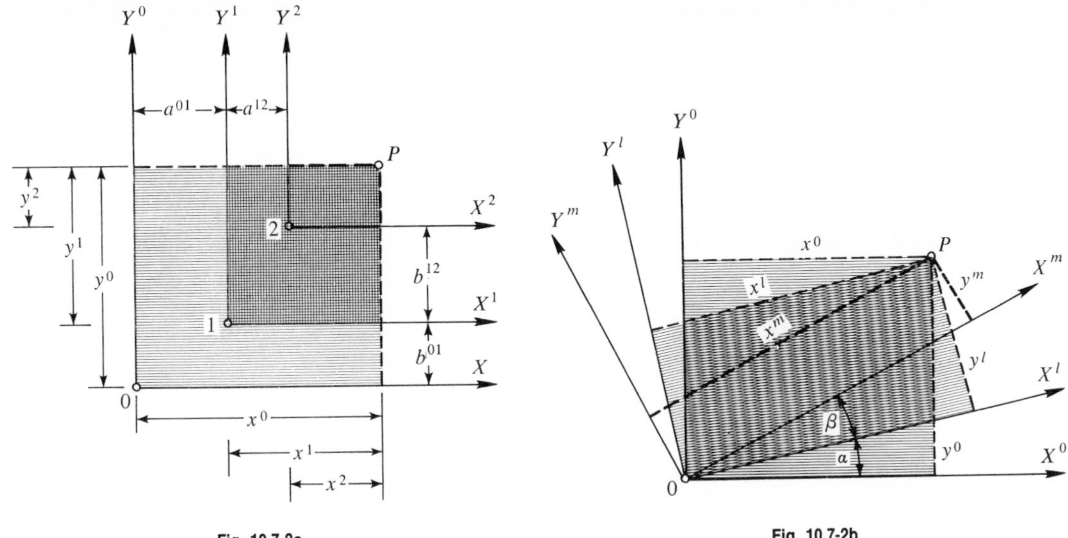

Fig. 10.7-2a

Fig. 10.7-2b

(2) Successive Transformations

(a) Translations. If the system of axes is successively translated as shown in Fig. 10.07–2*a*, then

$$\underbrace{\begin{bmatrix} x^0 \\ y^0 \end{bmatrix}}_{[s^0]} = \underbrace{\begin{bmatrix} a^{01} \\ b^{01} \end{bmatrix}}_{[d^{01}]} + \underbrace{\begin{bmatrix} x^1 \\ y^1 \end{bmatrix}}_{[s^1]} \qquad \underbrace{\begin{bmatrix} x^1 \\ y^1 \end{bmatrix}}_{[s^1]} = \underbrace{\begin{bmatrix} a^{12} \\ b^{12} \end{bmatrix}}_{[d^{12}]} + \underbrace{\begin{bmatrix} x^2 \\ y^2 \end{bmatrix}}_{[s^2]}$$

or, directly,

$$\underbrace{\begin{bmatrix} x^0 \\ y^0 \end{bmatrix}}_{[s^0]} = \underbrace{\begin{bmatrix} a^{01} \\ b^{01} \end{bmatrix}}_{[d^{01}]} + \underbrace{\begin{bmatrix} a^{12} \\ b^{12} \end{bmatrix}}_{[d^{12}]} + \underbrace{\begin{bmatrix} x^2 \\ y^2 \end{bmatrix}}_{[s^2]} = \underbrace{\begin{bmatrix} a^{02} \\ b^{02} \end{bmatrix}}_{[d^{02}]} + \underbrace{\begin{bmatrix} x^2 \\ y^2 \end{bmatrix}}_{[s^2]}$$

where $[s^0]$, $[s^1]$, $[s^2]$ are the column matrices of the coordinates of P in the 0, 1, 2 systems respectively and $[d^{01}]$, $[d^{12}]$, $[d^{02}]$ are the column matrices of the respective linear displacements of the origin.

Inversely,

$$[s^2] = [s^1] - [d^{12}] \qquad [s^1] = [s^0] - [d^{01}]$$

and

$$[s^2] = [s^0] - [d^{01}] - [d^{12}] = [s^0] - [d^{02}]$$

Both relations can be extended to include any number of translations.

(b) Rotations. If the system of axes is successively rotated as shown in Fig. 10.07–2*b*, then

$$\underbrace{\begin{bmatrix} x^0 \\ y^0 \end{bmatrix}}_{[s^0]} = \underbrace{\begin{bmatrix} \cos\alpha & -\sin\alpha \\ \sin\alpha & \cos\alpha \end{bmatrix}}_{[\pi^{0l}]} \underbrace{\begin{bmatrix} x^l \\ y^l \end{bmatrix}}_{[s^l]} \qquad \underbrace{\begin{bmatrix} x^l \\ y^l \end{bmatrix}}_{[s^l]} = \underbrace{\begin{bmatrix} \cos\beta & -\sin\beta \\ \sin\beta & \cos\beta \end{bmatrix}}_{[\pi^{lm}]} \underbrace{\begin{bmatrix} x^m \\ y \end{bmatrix}}_{[s^m]}$$

or directly,

$$\underbrace{\begin{bmatrix} x^0 \\ y^0 \end{bmatrix}}_{[s^0]} = \underbrace{\begin{bmatrix} \cos\alpha & -\sin\alpha \\ \sin\alpha & \cos\alpha \end{bmatrix}}_{[\pi^{0l}]} \underbrace{\begin{bmatrix} \cos\beta & -\sin\beta \\ \sin\beta & \cos\beta \end{bmatrix}}_{[\pi^{lm}]} \underbrace{\begin{bmatrix} x^m \\ y^m \end{bmatrix}}_{[s^m]} = \underbrace{\begin{bmatrix} \cos(\alpha+\beta) & -\sin(\alpha+\beta) \\ \sin(\alpha+\beta) & \cos(\alpha+\beta) \end{bmatrix}}_{[\pi^{0m}]} \underbrace{\begin{bmatrix} x^m \\ y^m \end{bmatrix}}_{[s^m]}$$

where $[s^0]$, $[s^l]$, $[s^m]$ are the coordinate matrices of P in the 0, l, m system and $[\pi^{0l}]$, $[\pi^{lm}]$, $[\pi^{0m}]$ are the respective angular transformation matrices.

Inversely,

$$[s^m] = [\pi^{ml}][s^l] \qquad [s^l] = [\pi^{l0}][s^0] \qquad \text{and} \qquad [s^m] = [\pi^{ml}][\pi^{l0}][s^0] = [\pi^{m0}][s^0]$$

Both relations can be extended to include any number of rotations. As in Sec. 10.07–1*b*, all π matrices are orthogonal.

11
SCALARS AND VECTORS

11.01 VECTOR ALGEBRA

(1) Basic Concepts

(a) Scalar. The quantity defined by the magnitude only and designated by ordinary letters such as $a, b, c, \ldots, A, B, C, \ldots, \alpha, \beta, \gamma$ is called the scalar. Typical scalars are length, time, temperature, and mass.

(b) Vector. The quantity defined by the magnitude and direction, and designated by boldface letters such as $\mathbf{a}, \mathbf{b}, \mathbf{c}, \ldots, \mathbf{A}, \mathbf{B}, \mathbf{C}$ is called the vector. Typical vectors are force, moment, displacement, velocity, and acceleration.

(c) Right-handed cartesian system. For the graphical representation of vectors two different rectangular (cartesian) coordinate systems can be used: the right-handed system and the left-handed coordinate system (Fig. 11.01–1). In this book the right-handed system is used consistently.

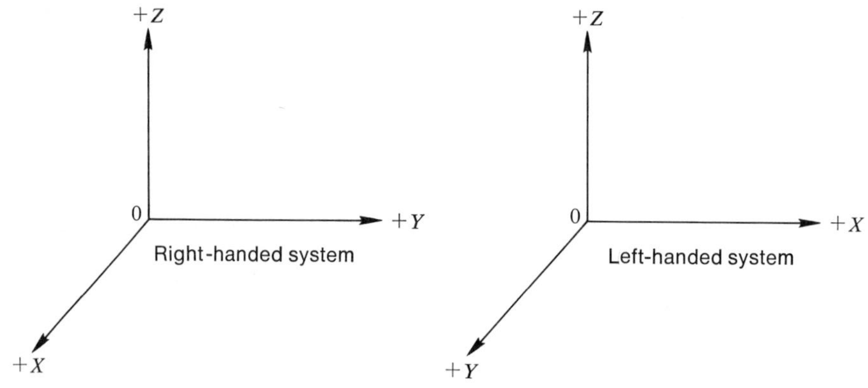

Fig. 11.01-1

(d) Representation. A vector $\mathbf{V}$ (any vector) can be represented graphically by a directed segment (Fig. 11.01–2) such as

$$\mathbf{V} = \overline{PQ}$$

where P is the initial point (origin of vector) and Q is the terminal point (terminus of vector).

(e) Magnitude of $\mathbf{V}$ (length of vector) is designated as V, or $|\mathbf{V}|$.

Analytically,

$$V = \sqrt{(x_P - x_Q)^2 + (y_P - y_Q)^2 + (z_P - z_Q)^2}$$

where x_P, y_P, z_P and x_Q, y_Q, z_Q are the coordinates of P and Q respectively.

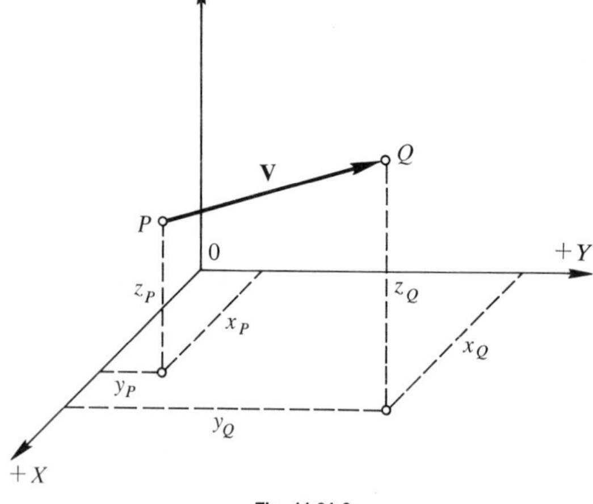

Fig. 11.01-2

(f) Unit vector. The ratio of $\mathbf{V}$ to V is called the unit vector $\mathbf{v}$ and consequently,

$$\mathbf{V} = V\mathbf{v}$$

(2) Addition and Subtraction

(a) Equal vectors of identical direction (Fig. 11.01–3). Two vectors $\mathbf{A} = A\mathbf{a}$ and $\mathbf{B} = B\mathbf{b}$ are equal and of identical direction if $A = B$ and $\mathbf{a} = \mathbf{b}$; that is,

$$\mathbf{A} = \mathbf{B} \qquad \text{or} \qquad A\mathbf{a} = B\mathbf{b}$$

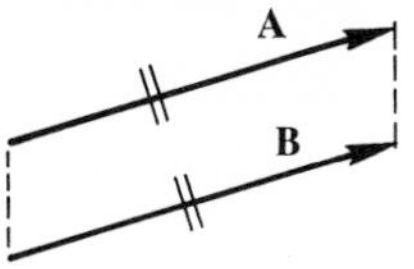

Fig. 11.01-3

(b) Equal vectors of opposite direction (Fig. 11.01–4). Two vectors $\mathbf{C} = C\mathbf{c}$ and $\mathbf{D} = D\mathbf{d}$ are equal but of opposite direction if $C = D$ and $\mathbf{c} = -\mathbf{d}$; that is,

$$\mathbf{C} = -\mathbf{D} \qquad \text{or} \qquad C\mathbf{c} = -D\mathbf{d}$$

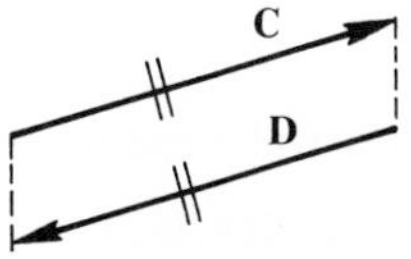

Fig. 11.01-4

(c) Sum of two vectors. The sum of two vectors **A** and **B** called their resultant **C** is obtained graphically by placing the origin of **B** on the origin of **A** and joining the terminus of **B** to the the terminus of **A** (Fig. 11.01–5). Analytically,

$$\mathbf{A} + \mathbf{B} = \mathbf{C}$$

(d) Difference of two vectors. The difference of vector **A** and **B** denoted as **D** is obtained graphically by placing the origin of **B** on the origin of **A** and joining the terminus of **B** to the terminus of **A** (Fig. 11.01–6). Analytically

$$\mathbf{A} - \mathbf{B} = \mathbf{D}$$

(e) Zero vector. If, in Fig. 11.01–6, $\mathbf{A} = \mathbf{B}$, then $\mathbf{D} = \mathbf{0}$ is called the zero vector. Analytically,

$$\mathbf{A} - \mathbf{B} = \mathbf{0}$$

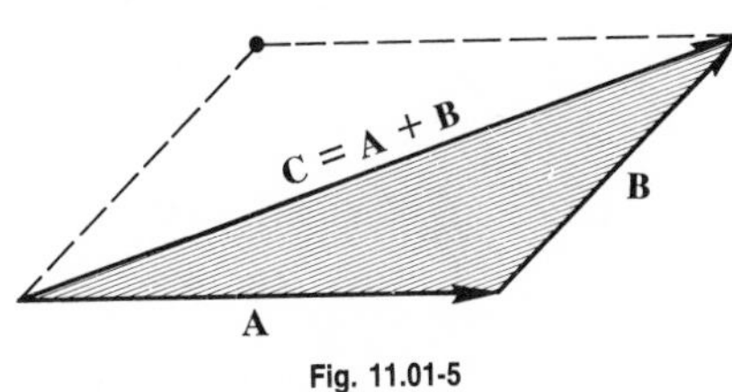

Fig. 11.01-5

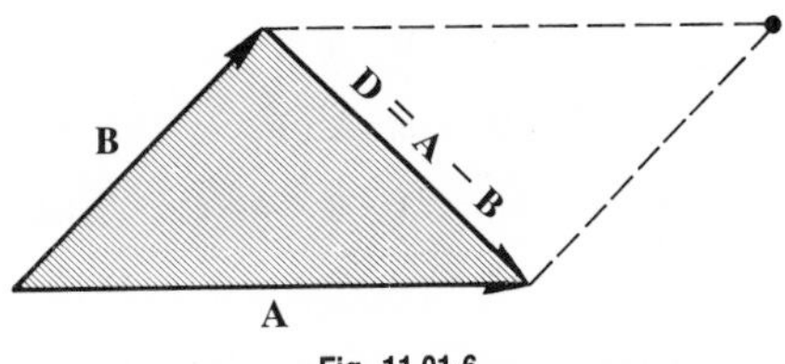

Fig. 11.01-6

(3) Vector Polygon

(a) Resultant. The resultant of two or several vectors is their vector sum **R**. Graphically (Fig. 11.01–7), **R** is the closing vector of the polygon obtained by placing the origin of the second vector on the terminus of the first vector, the origin of the third vector on the terminus of the second vector, and so on. Analytically

$$\mathbf{R} = \mathbf{A} + \mathbf{B} + \mathbf{C}$$

(b) Summation laws. The summation of vectors is commutative

$$\mathbf{R} = \mathbf{A} + \mathbf{B} + \mathbf{C} = \mathbf{B} + \mathbf{C} + \mathbf{A} = \mathbf{C} + \mathbf{A} + \mathbf{B}$$

and is also associative,

$$\mathbf{R} = (\mathbf{A} + \mathbf{B}) + \mathbf{C} = \mathbf{A} + (\mathbf{B} + \mathbf{C})$$

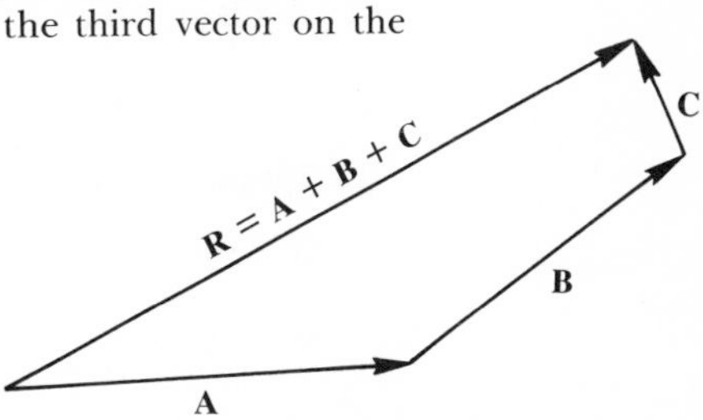

Fig. 11.01-7

(4) Scalar and Vector Relations

(a) Sum and difference. The sum or difference of a scalar m and vector $\mathbf{A}$ is not possible.

(b) Product laws.

$m\mathbf{A} = \mathbf{A}m$	commutative law
$m(n\mathbf{A}) = (mn)\mathbf{A} = n(m\mathbf{A})$	associative law
$(m+n)\mathbf{A} = m\mathbf{A} + n\mathbf{A}$	distributive law
$m(\mathbf{A}+\mathbf{B}) = m\mathbf{A} + m\mathbf{B}$	distributive law

where m, n are scalars.

(5) Dot Product of Two Vectors

(a) Definition. The dot product of two vectors $\mathbf{A} = A\mathbf{a}$ and $\mathbf{B} = B\mathbf{b}$ (also called the scalar product) denoted as $\mathbf{A}\cdot\mathbf{B}$ or $(\mathbf{AB})$ and read $\mathbf{A}$ dot $\mathbf{B}$ is defined as the product of their magnitudes and the cosine of the angle θ between them (Fig. 11.01–8). The result is a scalar. Analytically,

$$\mathbf{A}\cdot\mathbf{B} = AB\mathbf{a}\cdot\mathbf{b} = AB\cos\theta \qquad 0 \le \theta \le \pi$$

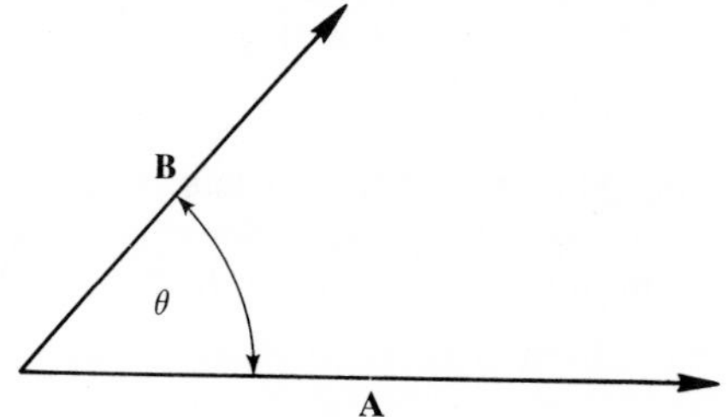

Fig. 11.01-8

(b) Special cases.

If $\theta = 0$ $\quad \mathbf{A}\cdot\mathbf{B} = AB$

If $\theta = \dfrac{\pi}{2}$ $\quad \mathbf{A}\cdot\mathbf{B} = 0$

If $\theta = \pi$ $\quad \mathbf{A}\cdot\mathbf{B} = -AB$

(c) Dot product laws.

$\mathbf{A}\cdot\mathbf{B} = \mathbf{B}\cdot\mathbf{A}$	commutative law
$m\mathbf{A}\cdot n\mathbf{B} = mn\mathbf{A}\cdot\mathbf{B}$	associative law
$\mathbf{A}\cdot(\mathbf{B}+\mathbf{C}) = \mathbf{A}\cdot\mathbf{B} + \mathbf{A}\cdot\mathbf{C}$	distributive law

(6) Cross Product of Two Vectors

(a) Definition. The cross product of two vectors $\mathbf{A} = A\mathbf{a}$ and $\mathbf{B} = B\mathbf{b}$ (also called the vector product) denoted as $\mathbf{A}\times\mathbf{B}$ or $[\mathbf{AB}]$ and read as $\mathbf{A}$ cross $\mathbf{B}$ is defined as the product of their magnitudes, the sine of the angle θ between them, and the unit vector $\mathbf{n}$ normal to their plane (Fig. 11.01–9). The result is a vector. Analytically,

$$\mathbf{A}\times\mathbf{B} = AB\mathbf{a}\times\mathbf{b} = AB\sin\theta\,\mathbf{n} \qquad 0 \le \theta \le \pi$$

(b) Special cases.

If $\theta = 0$ $\quad \mathbf{A}\times\mathbf{B} = 0$

If $\theta = \dfrac{\pi}{2}$ $\quad \mathbf{A}\times\mathbf{B} = AB\mathbf{n}$

If $\theta = \pi$ $\quad \mathbf{A}\times\mathbf{B} = 0$

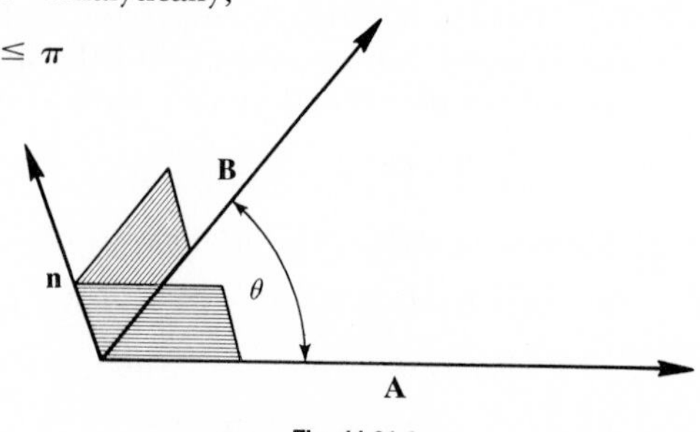

Fig. 11.01-9

(c) Cross product laws.

$$\mathbf{A} \times \mathbf{B} = -\mathbf{B} \times \mathbf{A} \qquad \text{commutative law fails}$$

$$(m\mathbf{A}) \times (n\mathbf{B}) = mn(\mathbf{A} \times \mathbf{B}) \qquad \text{associative law}$$

$$\mathbf{A} \times (\mathbf{B} + \mathbf{C}) = \mathbf{A} \times \mathbf{B} + \mathbf{A} \times \mathbf{C} \qquad \text{distributive law}$$

where m, n are scalars.

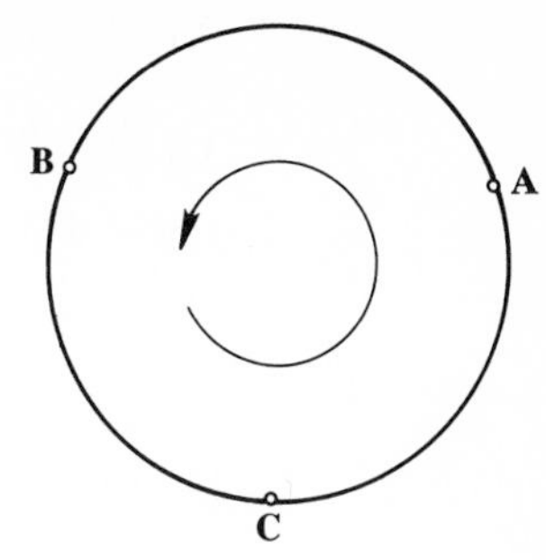

Fig. 11.01-10

(7) Scalar Triple Product

(a) General case. The scalar triple product designated as $[\mathbf{A} \quad \mathbf{B} \quad \mathbf{C}]$ is a dot product of two vectors one of which is a cross product of two other vectors. The result is a scalar. Analytically,

$$[\mathbf{A} \quad \mathbf{B} \quad \mathbf{C}] = \mathbf{A} \cdot (\mathbf{B} \times \mathbf{C}) = \mathbf{B} \cdot (\mathbf{C} \times \mathbf{A}) = \mathbf{C} \cdot (\mathbf{A} \times \mathbf{B})$$

$$= (\mathbf{A} \times \mathbf{B}) \cdot \mathbf{C} = (\mathbf{B} \times \mathbf{C}) \cdot \mathbf{A} = (\mathbf{C} \times \mathbf{A}) \cdot \mathbf{B}$$

where the dot and cross signs are interchangeable and **A**, **B**, **C** follow the sequence of Fig. 11.01–10.

(b) Special case. If **A**, **B**, **C** are three mutually perpendicular nonzero vectors (Fig. 11.01–11), then

$$[\mathbf{A} \quad \mathbf{B} \quad \mathbf{C}] = \pm ABC$$

is the positive (negative) volume of the parallelepiped having **A**, **B**, **C** as edges in the right-handed (left-handed) cartesian system.

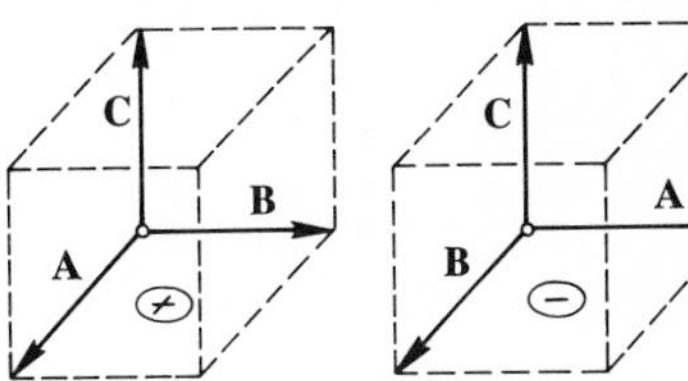

Fig. 11.01-11

(c) Unit vectors. If **a**, **b**, **c** are three mutually perpendicular unit vectors (Fig. 11.01–12), then

$$\mathbf{a} \cdot \mathbf{b} = \mathbf{b} \cdot \mathbf{c} = \mathbf{c} \cdot \mathbf{a} = 1 \qquad \text{and} \qquad [\mathbf{a} \quad \mathbf{b} \quad \mathbf{c}] = \pm 1$$

where the physical interpretation is that of Sec. 11.01–7*b*.

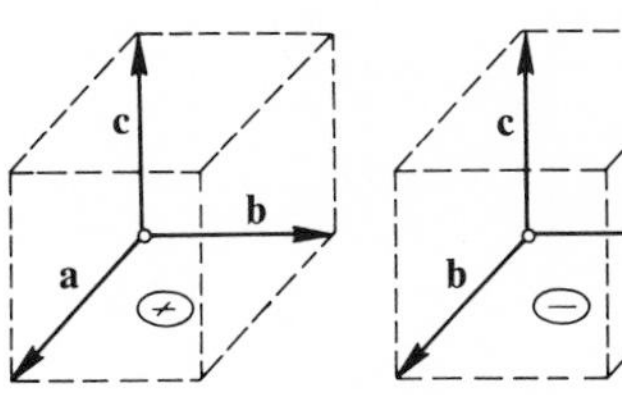

Fig. 11.01-12

(8) Vector Triple Product

(a) General case. The vector triple product is a cross product of two vectors, one of which is a cross product of two other vectors. The result is a vector. Analytically,

$$\mathbf{A} \times (\mathbf{B} \times \mathbf{C}) = (\mathbf{A} \cdot \mathbf{C})\mathbf{B} - (\mathbf{A} \cdot \mathbf{B})\mathbf{C}$$

$$(\mathbf{A} \times \mathbf{B}) \times \mathbf{C} = (\mathbf{A} \cdot \mathbf{C})\mathbf{B} - (\mathbf{B} \cdot \mathbf{C})\mathbf{A}$$

which shows that

$$\mathbf{A} \times (\mathbf{B} \times \mathbf{C}) \neq (\mathbf{A} \times \mathbf{B}) \times \mathbf{C}$$

(b) Special case. If **A**, **B**, **C** are three mutually perpendicular vectors (Fig. 11.01–11), then

$$\mathbf{A} \times (\mathbf{B} \times \mathbf{C}) = (\mathbf{A} \times \mathbf{B}) \times \mathbf{C} = 0$$

(c) Unit vectors. If **a**, **b**, **c** are three mutually perpendicular unit vectors in the right-handed system (Fig. 11.01–12), then

$$\mathbf{a} = \mathbf{b} \times \mathbf{c} \qquad \mathbf{b} = \mathbf{c} \times \mathbf{a} \qquad \mathbf{c} = \mathbf{a} \times \mathbf{b}$$

and also

$$\mathbf{a} \times (\mathbf{b} \times \mathbf{c}) = \mathbf{b} \times (\mathbf{c} \times \mathbf{a}) = \mathbf{c} \times (\mathbf{a} \times \mathbf{b}) = 0$$

11.02 CARTESIAN VECTORS

(1) Resolution

(a) Components. In the cartesian right-handed system (Fig. 11.01–1), a given vector **V** can be represented as the vector sum of its components $\mathbf{V}_x$, $\mathbf{V}_y$, $\mathbf{V}_z$ which are parallel to the respective coordinate axes (Fig. 11.02–1). Analytically,

$$\mathbf{V} = \mathbf{V}_x + \mathbf{V}_y + \mathbf{V}_z = V_x\mathbf{i} + V_y\mathbf{j} + V_z\mathbf{k}$$

where V_x, V_y, V_z are the magnitudes of the respective components and **i**, **j**, **k** are the unit vectors parallel to the X, Y, Z axes respectively (Fig. 11.02–2).

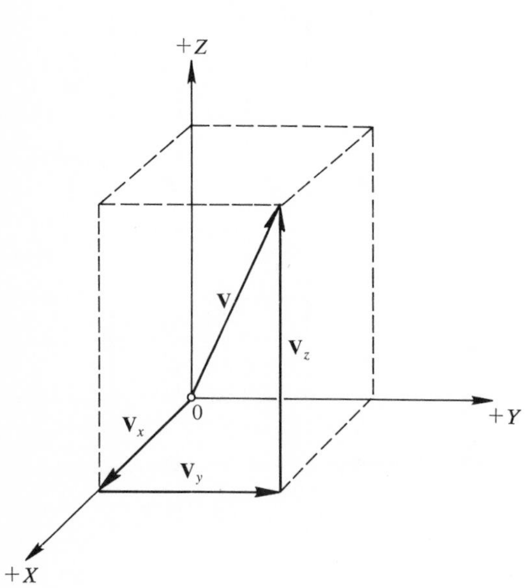

Fig. 11.02-1

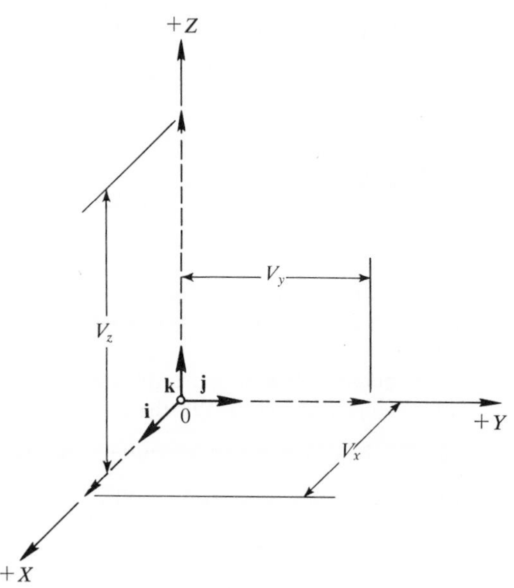

Fig. 11.02-2

(b) Magnitude of **V** is then the length of the diagonal of the rectangular parallelepiped having $\mathbf{V}_x$, $\mathbf{V}_y$, $\mathbf{V}_z$ as edges (Fig. 11.02–1). Analytically,

$$V = \sqrt{V_x^2 + V_y^2 + V_z^2}$$

(c) Unit vector. By definition, the unit vector **v** of **V** is

$$\mathbf{v} = \frac{\mathbf{V}}{V} = \frac{V_x}{V}\mathbf{i} + \frac{V_y}{V}\mathbf{j} + \frac{V_z}{V}\mathbf{k} = v_x\mathbf{i} + v_y\mathbf{j} + v_z\mathbf{k}$$

example:

The magnitude of $\mathbf{A} = 3\mathbf{i} + 4\mathbf{j} + 12\mathbf{k}$ is $A = \sqrt{9 + 16 + 144} = 13$ and its unit vector is $\mathbf{a} = \frac{3}{13}\mathbf{i} + \frac{4}{13}\mathbf{j} + \frac{12}{13}\mathbf{k}$.

(2) Geometry

(a) Direction cosines of the vector **V** are the cosines of the angles α, β, γ measured to **V** from the positive X, Y, Z axes (Fig. 11.02–3). Analytically,

$$\cos\alpha = \frac{V_x}{V} = v_x \qquad \cos\beta = \frac{V_y}{V} = v_y \qquad \cos\gamma = \frac{V_z}{V} = v_z$$

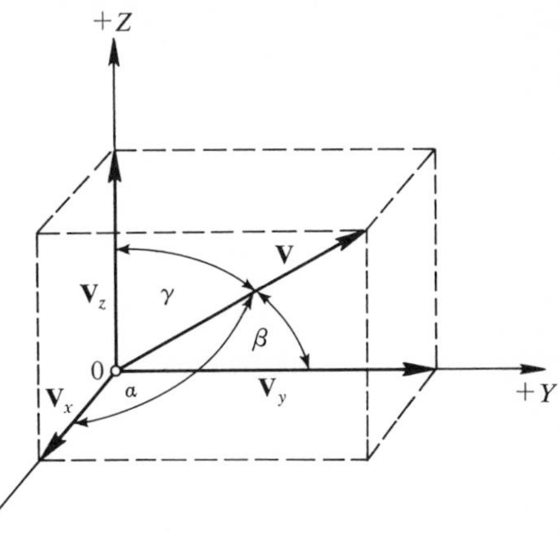

Fig. 11.02-3

(b) Relationship of the direction angles is given by

$$\cos^2\alpha + \cos^2\beta + \cos^2\gamma = 1$$

which shows that

$$v_x^2 + v_y^2 + v_z^2 = 1$$

(c) Independence. Out of three direction angles two and only two are independent and the third one is determined from the relationship of Sec. 11.02–2*b*. Consequently, the direction of a vector is defined completely and uniquely by two direction angles (any two).

example:

The direction cosines of $\mathbf{A} = 3\mathbf{i} + 4\mathbf{j} + 12\mathbf{k}$ are

$$\cos\alpha = \frac{A_x}{A} = \frac{3}{\sqrt{3^2 + 4^2 + 12^2}} = 0.23076$$

$$\cos\beta = \frac{A_y}{A} = \frac{4}{\sqrt{3^2 + 4^2 + 12^2}} = 0.30769$$

$$\cos\gamma = \frac{A_z}{A} = \frac{12}{\sqrt{3^2 + 4^2 + 12^2}} = 0.92307$$

and must satisfy

$$(0.23076)^2 + (0.30769)^2 + (0.92307)^2 = 1$$

(3) Addition and Subtraction

(a) Sum of two vectors $\mathbf{A} = A_x\mathbf{i} + A_y\mathbf{j} + A_z\mathbf{k}$ and $\mathbf{B} = B_x\mathbf{i} + B_y\mathbf{j} + B_z\mathbf{k}$ equals the vector sum of the sums of their respective cartesian components.

$$\mathbf{A} + \mathbf{B} = (A_z + B_x)\mathbf{i} + (A_y + B_y)\mathbf{j} + (A_z + B_z)\mathbf{k}$$

(b) Difference of the same two vectors A and **B** equals the vector sum of the differences of their respective cartesian components.

$$\mathbf{A} - \mathbf{B} = (A_x - B_x)\mathbf{i} + (A_y - B_y)\mathbf{j} + (A_z - B_z)\mathbf{k}$$

examples:

If $\mathbf{A} = -3\mathbf{i} + 10\mathbf{j} + 13\mathbf{k}$ and $\mathbf{B} = -8\mathbf{i} - 5\mathbf{j} + 13\mathbf{k}$, then

$$\mathbf{A} + \mathbf{B} = (-3-8)\mathbf{i} + (10-5)\mathbf{j} + (13+13)\mathbf{k} = -11\mathbf{i} + 5\mathbf{j} + 26\mathbf{k}$$

$$\mathbf{A} - \mathbf{B} = (-3+8)\mathbf{i} + (10+5)\mathbf{j} + (13-13)\mathbf{k} = 5\mathbf{i} + 15\mathbf{j}$$

(4) Operations with a Scalar Factor

(a) Scalar factor. The scalar factor of a vector is the largest common factor of all the magnitudes of its cartesian components.

example:

$$\mathbf{V} = 18\mathbf{i} - 27\mathbf{j} + 90\mathbf{k} = 9(2\mathbf{i} - 3\mathbf{j} + 10\mathbf{k})$$

(b) Product of a scalar and a vector is the vector sum of the products of the scalar m and the respective cartesian components of the vector $\mathbf{V}$.

example:

$$m\mathbf{V} = 3(18\mathbf{i} - 27\mathbf{j} + 90\mathbf{k}) = 54\mathbf{i} - 81\mathbf{j} + 270\mathbf{k}$$

(c) Quotient of a vector and a scalar is the vector sum of the quotients of the respective cartesian components of the vector $\mathbf{V}$ and the scalar n.

example:

$$\frac{\mathbf{V}}{n} = \frac{18\mathbf{i} - 27\mathbf{j} + 90\mathbf{k}}{9} = 2\mathbf{i} - 3\mathbf{j} + 10\mathbf{k}$$

(5) Dot Product of Two Vectors

(a) Orthogonal unit vectors. The dot products of the cartesian unit vectors **i**, **j**, **k** are special cases of the dot product defined in Sec. 11.01–5*a*.

$$\mathbf{i}\cdot\mathbf{i} = 1 \qquad \mathbf{i}\cdot\mathbf{j} = 0 \qquad \mathbf{i}\cdot\mathbf{k} = 0$$

$$\mathbf{j}\cdot\mathbf{i} = 0 \qquad \mathbf{j}\cdot\mathbf{j} = 1 \qquad \mathbf{j}\cdot\mathbf{k} = 0$$

$$\mathbf{k}\cdot\mathbf{i} = 0 \qquad \mathbf{k}\cdot\mathbf{i} = 0 \qquad \mathbf{k}\cdot\mathbf{k} = 1$$

(b) Orthogonal components. The dot product of two vectors $\mathbf{A} = A_x\mathbf{i} + A_y\mathbf{j} + A_z\mathbf{k}$ and $\mathbf{B} = B_x\mathbf{i} + B_y\mathbf{j} + B_z\mathbf{k}$ is

$$\mathbf{A}\cdot\mathbf{B} = A_xB_x + A_yB_y + A_zB_z$$

example:

If $\mathbf{A} = 2\mathbf{i} - 3\mathbf{j} + 4\mathbf{k}$ and $\mathbf{B} = 4\mathbf{i} + 5\mathbf{j} + 7\mathbf{k}$, then

$$\mathbf{A}\cdot\mathbf{B} = (2)(4) + (-3)(5) + (4)(7) = 21$$

(c) Angle θ between two vectors $\mathbf{A}$ and $\mathbf{B}$ defined above is given (Sec. 11.01–5*a*) by

$$\cos\theta = \frac{\mathbf{A}\cdot\mathbf{B}}{AB} = \frac{A_xB_x + A_yB_y + A_zB_z}{AB}$$

where A, B are the magnitudes of $\mathbf{A}$, $\mathbf{B}$ respectively.

example:

If **A**, **B** are the vectors of Sec. 11.02–5*b*, then

$$\cos\theta = \frac{21}{\sqrt{2^2 + (-3)^2 + 4^2}\sqrt{4^2 + 5^2 + 7^2}} = 0.41128$$

(d) Magnitude of the vector $\mathbf{A} = A_x\mathbf{i} + A_y\mathbf{j} + A_z\mathbf{k}$ is also defined as

$$A = \sqrt{\mathbf{A}\cdot\mathbf{A}} = \sqrt{A_x^2 + A_y^2 + A_z^2}$$

(6) Cross Product of Two Vectors

(a) Orthogonal unit vectors. The cross products of the cartesian unit vectors **i**, **j**, **k** are special cases of the cross product defined in Sec. 11.01–6*a*.

$$\mathbf{i}\times\mathbf{i} = 0 \qquad \mathbf{i}\times\mathbf{j} = \mathbf{k} \qquad \mathbf{i}\times\mathbf{k} = -\mathbf{j}$$
$$\mathbf{j}\times\mathbf{i} \times -\mathbf{k} \qquad \mathbf{j}\times\mathbf{j} = 0 \qquad \mathbf{j}\times\mathbf{k} = \mathbf{i}$$
$$\mathbf{k}\times\mathbf{i} = \mathbf{j} \qquad \mathbf{k}\times\mathbf{j} = -\mathbf{i} \qquad \mathbf{k}\times\mathbf{k} = 0$$

(b) Orthogonal components. The cross product of two vectors $\mathbf{A} = A_x\mathbf{i} + A_y\mathbf{j} + A_z\mathbf{k}$ and $B = B_x\mathbf{i} + B_y\mathbf{j} + B_z\mathbf{k}$ is

$$\mathbf{A}\times\mathbf{B} = \begin{vmatrix} \mathbf{i} & \mathbf{j} & \mathbf{k} \\ A_x & A_y & A_z \\ B_x & B_y & B_z \end{vmatrix} = \begin{vmatrix} \mathbf{i} & A_x & B_x \\ \mathbf{j} & A_y & B_y \\ \mathbf{k} & A_z & B_z \end{vmatrix}$$

$$= (A_yB_z - A_zB_y)\mathbf{i} - (A_xB_z - A_zB_x)\mathbf{j} + (A_xB_y - A_yB_x)\mathbf{k}$$

example:

If $\mathbf{A} = 2\mathbf{i} - 3\mathbf{j} + 4\mathbf{k}$ and $B = 4\mathbf{i} + 5\mathbf{j} + 7\mathbf{k}$, then

$$\mathbf{A}\times\mathbf{B} = \begin{vmatrix} \mathbf{i} & \mathbf{j} & \mathbf{k} \\ 2 & -3 & 4 \\ 4 & 5 & 7 \end{vmatrix}$$

$$= [(-3(7) - (4)(5)]\mathbf{i} - [(2)(7) - (4)(4)]\mathbf{j} + [(2)(5) - (-3)(4)]\mathbf{k}$$
$$= -41\mathbf{i} + 2\mathbf{j} + 22\mathbf{k}$$

and the magnitude of $\mathbf{A}\times\mathbf{B}$ is

$$|\mathbf{A}\times\mathbf{B}| = \sqrt{(-41)^2 + (2)^2 + (22)^2} = 46.57$$

(c) Angle θ between two vectors **A** and **B** defined above is given (Sec. 11.01–6*a*) by

$$\sin\theta = \frac{|\mathbf{A}\times\mathbf{B}|}{AB} = \frac{\sqrt{(A_yB_z - A_zB_y)^2 + (A_xB_z - A_zB_x)^2 + (A_xB_y - A_yB_x)^2}}{AB}$$

where A, B are the magnitudes of **A**, **B** respectively.

example:

If **A**, **B** are the vectors of Sec. 11.02–6*b*, then

$$\sin\theta = \frac{|\mathbf{A}\times\mathbf{B}|}{AB} = \frac{46.57}{\sqrt{(2)^2 + (-3)^2 + (4)^2}\sqrt{(4)^2 + (5)^2 + (7)^2}} = 0.91156$$

The result of this problem and that of problem of Sec. 11.02–5*c* must satisfy the relation $\sin^2\theta + \cos^2\theta = 1$.

(7) Triple Products

(a) Scalar triple product (Sec. 11.01–7a) of the vectors

$$\mathbf{A} = A_x\mathbf{i} + A_y\mathbf{j} + A_z\mathbf{k} \qquad \mathbf{B} = B_x\mathbf{i} + B_y\mathbf{j} + B_z\mathbf{k} \qquad \mathbf{C} = C_x\mathbf{i} + C_y\mathbf{j} + C_z\mathbf{k}$$

is

$$\mathbf{A}\cdot(\mathbf{B}\times\mathbf{C}) = (\mathbf{A}\times\mathbf{B})\cdot\mathbf{C} = \begin{vmatrix} A_x & A_y & A_z \\ B_x & B_y & B_z \\ C_x & C_y & C_z \end{vmatrix} = \begin{vmatrix} A_x & B_x & C_x \\ A_y & B_y & C_y \\ A_z & B_z & C_z \end{vmatrix}$$

$$= A_x(B_yC_z - B_zC_y) - A_y(B_xC_z - B_zC_x) + A_z(B_xC_y - B_yC_x)$$

(b) Vector triple product (Sec. 11.01–8a) of the vectors **A**, **B**, **C** defined above is

$$\mathbf{A}\times(\mathbf{B}\times\mathbf{C}) = \begin{vmatrix} \mathbf{B} & \mathbf{C} \\ \mathbf{A}\cdot\mathbf{B} & \mathbf{A}\cdot\mathbf{C} \end{vmatrix} = (\mathbf{A}\cdot\mathbf{C})\mathbf{B} - (\mathbf{A}\cdot\mathbf{B})\mathbf{C}$$

$$(\mathbf{A}\times\mathbf{B})\times\mathbf{C} = \begin{vmatrix} \mathbf{A}\cdot\mathbf{C} & \mathbf{B}\cdot\mathbf{C} \\ \mathbf{A} & \mathbf{B} \end{vmatrix} = (\mathbf{A}\cdot\mathbf{C})\mathbf{B} - (\mathbf{B}\cdot\mathbf{C})\mathbf{A}$$

where $\mathbf{A}\cdot\mathbf{B} = A_xB_x + A_yB_y + A_zB_z$, $\mathbf{B}\cdot\mathbf{C} = B_xC_x + B_yC_y + B_zC_z$, and $\mathbf{A}\cdot\mathbf{C} = A_xC_x + A_yC_y + A_zC_z$.

11.03 VECTOR CALCULUS

(1) Vector Derivatives

(a) Vector function. If to each value of a scalar t there exists a vector $\mathbf{F}(t)$ then $\mathbf{F}(t)$ is said to be the vector function of the single scalar variable t. In the cartesian system,

$$\mathbf{F}(t) = \mathbf{F}_x(t) + \mathbf{F}_y(t) + \mathbf{F}_z(t) = F_x(t)\mathbf{i} + F_y(t)\mathbf{j} + F_z(t)\mathbf{k}$$

where $F_x(t)$, $F_y(t)$, $F_z(t)$ are the magnitudes of the vector function's components $\mathbf{F}_x(t)$, $\mathbf{F}_y(t)$, $\mathbf{F}_z(t)$, respectively.

(b) Limit, continuity, and derivatives of the vector function $\mathbf{F}(t)$ follow rules similar to those of the scalar function $F(t)$.

(c) First derivative of $\mathbf{F}(t)$ with respect to t is

$$\frac{d[\mathbf{F}(t)]}{dt} = \frac{d[F_x(t)]}{dt}\mathbf{i} + \frac{d[F_y(b)]}{dt}\mathbf{j} + \frac{d[F_z(t)]}{dt}\mathbf{k}$$

or, simply,

$$\dot{\mathbf{F}} = \dot{F}_x\mathbf{i} + \dot{F}_y\mathbf{j} + \dot{F}_z\mathbf{k}$$

where the dot above the symbols indicates the first derivative with respect to t and **i**, **j**, **k** are unaffected by the differentiation process.

(d) Second derivative of $\mathbf{F}(t)$ with respect to t is

$$\ddot{\mathbf{F}} = \ddot{F}_x\mathbf{i} + \ddot{F}_y\mathbf{j} + \ddot{F}_z\mathbf{k}$$

where the double dot above the symbols indicates the second derivative with respect to t.

example:

The first and second derivatives of $\mathbf{F} = \sin\omega t\,\mathbf{i} + \cos\omega t\,\mathbf{j} + \omega t\,\mathbf{k}$ with respect to t are

$$\dot{\mathbf{F}} = \omega\cos\omega t\,\mathbf{i} - \omega\sin\omega t\,\mathbf{j} + \omega\mathbf{k}$$

$$\ddot{\mathbf{F}} = -\omega^2\sin\omega t\,\mathbf{i} - \omega^2\cos\omega t\,\mathbf{j}$$

where ω is constant ($d\omega/dt = 0$).

(e) Derivatives of sum and products of two vector functions $\mathbf{A} = \mathbf{A}(t)$ and $\mathbf{B} = \mathbf{B}(t)$ with respect to their scalar variable t are

$$\frac{d}{dt}(\mathbf{A}+\mathbf{B}) = \dot{\mathbf{A}}+\dot{\mathbf{B}}$$

$$\frac{d}{dt}(\mathbf{A}\cdot\mathbf{B}) = \mathbf{A}\cdot\dot{\mathbf{B}}+\dot{\mathbf{A}}\cdot\mathbf{B}$$

$$\frac{d}{dt}(\mathbf{A}\times\mathbf{B}) = \mathbf{A}\times\dot{\mathbf{B}}+\dot{\mathbf{A}}\times\mathbf{B}$$

where $\dot{\mathbf{A}}$ and $\dot{\mathbf{B}}$ are the first derivatives of $\mathbf{A}$ and $\mathbf{B}$ respectively.

(f) Second derivatives of these vector functions with respect to t are

$$\frac{d^2}{dt^2}(\mathbf{A}+\mathbf{B}) = \ddot{\mathbf{A}}+\ddot{\mathbf{B}}$$

$$\frac{d^2}{dt^2}(\mathbf{A}\cdot\mathbf{B}) = \mathbf{A}\cdot\ddot{\mathbf{B}}+2\dot{\mathbf{A}}\cdot\dot{\mathbf{B}}+\ddot{\mathbf{A}}\cdot\mathbf{B}$$

$$\frac{d^2}{dt^2}(\mathbf{A}\times\mathbf{B}) = \mathbf{A}\times\ddot{\mathbf{B}}+2\dot{\mathbf{A}}\times\dot{\mathbf{B}}+\ddot{\mathbf{A}}\times\mathbf{B}$$

(2) Indefinite Vector Integrals

(a) Definition of the indefinite integral of the vector function $\mathbf{f}(t)$ is formally identical to that of the indefinite integral of its scalar equivalent $f(t)$. Analytically,

$$\int \mathbf{f}(t)\,dt = \int \frac{d}{dt}[\mathbf{F}(t)]\,dt = \mathbf{F}(t)+\mathbf{C}$$

where $\mathbf{C}$ is the vector constant of integration which has a similar meaning as its scalar counterpart in Sec. 9.01–1*c*.

(b) Cartesian components. If $\mathbf{f}(t) = f_x(t)\mathbf{i}+f_y(t)\mathbf{j}+f_z(t)\mathbf{k}$, then

$$\int \mathbf{f}(t)\,dt = \mathbf{i}\int f_x(t)\,dt+\mathbf{j}\int f_y(t)\,dt+\mathbf{k}\int f_z(t)\,dt = F_x(t)\mathbf{i}+F_y(t)\mathbf{j}+F_z(t)\mathbf{k}+\mathbf{C}$$

where $F_x(t)$, $F_y(t)$, $F_z(t)$ are the indefinite integrals of $f_x(t)$, $f_y(t)$, $f_z(t)$ respectively, $\mathbf{i}$, $\mathbf{j}$, $\mathbf{k}$ are constants unaffected by the integration process and

$$\mathbf{C} = C_x\mathbf{i}+C_y\mathbf{j}+C_z\mathbf{k}$$

is the vector constant of integration defined by three components of magnitudes C_x, C_y, C_z.

example:

$$\int (5t\mathbf{i}+\cos t\mathbf{j}+\sin t\mathbf{k})\,dt = \left(\frac{5t^2}{2}\right)\mathbf{i}+(\sin t)\mathbf{j}-(\cos t)\mathbf{k}+\mathbf{C}$$

(3) Definite Vector Integrals

(a) Definition of the definite integral of a vector function $\mathbf{f}(t)$ is also formally identical to that of the definite integral of its scalar equivalent $f(t)$. Analytically,

$$\int_{t_1}^{t_2} \mathbf{f}(t)\,dt = \int_{t_1}^{t_2} \frac{d}{dt}[\mathbf{F}(t)]\,dt = \mathbf{F}(t_2) - \mathbf{F}(t_1)$$

where $t_1 \le t \le t_2$ is the closed interval of integration and t_1, t_2 are the lower and upper limits of integration respectively.

(b) Cartesian components. If $\mathbf{f}(t) = f_x(t)\mathbf{i} + f_y(t)\mathbf{j} + f_z(t)\mathbf{k}$, then

$$\int_{t_1}^{t_2} \mathbf{f}(t)\,dt = [F_x(t_2) - F_x(t_1)]\mathbf{i} + [F_y(t_2) - F_y(t_1)]\mathbf{j} + [F_z(t_2) - F_z(t_1)]\mathbf{k}$$

where $F_x(t_2)$, $F_y(t_2)$, $F_z(t_2)$, and $F_x(t_1)$, $F_y(t_1)$, $F_z(t_1)$ are the integrals of $f_x(t)$, $f_y(t)$, $f_z(t)$ in terms of t_2 and t_1 respectively.

example:

$$\int_1^2 (2t\mathbf{i} + 9t^2\mathbf{j} + 12t^3\mathbf{k})\,dt = \left[\frac{2t^2}{2}\mathbf{i} + \frac{9t^3}{3}\mathbf{j} + \frac{12t^4}{4}\mathbf{k}\right]_1^2$$

$$= \left[\frac{(2)(2)^2}{2} - \frac{(2)(1)^2}{2}\right]\mathbf{i} + \left[\frac{(9)(2)^3}{3} - \frac{(9)(1)^3}{3}\right]\mathbf{j} + \left[\frac{(12)(2)^4}{4} - \frac{(12)(1)^4}{4}\right]\mathbf{k}$$

$$= 3\mathbf{i} + 21\mathbf{j} + 45\mathbf{k}$$

12
NUMERICAL PROCEDURES

12.01 DECIMAL SYSTEM

(1) Definitions

(a) Decimal notation for expressing numbers is called a place system of base 10 in which each digit is a multiple of the respective position power of 10. *Digits* of the decimal system are 0, 1, 2, 3, 4, 5, 6, 7, 8, 9.

(b) Decimal point is the reference point of position. Digits on the left side of this point are multiples of $\ldots, 10^2, 10^1, 10^0$ respectively, and digits on the right side of this point are multiples of $10^{-1}, 10^{-2}, 10^{-3}, \ldots,$ respectively. The digits on the right side of the decimal point are called the *decimal places.*

examples:

$$2{,}345 = 2\times 10^3 + 3\times 10^2 + 4\times 10^1 + 5\times 10^0$$

$$234.5 = 2\times 10^2 + 3\times 10^1 + 4\times 10^0 + 5\times 10^{-1}$$

$$23.45 = 2\times 10^1 + 3\times 10^0 + 4\times 10^{-1} + 5\times 10^{-2}$$

$$2.345 = 2\times 10^0 + 3\times 10^{-1} + 4\times 10^{-2} + 5\times 10^{-3}$$

(c) Decimal fraction is the quotient of a number and a power of 10.

examples:

$$2{,}345 = \frac{2{,}345}{10^0} \qquad 2.345 = \frac{2{,}345}{10^3}$$

$$234.5 = \frac{2{,}345}{10^1} \qquad 0.2345 = \frac{2{,}345}{10^4}$$

$$23.45 = \frac{2{,}345}{10^2} \qquad 0.02345 = \frac{2{,}345}{10^5}$$

(d) Every fraction can be converted to a decimal (number or fraction) by performing the indicated division.

(e) Terminating decimal. When the division process yields a finite number of digits, the quotient is called terminating decimal.

examples:

$\frac{2}{4} = 0.50 \qquad \frac{16}{64} = 0.25 \qquad \frac{1}{8} = 0.125$

(f) Endless decimal. When the division process does not terminate, the quotient is called endless decimal.

examples:

$\frac{43}{57} = 0.7543859\ldots$

$\frac{1}{3} = 0.3333333\ldots = 0.\dot{3}$

where the last decimal $0.\dot{3}$ is called *periodic number.*

(2) Four Fundamental Operations

(a) Addition of decimals. First all decimals are written in a vertical column aligned on their decimal points. Then the columns of digits are added in a sum with the decimal point placed under the other decimal points. It is desirable to add once going upward and then downward to check the result.

examples:

	701.2		0.373
	10.46		0.002
	1.131		1.719
	2712.3		20.103
Sum =	3425.091	Sum =	22.197

(b) Subtraction of decimals. First the subtrahend is placed below the minuend with the decimal points aligned. Then the digits of each column are subtracted with the decimal point placed under the other decimal points. To check, the sum of the difference and the subtrahend must equal the minuend.

examples:

Subtraction:		Check:	
Minuend	127.830	Difference	68.858
Subtrahend	− 58.972	Subtrahend	+ 58.972
Difference	68.858	Minuend	127.830

(c) Multiplication of decimals. First the product of the factors is found, disregarding the presence of decimal points in both factors. Then the number of decimal places of both factors are counted and assigned to the product. Finally the result is checked by multiplying the factors in reversed order.

example:

Multiplication:

31.64	= multiplicand	⟵	2 decimal places
× 2.783	= multiplier	⟵	3 decimal places
9492	= (3 × 3164)		
25312	= (8 × 3164)		
22148	= (7 × 3164)		
6328	= (2 × 3164)		
88.05412	= product	⟵	5 decimal places

Check:

2.783	= multiplicand	⟵	3 decimal places
× 31.64	= multiplier	⟵	2 decimal places
11132	= (4 × 2783)		
16698	= (6 × 2783)		
2783	= (1 × 2783)		
8349	= (3 × 2783)		
88.05412	= product	⟵	5 decimal places

(d) Division of decimals. First the decimal point is removed from the divisor by moving the decimal point in the dividend an equal number of places (if the dividend has fewer decimal places than the divisor, then the respective number of zeros must be added). Then the division process is carried out with the decimal point of the quotient aligned above the decimal point of the dividend. Finally the product of the quotient and of the divisor plus the remainder (if there is one) must equal the dividend, which is a useful and necessary check.

example:

Removal of decimal point in divisor:

Case	Initial position of decimal points	Shifted position of decimal points
1	56.789 : 1.234	56.789. : 1.234.
2	56.789 : 12.34	56.78.9 : 12.34.
3	56.789 : 123.4	56.7.89 : 123.4.

For solution see cases 1, 2, and 3 below

example:

Case 1:

Divisor = 1234

```
    46.02   = quotient
  -------
   56789    = dividend
  -4936     = (4 × 1234)
  -------
    7429
  -7404     = (6 × 1234)
  -------
     25 0
    -00 0   = (0 × 1234)
  -------
     25 00
    -24 68  = (2 × 1234)
  -------
        32  = remainder
```

Check: $\underbrace{(46.02)}_{\text{quotient}} \times \underbrace{(1234)}_{\text{divisor}} + \underbrace{0.32}_{\text{remainder}} = \underbrace{56789}_{\text{dividend}}$

example:

Case 2:

Divisor = 1234

```
    4.602   = quotient
  -------
   5678.9   = dividend
  -4936     = (4 × 1234)
  -------
    742 9
   -740 4   = (6 × 1234)
  -------
      2 50
     -0 00  = (0 × 1234)
  -------
      2 500
     -2 468 = (2 × 1234)
  -------
         32 = remainder
```

Check: $\underbrace{(4.602)}_{\text{quotient}} \times \underbrace{(1234)}_{\text{divisor}} + \underbrace{0.032}_{\text{remainder}} = \underbrace{56789}_{\text{dividend}}$

example:

Case 3:

By inspection, quotient = 0.4602 and remainder = 0.0032

(3) Special Rules

(a) Divisibility of numbers. A decimal N consisting of a finite number of digits is perfectly divisible by:

2, if its last digit is an even number
3, if the sum of all its digits is divisible by 3
4, if the number made up of its last two digits is divisible by 4
5, if the last digit is 0 or 5
6, if it is even and divisible by 3
8, if the number made up of its last three digits is divisible by 8
9, if the sum of all its digits is divisible by 9
10, if the last digit is zero
11, if the difference of the sum of odd-place digits and of the sum of even-place digits is divisible by 11 or is zero
12, if it is divisible by 3 and 4
25, if the last two digits are 00, 25, 50, or 75
50, if the last two digits are 00 or 50
100, if the last two digits are 00

No rule is available for the divisibility of a number *by 7.*

example:

The number 443,520 is divisible by:

2, since the last digit is 0 (even)
3, since $4+4+3+5+2+0 = 18$ is divisible by 3
4, since 20 is divisible by 4
5, since the last digit is zero
6, since it is an even number divisible by 3
8, since 520 is divisible by 8
9, since $4+4+3+5+2+0 = 18$ is divisible by 9
10, since the last digit is 0
11, since $(4+5+0)-(4+3+2) = 0$
12, since it is divisible by 3 and 4

(b) Special products. A decimal N consisting of a finite number of digits can be conveniently multiplied by:

5, if it is multiplied by 10 and divided by 2
25, if it is multiplied by 100 and divided by 4
125, if it is multiplied by 1,000 and divided by 8

example:

$$1{,}234 \times 5 = \frac{12{,}340}{2} = 6{,}170$$

$$1{,}234 \times 25 = \frac{123{,}400}{4} = 30{,}850$$

$$1{,}234 \times 125 = \frac{1{,}234{,}000}{8} = 154{,}250$$

(c) Special quotients. A decimal N consisting of a finite number of digits can be conveniently divided by:

5, if it is multiplied by 2 and divided by 10
25, if it is multiplied by 4 and divided by 100
125, if it is multiplied by 8 and divided by 100

examples:

$$29{,}375 : 5 = 2937.5 \times 2 = 5{,}875$$

$$29{,}375 : 25 = 293.75 \times 4 = 1{,}175$$

$$29.375 : 125 = 29.375 \times 8 = 235$$

12.02 Binary and Other Systems

(1) Definitions

(a) Binary notation for expressing numbers is called a place system of base 2 in which each digit is a multiple of the respective position power of 2. Digits of the binary system are 0 and 1.

(b) Binary point is the reference point of position. Digits on the left side of this point are multiples of $\ldots, 2^2, 2^1, 2^0$, respectively, and digits on the right side of this point are multiples of $2^{-1}, 2^{-2}, 2^{-3}, \ldots$, respectively.

examples:

$$1\,101 = 1\times 2^3 + 1\times 2^2 + 0\times 2^1 + 1\times 2^0 \qquad 11.01 = 1\times 2^1 + 1\times 2^0 + 0\times 2^{-1} + 1\times 2^{-2}$$

$$110.1 = 1\times 2^2 + 1\times 2^1 + 0\times 2^0 + 1\times 2^{-1} \qquad 1.101 = 1\times 2^0 + 1\times 2^{-1} + 0\times 2^{-2} + 1\times 2^{-3}$$

(c) Binary fraction is the quotient of a number and a power of 2.

examples:

$$1\,101 = \frac{1\,101}{2^0} \qquad 1.101 = \frac{1\,101}{2^3}$$

$$110.1 = \frac{1\,101}{2^1} \qquad 0.1101 = \frac{1\,101}{2^4}$$

$$11.01 = \frac{1\,101}{2^2} \qquad 0.01101 = \frac{1\,101}{2^5}$$

(d) Advantage of binary notation is that only two different symbols (0 and 1) are involved in the representation of any number which in turn requires only two devices of transmission.

examples:

On a punch card, 1 can be represented by a hole and 0 by the absence of a hole.

On a magnetic tape, 1 can be represented by a magnetized spot and 0 by the absence of a magnetized spot.

(e) Disadvantage of binary notation is the large number of places required to express a number.

example:

$1\times 2^4 + 0\times 2^3 + 0\times 2^2 + 0\times 2^1 \times 0\times 2^0 \equiv 10\,000$ in the binary system and equals $1\times 10^1 + 6\times 10^0 \equiv 16$ in the decimal system.

(f) General notation for expressing numbers is called a place system of base B in which each digit is a multiple of the respective position power of B. Digits of the general system are $0, 1, 2, \ldots, B-1$.

examples:

$B = 3$: $\quad 21.20 = 2\times 3^1 + 1\times 3^0 + 2\times 3^{-1} + 0\times 3^{-2}$

$B = 7$: $\quad 534.6 = 5\times 7^2 + 3\times 7^1 + 4\times 7^0 + 6\times 7^{-1}$

(2) Conversions

(a) Binary to decimal conversion is performed by successive summation of the respective multiples of powers of 2.

example:

Binary number 101.11 equals decimal number 9.75; that is

$$1\times 2^3 = 8 = 8.00$$
$$0\times 2^2 = 0 = 0.00$$
$$1\times 2^0 = 1 = 1.00$$
$$1\times 2^{-1} = \tfrac{1}{2} = 0.50$$
$$1\times 2^{-2} = \tfrac{1}{4} = 0.25$$
$$\text{Decimal number} = 9.75$$

(b) Decimal to binary conversion is performed by means of Table 12.02–1 of the powers of 2.

example:

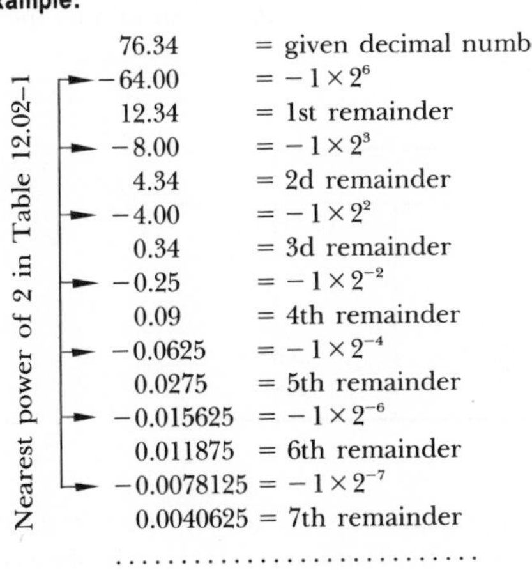

	76.34	= given decimal number
→	−64.00	$= -1\times 2^6$
	12.34	= 1st remainder
→	−8.00	$= -1\times 2^3$
	4.34	= 2d remainder
→	−4.00	$= -1\times 2^2$
	0.34	= 3d remainder
→	−0.25	$= -1\times 2^{-2}$
	0.09	= 4th remainder
→	−0.0625	$= -1\times 2^{-4}$
	0.0275	= 5th remainder
→	−0.015625	$= -1\times 2^{-6}$
	0.011875	= 6th remainder
→	−0.0078125	$= -1\times 2^{-7}$
	0.0040625	= 7th remainder
	....	

(Arrows labelled: Nearest power of 2 in Table 12.02–1)

TABLE 12.02–1 Powers of 2

	$2^0 = 1$
$2^1 = 2$	$2^{-1} = 0.5$
$2^2 = 4$	$2^{-2} = 0.25$
$2^3 = 8$	$2^{-3} = 0.125$
$2^4 = 16$	$2^{-4} = 0.062\,5$
$2^5 = 32$	$2^{-5} = 0.031\,25$
$2^6 = 64$	$2^{-6} = 0.015\,625$
$2^7 = 128$	$2^{-7} = 0.007\,812\,5$
$2^8 = 256$	$2^{-8} = 0.003\,906\,25$
$2^9 = 512$	$2^{-9} = 0.001\,953\,125$
$2^{10} = 1\,024$	$2^{-10} = 0.000\,976\,562\,5$
$2^{11} = 2\,048$	$2^{-11} = 0.000\,488\,231\,25$
$2^{12} = 4\,096$	$2^{-12} = 0.000\,244\,115\,625$
$2^{13} = 8\,192$	$2^{-13} = 0.000\,122\,057\,812\,5$
$2^{14} = 16\,384$	$2^{-14} = 0.000\,061\,028\,906\,25$
$2^{15} = 32\,768$	$2^{-15} = 0.000\,030\,514\,453\,125$

This process can be repeated to a desired accuracy. If the process stops with the 7th remainder, as shown above, then the binary equivalent of the given decimal number 76.34 is

$$1\times 2^6 + 0\times 2^5 + 0\times 2^4 + 1\times 2^3 + 1\times 2^2 + 0\times 2^1 + 0\times 2^0 + 0\times 2^{-1} + 1\times 2^{-2} + 0\times 2^{-3}$$
$$+ 1\times 2^{-4} + 0\times 2^{-5} + 1\times 2^{-6} + 1\times 2^{-7} + \cdots = 1\,001\,100.010\,101\,1\cdots$$

12.03 APPROXIMATE COMPUTATIONS

(1) Significant Figures

(a) Measured data inherently are not exact and if recorded in decimal notation they consist of a finite set of decimal digits called *significant figures*, the last of which is called the *doubtful digit.*

(b) Zero is a significant figure if surrounded by other digits, or if specifically designated as such; otherwise the zeros are used to fix the decimal point and hence are not significant.

examples:

570.2 has four significant figures: 5,7,0,2. 0.00134 has three significant figures, 1,3,4, and the zeros are used to fix the decimal point.

210,00 has five significant figures, 2,1,0,0,0, where the zeros are designated as significant; or has two significant figures, 2,1, if the zeros are not designated as significant.

(c) Scientific notation. A large or a small number N can be expressed in the form

$$N = t \times 10^n$$

where t is a number between 1 and 10 and n is an integer.

examples:

$400{,}000 = 4 \times 10^5$ $\quad 8{,}200 = 8.2 \times 10^3$ $\quad 0.025 = 2.5 \times 10^{-2}$

(2) Retention of Significant Figures

(a) Rounding off numbers. Frequently the number is rounded off by rejecting one or several of its last figures (digits). In rejecting figures, the last figure retained should be increased by 1 if the figure rejected (dropped) is 5 or greater.

example:

Given number	Rounded off to		
	Four figures	Three figures	Two figures
3.1416	3.142	3.14	3.1
2.7183	2.718	2.72	2.7
347.25	347.3	347	350
10,347	10,350	10,300	10,000

(b) Addition and subtraction. The number of significant figures of the sum or difference should be rounded off by eliminating any digit resulting from operations on broken column on the right as shown below.

examples:

$$\begin{array}{r} 201.3 \\ 1.05 \\ 21.76 \\ 0.0013 \\ \hline 224.1113 \doteq 224.1 \end{array} \qquad \begin{array}{r} 201.3 \\ -1.05 \\ -21.76 \\ -0.0013 \\ \hline 178.4887 \doteq 178.5 \end{array}$$

(c) Multiplication and division. The number of significant figures of the product or of the quotient should be rounded off to a number of significant figures equal to that of the least accurate term involved in the calculation.

examples:

$$3.14159 \times 21.13 = 66.38179 \doteq 66.38$$

$$3.14159 : 21.13 = 0.14868 \doteq 0.1487$$

where 21.13 is the least accurate number (4 significant figures) and therefore both results are rounded off to four significant figures.

(d) Squares and cubes. The number of significant figures of a square or of a cube of a number N should be rounded off to the number of significant figures of N.

examples:

$2.13^2 = 4.5369 \doteq 4.54 \qquad 2.13^3 = 9.6636 \doteq 9.66$

(e) Square roots and cube roots. The number of significant figures of a square root or of a cube root of a number N should be rounded off to the number of significant figures of N.

examples:

$$\sqrt{2.13} = 1.4595 \doteq 1.46$$

$$\sqrt[3]{2.13} = 1.2866 \doteq 1.29$$

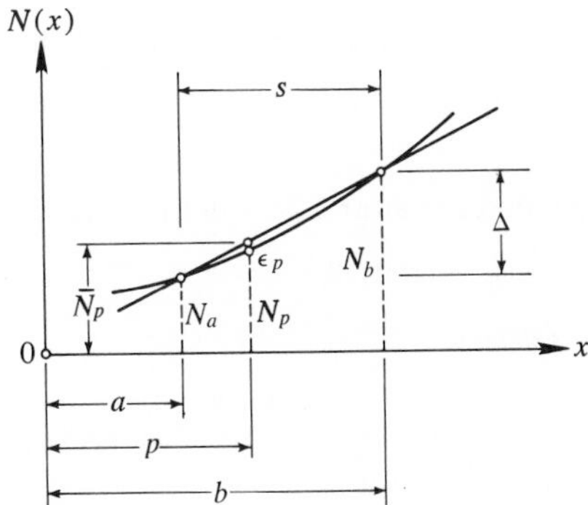

Fig. 12.03-1

(3) Interpolation of Tabulated Values

(a) Definition. The process of finding a value of a function between two tabulated values by a procedure other than the evaluation of the function generating these values is called interpolation.

(b) Linear interpolation. If N_a and N_b are two tabulated values representing the ordinates of $N(x)$ at $x = a$ and $x = b$ in Fig. 12.03–1, then the approximate value of N_p at $x = p$ is

$$N_p = N_a + \frac{N_b - N_a}{b - a}(p - a) = N_a + f\Delta$$

where $N_b - N_a = \Delta$ is the tabular difference, $b - a = s$ is the interval of interpolation, and $(p - a)/s = f$ is the interpolation factor (decimal fraction).

example:

The circumference of a circle of diameter $D = 4$ is $C_4 = 4\pi = 12.5664$ and of a circle of diameter $D = 5$ is $C_5 = 5\pi = 15.7080$. Then by linear interpolation the circumference of a circle of diameter $D = 4.25$ is

$$D = 12.5664 + \frac{15.7080 - 12.5664}{5 - 4}(4.25 - 4) = 13.3518$$

which is in a good agreement with the exact value $4.25\pi = 13.3518$.

(c) Limitations. The application of linear interpolation is limited to densely spaced values which can be represented graphically by a smooth curve.

example:

If $\sin 20°10' = 0.3448$ and $\sin 20°20' = 0.3475$, then $\Delta = 0.3475 - 0.3448 = 0.0027$, and for

$$f = 0.1: \quad \sin 20°11' = 0.3448 + 0.0027 \times 0.1 = 0.3451$$

$$f = 0.2: \quad \sin 20°12' = 0.3448 + 0.0027 \times 0.2 = 0.3453$$

. . . .

$$f = 0.8: \quad \sin 20°18' = 0.3448 + 0.0027 \times 0.8 = 0.3470$$

$$f = 0.9: \quad \sin 20°19' = 0.3448 + 0.0027 \times 0.9 = 0.3473$$

where the last digit was rounded off in all results.

(4) Numerical Errors

(a) Absolute error ϵ is the difference between the true (correct) value N (assumed to be known) and the approximate value $\bar{N}$ (obtained by measurements or calculations).

$$\epsilon = N - \bar{N}$$

(b) Relative error $\bar{\epsilon}$ is the ratio of the absolute error ϵ to the true value N.

$$\bar{\epsilon} = \frac{\epsilon}{N}$$

(c) Percent error $\bar{\epsilon}\%$ is defined as

$$\bar{\epsilon}\% = (100\bar{\epsilon})\%$$

example:

If the correct value $N = 200m$ and the measured value $\bar{N} = 199.9m$, then the absolute error $\epsilon = 200 - 199.9 = 0.1m$, the relative error $\bar{\epsilon} = 0.1/200 = 0.0005$, and the percent error $\bar{\epsilon}\% = (100)(0.0005) = 0.05\%$.

(d) Types of errors. Aside from possible outright errors (blunders), four types of basic errors occur in numerical calculations: (1) inherent errors, (2) truncation errors, (3) round-off errors, (4) interpolation errors.

(e) Inherent error. The error in the initial data based on inaccuracy of measurements, observation, or recording is called the inherent error, ϵ_i.

(f) Truncation error. The error due to approximation of a function by a series of few terms is called the truncation error, ϵ_t.

example:

The correct value of $N = \sin\frac{\pi}{2} = 1.00000$

The approximate value of N computed by series expansion (Sec. 8.06–2c) is

$$\bar{N} = \frac{\pi}{2} - \frac{(\pi/2)^3}{3!} + \frac{(\pi/2)^5}{5!} - \frac{(\pi/2)^7}{7!} + \cdots$$

If only the first term of the series is used, $\epsilon_t = N - \bar{N} = 1.00000 - \frac{\pi}{2} = -0.57080 \quad (-57\%)$

If the first two terms are used, $\epsilon_t = N - \bar{N} = 1.00000 - \frac{\pi}{2} + \frac{(\pi/2)^3}{3!} = +0.07516 \quad (+7.5\%)$

If the first three terms are used, $\epsilon_t = N - \bar{N} = 1.00000 - \frac{\pi}{2} + \frac{(\pi/2)^3}{3!} - \frac{(\pi/2)^5}{5!} = -0.00453 \quad (-0.5\%)$

Finally, if the first four terms of the series are used, $\epsilon_t = 0.00015$ and the truncation error becomes insignificant.

(g) Round-off error. The error due to finite number of digits expressing a transcendent number, a periodic number, or a number with rejected figures (Sec. 12.03–2) is called the round-off error, ϵ_r.

(h) Interpolation error. The error due to approximation of a number by its interpolative equivalent (Sec. 12.03–3) is called the interpolation error, ϵ_p (Fig. 12.03–1).

(j) Numerical accuracy. The accuracy of numerical calculations is governed by all these errors, and the results cannot be more accurate than the given data. Since the technical data are seldom known with an accuracy greater than ±0.2 percent, the computations performed should remain within the same range of error.

12.04 APPLICATION OF TABLES OF NUMERICAL CONSTANTS

(1) Square Roots and Cube Roots of Fractions

(a) Direct values. Tables A.01 and A.02 give directly the numerical values of $\sqrt{N/n}$ and $\sqrt[3]{N/n}$ for $N = 1, 2, 3, \pi, 4, 5, 6, 2\pi, 7, 8, 9, 3\pi$ and $n = 1, 2, \ldots, 9$.

examples:

From Table A.01:

$$\sqrt{\tfrac{2}{3}} = 0.81650$$

$$\sqrt{\frac{2\pi}{5}} = 1.12100$$

From Table A.02:

$$\sqrt[3]{\tfrac{2}{3}} = 0.87358$$

$$\sqrt[3]{\frac{2\pi}{5}} = 1.07912$$

(b) Resolution of radicand. The same tables can be used in cases where N/n can be resolved into a product and/or a quotient of two or several fractions whose roots are given in the tables.

examples:

From Table A.01,

$$\sqrt{\frac{2\pi}{27}} = \sqrt{\frac{2\pi}{9}} \times \sqrt{\frac{1}{3}} = 0.83554 \times 0.57735 = 0.48240$$

$$\sqrt{\frac{15}{2\pi}} = \sqrt{\frac{5}{2}} : \sqrt{\frac{\pi}{3}} = 1.58133 : 1.02333 = 1.54528$$

From Table A.02,

$$\sqrt[3]{\frac{21\pi}{40}} = \sqrt[3]{\frac{3\pi}{8}} \times \sqrt[3]{\frac{7}{5}} = 1.05615 \times 1.11869 = 1.18150$$

$$\sqrt[3]{\frac{20}{3\pi}} = \sqrt[3]{10} : \sqrt[3]{\frac{3\pi}{2}} = 2.15444 : 1.67654 = 0.68858$$

where $\sqrt[3]{10}$ is given in Table A.03.

(c) Resolution of exponent. The same tables can be used in cases where the exponent of the fraction can be resolved into sums, differences, or multiples of $\frac{1}{2}$ and $\frac{1}{3}$.

examples:

$$(N/n)^{3/2} = (N/n) \times \sqrt{N/n} \qquad (N/n)^{2/3} = \sqrt[3]{N/n} \times \sqrt[3]{N/n}$$

$$(N/n)^{5/2} = (N/n)^2 \times \sqrt{N/n} \qquad (N/n)^{4/3} = (N/n) \times \sqrt[3]{N/n}$$

$$(N/n)^{5/6} = \sqrt{N/n} \times \sqrt[3]{N/n} \qquad (N/n)^{1/6} = \sqrt{N/n} : \sqrt[3]{N/n}$$

(d) Linear interpolation. Ten-point linear interpolation between two adjacent tabulated values in Tables A.01 and A.02 leads to large errors and is not admissible.

(2) Squares, Cubes, and Roots of Integers

(a) Direct values. Appendix Tables A.03 to A.22 give directly the numerical values of N^2, N^3, $\sqrt{N}$, $\sqrt[3]{N}$, $\sqrt[4]{N}$, and $\sqrt[5]{N}$ for $N = 1, 2, \ldots, 999, 1{,}000$.

example:

From Table A.05, for $N = 123$,

$$123^2 = 15{,}129 \qquad 123^3 = 1{,}860{,}867 \qquad \sqrt{123} = 11.09054$$

$$\sqrt[3]{123} = 4.97319 \qquad \sqrt[4]{123} = 3.33025 \qquad \sqrt[5]{123} = 2.61807$$

(b) Resolution of ***N*****.** The same tables can be used where N can be resolved into a product and/or a quotient of integers whose squares, cubes, and/or roots are given in the tables.

examples:

From Table A.05:

$$123{,}000^2 = (123 \times 1000)^2 = 123^2 \times 1{,}000^2 = 15{,}129 \times 1{,}000{,}000 = 15{,}129{,}000{,}000$$

From Tables A.05 and A.03:

$$\sqrt{1{,}230} = \sqrt{123 \times 10} = \sqrt{123} \times \sqrt{10} = 11.09054 \times 3.16228 = 35.07139$$

From Tables A.05 and A.04:

$$\sqrt[3]{1.23} = \sqrt[3]{\frac{123}{100}} = \frac{\sqrt[3]{123}}{\sqrt[3]{100}} = \frac{4.97319}{4.64159} = 1.07144$$

(c) Resolution of exponent. The same tables can be used in cases where the exponent of the number can be resolved into sums, differences, products, or quotients of 2, 3, $\frac{1}{2}$, $\frac{1}{3}$, $\frac{1}{4}$, and $\frac{1}{5}$.

examples:

$$\sqrt{N^3} = N^{3/2} = N^{1+(1/2)} = N \times \sqrt{N} \qquad \sqrt[3]{N^4} = N^{4/3} = N^{1+(1/3)} = N \times \sqrt[3]{N}$$

$$\sqrt[3]{N^2} = N^{2/3} = N^{1-(1/3)} = N : \sqrt[3]{N} \qquad \sqrt[4]{N^3} = N^{3/4} = N^{1-(1/4)} = N : \sqrt[4]{N}$$

$$\sqrt[4]{N^5} = N^{5/4} = N^{1+(1/4)} = N \times \sqrt[4]{N} \qquad \sqrt[5]{N^6} = N^{6/5} = N^{1+(1/5)} = N \times \sqrt[5]{N}$$

$$\sqrt[5]{N^4} = N^{4/5} = N^{1-(1/5)} = N : \sqrt[5]{N} \qquad \sqrt[6]{N^5} = N^{5/6} = N^{(1/2)+(1/3)} = \sqrt{N} \times \sqrt[3]{N}$$

(d) Linear interpolation. Ten-point linear interpolation of two adjacent tabulated values in Tables A.03 to A.22 is admissible and the error never exceeds one unit in the fifth significant figure.

examples:

From Table A.09:

$\sqrt{326.6} = \sqrt{326} + 0.6\Delta = 18.05547 + 0.6 \times 0.02767 = 18.07207$

where $\Delta = \sqrt{327} - \sqrt{326} = 18.08314 - 18.05547 = 0.02767$ and the result must satisfy $18.072^2 = 326.597 \doteq 326.6$.

From Tables A.09 and A.03:

$\sqrt{3{,}266} = \sqrt{10} \times \sqrt{326.6} = 3.16228 \times 18.07207 = 57.14895$

where $\sqrt{326.6}$ is taken from the first example, and the result must satisfy $57.149^2 = 3{,}266$.

12.05 APPLICATION OF COMMON LOGARITHMS

(1) Common Logarithm of N

(a) Common logarithm of N (decadic or Brigg's logarithm) denoted as $\log N$ is the exponent of the power of 10 so that

$$10^{\log N} = N$$

(b) Properties. Every positive number has a common logarithm which can be expressed as a decimal number with accuracy to a definite number of significant figures. The integral part of the logarithm is called the *characteristic c* and the decimal fraction (usually endless) is called the *mantissa m*.

examples:

If $10^3 = 1{,}000$, then $\log 1{,}000 = 3.0000$ where 3 is the characteristic and 0.0000 is the mantissa.
If $10^{2.7582} = 573$, then $\log 573 = 2 + 0.7582 = 2.7582$ where 2 is the characteristic and 0.7582 is the mantissa.

(c) Characteristic of $\log N$ is determined according to the following rules:

(α) If $N > 1$ (N is greater than 1) the characteristic of $\log N$ is the positive number of digits before the decimal point diminished by 1.
(β) If $N < 1$ (N is less than one) the characteristic of $\log N$ is the negative number of zeros after the decimal point diminished by 1.

examples:

$$\log 573 = 2 + 0.7582 = 2.7582$$

$$\log 57.3 = 1 + 0.7582 = 1.7582$$

$$\log 5.73 = 0 + 0.7582 = 0.7582$$

$$\log 0.573 = -1 + 0.7582 = 9.7582 - 10$$

$$\log 0.0573 = -2 + 0.7582 = 8.7582 - 10$$

$$\log 0.00573 = -3 + 0.7582 = 7.7582 - 10$$

where the last three logarithms have been expressed in a more convenient form by adding and subtracting 10.

(d) Mantissa of $\log N$ is a tabulated value (Tables A.23 and A.24) which is independent of the decimal point.

example:

log 573, log 57.3, log 5.73, ... have all the same mantissa 0.7582 as shown in (c).

(2) Finding of $\log N$

(a) Direct values. The characteristic of $\log N$ is determined by the rules of Sec. 12.05–1c and the mantissa for any number defined by three significant figures can be found directly in Tables A.23 and A.24.

examples:

$\log 83.1 = 1.9196$ $\qquad \log 2.75 = 0.4393$

$\log 0.112 = 9.0492 - 10$ $\qquad \log 0.013 = 8.1139 - 10$

where the last mantissa is taken from Table A.23 under 130.

(b) Special values. It is useful to remember (Sec. 2.13–2) that

$\log 10 = 1.000$	$\log 0 = -\infty$
$\log 1 = 0.0000$	$\log \infty = \infty$

Several other special values of $\log N$ frequently occurring in technical calculations are given in Table A.24.

(c) Interpolated values. For N defined by more than three significant figures the mantissa of $\log N$ can be found approximately by the linear interpolation (Sec. 12.03–3b) of two adjacent tabulated values in Tables A.23 and A.24. To facilitate the interpolation procedure the average tabular difference $\bar{\Delta}$ is given in the last column of each table. The exact Δ is the difference of two adjacent tabulated values.

example:

$$\log 8{,}736 = \log 8{,}730 + 0.6\Delta = 3.9410 + 0.6 \times 5 \times 10^{-4} \doteq 3.9413$$

where in this case the average difference $\bar{\Delta}$ is the exact difference $\Delta = 5 \times 10^{-4}$.

(3) Finding of Antilog of $\log N$

(a) Direct values. The antilog of $\log N$ is N. If the mantissa of $\log N$ is a tabulated value, then the antilog is the corresponding N of the table.

example:

From Table A.24,

Antilog of $3.8267 = 6{,}710$ $\qquad$ antilog of $(8.8267 - 10) = 0.0671$

(b) Special values. Using inversely the relations of Sec. 12.05–2b,

antilog of $1.0000 = 10$	antilog of $(-\infty) = 0$
antilog of $0.0000 = 1$	antilog of $(+\infty) = +\infty$

(c) Interpolated values. If the mantissa of log N cannot be found exactly in Table A.23 or A.24, then the following linear interpolation gives a good approximation of the desired N.

$$N = \bar{N} + \frac{(m - \bar{m})10^{c+2}}{\Delta}$$

where m is the mantissa of log N, $\bar{m}$ is the nearest smaller mantissa, $\bar{N}$ is the number corresponding to $\bar{m}$, Δ is the tabular difference of the nearest higher and nearest lower mantissa, and c is the characteristic of log N.

example:

From Table A.24,

$$\text{antilog of } 3.9413 = 8{,}730 + \frac{(0.9413 - 0.9410) \times 10^5}{5} = 8{,}736$$

where $c + 2 = 3 + 2 = 5$.

(4) Logarithmic Calculations

(a) Purpose. The logarithmic calculations are used for the evaluation of complex numerical expressions consisting of products, quotients, powers, and roots of numbers but are *not applicable in the evaluation of sums or differences.*

(b) Logarithms of products and quotients. According to Sec. 2.13–3,

$$\log (A \times B) = \log A + \log B$$

$$\log \frac{A}{B} = \log A - \log B$$

$$\log \frac{A \times B}{C \times D} = \log A + \log B - \log C - \log D$$

where A, B, C, D are given numbers.

examples:

$N = 127.5 \times 0.0023 \times 17.6$	$\log 127.5 = 2.1055$ $\log 0.0023 = 7.3617 - 10$ $\log 17.6 = 1.2455$ $\log N = 10.7127 - 10 = 0.7127$
$N = 5.16 + \frac{(0.7127 - 0.7126) \times 10^2}{9} = 5.161$	where $c + 2 = 0 + 2 = 2$

$N = \frac{127.5 \times 0.0023}{17.6}$	$\log 127.5 = 2.1055$ $\log 0.0023 = 7.3617 - 10$ $-\log 17.6 = -1.2455$ $\log N = 8.2217 - 10$
$N = 0.0166 + \frac{(0.2217 - 0.2201) \times 10^0}{26} = 0.01667$	where $c + 2 = -2 + 2 = 0$

(c) Logarithms of powers and roots. According to Sec. 2.13–3,

$$\log A^m = m \log A \qquad \log \frac{1}{A^m} = -m \log A$$

$$\log \sqrt[n]{A} = \frac{1}{n} \log A \qquad \log \frac{1}{\sqrt[n]{A}} = -\frac{1}{n} \log A$$

$$\log \sqrt[n]{A^m} = \frac{m}{n} \log A \qquad \log \frac{1}{\sqrt[n]{A^m}} = -\frac{m}{n} \log A$$

where A is a given number and m, n are given integers (or decimal numbers).

examples:

$N = \sqrt[3]{0.148}$	$\log N = \frac{1}{3} \log 0.148$ $= (\frac{1}{3})(9.1703 - 10)$ $= (\frac{1}{3})(29.1703 - 30)$ $= 9.7234 - 10$
$N = 0.528 + \dfrac{(0.7234 - 0.7226) \times 10}{9} = 0.5289$ where $c = -1 + 2 = 1$	

It should be noted that the second factor in the product $(\frac{1}{3})(9.1703–10)$ can be conveniently modified by adding and subtracting any multiple of 10 without affecting the result.

$N = \sqrt[5]{12.57^2}$	$\log N = \frac{2}{5} \log 12.57$ $= 0.4 \times 1.0994$ $= 0.4398$
$N = 2.75 + \dfrac{(0.4398 - 0.4393) \times 10^2}{16} = 2.753$ where $c = 0 + 2 = 2$	

12.06 APPLICATION OF NATURAL LOGARITHMS

(1) Natural Logarithm of N

(a) Natural logarithm of N (Napier's logarithm) denoted as $\ln N$ is the exponent of the power of e so that

$$e^{\ln N} = N$$

where

$$e = \lim_{n \to \infty} \left(1 + \frac{1}{n}\right)^n = 1 + \frac{1}{1!} + \frac{1}{2!} + \frac{1}{3!} + \cdots = 2.71828\ldots$$

is called the base of natural logarithms.

(b) Properties. Every positive number has a natural logarithm which again can be expressed as a decimal number with accuracy to a definite number of significant figures. The integral part of the logarithm is again called the *characteristic* c and the decimal fraction (usually endless) is called also the *mantissa* m.

example:

If $e^{c+m} = N$, then $\ln N = c + m$, where the characteristic does not define directly the position of the decimal point but is a tabulated value.

(c) Transformations. The relations between $\log N$ and $\ln N$ (given in Sec. 2.13–1d) are:

$\ln N = (2.30258\ldots)\log N$	$\log N = (0.43429\ldots)\ln N$

examples:

If $\log 573 = 2.7582$, then $\ln 573 = 2.3026 \times 2.7582 = 6.3510$.

If $\log 0.573 = 9.7582 - 10 = -0.2418$, then $\ln 0.573 = 2.3026 \times (-0.2418) = -0.5568$, where the transformation error in the last digit may be ± 1 (see Sec. 12.06–2a).

(2) Finding of $\ln N$

(a) Direct values. The characteristic and the mantissa of $\ln N$ for $N = 1.00 - 9.99$ can be found directly in Tables A.25 and A.26.

For $N < 1$, N is expressed as $t/10^n$, and

$$\ln N = \ln \frac{t}{10^n} = \ln t - n \ln 10$$

For $N > 10$, N is expressed as $t \times 10^n$ and

$$\ln N = \ln (t \times 10^n) = \ln t + n \ln 10$$

where $\ln t$ is one of the tabulated values, n is an integer, and $\ln 10 = 2.3026$.

examples:

In Table A.25:

$$\ln 5.73 = 1.7457$$

$$\ln 57.3 = \ln (5.73 \times 10) = \ln 5.73 + \ln 10 = 1.7457 + 2.3026 = 4.0483$$

$$\ln 573 = \ln (5.73 \times 10^2) = \ln 5.73 + 2 \ln 10 = 1.7457 + 4.6052 = 6.3509$$

$$\ln 0.573 = \ln \frac{5.73}{10} = \ln 5.73 - \ln 10 = 1.7457 - 2.3026 = -0.5569$$

(b) Special values. It is again useful to remember that

$\ln e = 1.0000$	$\ln 0 = -\infty$
$\ln 1 = 0.0000$	$\ln \infty = \infty$

(c) Interpolated values. For N defined by more than three significant figures the natural logarithm of N can be found approximately by the linear interpolation similar to that of Sec. 12.05–2*c*.

example:

$$\ln 34.56 = \ln (3.456 \times 10) = \ln 3.456 + \ln 10 = 3.5458$$

where $\ln 3.456 = \ln 3.45 + 0.6\Delta = 1.2413 + 0.6 \times 0.0029 = 1.2432$

$\ln 10 = 2.3026$

12.07 APPLICATION OF TABLES OF TRIGONOMETRIC FUNCTIONS

(1) Natural Trigonometric Functions in Degrees and Minutes

(a) Direct values. Tables A.27 to A.31 give directly the numerical values of $\sin \alpha$, $\cos \alpha$, $\tan \alpha$, and $\cot \alpha$ for $\alpha = 0° - 90°$ in intervals of 10 minutes. For $\alpha = 0° - 45°$, the respective value is found referring to the top boxhead and moving down along the left stub of α. For $\alpha = 45° - 90°$, the respective value is found referring to the bottom boxhead and moving up along the right stub of α.

examples:

$\sin 12°20' = 0.2136$ is found in Table A.28, under the top boxhead "nat. sine," left stub, $\alpha = 12°20'$.

$\sin 58°30' = 0.8526$ is found in Table A.30, above the bottom boxhead "nat. sine," right stub $\alpha = 58°30'$.

(b) Large angles. If $\alpha > 90°$, the argument is reduced according to Sec. 5.02–4*c*, *d* (before entering the respective tables).

examples:

$$\cos 293°40' = \cos (270° + 23°40') = \sin 23°40' = 0.4014$$

$$\tan 112°10' = \tan (90° + 22°10') = -\cot 22°10' = -2.4545$$

(c) Negative angle. If $\alpha < 0°$, the given function is transformed according to Sec. 5.02–4*a* (before entering the respective table).

examples:

$$\sin (-60°20') = -\sin 60°20' = -0.8089$$

$$\tan (-112°10') = -\tan (90° + 22°10') = \cot 22°10' = 2.4545$$

(d) Secant and cosecant. The values of $\sec \alpha$ and $\csc \alpha$ are taken as the reciprocal values of $\cos \alpha$ and $\sin \alpha$ respectively.

examples:

$$\sec 72°50' = \frac{1}{\cos 72°50'} = \frac{1}{0.2952} = 3.3875$$

$$\csc 72°50' = \frac{1}{\sin 72°50'} = \frac{1}{0.9555} = 1.0466$$

(e) Interpolated values. The linear-interpolation procedure described in Sec. 12.03–3*b* is directly applicable in Tables A.27 to A.31.

examples:

$$\sin 12°24' = \sin 12°20' + 0.4\Delta = 0.2136 + 0.4 \times 0.0028 = 0.2147$$

where $\Delta = \sin 12°30' - \sin 12°20' = 0.2164 - 0.2136 = 0.0028$

$$\cos 12°24' = \cos 12°20' + 0.4\Delta = 0.9769 + 0.4 \times (-0.0006) = 0.9767$$

where $\Delta = \cos 12°30' - \cos 12°20' = 0.9763 - 0.9769 = -0.0006$

It is important to note that for $\sin \alpha$ and $\tan \alpha$, Δ is positive and for $\cos \alpha$ and $\cot \alpha$, Δ is negative.

(2) Common Logarithms of Trigonometric Functions

(a) Direct values. Tables A.27 to A.31 give directly the numerical values of $\log \sin \alpha$, $\log \cos \alpha$, $\log \tan \alpha$, and $\log \cot \alpha$ for $\alpha = 0° - 90°$ in intervals of 10 minutes. For $\alpha = 0° - 45°$, the respective value is found referring to the top boxhead and moving down along the left stub α. For $\alpha = 45° - 90°$, the respective value is found referring to the bottom boxhead and moving up along the right stub α. The tabulated value includes the characteristic and the mantissa of the logarithm. If the characteristic is 7, 8, or 9, the tabulated value must be supplemented by -10.

examples:

$\log \sin 12°20' = 9.3296 - 10$ is found in Table A.28, under the top boxhead "log sine," left stub, $\alpha = 12°20'$.

$\log \cot 12°20' = 0.6603$ is found in Table A.28, under the top boxhead "log cotangent," left stub, $\alpha = 12°20'$.

(b) log secant and log cosecant. The logarithms of $\sec \alpha = 1/\cos \alpha$ and $\csc \alpha = 1/\sin \alpha$ are

$$\log \sec \alpha = -\log \cos \alpha \qquad \log \csc \alpha = -\log \sin \alpha$$

where $\log 1 = 0$.

examples:

$$\log \sec 12°20' = -\log \cos 12°20' = -(9.9899 - 10) = 0.0101$$

$$\log \csc 12°20' = -\log \sin 12°20' = -(9.3296 - 10) = 0.6704$$

(c) Interpolated values. The linear-interpolation procedure described in Sec. 12.03–3*b* is directly applicable for the values of $\log \sin \alpha$ and $\log \tan \alpha$ in the range $\alpha = 3° - 90°$ and for the values of $\log \cos \alpha$ and $\log \cot \alpha$ in the range $\alpha° = 0° - 87°$. For the excluded values of α, the linear interpolation of tabulated values is not accurate.

examples:

$$\log \sin 12°24' = \log \sin 12°20' + 0.4\Delta = 9.3296 - 10 + 0.4 \times 0.0057 = 9.3319 - 10$$

where

$$\Delta = \log \sin 12°30' - \log \sin 12°20' = (9.3353 - 10) - (9.3296 - 10) = 0.0057$$

$$\log \cos 12°24' = \log \cos 12°20' + 0.4\Delta = 9.9899 - 10 + 0.4 \times -0.0003 = 9.9898 - 10$$

where

$$\Delta = \log \cos 12°30' - \log \cos 12°20' = (9.9896 - 10) - (9.9898 - 10) = -0.0003$$

It is important to note that for log sin α and log tan α, Δ is positive and for log cos α and log cot α, Δ is negative.

(d) Antilogarithms. The antilog of the given logarithm of a trigonometric function is found by a procedure similar to that of Sec. 12.06–3 from Tables A.27 to A.31 as shown below.

example:

If $\log \tan \alpha = 9.9736 - 10$, find α.

Referring to Table A.31, the nearest higher and lower logarithms in the column "log tangent" are respectively,

$$\log \tan 43°20' = 9.9747 - 10 = \log \tan (\bar{\alpha} + 10')$$

$$\log \tan 43°10' = 9.9722 - 10 = \log \tan \bar{\alpha}$$

from which $\Delta = 0.0025$ is their tabular difference.

In the same table,

$$\log \tan \alpha = 9.9736 - 10$$

$$\log \tan \bar{\alpha} = 9.9722 - 10$$

from which $m - \bar{m} = 0.0014$ is the difference of their mantissas.

Then,

$$\alpha = \bar{\alpha} + \frac{m - \bar{m}}{\Delta} 10' = 43°10' + \frac{0.0014}{0.0025} 10' = 43°16'$$

but also $\alpha = 180° + 43°16', 360° + 43°16', \ldots.$

12.08 APPLICATION OF TABLES OF ELEMENTARY FUNCTIONS

(1) Natural Trigonometric Functions in Radians

(a) Direct values. Tables A.32 to A.38 give directly the numerical values of sin x, cos x, and tan x for $x = 0.01$ to 3.00 radians in intervals of 0.01 radian and for $x = 3.00$ to 5.00 radians in intervals of 0.05 radian.

examples:

From Table A.34:

$\sin 1.41 = 0.98710$ $\cos 1.41 = 0.16010$ $\tan 1.41 = 6.16536$

(b) Indirect values. By their definitions, the numerical values of sec x, csc x, and cot x can be obtained from the same tables as the respective reciprocal values of cos x, sin x, and tan x.

examples:

From Table A.34: $\sec 1.41 = \dfrac{1}{\cos 1.41} = 6.24610$ $\qquad \csc 1.41 = \dfrac{1}{\sin 1.41} = 1.01307$

(c) Interpolated values. The linear-interpolation procedure described in Sec. 12.03–3*b* is directly applicable for the values of all trigonometric functions in Tables A.32 to A.38.

(d) Large arguments. If x is greater than 5.00 but less than 10, the argument x can be resolved into the sum of any two numbers and inserted in the respective formula of Sec. 5.04–1*a* which can be then evaluated in terms of the tabulated values. If x is greater than 10, then the argument x can be resolved into the sum of a multiple of 2π (such as $2\pi, 4\pi, 6\pi, \ldots$) and a remainder and inserted in the reduction formula of Sec. 5.02–4*d* which then again can be evaluated by means of the tabulated values.

examples:

The numerical value of cos 7.26 is found by means of

$\cos(\alpha + \beta) = \cos\alpha\cos\beta - \sin\alpha\sin\beta$ (see Sec. 5.04–1*a*)

which for $\alpha = 5.00$, $\beta = 2.26$, in terms of Tables A.36 and A.38 yields

$$\cos 7.26 = \cos 5.00 \cos 2.26 - \sin 5.00 \sin 2.26$$
$$= 0.28366 \times (-0.63592) - (-0.95892) \times 0.77175 = 0.55966$$

The numerical value of cos 51.73 is found by means of

$\cos(2n\pi + \alpha) = \cos\alpha$ (see Sec. 5.02–4*d*)

where $2n\pi$ in this case is $16\pi = 50.265$ and $\alpha = 51.73 - 50.265 = 1.488$.

Then $\cos 51.73 = \cos 1.488 = \cos 1.480 + 0.8\Delta = 0.08270$ where, from Table A.34,

$\Delta = \cos 1.490 - \cos 1.480 = 0.08071 - 0.09067 = -0.00797$

(2) Exponential Functions of Base $e = 2.718281\ldots$

(a) Direct values. Tables A.32 to A.38 give directly the numerical values of e^x and e^{-x} for $x = 0.01 - 3.00$ in intervals of 0.01 and for $x = 3.00 - 5.00$ in intervals of 0.05.

examples:

From Table A.34: $e^{1.41} = 4.09596$ $\qquad e^{-1.41} = 0.24414$

(b) Interpolated values. The linear-interpolation procedure of Sec. 12.03–3*b* is directly applicable for the values of e^x, e^{-x} in Tables A.32 to A.38 if $x = 0.01$ to 3.00. For larger arguments this procedure is inaccurate and one of the procedures of the subsequent section must be used.

(c) Large arguments. If x is greater than 3 but less than 10, the exponential expression can be resolved as follows.

$$e^x = e^{x_1}e^{x_2} \qquad e^{-x} = e^{-x_1}e^{-x_2}$$

where $x = x_1 + x_2$, and e^{x_1}, e^{x_2} or e^{-x_1}, e^{-x_2} are values given in tables.

If $x > 10$, the logarithmic transformation given below is more convenient.

$$e^x = e^{[x/(\ln 10)]\ln 10} = e^{(a+b)\ln 10} = 10^a e^c$$

where a is the integer of $x/(\ln 10)$, b is the decimal of $x/(\ln 10)$, $c = b \ln 10 = 2.30258b$, and e^c is a value given in tables.

examples:

From Tables A.34 and A.38:

$$e^{6.41} = e^{5.00}e^{1.41} = 148.41316 \times 4.09596 = 607.89436$$

For the evaluation of e^{72} introduce

$$\frac{72}{\ln 10} = \frac{72}{2.30259} = 31.26914 \qquad a = 31 \qquad b = 0.26914 \qquad c = 0.61972$$

and $$e^{72} = 10^{31}e^{0.61972} \approx 10^{31}e^{0.620} = 1.85893 \times 10^{31}$$

(3) Hyperbolic Functions

(a) Direct values. Tables A.32 to A.38 give directly the numerical values of sinh x, cosh x, and tanh x for $x = 0.01$ to 3.00 in intervals of 0.01 and for $x = 3.00$ to 5.00 in intervals of 0.05.

examples:

From Table A.34:

$$\sinh 1.41 = 1.92591 \qquad \cosh 1.41 = 2.17005 \qquad \tanh 1.41 = 0.88749$$

(b) Indirect values. By their definitions, the numerical values of sech x, csch x, and coth x can be obtained from the same tables as the respective reciprocal values of cosh x, sinh x, and tanh x.

examples:

$$\operatorname{sech} 1.41 = \frac{1}{\cosh 1.41} = 0.46082 \qquad \operatorname{csch} 1.41 = \frac{1}{\sinh 1.41} = 0.51924$$

(c) Interpolated values. The linear-interpolation procedure of Sec. 12.03–3*b* is directly applicable for the values of hyperbolic functions in Tables A.32 to A.38 in the range of $x = 0.01$ to 3.00. For larger arguments this procedure is inaccurate and one of the procedures of the subsequent section must be used.

(d) Large arguments. If x is greater than 3 but less than 10, the exponential resolution of Sec. 12.08–2*d* should be used. If x is greater than 10, then

$$\sinh x \approx \cosh x \approx \frac{e^x}{2} \qquad \tanh x \approx 1 - 2e^{-2x} \qquad \coth x \approx 1 + 2e^{-2x}$$

examples:

$$\sinh 6.41 = \frac{e^{6.41} - e^{-6.41}}{2} = \frac{e^{5.00}e^{1.41} - e^{-5.00}e^{-1.41}}{2} = 303.94636$$

where

$$e^{5.00}e^{1.41} = 607.89436 \qquad \text{and} \qquad e^{-5.00}e^{-1.41} = \frac{1}{607.89436} = 0.00164$$

are taken from example of Sec. 12.08–2*c*.

$$\sinh 72 = \frac{e^{72}}{2} = 0.92946 \times 10^{31}$$

where e^{72} is taken from example of Sec. 12.08–2*c*.

(4) Inverse Functions

(a) Direct values. Tables A.32 to A.38 can be used directly for finding the inverse trigonometric and hyperbolic functions in limits of the respective table as shown below.

example:

From Table A.32: $\sin^{-1} 0.35227 = 0.36 \qquad \sinh^{-1} 0.36783 = 0.36$

(b) Interpolated values. The linear-interpolation procedure of Sec. 12.03–3*b* is directly applicable in Tables A.32 to A.38 for $\sin^{-1} y$, $\cos^{-1} y$, $\tan^{-1} y$, and for $\sinh^{-1} y$, $\cosh^{-1} y$, $\tanh^{-1} y$.

examples:

In Table A.34:

$$\sin^{-1} 0.90000 = 1.11 + \frac{0.90000 - 0.89570}{0.90010 - 0.89570} 0.01 = 1.11977$$

where $0.90000 - 0.89570$ is the difference between the given argument and the nearest lower tabulated value, $0.90010 - 0.89570$ is the table difference, and 0.01 is the decimal position factor.

In Table A.33:

$$\sinh^{-1} 0.90000 = 0.80 + \frac{0.90000 - 0.88811}{0.90152 - 0.88811} 0.01 = 0.80887$$

where the interpolative correction is computed similarly as above.

12.09 COMPUTATIONS WITH COMPLEX NUMBERS

(1) Geometric Representation

(a) Cartesian form. A complex number $p = a + bi$ introduced in Sec. 2.12–1*c* can be represented as the point P in the rectangular (cartésian) coordinate system of Fig. 12.09–1, in which the X axis is referred to as the real axis, the Y axis as the imaginary axis, and the X, Y plane is called the complex plane (Agrand or Gauss plane).

(b) Polar form. Employing polar coordinates in the complex plane

$$a = r \cos \theta \qquad b = r \sin \theta$$

the complex number

$$p = a + bi = r(\cos \theta + i \sin \theta) = r[\cos(\theta + 2k\pi) + i \sin(\theta + 2k\pi)]$$

where

$$r = \sqrt{a^2 + b^2} \qquad \theta = \tan^{-1} \frac{a}{b} \qquad \frac{b}{a} = \ldots -2, -1, 0, 1, 2, \ldots$$

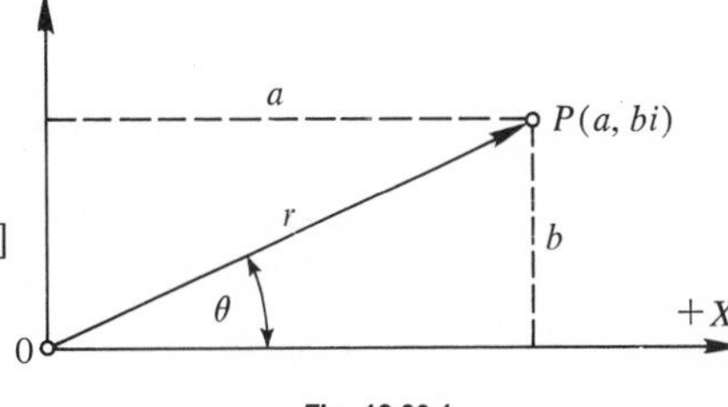

Fig. 12.09-1

(c) Vector form. In terms of vector notation, the position vector of P in Fig. 12.09–1 is

$$\mathbf{p} = a\mathbf{e}_1 + (bi)\mathbf{e}_2$$

where the magnitude of $\mathbf{p}$ is r and $\mathbf{e}_1$, $\mathbf{e}_2$ are the unit vectors along the X and Y axis respectively. The notation $\mathbf{e}_1$, $\mathbf{e}_2$ instead of $\mathbf{i}$, $\mathbf{j}$ is used to avoid a confusion of the symbol $i = \sqrt{-1}$ and $\mathbf{i}$ = unit vector along the X axis.

example:

If $p = 3-4i$, then

$$r = \sqrt{(3)^2+(-4)^2} = 5 \qquad \tan\theta = \frac{-4}{3} \qquad \theta = 306°52'$$

In polar form, $p = 5(\cos 306°52' + i \sin 306°52')$ but also

$$p = 5\cos 666°52' + i \sin 666°52'$$
$$= 5\cos 1{,}026°52' + i \sin 1{,}026°52')$$

. . . .

(d) Product and quotient. If p_1 and p_2 are two complex numbers defined as

$$p_1 = r_1(\cos\theta_1 + i\sin\theta_1) \qquad \text{and} \qquad p_2 = r_2(\cos\theta_2 + i\sin\theta_2)$$

then their product is

$$p_1p_2 = r_1r_2[\cos(\theta_1+\theta_2) + i\sin(\theta_1+\theta_2)]$$

and their quotient is

$$\frac{p_1}{p_2} = \frac{r_1}{r_2}[\cos(\theta_1-\theta_2) + i\sin(\theta_1-\theta_2)]$$

examples:

If $p_1 = 10(\cos 45° + i\sin 45°)$ and $p_2 = 20(\cos 30° + i\sin 30°)$, then

$$p_1p_2 = 200(\cos 75° + i\sin 75°) = 51.764 + 193.186i$$

$$\frac{p_1}{p_2} = \tfrac{1}{2}(\cos 15° + i\sin 15°) = 0.48297 + 0.12941i$$

(e) Powers and roots. If $p = r(\cos\theta + i\sin\theta)$ and m, n are integers, then

$$p^m = r^m(\cos\theta + i\sin\theta)^m = r^m(\cos m\theta + i\sin m\theta)$$

$$\sqrt[n]{p} = \sqrt[n]{r}\sqrt[n]{\cos\theta + i\sin\theta} = \sqrt[n]{r}\left(\cos\frac{\theta+2k\pi}{n} + i\sin\frac{\theta+2k\pi}{n}\right)$$

$$\sqrt[n]{p^m} = \sqrt[n]{r^m}\sqrt[n]{(\cos\theta + i\sin\theta)^m} = \sqrt[n]{r^m}\left[\cos\frac{(\theta+2k\pi)m}{n} + i\sin\frac{(\theta+2k\pi)m}{n}\right]$$

where $k = 0, 1, 2, \ldots, n-1$ and there are n roots of p. These relations are known as De Moivre's theorems.

example:

If $p = 10(\cos 45° + i\sin 45°)$, then

$$p^2 = 100(\cos 90° + i\sin 90°) = 100i$$

$$\sqrt{p} = \begin{cases} \sqrt{10}(\cos 22°30' + i\sin 22°30') = 2.92159 + 1.21015i \\ \sqrt{10}(\cos 112°30' + i\sin 112°30') = -1.21015 + 2.92159i \end{cases}$$

(2) Exponential representation

(a) Exponential form. By definition of $e^{i\theta}$ (Sec. 8.07–1a, $x = \theta$)

$$e^{i\theta} = \cos\theta + i\sin\theta$$

and consequently

$$e^{i(\theta+2k\pi)} = \cos(\theta+2k\pi) + i\sin(\theta+2k\pi)$$

where $k = \ldots, -2, -1, 0, 1, 2, \ldots$.

Then the polar form of p (Sec. 12.09–1b) written in terms of these relations becomes

$$p = a + bi = r(\cos\theta + i\sin\theta) = r[\cos(\theta+2k\pi) + i\sin(\theta+2k\pi)] = re^{i\theta} = re^{i(\theta+2k\pi)}$$

and is called the exponential form (Euler's formula).

example:

If $p = 3 - 4i$, then from Sec. 12.09–1c,

$$r = 5 \qquad \theta = 306°52' = 5.35584 \text{ rad} \qquad p = 5e^{5.35584i}$$

which in electrical technology is frequently abbreviated as

$$p = 5\underline{/306°52'}$$

(b) Products and quotients. If p_1 and p_2 are two complex numbers defined as

$$p_1 = r_1 e^{i\theta_1} \qquad \text{and} \qquad p_2 = r_2 e^{i\theta_2}$$

then their product is

$$p_1 p_2 = r_1 r_2 e^{i(\theta_1+\theta_2)}$$

and their quotient is

$$\frac{p_1}{p_2} = \frac{r_1}{r_2} e^{i(\theta_1-\theta_2)}$$

examples:

If $p_1 = 10e^{\pi i/4}$ and $p_2 = 20e^{\pi i/6}$, then

$$p_1 p_2 = 200e^{5\pi i/12} \qquad \text{and} \qquad \frac{p_1}{p_2} = \tfrac{1}{2} e^{\pi i/12}$$

(c) Powers and roots. If $p = re^{i\theta}$ and m, n are integers, then

$$p^m = r^m e^{im\theta} = r^m e^{im(\theta+2k\pi)}$$

$$\sqrt[n]{p} = \sqrt[n]{r} e^{(\theta+2k\pi)i/n}$$

$$\sqrt[n]{p^m} = \sqrt[n]{r^m} e^{(\theta+2k\pi)mi/n}$$

example:

If $p = 10e^{\pi i/4}$, then

$$p^4 = 10{,}000e^{\pi i} = -10{,}000$$

where $e^{\pi i} = \cos\pi + i\sin\pi = -1$ (Sec. 8.07–1a).

(d) Special values. The relations of Sec. 12.09–2c yield the following special values:

$e^{\pi i/2} = \cos\frac{\pi}{2} + i\sin\frac{\pi}{2} = i$	$e^{-\pi i/2} = \cos\frac{\pi}{2} + i\sin -\frac{\pi}{2} = -i$
$e^{2\pi i/2} = \cos\pi + i\sin\pi = -1$	$e^{-2\pi i/2} = \cos -\pi + i\sin -\pi = -1$
$e^{3\pi i/2} = \cos\frac{3\pi}{2} + i\sin\frac{3\pi}{2} = -i$	$e^{-3\pi i/2} = \cos -\frac{3\pi}{2} + i\sin -\frac{3\pi}{2} = i$
$e^{4\pi i/2} = \cos\frac{4\pi}{2} + i\sin\frac{4\pi}{2} = 1$	$e^{-4\pi i/2} = \cos -\frac{4\pi}{2} + i\sin -\frac{4\pi}{2} = 1$

and in general for $k = \ldots -2, -1, 0, 1, 2, \ldots$

$e^{2k\pi i} = \cos 2k\pi + i\sin 2k\pi = 1$	$e^{(2k+1)\pi i} = \cos(2k+1)\pi + i\sin(2k+1)\pi = -1$
$e^{(\theta+2k\pi)i} = \cos(\theta+2k\pi) + i\sin(\theta+2k\pi) = e^{i\theta}$	$e^{[\theta+(2k+1)\pi]i} = \cos[\theta+(2k+1)\pi] + i\sin[\theta+(2k+1)\pi] = -e^{-i\theta}$
$e^{\theta+2k\pi i} = e^{\theta}(\cos 2k\pi + i\sin 2k\pi) = e^{\theta}$	$e^{\theta+(2k+1)\pi i} = e^{\theta}[\cos(2k+1)\pi + i\sin(2k+1)\pi] = -e^{\theta}$

(3) Extraction of Roots of Numbers

(a) General case. The n roots of equation $x^n = a$ are

$$x_{1,2,\ldots,n} = \sqrt[n]{a}\left(\cos\frac{2k\pi}{n} + i\sin\frac{2k\pi}{n}\right)$$

and of equation $x^n = -a$ are

$$x_{1,2\ldots,n} = \sqrt[n]{a}\left[\cos\frac{(2k+1)\pi}{n} + i\sin\frac{(2k+1)\pi}{n}\right]$$

where $k = 0, 1, 2, \ldots, n-1$ and $a > 0$.

(b) Special cases. The roots of $x^n = a$ for $a = 1$, $a = -1$, and $n = 2, 3, 4$ are represented by circles of radius $r = 1$ in the complex plane in Fig. 12.09–2.

example:

The roots of $x^3 = 27$ are

$$x_1 = \sqrt[3]{27}\left[\cos\frac{(2)(0)\pi}{3} + i\sin\frac{(2)(0)\pi}{3}\right] = 3(\cos 0° + i\sin 0°) = 3$$

$$x_2 = \sqrt[3]{27}\left[\cos\frac{(2)(1)\pi}{3} + i\sin\frac{(2)(1)\pi}{3}\right] = 3(\cos 120° + i\sin 120°) = 3\left(-\frac{1}{2} + \frac{i\sqrt{3}}{2}\right)$$

$$x_3 = \sqrt[3]{27}\left[\cos\frac{(2)(2)\pi}{3} + i\sin\frac{(2)(2)\pi}{3}\right] = 3(\cos 240° + i\sin 240°) = 3\left(-\frac{1}{2} - \frac{i\sqrt{3}}{2}\right)$$

where the factors of 3 are those shown in Fig. 12.09–2 for $x^3 = 1$.

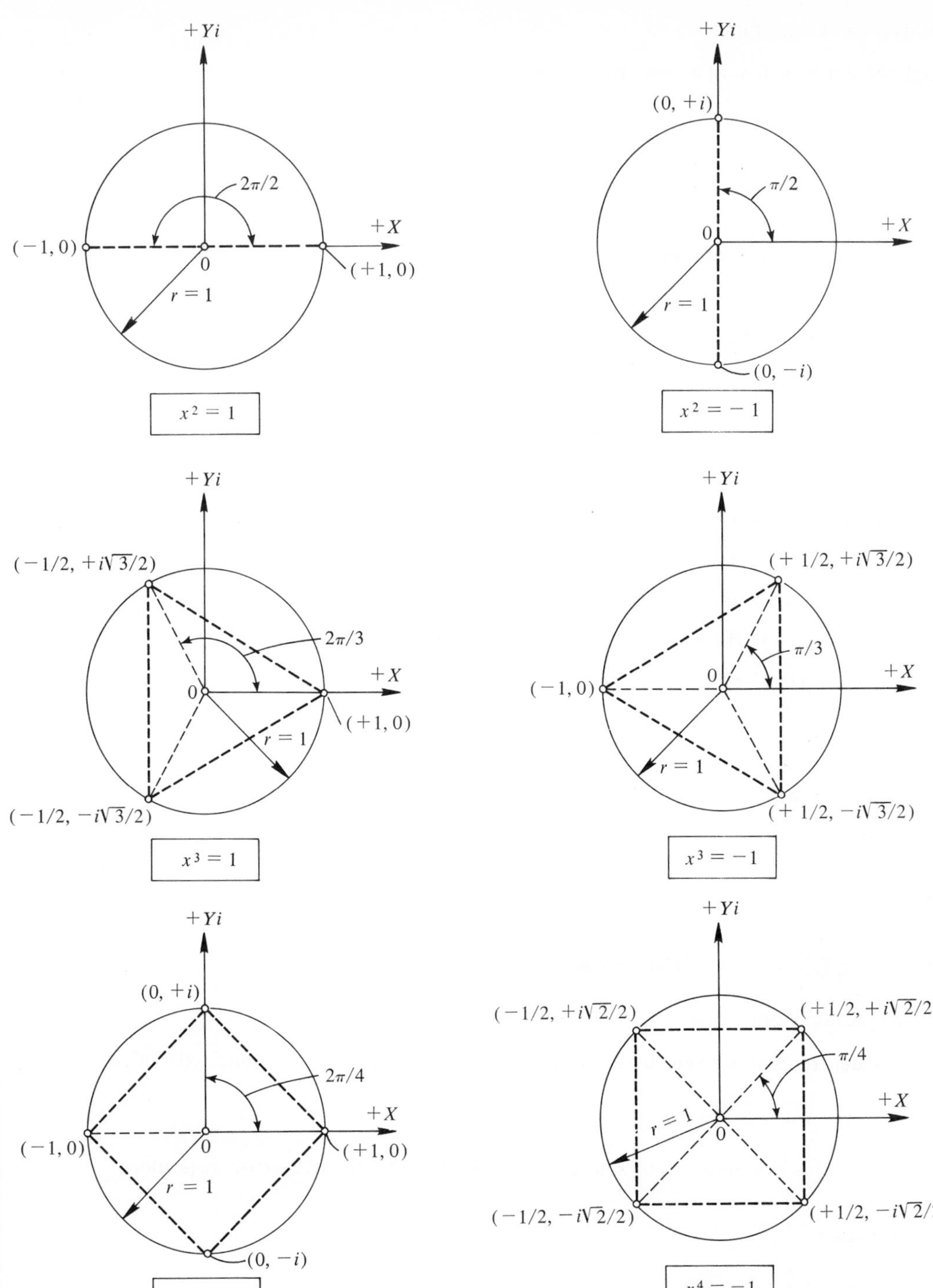
+Yi
+X
(−1, 0)
(+1, 0)
0
2π/2
r = 1
x² = 1
+Yi
(0, +i)
π/2
0
+X
r = 1
(0, −i)
x² = − 1
+Yi
(−1/2, +i√3/2)
2π/3
+X
0
(+1, 0)
r = 1
(−1/2, −i√3/2)
x³ = 1
+Yi
(+ 1/2, +i√3/2)
π/3
0
+X
(−1, 0)
r = 1
(+ 1/2, −i√3/2)
x³ = −1
+Yi
(0, +i)
2π/4
+X
(−1, 0)
0
(+1, 0)
r = 1
(0, −i)
x⁴ = 1
+Yi
(−1/2, +i√2/2)
(+1/2, +i√2/2)
π/4
+X
r = 1
0
(−1/2, −i√2/2)
(+1/2, −i√2/2)
x⁴ = −1

Fig. 12.09-2

(4) Natural Logarithms

(a) Natural logarithm of a complex number $a + bi$.

$$\ln(a+bi) = \ln[re^{i(\theta+2k\pi)}]$$
$$= \ln r + i(\theta + 2k\pi)\ln e = \ln r + i(\theta + 2k\pi)$$

where $k = \ldots, -2, -1, 0, 1, 2, \ldots$ and $\ln e = 1$.

(b) Natural logarithm of a pure imaginary number bi.

If $a = 0$, $b > 0$, then $r = b$, $\tan\theta = b/0 = +\infty$, $\theta = \pi/2$, and

$$\ln bi = \ln b + i\left(\frac{\pi}{2} + 2k\pi\right)$$

If $a = 0$, $b < 0$, then $r = b$, $\tan\theta = -b/0 = -\infty$, $\theta = 3\pi/2$, and

$$\ln bi = \ln(-b) + i\left(\frac{3\pi}{2} + 2k\pi\right)$$

(c) Natural logarithm of a real number, $a \neq 0$.

If $a > 0$, $b = 0$, then $r = a$, $\tan\theta = 0/a = 0$, $\theta = 0$, and

$$\ln a = \ln a + 2k\pi i$$

If $a < 0$, $b = 0$, then $r = -a$, $\tan\theta = 0/-a = 0$, $\theta = \pi$, and

$$\ln a = \ln(-a) + (\pi + 2k\pi)i$$

examples:

$$\ln(+1) = 2\pi ki \qquad \ln i = \left(\frac{\pi}{2} + 2k\pi\right)i$$

$$\ln(-1) = (\pi + 2\pi k)i \qquad \ln(-i) = \left(\frac{3\pi}{2} + 2k\pi\right)i$$

12.10 USEFUL APPROXIMATIONS

(1) Algebraic Approximations

(a) Square roots of large numbers. If $N = a^2 + \Delta$ and Δ is very small compared to a^2, then

$$\sqrt{N} = \sqrt{a^2 + \Delta} \approx a + \frac{\Delta}{2a}$$

where a^2 is a number to be selected from Tables A.03 to A.22 or by inspection.

example:

$$\sqrt{234{,}567} = \sqrt{484^2 + 311} \approx 484 + \frac{311}{(2)(484)} = 484.321$$

which should satisfy (approximately)

$$484.321^2 = 234{,}566.967 \doteq 234{,}567$$

(b) *n*th roots of large numbers. If $N = a^n + \Delta$ and Δ is very small compared to a^n, then

$$\sqrt[n]{N} = \sqrt[n]{a^n + \Delta} \approx a + \frac{\Delta}{na^{n-1}}$$

where a^n is a number to be selected from Tables A.03 to A.22.

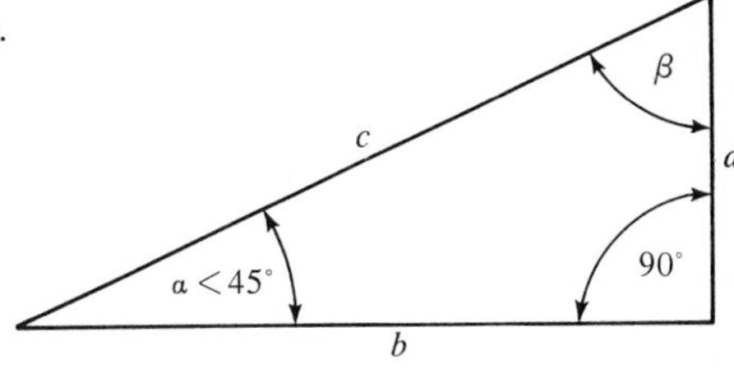

Fig. 12.10-1

example:

$$\sqrt[3]{234{,}567} = \sqrt[3]{61.6^3 + 822} \simeq 61.6 + \frac{822}{(3)(61.6^2)} = 61.6722$$

which should satisfy (approximately)

$61.6722^3 = 234{,}567.76 \doteq 234{,}568$

(2) Geometric Approximations

(a) Smallest angle in right triangle. If α is the smallest angle in the right triangle of Fig. 12.10–1, then

$$\alpha \simeq \frac{3a}{b+2c} \quad \text{in radians} \qquad \text{or} \qquad \alpha \simeq \frac{172a}{b+2c} \quad \text{in degrees}$$

where a, b, c are the known sides.

examples:

The errors of this formula are:

$\alpha = 5°$	$\alpha = 4.975°$	$\epsilon = 1.50'$	$\alpha = 30°$	$\alpha = 29.983°$	$\epsilon = 1.02'$
$\alpha = 10°$	$\alpha = 9.991°$	$\epsilon = 0.53'$	$\alpha = 40°$	$\alpha = 39.952°$	$\epsilon = 2.88'$
$\alpha = 20°$	$\alpha = 20.015°$	$\epsilon = 0.90'$	$\alpha = 45°$	$\alpha = 44.923°$	$\epsilon = 4.68'$

(b) Length of a circular arc of Fig. 12.10–2. For $\omega < 45°$,

$$L \approx \frac{6Rl}{3R - h} = \frac{6l}{3 - n}$$

where r, l, h are known and $n = h/R$.

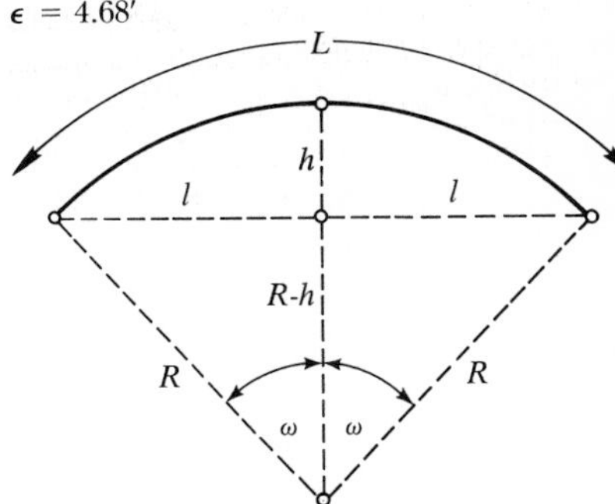

Fig. 12.10-2

example:

If $R = 10$, $h = 4$, then $l = \sqrt{100 - 36} = 8$, $n = 4/10 = 0.4$, and

$$L \approx \frac{(6)(8)}{3 - 0.4} = 18.61$$

The range of errors of this formula is the same as in Sec. 12.10–2*a*.

(c) Length of ellipse.

$$L \simeq \pi[1.5(a+b) - \sqrt{ab}]$$

where a, b are the semiaxes.

(d) Length of a parabolic arc of Fig. 12.10–3. For $x/y = \lambda < \frac{1}{5}$,

$$L \simeq y(1 + \tfrac{2}{3}\lambda^2 - \tfrac{2}{5}\lambda^4)$$

where x, y are the coordinates of P.

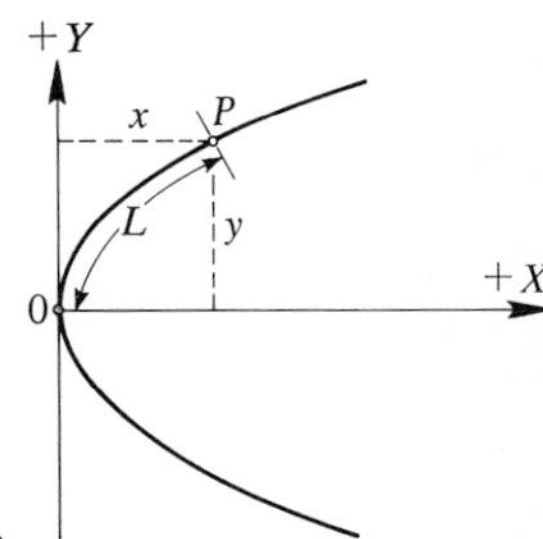

Fig. 12.10-3

(3) Transcendental Approximations

(a) Trigonometric functions and their inverses.

Approximation*	Interval	Conversion
$\sin x \approx x$ $\sin^{-1} x \approx x$	$[-0.11, 0.11]$	$0.11 \text{ rad} \equiv 6.30°$
$\cos x \approx 1$ $\cos^{-1} 1 \approx x$	$[-0.06, 0.06]$	$0.06 \text{ rad} \equiv 3.44°$
$\tan x \approx x$ $\tan^{-1} x \approx x$	$[-0.08, 0.08]$	$0.08 \text{ rad} \equiv 4.58°$

*Error of less than ±0.2%

(b) Hyperbolic functions and their inverses

Approximation*	Interval	Conversion
$\sinh x \approx x$ $\sinh^{-1} x \approx x$	$[-0.22, 0.22]$	$0.22 = 0.070\pi$
$\cosh x \approx 1$ $\cosh^{-1} 1 \approx x$	$[-0.06, 0.06]$	$0.06 = 0.019\pi$
$\tanh x \approx x$ $\tanh^{-1} x \approx x$	$[-0.18, 0.18]$	$0.18 = 0.057\pi$

*Error of less than ±0.2%

Fig. 12.10-4

(4) Area Approximations

(a) Area of circular segment of Fig. 12.10–4 is

$$A \approx \frac{2R^2\omega^3}{3}(1 - 0.2\omega^2 + 0.02\omega^4)$$

where 2ω is the central angle in radians. For $2\omega = \pi$(half circle) the error of this approximation is 3.3 percent, and it decreases rapidly as 2ω decreases. For $2\omega < \pi/2$ the error is less than 0.1 percent.

(b) Area of elliptical segment of Fig. 12.10–5 for $x < a/2$ is

$$A \approx ab\left(\frac{\pi}{2} - \frac{x}{a} - \frac{x^3}{6a^3}\right) - xy$$

where x, y are the coordinates of P. For $x > a/2$ use the exact formula,

$$A = ab \cos^{-1}\frac{x}{a} - xy$$

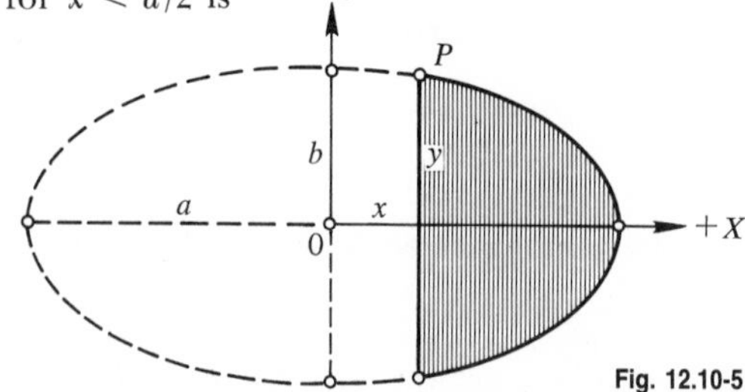

Fig. 12.10-5

(c) Areas of hyperbolic segment of Fig. 12.10–6 for $x > 2a$ is

$$A \approx xy - ab\left(\ln\frac{2x}{a} - \frac{a^2}{4x^2}\right)$$

where x, y are the coordinates of P. For $x < 2a$ use the exact formula,

$$A = xy - ab\cosh^{-1}\frac{x}{a}$$

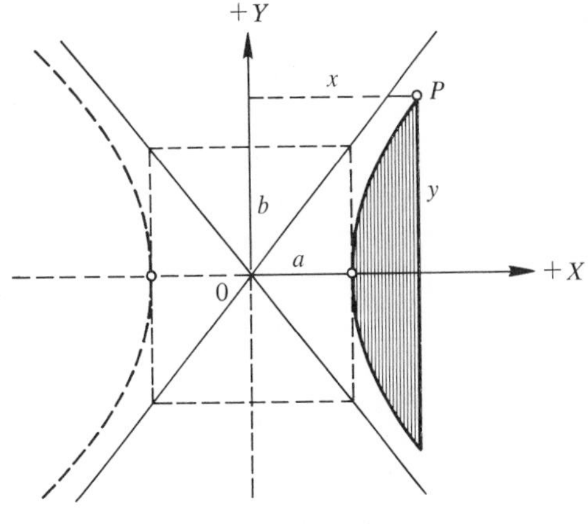

Fig. 12.10-6

(d) Area of parabolic segment of Fig. 12.10–7 is exactly

$$A = \frac{(x_1 - x_2)(y_1 - y_2)^2}{6(y_1 + y_2)}$$

where x_1, y_1 and x_2, y_2 are the coordinates of P_1 and P_2, respectively, $(x_1 \neq x_2)$.

If $x_1 = x_2$, then

$$A = \frac{4x_1y_1}{3}$$

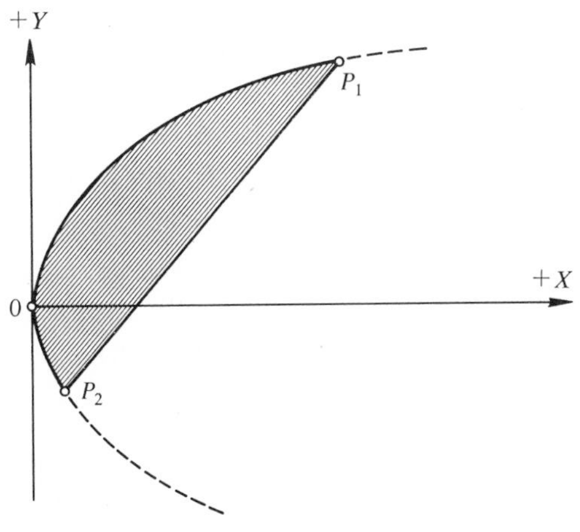

Fig. 12.10-7

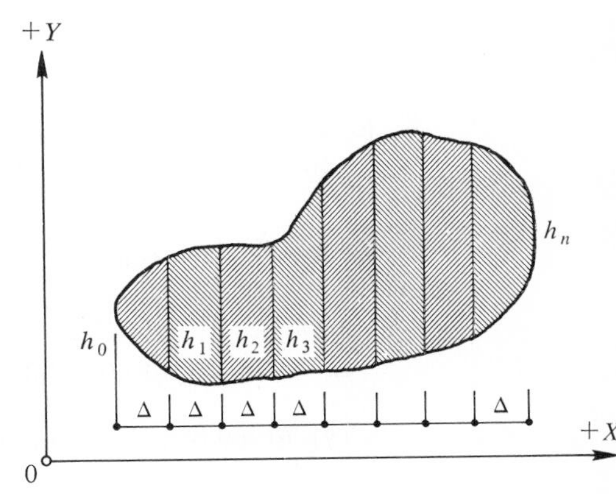

Fig. 12.10-8

(e) Irregular area of Fig. 12.10–8 is first divided into n vertical strips by equidistant parallel chords of lengths $h_0, h_1, h_2, \ldots, h_n$ and then evaluated by one of the approximate formulas given below.

(α) *Trapezoidal rule* (n = even or odd):

$$A \approx \Delta\left(\frac{h_0}{2} + h_1 + h_2 + \cdots + \frac{h_n}{2}\right)$$

(β) *Simpson's rule* (n = even):

$$A \approx \frac{\Delta}{3}(h_0 + 4h_1 + 2h_2 + 4h_3 + 2h_4 + \cdots + h_n)$$

where h_0 and/or h_n may be zero.

12.11 INTEREST AND ANNUITIES

(1) Interest

(a) Definition. The money paid for the use of another's money (principal P) is called the interest p. The interest is paid at the ends of specified intervals of time (annually, semiannually, etc.).

(b) Rate. The ratio of the amount of interest paid for a given period of time to the principal is called the rate of interest and is usually expressed as a percentage.

example:

If the principal $P = \$10{,}000$ is used for a period of one year for the interest of $p = \$600$, then the rate of interest is said to be

$$i = \frac{p}{P} = \frac{600}{10{,}000} = 0.06 \qquad \text{or } 6\%$$

(c) Simple interest is the money paid for the use of principal P for the number of periods n.

$$I = Pni$$

example:

The total simple interest paid for use of $P = \$10{,}000$ during $n = 5$ periods (years) at rate of $i = 6\%$ per period is

$$I = Pni = (10{,}000)(5)(0.06) = \$3{,}000$$

(d) Compound interest is the amount of money paid by the end of each period on the principal and on the accumulated interest.

(e) Accumulated amount A_n. The principal placed for n periods accumulates:

(α) At simple interest i to: $\quad A_n = P(1+ni)$

(β) At interest i compounded at the end of each period to: $\quad A_n = P(1+i)^n$

(γ) At interest i/q compounded q times during each period to: $\quad A_n = P\left(1+\frac{i}{q}\right)^{nq}$

(f) Table of $(1+i)^n$. The numerical values of $(1+i)^n$ for $n = 1, 2, \ldots, 50$ and $i = 1, 2, \ldots, 8$ are given in Table A.39.

examples:

If the principal $P = \$10{,}000$ is deposited for 10 years at the annual interest $i = 4\%$ compounded annually, then from Table A.39,

$$A_n = P(1+i)^n = 10{,}000(1+0.04)^{10} = 10{,}000(1.48024) = \$14{,}480.24$$

If the same principal is deposited for the same number of years at the same annual interest which is, however, compounded quarterly, then $i = 4/4 = 1\%$, $n = 10 \times 4 = 40$, and from Table A.39,

$$A_n = 10{,}000(1+0.01)^{40} = 10{,}000(1.48886) = \$14{,}888.60$$

(g) Present amount P. The present value P which during n periods accumulates to A_n:

(α) At simple interest i is: $P = A_n(1+ni)^{-1}$

(β) At interest i compounded at the end of each period is: $P = A_n(1+i)^{-n}$

(γ) At interest i/q compounded q times during each period is: $P = A_n\left(1+\dfrac{i}{q}\right)^{-nq}$

(h) Table of $(1+i)^{-n}$. The numerical values of $(1+i)^{-n}$ for $n = 1, 2, \ldots, 50$ and $i = 1, 2, \ldots, 8$ are given in Table A.40.

examples:

The amount required to be deposited to accumulate to \$10,000 at the end of 10 years at the annual rate $i = 4\%$ compounded annually is (Table A.40)

$$P = A_n(1+i)^{-n} = 10{,}000(1+0.04)^{-10} = 10{,}000(0.67556) = \$6{,}755.60$$

If the interest is compounded quarterly, then $i = 4/4 = 1\%$, $n = 10 \times 4 = 40$, and (Table A.40)

$$P = 10{,}000(1+0.01)^{-40} = 10{,}000(0.67165) = \$6{,}716.50$$

(2) Annuities

(a) Definition. The annuity is a sequence of equal payments made at equal periods of time.

(b) Amount of annuity $S_{\overline{n}|i}$. If a payment R is deposited at the end of each successive period for n periods and the interest i is compounded at the end of each period, then the accumulated amount at the end of the nth period is

$$S_{\overline{n}|i} = R\frac{(1+i)^n - 1}{i}$$

If the periodic payment R is made at the beginning of each period, then

$$S_{\overline{n}|i} = R(1+i)\left[\frac{(1+i)^n - 1}{i}\right]$$

(c) Table of $[(1+i)^n + 1]/i$. The numerical values of $[(1+i)^n + 1]/i$ for $n = 1, 2, \ldots, 50$ and $i = 1, 2, \ldots, 8$ are given in Table A.41.

example:

If the payment $R = \$1{,}000$ is made at the end of each year for $n = 10$ years and the interest $i = 5\%$ is compounded annually, then at the end of the tenth year the accumulated amount is (Table A.41)

$$S_{\overline{10}|5} = R\frac{(1+i)^n - 1}{i} = 1{,}000\frac{(1+0.05)^{10} - 1}{0.05} = 1{,}000(12.57789) = \$12{,}577.89$$

(d) Present value of annuity $A_{\overline{n}|i}$. The present amount required to produce a periodic payment R at the end of each successive period for n periods during which the interest i is compounded on the outstanding balance at the end of each period is

$$A_{\overline{n}|i} = R\frac{1-(1+i)^{-n}}{i}$$

If the periodic payment is made at the beginning of each period, then

$$A_{\overline{n}|i} = R(1-i)\left[\frac{1-(1+i)^{-n}}{i}\right]$$

(e) Table of $[1-(1+i)^{-n}]/i$. The numerical values of $[1-(1+i)^{-n}]/i$ for $n = 1, 2, \ldots, 50$ and $i = 1, 2, \ldots, 8$ are given in Table A.42.

example:

The present value of annuity which will pay $R = \$1{,}000$ annually at the end of each year for a period of 10 years at compound interest $i = 5\%$ is (Table A.42)

$$A_{\overline{10}|5} = R\frac{1-(1+i)^{-n}}{i} = 1{,}000\frac{1-(1+0.05)^{-10}}{0.05} = 1{,}000(7.72173) = \$7{,}721.73$$

13
UNIT SYSTEMS AND THEIR RELATIONSHIPS

13.01 QUANTITATIVE MEASUREMENTS

(1) Definitions

(a) Physical quantities (such as length, mass, temperature, etc.) are measured by comparison to the quantities of the same kind designated as units of measure.

(b) Concepts of measurement. Thus the statement of a physical quantity X involves two concepts:

(α) *Numerical factor* n (dimensionless number) defining the number of units
(β) *Symbol of unit* u designating the dimensions

Analytically,

$$X = nu \qquad \text{or} \qquad n = X/u$$

examples:

Equatorial radius of earth	6.378×10^3 km	3.963×10^3 miles
Polar radius of earth	6.356×10^3 km	3.950×10^3 miles
Volume of earth	1.083×10^{21} m^3	3.825×10^{22} ft^3
Mean density of earth	5.522×10^3 kg/m^3	3.447×10^2 lb/ft^3
Average linear velocity of earth	2.977×10 km/s	1.850×10 miles/sec
Average angular velocity of earth	7.272×10^{-5} rad/s	2.618×10^{-1} rad/h
Acceleration due to gravity	9.80665 m/s^2	3.21740×10 ft/sec^2
Speed of sound at 0°C	3.314×10^2 m/s	1.087 ft/sec
Speed of light in vacuum	3.000×10^8 m/s	9.843×10^8 ft/sec
Density of water at 4°C	1.000×10^{-3} kg/cm^3	
Density of dry air at 0°C	1.293×10^{-6} kg/cm^3	
Density of mercury at 0°C	1.359×10^{-2} kg/cm^3	
Mass of electron	0.91091×10^{-27} g	
Mass of proton	1.67252×10^{-24} g	
Mass of neutron	1.67548×10^{-24} g	

(2) Basic and Derived Units

(a) Origin of units. The various units of measure (such as meter, kilogram, degree Celsius, etc.) are by no means prescribed by nature but are products of human selection (national or international conventions).

(b) Classification. As the physical quantities are of the basic type (length, mass, time, temperature, etc.) and of the derived type (volume, velocity, work, etc.) their units are also designated as the basic and the derived units.

(c) Systems of units. Out of several systems of units, three systems are introduced in Tables 13.02–1 through 13.02–5; their relationships in terms of their conversion factors are given in Appendix B.

13.02 INTERNATIONAL SYSTEM OF UNITS

(1) Establishment of the System

(a) Convention. At the Eleventh General Conference on Weights and Measures of 1960, the *Metric System* (with some modifications) was given the name *International System of Units* and the abbreviation *SI* in all languages.*

(b) Recommendation. The members of this conference recommended the adaptation of the SI system for all scientific, technical, practical, and teaching purposes.

(2) Basic Units of the SI System

(a) Six units defined in this section are the basic units of the SI system. Their definitions are extracted from the records of the general conference.

(b) Unit of length. "The *meter* (m) is the length of exactly 1 650 763.73 wavelengths of the radiation in vacuum corresponding to the unperturbed transition between the levels $2p_{10}$ and $5d_5$ of the atom of Krypton 86, the orange-red line."

(c) Unit of mass. "The *kilogram* (kg) is the mass of a particular cylinder of platinum-iridium alloy, called the International Prototype Kilogram, which is preserved in a vault at Sevres, France by the International Bureau of Weights and Measures."

(d) Unit of time. "The *ephemeris second* (s) is exactly 1/31 556 925.974 7 of the tropical year of 1900, January, 0 days and 12 hours ephemeris time."

(e) Unit of electric current. "The *ampere* (A) is the constant current which, if maintained in two straight parallel conductors of infinite length, of negligible circular sections, and placed 1 meter apart in a vacuum, will produce between these conductors a force equal to 2×10^{-7} newton† per meter of length."

(f) Unit of temperature. "The *thermodynamic Kelvin degree* (°K) is the unit of temperature determined by the Carnot cycle with the triple-point temperature of water defined as exactly 273.16°K." The International Practical Kelvin Temperature Scale of 1960 and the International Practical Celsius Temperature Scale of 1960 are defined by a set of interpolation equations based on the following reference temperatures:

State	°K	°C
Oxygen, liquid-gas equilibrium	90.18	– 182.97
Water, solid-liquid equilibrium	273.15	0.00
Water, solid-liquid-gas equilibrium	273.16	0.01
Water, liquid-gas equilibrium	373.15	100.00
Zinc, solid-liquid equilibrium	692.655	419.505
Sulfur, liquid-gas equilibrium	717.75	444.6
Silver, solid-liquid equilibrium	1233.95	960.8
Gold, solid-liquid equilibrium	1336.15	1063.0

(g) Unit of luminous intensity. "The *candela* (cd) is of such a value that the luminous intensity of a full radiator at the melting temperature of platinum is 60 candela per centimeter squared."

*Recent revisions of units of time, temperature, and luminous intensity are given in Sec. 13.02–6 on p. 288.
†For the definition of 1 newton, see Sec. 13.02–3*b*.

(h) Multiples and fractions of these units defined by their factor, prefix, and symbol are given in Table 13.02–2.

(i) Special multiples and fractions of 1 meter frequently used in technology are:

1 kilometer = 1 km = 10^3 m 1 centimeter = 1 cm = 10^{-2} m 1 millimeter = 1 mm = 10^{-3} m	1 micrometer = 1 μm = 10^{-6} m 1 nanometer = 1 nm = 10^{-9} m 1 angstrom = 1 Å = 10^{-10} m
1 astronomical unit = 1.49598×10^{11} m, 1 light year = 9.46055×10^{15} m, 1 parsec = 3.08374×10^{16} m	

(3) Derived Units of the SI System

(a) Ten units defined in this section are the most important derived units of the SI system. Their definitions have been extracted from the records of the general conference.

(b) Unit of force. "The *newton* (N) is that force which gives to a mass of 1 kilogram an acceleration of 1 meter per second per second."

(c) Unit of energy. "The *joule* (J) is the work done when the point of application of 1 newton is displaced a distance of 1 meter in the direction of the force."

(d) Unit of power. "The *watt* (W) is the power which gives rise to the production of energy at the rate of 1 joule per second."

(e) Unit of electric quantity. "The *coulomb* (C) is the quantity of electricity transported in 1 second by a current of 1 ampere."

TABLE 13.02–1 Systems of Basic Units

Designation	Dimension	System		
		English (FPS)	Metric (MKS)	International (SI)
Length	L	foot (ft)	meter (m)	meter (m)
Mass	M	pound (lb)	kilogram (kg)	kilogram (kg)
Time	t	second (sec)	second (s)	second (s)
Electric current	A	ampere (A)	ampere (A)	ampere (A)
Temperature	T	degree Fahrenheit (°F)	degree Celsius (°C)	degree Kelvin (°K)
Luminous intensity	l	candela (cd)	candela (cd)	candela (cd)

TABLE 13.02–2 Decimal Multiples and Fractions of Units

Factor	Prefix	Symbol	Factor	Prefix	Symbol
10^1	deka	D*	10^{-1}	deci	d
10^2	hecto	h	10^{-2}	centi	c
10^3	kilo	k	10^{-3}	milli	m
10^6	mega	M	10^{-6}	micro	μ
10^9	giga	G	10^{-9}	nano	n
10^{12}	tera	T	10^{-12}	pico	p
10^{15}	femta	F	10^{-15}	femto	f
10^{18}	atta	A	10^{-18}	atto	a

*In some literature, the symbol "da" is used for deka.

(f) Unit of electric potential difference and electromotive force. "The *volt* (V) is the difference of electric potential between two points of a conducting wire carrying a constant current of 1 ampere, when the power dissipated between these points is equal to 1 watt."

(g) Unit of electric resistance. "The ohm (Ω) is the electric resistance between two points of a conductor when a constant difference of potential of 1 volt, applied between these two points, produces in this conductor a current of 1 ampere, this conductor not being the source of any electromotive force.".

(h) Unit of electric capacitance. "The farad (F) is the capacitance of a capacitor between the plates of which there appears a difference of potential of 1 volt when it is charged by a quantity of electricity equal to 1 coulomb."

(i) Unit of electric inductance. "The henry (H) is the inductance of a closed circuit in which an electromotive force of 1 volt is produced when the electric current in the circuit varies uniformly at a rate of 1 ampere per second."

(j) Unit of magnetic flux. "The weber (Wb) is the magnetic flux which, linking a circuit of one turn, produces in it an electromotive force of 1 volt as it is reduced to zero at a uniform rate in 1 second."

(k) Unit of luminous flux. "The lumen (lm) is the luminous flux emitted in a solid angle of 1 steradian by a uniform point source having an intensity of 1 candela."

TABLE 13.02–3 Systems of Derived Units, Geometry

Designation	Dimensions	System		
		English (FPS)	Metric (MKS)	International (SI)
Area	L^2	ft^2	m^2	m^2
Static moment of area	L^3	ft^3	m^3	m^3
Moment of inertia of area	L^4	ft^4	m^4	m^4
Product of inertia of area	L^4	ft^4	m^4	m^4
Polar moment of inertia of area	L^4	ft^4	m^4	m^4
Volume	L^3	ft^3	m^3	m^3
Static moment of volume	L^4	ft^4	m^4	m^4
Moment of inertia of volume	L^5	ft^5	m^5	m^5
Product of inertia of volume	L^5	ft^5	m^5	m^5
Polar moment of inertia of volume	L^5	ft^5	m^5	m^5
Mass = $M = W/g$*	M	lb	kg	kg
Static moment of mass	ML	lb·ft	kg·m	kg·m
Moment of inertia of mass	ML^2	$lb \cdot ft^2$	$kg \cdot m^2$	$kg \cdot m^2$
Product of inertia of mass	ML^2	$lb \cdot ft^2$	$kg \cdot m^2$	$kg \cdot m^2$
Polar moment of inertia of mass	ML^2	$lb \cdot ft^2$	$kg \cdot m^2$	$kg \cdot m^2$
Curvature of a curve	L^{-1}	1/ft	1/m	1/m
Torsion of a curve	L^{-1}	1/ft	1/m	1/m
Plane angle†	R	rad	rad	rad
Solid angle†	S	sr	sr	sr

*In the English system (FPS) and in the metric system (MKS), the mass M is a derived unit defined as the weight W divided by the acceleration due to gravity g. lb = pound mass; kg = kilogram mass.
†The unit of plane angle called radian (rad) and the unit of solid angle called steradian (sr) are supplementary units.

(4) Supplementary Units

(a) Two units defined in this section are the supplementary units of the SI system. Their definitions are extracted from the records of the General Conference.

(b) Unit of plane angle. The *radian* (rad) is the plane angle subtended at the center of a circle by an arc whose length is equal to the radius (Fig. 13.02–1).

(c) Unit of solid angle. The steradian (sr) is the space angle subtended at the vertex of spherical sector whose spherical part of the surface is equal to the square of the radius of the sphere (Fig. 13.02–2).

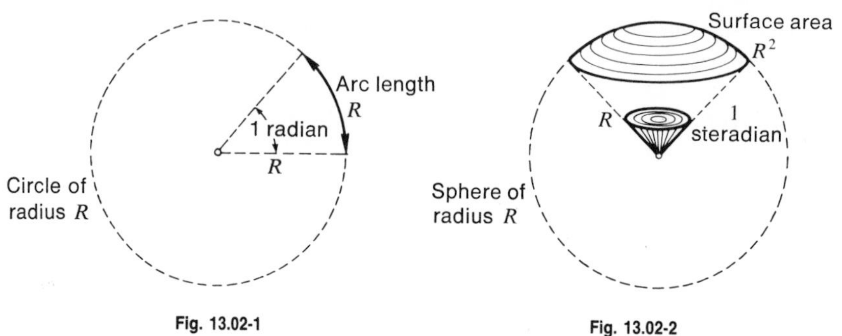

Fig. 13.02-1

Fig. 13.02-2

TABLE 13.02–4 Systems of Derived Units, Mechanics

Designation	Dimension	System		
		English (FPS)	Metric (MKS)	International (SI)
Linear velocity	Lt^{-1}	ft/sec	m/s	m/s
Angular velocity	Rt^{-1}	rad/sec	rad/s	rad/s
Linear acceleration	Lt^{-2}	ft/sec^2	m/s^2	m/s^2
Angular acceleration	Rt^{-2}	rad/sec^2	rad/s^2	rad/s^2
Linear momentum*	MLt^{-1}	lb·ft/sec	kg·m/s	N·s
Angular momentum	ML^2t^{-1}	lb·ft^2/sec	kg·m^2/s	N·m/s
Force†	MLt^{-2}	lb_f	kg_f	N
Moment of force	ML^2t^{-2}	lb_f·ft	kg_f·m	N·m
Linear impulse	MLt^{-1}	lb_f·sec	kg_f·s	N·s
Angular impulse	ML^2t^{-1}	lb_f·m·sec	kg_f·m·s	N·m·s
Stress (pressure)	$ML^{-1}t^{-2}$	lb_f/ft^2	kg_f/m^2	N/m^2
Work (energy)‡	ML^2t^{-2}	ft·lb_f	m·kg_f	J
Linear power§	ML^2t^{-3}	lb_f·ft/sec	kg_f·m/s	W
Angular power	ML^2Rt^{-3}	lb_f·ft·rad/sec	kg_f·m·rad/s	W rad
Linear stiffness	Mt^{-2}	lb_f/ft	kg_f/m	N/m
Angular stiffness	$ML^2R^{-1}t^{-2}$	lb_f·ft/rad	kg_f·m/rad	N·m/rad
Linear flexibility	$M^{-1}t^{-2}$	ft/lb_f	m/kg_f	m/N
Angular flexibility	$M^{-1}L^{-2}Rt^{-2}$	rad/lb_f·ft	rad/kg_f·m	rad/N·m
Viscosity	$ML^{-1}t^{-1}$	lb/ft·sec	kg/m·s	N·s/m^2
Frequency	t^{-1}	1/sec	1/s	1/s

*N = newton = kg·m/s^2.

†lb_f = lb·(32.174 ft/s^2), 1 kg_f = kg (9.80665 m/s^2).

‡J = joule = N·m = kg·m^2/s^2 = 10^7 ergs.

§Watt = N·m/s = kg·m^2/s^3 = J/s.

(5) Related Units of the SI System

(a) Ten additional units related to the units of the SI system are: the liter, the kilogram-force, the dyne, the bar, the technical atmosphere, the torr, the physical atmosphere, the erg, the technical calorie, and the (metric) horsepower.

(b) Liter (l) is now a special name for the cubic decimeter. $1\,l = 1\,dm^2$. Formerly the liter was defined as the volume occupied by a 1-kg mass of pure water at 4°C (39.2°F) and standard atmospheric pressure (1 atm). The weight of 1 liter of water ≅ 1 kilogram-force.

(c) Kilogram-force (kg_f) is that force which, when applied to a body having a mass of 1 kg, gives it an acceleration of 9.806 650 meters per second per second at latitude 45° and sea level. $1\,kg_f = 9.806\,650$ N.

(d) Dyne (dyn) is that force which, when applied to a body having a mass of 1 gram, gives it an acceleration of 1 centimeter per second per second. 10^5 dyn = 1 N = 0.101 971 6 kg_f.

(e) Bar (bar) is the pressure equal to $10^5\,N/m^2$. 1 bar = 10 N/cm^2 = 1.019716 kg_f/cm^2 = 750.062 torr.

(f) Technical atmosphere (at) is the pressure resulting from the uniform action of 1 kg_f over an area of 1 cm^2. 1 at = 1 kg_f/cm^2 = $10^4\,kg_f/m^2$ = 9.806 650 N/cm^2.

(g) Torr (torr) is the pressure of a 1-mm column of mercury. (This value is accurate to 1 part in 7 million.) 1 torr = 1.33322 millibars (mb) = $1.333\,220 \times 10^3\,dyn/cm^2$ = 133.322 N/m^2 = $1.359\,506 \times 10^{-3}\,kg_f/cm^2$ = 1/760 atm.

TABLE 13.02–5 Systems of Derived Units Heat, Electricity, Magnetism

Designation	Dimension	System: English (FPS)	Metric (MKS)	International (SI)
Quantity of heat[a]	ML^2t^{-2}	ft·lb_f	m·kg_f	J
Heat capacity	$ML^2t^{-2}T^{-1}$	ft·lb_f/°F	m·kg_f/°C	J/°K
Specific heat	$L^2t^{-2}T^{-1}$	ft·lb_f/lb·°F	m·kg_f/kg·°C	J/kg·°K
Heat flow[b]	ML^2t^{-3}	ft·lb_f/sec	m·kg_f/s	W
Heat conductivity	$MLt^{-3}T^{-1}$	lb_f/sec·°F	kg_f/s·°C	W/m·°K
Heat transfer coef.	$Mt^{-3}T^{-1}$	lb_f/ft·sec·°F	kg_f/m·s·°C	W/m^2·°K
Electric charge[c]	At	C	C	C
Voltage potential[d]	$MA^{-1}L^2 \cdot t^{-3}$	V	V	V
Electric resistance[e]	$MA^{-2}L^2 \cdot t^{-2}$	Ω	Ω	Ω
Electric capacitance[f]	$M^{-1}A^2L^{-2} \cdot t^4$	F	F	F
Electric inductance[g]	$MA^{-2}L^2 \cdot t^{-2}$	H	H	H
Electric field strength	$MA^{-1}L \cdot t^{-3}$		V/m	V/m
Dielectric constant	$M^{-1}A^2L^{-3} \cdot t^4$		F/m	F/m
Permeability	$MA^{-2}L \cdot t^{-2}$		H/m	H/m
Magnetic intensity	$A \cdot L^{-1}$		A/m	A/m
Magnetic flux[h]	$MA^{-1}L^2 \cdot t^{-2}$	Wb	Wb	Wb
Magnetic flux density[i]	$MA^{-1} \cdot t^{-2}$	T	T	T
Luminous flux[j]	(cd)(sr)	lm	lm	lm
Luminance	$(cd)(L)^{-2}$	cd/ft²	cd/m²	cd/m²
Illumination[k]	$(cd)(sr)(L)^{-2}$	lm/ft²	lx	lx

[a] J = joule = N·m = $kg \cdot m^2/s^2$ = 10^7 erg
[b] W = watt = J/s = N·m/s = $kg \cdot m^2/s^3$
[c] C = coulomb = A·s = ampere·second
[d] V = volt = W/A = N·m/A·s = $kg \cdot m^2/A \cdot s^3$
[e] Ω = ohm = V/A = $N \cdot m/A^2 \cdot s$ = $kg \cdot m^2/A^2 \cdot s^3$
[f] F = farad = A·s/V = $A^2 \cdot s^2/N \cdot m$ = $A^2 \cdot s^4/kg \cdot m^2$
[g] H = henry = V·s/A = $N \cdot m/A^2$ = $kg \cdot m^2/A^2 \cdot s^2$
[h] Wb = weber = V·s = N·m/A = $kg \cdot m^2/A \cdot s^2$
[i] T = tesla = Wb/m^2 = N/A·m = $kg/A \cdot s^2$
[j] lm = lumen = cd·sr
[k] lx = lux = lm/m^2

(h) Physical (standard) atmosphere (atm) is the pressure equal to the weight of a 760-mm column of mercury at latitude 45°, sea level, and 0°C. 1 atm = 760 torr = 1.033 228 kg_f/cm^2 = 1.033 228 at.

(i) Erg (erg) is equal to the work done by a force of 1 dyn moving a particle a distance of 1 cm. 10^7 ergs = 1 N·m = 0.101 971 6 kg_f·m = 1 J.

(j) Technical calorie (cal) is the quantity of heat required to raise the temperature of 1 gram of water from 14.5 to 15.5°C. 1 cal = 4.1865 J = 4.1865×10^7 ergs.

(k) Horsepower (metric) (hp) is the work done by 75 kg_f moving a particle a distance of 1 meter in 1 second. 1 hp = 735.497 500 J/s = 735.497 500 W.

(6) Revised and Additional Units of the SI System

(a) Definitions of three basic units (second, kelvin, candela) were revised at the Thirteenth General Conference on Weights and Measures in 1967. These new definitions have no quantitative effect on the corresponding units defined in Sec. 13.02–2.

(b) Unit of time. "The *second* (s) is the duration of 9 192 631 770 periods of the radiation corresponding to the transition between the two hyperfine levels of the fundamental [ground] state of the atom of cesium 133."

(c) Unit of thermodynamic temperature. "The *Kelvin** (K) is the fraction 1/273.16 of the thermodynamic temperature of the triple point of water."

(d) Unit of luminous intensity. "The *candela* (cd) is the luminous intensity, in the direction of the normal, of a blackbody surface 1/600 000 square meter in area, at the temperature of solidification of platinum (2042°K) under the pressure of 1 physical atmosphere (1 atm = 101 325 pascals).

(e) One additional basic unit and two derived units (mole, pascal, siemens) were adopted at the Fourteenth General Conference on Weights and Measures in 1971.

(f) Unit of amount of substance. "The *mole* (mol) is the amount of substance containing the same number of particles as there are atoms in 0.012 kilogram of the pure carbon nuclide ^{12}C."

(g) Unit of pressure or stress. "The *pascal* (Pa) is the pressure or stress of one newton per square meter."

(h) Unit of electrical conductance. "The *siemens* (S) is the electrical conductance of a conductor in which a current of one ampere is produced by an electric potential difference of one volt."

*In this book, the notation 1 degree Kelvin = °K is used to eliminate the potential conflict with the FPS unit of force, 1 Kip = 1000 pound-force = 1 K.

Appendix A
NUMERICAL TABLES

A.01 SQUARE ROOTS OF N/n*

$N = 1\text{–}3\pi$ $n = 2\text{–}9$

N	$\sqrt{N/2}$	$\sqrt{N/3}$	$\sqrt{N/4}$	$\sqrt{N/5}$	$\sqrt{N/6}$	$\sqrt{N/7}$	$\sqrt{N/8}$	$\sqrt{N/9}$
1	0.70711	0.57735	0.50000	0.44721	0.40825	0.37796	0.35355	0.33333
2	1.00000	0.81650	0.70711	0.63246	0.57735	0.53452	0.50000	0.47140
3	1.22475	1.00000	0.86603	0.77460	0.70711	0.65465	0.61237	0.57735
π	1.25331	1.02333	0.88623	0.79267	0.72360	0.66992	0.62666	0.59082
4	1.41421	1.15470	1.00000	0.89443	0.81649	0.75593	0.70711	0.66660
5	1.58133	1.29099	1.11803	1.00000	0.91287	0.84515	0.79057	0.74536
6	1.73205	1.41421	1.22474	1.09545	1.00000	0.92582	0.86603	0.81658
2π	1.77245	1.44720	1.25331	1.12100	1.02332	0.94742	0.90180	0.83554
7	1.87082	1.52753	1.32288	1.18322	1.08011	1.00000	0.93541	0.88192
8	2.00000	1.63299	1.41421	1.26491	1.15470	1.06904	1.00000	0.94281
9	2.12132	1.73205	1.50000	1.34164	1.22474	1.13389	1.06066	1.00000
3π	2.17080	1.77245	1.50000	1.37294	1.25331	1.16034	1.08540	1.02333

*For applications see Sec. 12.04–1.

A.02 CUBE ROOTS OF N/n*

$N = 1\text{–}3\pi$ $n = 2\text{–}9$

N	$\sqrt[3]{N/2}$	$\sqrt[3]{N/3}$	$\sqrt[3]{N/4}$	$\sqrt[3]{N/5}$	$\sqrt[3]{N/6}$	$\sqrt[3]{N/7}$	$\sqrt[3]{N/8}$	$\sqrt[3]{N/9}$
1	0.79370	0.69336	0.62996	0.58480	0.55032	0.52276	0.50000	0.48075
2	1.00000	0.87358	0.79370	0.73681	0.69336	0.65863	0.62996	0.60571
3	1.14471	1.00000	0.90856	0.84343	0.79370	0.75395	0.72112	0.69336
π	1.16245	1.01549	0.92264	0.85650	0.80600	0.76563	0.73230	0.70410
4	1.25992	1.10064	1.00000	0.92832	0.87358	0.82983	0.79370	0.76314
5	1.35721	1.18563	1.07722	1.00000	0.94104	0.89390	0.85499	0.82207
6	1.44225	1.26061	1.14471	1.06266	1.00000	0.94991	0.90856	0.87358
2π	1.46470	1.27944	1.16261	1.07912	1.01549	0.96463	0.92264	0.88724
7	1.51829	1.32635	1.20507	1.11869	1.05273	1.00000	0.95647	0.91964
8	1.58740	1.38672	1.25992	1.16961	1.10064	1.04552	1.00000	0.96150
9	1.65096	1.44225	1.31037	1.21761	1.14471	1.08738	1.04004	1.00000
3π	1.67654	1.46459	1.33067	1.23528	1.16245	1.10423	1.05615	1.01549

*For applications see Sec. 12.04–1.

A.03 SQUARES, CUBES, AND ROOTS OF N*

N = 1–50

N	N^2	N^3	$\sqrt{N}$	$\sqrt[3]{N}$	$\sqrt[4]{N}$	$\sqrt[5]{N}$
1	1	1	1.000 000	1.000 000	1.000 000	1.000 000
2	4	8	1.414 214	1.259 921	1.189 207	1.148 698
3	9	27	1.732 051	1.442 250	1.316 074	1.245 731
4	16	64	2.000 000	1.587 401	1.414 214	1.319 508
5	25	125	2.236 068	1.709 976	1.495 349	1.379 730
6	36	216	2.449 490	1.817 121	1.565 085	1.430 969
7	49	343	2.645 751	1.912 931	1.626 576	1.475 773
8	64	512	2.828 427	2.000 000	1.681 793	1.515 717
9	81	721	3.000 000	2.080 084	1.732 051	1.551 846
10	100	1 000	3.162 278	2.154 435	1.778 279	1.584 893
11	121	1 331	3.316 625	2.223 980	1.821 160	1.615 394
12	144	1 728	3.464 102	2.289 428	1.861 210	1.643 752
13	169	2 179	3.605 513	2.351 335	1.898 829	1.670 278
14	196	2 744	3.741 657	2.410 142	1.934 336	1.695 218
15	225	3 375	3.872 983	2.466 212	1.967 990	1.718 772
16	256	4 096	4.000 000	2.519 842	2.000 000	1.741 141
17	289	4 913	4.123 106	2.571 282	2.030 543	1.762 340
18	324	5 832	4.242 641	2.620 741	2.059 767	1.782 602
19	361	6 859	4.358 899	2.668 402	2.087 798	1.801 983
20	400	8 000	4.472 136	2.734 418	2.114 743	1.820 564
21	441	9 261	4.582 577	2.758 924	2.140 695	1.838 416
22	484	10 648	4.690 416	2.802 039	2.165 737	1.855 601
23	529	12 167	4.795 832	2.843 867	2.189 939	1.872 171
24	576	13 824	4.898 979	2.884 499	2.213 364	1.888 175
25	625	15 625	5.000 000	2.924 018	2.236 068	1.903 654
26	676	17 576	5.099 020	2.962 496	2.258 101	1.918 645
27	729	19 683	5.196 152	3.000 000	2.279 507	1.933 182
28	784	21 952	5.291 503	3.036 589	2.300 327	1.947 294
29	841	24 389	5.385 165	3.072 317	2.320 596	1.961 009
30	900	27 000	5.477 226	3.107 236	2.340 347	1.974 350
31	961	29 791	5.567 654	3.141 381	2.359 611	1.987 341
32	1 024	32 768	5.656 854	3.174 802	2.378 414	2.000 000
33	1 089	35 937	5.744 563	3.207 534	2.396 782	2.012 347
34	1 156	39 304	5.830 952	3.239 612	2.414 736	2.024 397
35	1 225	42 875	5.916 080	3.271 066	2.432 299	2.036 168
36	1 296	46 656	6.000 000	3.301 927	2.449 490	2.047 673
37	1 369	50 653	6.082 763	3.332 222	2.466 326	2.058 924
38	1 444	54 872	6.164 414	3.361 975	2.482 828	2.069 925
39	1 521	59 319	6.244 998	3.391 211	2.498 999	2.080 719
40	1 600	64 000	6.324 555	3.419 952	2.514 867	2.091 279
41	1 681	68 921	6.403 124	3.448 217	2.530 440	2.101 632
42	1 764	74 088	6.480 741	3.476 027	2.545 730	2.111 786
43	1 849	79 507	6.557 439	3.503 398	2.560 750	2.121 747
44	1 936	85 184	6.633 250	3.530 348	2.575 506	2.131 526
45	2 025	91 125	6.708 204	3.556 893	2.590 020	2.141 127
46	2 116	97 336	6.782 329	3.583 048	2.604 291	2.150 560
47	2 209	103 823	6.855 655	3.608 826	2.618 330	2.159 830
48	2 304	110 592	6.928 203	3.634 241	2.632 148	2.168 944
49	2 401	117 649	7.000 000	3.659 306	2.645 751	2.177 906
50	2 500	125 000	7.071 068	3.684 031	2.659 148	2.186 724

*For applications of Tables A.03 to A.22 see Sec. 12.04–2.

A.04 SQUARES, CUBES, AND ROOTS OF N

N = 51–100

N	N^2	N^3	$\sqrt{N}$	$\sqrt[3]{N}$	$\sqrt[4]{N}$	$\sqrt[5]{N}$
51	2 601	132 651	7.141 428	3.708 430	2.672 345	2.195 402
52	2 704	140 608	7.211 103	3.732 511	2.685 350	2.203 945
53	2 809	148 877	7.280 110	3.756 286	2.698 168	2.212 357
54	2 916	157 464	7.348 469	3.779 766	2.710 806	2.220 643
55	3 025	166 375	7.416 198	3.802 952	2.723 270	2.228 807
56	3 136	175 616	7.483 315	3.825 862	2.735 565	2.236 864
57	3 249	185 193	7.549 834	3.848 501	2.747 696	2.244 786
58	3 364	195 112	7.615 773	3.870 877	2.759 669	2.252 608
59	3 481	205 379	7.681 146	3.892 996	2.771 488	2.260 322
60	3 600	216 000	7.745 967	3.914 868	2.783 158	2.267 933
61	3 721	226 981	7.819 250	3.936 497	2.794 682	2.275 443
62	3 844	238 328	7.874 008	3.957 892	2.806 066	2.282 855
63	3 969	250 047	7.937 254	3.979 057	2.817 313	2.290 172
64	4 096	262 144	8.000 000	4.000 000	2.828 427	2.297 397
65	4 225	274 625	8.062 258	4.020 726	2.839 412	2.304 532
66	4 356	287 496	8.124 038	4.041 240	2.850 270	2.311 579
67	4 489	300 763	8.185 353	4.061 548	2.861 006	2.318 542
68	4 624	314 432	8.246 211	4.081 655	2.871 622	2.325 422
69	4 761	328 509	8.306 624	4.101 566	2.882 121	2.332 222
70	4 900	343 000	8.366 600	4.121 285	2.892 508	2.338 943
71	5 041	357 911	8.426 150	4.140 818	2.902 783	2.345 588
72	5 184	372 248	8.485 281	4.160 168	2.912 951	2.352 158
73	5 329	389 017	8.544 004	4.179 339	2.923 013	2.358 656
74	5 476	405 224	8.602 325	4.198 336	2.932 972	2.365 083
75	5 625	421 875	8.660 254	4.217 163	2.942 831	2.371 441
76	5 776	438 976	8.717 798	4.235 824	2.952 592	2.377 731
77	5 929	456 533	8.774 964	4.254 321	2.962 257	2.383 956
78	6 084	474 552	8.831 761	4.272 659	2.971 828	2.390 116
79	6 241	493 039	8.888 194	4.290 840	2.981 308	2.396 213
80	6 400	512 000	8.944 272	4.308 869	2.990 698	2.402 249
81	6 561	531 441	9.000 000	4.326 749	3.000 000	2.408 225
82	6 724	551 368	9.055 385	4.344 481	3.009 217	2.414 142
83	6 889	571 787	9.110 434	4.362 071	3.018 349	2.420 001
84	7 056	592 704	9.165 151	4.379 519	3.027 400	2.425 805
85	7 225	614 125	9.219 544	4.396 830	3.036 370	2.431 553
86	7 396	636 056	9.273 618	4.414 005	3.045 262	2.437 248
87	7 569	658 503	9.327 379	4.431 048	3.054 076	2.442 890
88	7 740	681 472	9.380 832	4.447 960	3.062 814	2.448 480
89	7 921	704 969	9.433 981	4.464 745	3.071 479	2.454 019
90	8 100	729 000	9.486 833	4.481 405	3.080 070	2.359 509
91	8 281	753 571	9.539 392	4.497 941	3.088 591	2.464 951
92	8 464	778 688	9.591 663	4.514 357	3.097 041	2.470 345
93	8 649	804 357	9.643 651	4.530 655	3.105 423	2.475 692
94	8 836	830 584	9.695 360	4.546 836	3.113 737	2.480 993
95	9 025	857 375	9.746 794	4.562 903	3.121 987	2.486 250
96	9 216	884 736	9.797 959	4.578 857	3.130 169	2.491 462
97	9 409	912 673	9.848 858	4.594 701	3.138 289	2.496 631
98	9 604	941 192	9.899 495	4.610 436	3.146 346	2.501 758
99	9 801	970 299	9.949 874	4.626 065	3.154 342	2.506 842
100	10 000	1 000 000	10.000 000	4.641 589	3.162 278	2.511 887

A.05 SQUARES, CUBES, AND ROOTS OF N

N	N	N^3	$\sqrt{N}$	$\sqrt[3]{N}$	$\sqrt[4]{N}$	$\sqrt[5]{N}$
101	10 201	1 030 301	10.049 876	4.657 010	3.170 154	2.516 890
102	10 404	1 061 208	10.099 505	4.672 329	3.177 972	2.521 855
103	10 609	1 092 727	10.148 892	4.687 548	3.185 733	2.526 780
104	10 816	1 124 864	10.198 039	4.702 669	3.193 437	2.531 668
105	11 025	1 157 625	10.246 951	4.717 694	3.201 086	2.536 517
106	11 236	1 191 016	10.295 630	4.732 623	3.208 680	2.541 330
107	11 447	1 225 043	10.344 080	4.747 459	3.216 108	2.546 108
108	11 664	1 259 712	10.392 305	4.762 203	3.223 710	2.550 849
109	11 881	1 295 029	10.440 307	4.776 856	3.231 146	2.555 555
110	12 100	1 331 000	10.488 088	4.791 420	3.238 532	2.560 227
111	12 321	1 367 631	10.535 654	4.805 896	3.245 867	2.564 865
112	12 544	1 404 928	10.583 005	4.820 285	3.253 153	2.569 470
113	12 769	1 442 897	10.630 146	4.834 588	3.260 390	2.570 042
114	12 996	1 481 544	10.677 078	4.848 808	3.267 580	2.578 582
115	13 225	1 520 875	10.723 805	4.862 944	3.274 722	2.583 090
116	13 456	1 560 896	10.770 330	4.876 999	3.281 818	2.587 567
117	13 689	1 601 613	10.816 654	4.890 973	3.288 868	2.592 013
118	13 924	1 643 032	10.862 780	4.904 868	3.295 873	2.596 429
119	14 161	1 685 159	10.908 712	4.918 685	3.302 834	2.600 814
120	14 400	1 728 000	10.954 451	4.932 424	3.309 751	2.605 171
121	14 641	1 771 561	11.000 000	4.946 087	3.316 625	2.609 499
122	14 884	1 815 848	11.045 361	4.959 676	3.323 456	2.613 798
123	15 129	1 860 867	11.090 537	4.973 190	3.330 246	2.618 069
124	15 376	1 906 624	11.135 529	4.986 631	3.336 994	2.622 312
125	15 625	1 953 125	11.180 340	5.000 000	3.343 702	2.626 528
126	15 876	2 000 376	11.224 972	5.013 298	3.350 369	2.630 717
127	16 129	2 048 383	11.269 428	5.026 526	3.356 997	2.634 879
128	16 384	2 097 152	11.313 709	5.039 684	3.363 586	2.639 016
129	16 641	2 146 689	11.357 817	5.052 774	3.370 136	2.643 126
130	16 900	2 197 000	11.401 754	5.065 797	3.376 648	2.647 212
131	17 161	2 248 091	11.445 523	5.078 753	3.383 123	2.651 272
132	17 474	2 299 969	11.489 125	5.091 643	3.389 561	2.655 307
133	17 689	2 352 637	11.532 563	5.104 469	3.395 963	2.659 318
134	17 956	2 406 104	11.575 837	5.117 230	3.402 328	2.663 305
135	18 225	2 460 375	11.618 950	5.129 928	3.408 658	2.667 269
136	18 496	2 515 456	11.661 904	5.142 563	3.414 953	2.671 208
137	18 769	2 571 353	11.704 700	5.155 137	3.421 213	2.675 125
138	19 044	2 628 072	11.747 340	5.167 649	3.427 439	2.679 019
139	19 321	2 685 619	11.789 826	5.180 101	3.433 632	2.682 891
140	19 600	2 744 000	11.832 160	5.192 494	3.439 791	2.686 740
141	19 881	2 803 221	11.874 342	5.204 828	3.445 917	2.690 567
142	20 164	2 863 288	11.916 375	5.217 103	3.452 010	2.694 373
143	20 449	2 924 207	11.958 261	5.229 315	3.458 010	2.698 157
144	20 736	2 985 984	12.000 000	5.241 483	3.464 102	2.701 920
145	21 025	3 048 625	12.041 595	5.253 588	3.470 100	2.705 662
146	21 316	3 121 856	12.083 046	5.265 637	3.476 067	2.709 384
147	21 609	3 176 523	12.124 356	5.277 632	3.482 005	2.713 085
148	21 904	3 241 792	12.165 525	5.289 572	3.487 911	2.716 767
149	22 201	3 307 949	12.206 556	5.301 459	3.493 788	2.720 428
150	22 500	3 375 000	12.247 449	5.313 293	3.499 636	2.724 070

A.06 SQUARES, CUBES, AND ROOTS OF N

N	N^2	N^3	$\sqrt{N}$	$\sqrt[3]{N}$	$\sqrt[4]{N}$	$\sqrt[5]{N}$
151	22 801	3 442 951	12.288 206	5.325 074	3.505 454	2.727 692
152	23 104	3 511 808	12.328 828	5.336 804	3.511 243	2.731 296
153	23 409	3 581 577	12.369 317	5.348 481	3.517 004	2.734 880
154	23 716	3 652 264	12.409 674	5.360 108	3.522 737	2.738 446
155	24 025	3 723 875	12.449 900	5.371 685	3.528 442	2.741 993
156	24 336	3 796 416	12.489 960	5.383 213	3.534 119	2.745 522
157	24 649	3 869 893	12.529 964	5.394 691	3.539 769	2.749 033
158	24 964	3 944 312	12.569 805	5.406 120	3.545 392	2.752 526
159	25 281	4 019 679	12.609 520	5.417 502	3.550 989	2.756 001
160	25 600	4 096 000	12.649 115	5.428 835	3.556 559	2.759.459
161	25 921	4 173 281	12.688 578	5.440 122	3.562 103	2.762 900
162	26 244	4 251 528	12.727 922	5.451 362	3.567 621	2.766 324
163	26 569	4 330 747	12.767 145	5.462 556	3.573 114	2.769 731
164	26 896	4 410 944	12.806 248	5.473 704	3.578 582	2.773 121
165	27 225	4 492 125	12.845 233	5.484 807	3.584 025	2.776 494
166	27 556	4 574 296	12.884 099	5.495 865	3.589 443	2.779 852
167	27 889	4 657 463	12.922 848	5.506 878	3.594 836	2.783 193
168	28 224	4 741 632	12.961 481	5.517 848	3.600 206	2.786 518
169	28 561	4 826 809	13.000 000	5.528 775	3.605 551	2.789 827
170	28 900	4 913 000	13.038 405	5.539 658	3.610 873	2.793 121
171	29 241	5 000 211	13.076 697	5.550 499	3.616 172	2.796 400
172	29 584	5 088 448	13.114 877	5.561 298	3.621 447	2.799 663
173	29 929	5 177 717	13.152 946	5.572 055	3.626 699	2.802 910
174	30 276	5 268 024	13.190 906	5.582 770	3.631 929	2.806 143
175	30 625	5 359 375	13.228 757	5.593 445	3.637 136	2.809 361
176	30 976	5 451 776	13.266 499	5.604 079	3.642 321	2.812 565
177	31 329	5 545 233	13.304 135	5.614 672	3.647 483	2.815 754
178	31 684	5 639 752	13.341 664	5.625 226	3.652 624	2.818 928
179	32 041	5 735 339	13.379 088	5.635 741	3.657 744	2.822 088
180	32 400	5 832 000	13.416 408	5.646 216	3.662 842	2.825 235
181	32 761	5 929 741	13.453 624	5.656 653	3.667 918	2.828 367
182	33 124	6 028 568	13.490 738	5.667 051	3.672 974	2.831 485
183	33 489	6 128 487	13.527 749	5.677 411	3.678 009	2.834 590
184	33 856	6 229 504	13.564 660	5.687 734	3.683 023	2.837 681
185	34 225	6 331 625	13.601 471	5.698 019	3.688 017	2.840 759
186	34 596	6 434 856	13.638 181	5.708 267	3.692 991	2.843 823
187	34 969	6 539 203	13.674 794	5.718 479	3.697 945	2.846 874
188	35 344	6 644 672	13.711 309	5.728 654	3.702 879	2.849 913
189	35 721	6 751 269	13.747 727	5.738 794	3.707 793	2.852 938
190	36 100	6 859 000	13.784 049	5.748 897	3.712 688	2.855 951
191	36 481	6 967 871	13.820 275	5.758 965	3.717 563	2.858 951
192	36 864	7 077 888	13.856 406	5.768 998	3.772 419	2.861 938
193	37 249	7 189 057	13.892 444	5.778 997	3.727 257	2.864 913
194	37 636	7 301 384	13.928 388	5.788 961	3.732 076	2.867 876
195	38 025	7 414 875	13.964 240	5.798 890	3.736 876	2.870 826
196	38 416	7 529 536	14.000 000	5.808 786	3.741 657	2.873 765
197	38 809	7 645 373	14.035 669	5.818 648	3.746 421	2.876 691
198	39 204	7 762 393	14.071 247	5.828 477	3.751 166	2.879 606
199	39 601	7 880 599	14.106 736	5.838 272	3.755 893	2.882 509
200	40 000	8 000 000	14.142 136	5.848 035	3.760 603	2.885 400

A.07 SQUARES, CUBES, AND ROOTS OF N

N	N^2	N^3	$\sqrt{N}$	$\sqrt[3]{N}$	$\sqrt[4]{N}$	$\sqrt[5]{N}$
201	40 401	8 120 601	14.177 447	5.857 766	3.765 295	2.888 279
202	40 804	8 242 408	14.212 670	5.867 464	3.769 970	2.891 148
203	41 209	8 365 427	14.247 807	5.877 131	3.774 627	2.894 005
204	41 616	8 489 664	14.282 857	5.886 765	3.779 267	2.896 850
205	42 025	8 615 125	14.317 821	5.896 369	3.783 890	2.899 685
206	42 436	8 741 816	14.352 700	5.905 941	3.788 850	2.902 508
207	42 849	8 869 743	14.387 495	5.915 482	3.793 085	2.905 321
208	43 264	8 998 912	14.422 205	5.924 992	3.797 658	2.908 122
209	43 681	9 129 329	14.456 839	5.934 472	3.802 214	2.910 913
210	44 100	9 261 000	14.491 377	5.943 922	3.806 754	2.913 693
211	44 521	9 393 931	14.525 839	5.953 342	3.811 278	2.916 463
212	44 944	9 528 128	14.560 220	5.962 732	3.815 786	2.919 222
213	45 369	9 663 597	14.594 520	5.972 093	3.820 277	2.921 971
214	45 796	9 800 344	14.628 739	5.981 424	3.824 753	2.924 710
215	46 225	9 938 375	14.662 878	5.990 727	3.829 214	2.927 438
216	46 656	10 077 696	14.696 939	6.000 000	3.833 659	2.930 156
217	47 089	10 218 313	14.730 920	6.009 245	3.838 088	2.932 864
218	47 524	10 360 232	14.764 823	6.018 462	3.842 502	2.935 562
219	47 961	10 503 459	14.798 649	6.027 650	3.846 901	2.938 251
220	48 400	10 648 000	14.832 397	6.036 811	3.851 128	2.940 929
221	48 841	10 793 861	14.866 069	6.045 944	3.855 654	2.943 598
222	49 284	10 941 048	14.899 664	6.055 049	3.860 008	2.946 257
223	49 729	11 089 567	14.933 185	6.064 127	3.864 348	2.948 906
224	50 176	11 239 424	14.966 630	6.073 178	3.868 673	2.951 546
225	50 625	11 390 625	15.000 000	6.082 202	3.872 983	2.954 177
226	51 076	11 543 176	15.033 296	6.091 199	3.877 280	2.956 798
227	51 529	11 697 083	15.066 519	6.100 170	3.881 561	2.959 410
228	51 984	11 852 352	15.099 669	6.109 147	3.885 829	2.962 013
229	52 441	12 008 989	15.132 746	6.118 033	3.890 083	2.964 607
230	52 900	12 167 000	15.165 751	6.126 926	3.894 323	2.967 191
231	53 361	12 326 391	15.198 648	6.135 792	3.898 549	2.969 767
232	53 824	12 487 168	15.231 546	6.144 634	3.902 761	2.972 334
233	54 289	12 649 337	15.264 338	6.153 449	3.906 960	2.974 892
234	54 756	12 812 904	15.297 059	6.162 240	3.911 145	2.977 441
235	55 225	12 977 875	15.329 710	6.171 006	3.915 317	2.979 982
236	55 696	13 144 256	15.362 915	6.179 747	3.919 476	2.982 513
237	56 169	13 312 053	15.394 804	6.188 463	3.923 621	2.985 037
238	56 644	13 481 272	15.427 249	6.197 154	3.922 754	2.987 551
239	57 121	13 615 919	15.459 625	6.205 822	3.931 873	2.990 058
240	57 600	13 824 000	15.491 933	6.214 465	3.935 979	2.992 556
241	58 081	13 997 521	15.524 175	6.223 084	3.940 073	2.995 045
242	58 564	14 172 488	15.556 349	6.231 680	3.944 154	2.997 527
243	59 049	14 348 907	15.588 457	6.240 251	3.948 220	3.000 000
244	59 536	14 526 784	15.620 499	6.248 800	3.952 278	3.002 465
245	60 025	14 706 125	15.652 476	6.257 325	3.956 321	3.004 922
246	60 516	14 886 936	15.684 387	6.265 827	3.960 352	3.007 371
247	61 009	15 069 223	15.716 234	6.274 305	3.964 371	3.009 812
248	61 504	15 252 992	15.748 016	6.282 761	3.968 377	3.012 245
249	62 001	15 438 249	15.779 734	6.291 195	3.972 371	3.014 671
250	62 500	15 625 000	15.811 388	6.299 605	3.976 354	3.017 088

A.08 SQUARES, CUBES, AND ROOTS OF N

N = 251–300

N	N^2	N^3	$\sqrt{N}$	$\sqrt[3]{N}$	$\sqrt[4]{N}$	$\sqrt[5]{N}$
251	63 001	15 813 251	15.842 980	6.307 994	3.980 324	3.019 498
252	63 504	16 003 008	15.874 508	6.316 360	3.984 283	3.021 900
253	64 009	16 194 277	15.905 974	6.324 704	3.988 229	3.024 295
254	64 516	16 387 064	15.937 377	6.333 026	3.992 165	3.026 682
255	65 025	16 581 375	15.968 719	6.341 326	3.996 088	3.029 061
256	65 536	16 777 216	16.000 000	6.349 604	4.000 000	3.031 433
257	66 049	16 974 593	16.031 220	6.357 861	4.003 901	3.033 798
258	66 564	17 173 512	16.062 378	6.366 097	4.007 790	3.036 155
259	67 081	17 373 979	16.093 477	6.374 311	4.011 668	3.038 505
260	67 600	17 576 000	16.124 516	6.382 504	4.015 534	3.040 848
261	68 121	17 779 581	16.155 494	6.390 677	4.019 390	3.043 183
262	68 644	17 984 728	16.186 414	6.398 828	4.023 234	3.045 512
263	69 169	18 191 447	16.217 275	6.406 959	4.027 068	3.048 784
264	69 696	18 399 744	16.248 077	6.415 069	4.030 890	3.050 147
265	70 225	18 609 625	16.278 821	6.423 158	4.034 702	3.052 454
266	70 756	18 821 096	16.309 506	6.431 228	4.038 503	3.054 755
267	71 289	19 034 163	16.340 135	6.439 277	4.042 293	3.057 048
268	71 824	19 248 832	16.370 706	6.447 306	4.046 073	3.059 334
269	72 361	19 465 109	16.401 219	6.455 315	4.049 842	3.061 614
270	72 900	19 683 000	16.431 677	6.466 304	4.053 600	3.063 887
271	73 441	19 902 511	16.462 078	6.471 274	4.057 349	3.066 153
272	73 984	20 123 648	16.492 423	6.479 224	4.061 086	3.068 413
273	74 529	20 346 417	16.522 712	6.487 154	4.064 814	3.070 666
274	75 076	20 570 824	16.552 945	6.495 065	4.068 531	3.072 912
275	75 625	20 796 875	16.583 124	6.502 957	4.072 238	3.075 152
276	76 176	21 024 576	16.613 248	6.510 830	4.075 935	3.077 385
277	76 729	21 253 933	16.643 317	6.518 684	4.079 622	3.079 612
278	77 284	21 484 952	16.673 332	6.526 519	4.083 299	3.081 832
279	77 841	21 717 639	16.703 293	6.534 335	4.086 966	3.084 046
280	78 400	21 952 000	16.733 201	6.542 133	4.090 623	3.086 254
281	78 961	22 188 041	16.763 055	6.549 912	4.094 271	3.088 455
282	79 524	22 425 768	16.792 856	6.557 672	4.097 909	3.090 650
283	80 089	22 665 187	16.822 604	6.565 414	4.101 537	3.092 839
284	80 656	22 906 304	16.852 300	6.573 138	4.105 155	3.095 021
285	81 225	23 149 125	16.881 943	6.580 844	4.108 764	3.097 198
286	81 796	23 393 656	16.911 535	6.588 532	4.112 364	3.099 368
287	82 369	23 639 903	16.941 074	6.596 202	4.115 954	3.101 533
288	82 944	23 887 872	16.970 563	6.603 854	4.119 534	3.103 691
289	83 521	24 137 569	17.000 000	6.611 489	4.123 106	3.105 844
290	84 100	24 389 000	17.029 386	6.619 106	4.126 678	3.107 990
291	84 681	24 642 171	17.058 722	6.626 705	4.130 221	3.110 130
292	85 264	24 897 088	17.088 007	6.634 287	4.133 764	3.112 265
293	85 849	25 153 757	17.117 243	6.641 852	4.137 299	3.114 394
294	86 436	25 412 184	17.146 428	6.649 400	4.140 825	3.116 517
295	87 025	25 672 375	17.175 564	6.656 930	4.144 341	3.118 634
296	87 616	25 934 336	17.204 651	6.664 444	4.147 849	3.120 745
297	88 209	26 198 073	17.233 688	6.671 940	4.151 348	3.122 851
298	88 804	26 463 592	17.262 677	6.679 200	4.154 838	3.124 951
299	89 401	26 730 899	17.291 616	6.686 883	4.158 319	3.127 046
300	90 000	27 000 000	17.320 508	6.694 330	4.161 791	3.129 135

A.09 SQUARES, CUBES, AND ROOTS OF N

N	N^2	N^3	$\sqrt{N}$	$\sqrt[3]{N}$	$\sqrt[4]{N}$	$\sqrt[5]{N}$
301	90 601	27 270 901	17.349 352	6.701 759	4.165 255	3.131 218
302	91 204	27 543 608	17.378 147	6.709 173	4.168 710	3.133 296
303	91 809	27 818 127	17.406 895	6.716 570	4.172 157	3.135 368
304	92 416	28 094 464	17.435 596	6.723 951	4.175 595	3.137 435
305	93 025	28 372 625	17.464 249	6.731 315	4.179 025	3.139 496
306	93 636	28 652 616	17.492 856	6.738 664	4.182 446	3.141 552
307	94 249	28 934 443	17.521 415	6.745 997	4.185 859	3.143 603
308	94 864	29 218 112	17.549 929	6.753 313	4.189 264	3.145 648
309	95 481	29 503 629	17.578 396	6.760 614	4.192 660	3.147 688
310	96 100	29 791 000	17.606 817	6.767 899	4.196 048	3.149 723
311	96 721	30 080 231	17.635 192	6.775 169	4.199 428	3.151 752
312	97 344	30 371 328	17.663 522	6.782 423	4.202 799	3.153 777
313	97 969	30 664 297	17.691 806	6.789 661	4.206 163	3.155 796
314	98 596	30 959 144	17.720 045	6.796 884	4.209 518	3.157 810
315	99 225	31 255 875	17.748 239	6.804 092	4.212 866	3.159 818
316	99 856	31 554 496	17.776 389	6.811 285	4.216 206	3.161 822
317	100 489	31 855 013	17.804 494	6.818 462	4.219 537	3.163 821
318	101 124	32 157 432	17.832 555	6.825 624	4.222 861	3.165 814
319	101 761	32 461 759	17.860 571	6.832 771	4.226 177	3.167 803
320	102 400	32 768 000	17.888 544	6.833 904	4.229 485	3.169 786
321	103 041	33 076 161	17.916 473	6.847 021	4.232 785	3.171 765
322	103 684	33 386 248	17.944 358	6.854 124	4.236 078	3.173 739
323	104 329	33 698 267	17.972 201	6.861 212	4.239 363	3.175 708
324	104 976	34 012 224	18.000 000	6.868 285	4.242 641	3.177 672
325	105 625	34 328 125	18.027 756	6.875 344	4.245 911	3.179 631
326	106 276	34 645 976	18.055 470	6.882 389	4.249 173	3.181 585
327	106 929	34 965 783	18.083 141	6.889 419	4.252 428	3.183 534
328	107 584	35 287 552	18.110 770	6.896 434	4.255 675	3.185 479
329	108 241	35 611 289	18.138 357	6.903 436	4.258 915	3.187 419
330	108 900	35 937 000	18.165 902	6.910 423	4.262 148	3.189 354
331	109 561	36 264 691	18.193 405	6.917 396	4.265 373	3.191 285
332	110 224	36 594 368	18.220 867	6.924 356	4.268 591	3.193 211
333	110 889	36 926 037	18.248 288	6.931 301	4.271 801	3.195 132
334	111 556	37 259 704	18.275 667	6.938 232	4.275 005	3.197 049
335	112 225	37 595 375	18.303 005	6.945 150	4.278 201	3.198 961
336	112 896	37 933 056	18.330 303	6.952 053	4.281 390	3.200 869
337	113 569	38 272 753	18.357 560	6.958 943	4.284 572	3.202 772
338	114 244	38 614 472	18.384 776	6.965 820	4.287 747	3.204 670
339	114 921	38 958 219	18.411 953	6.972 683	4.290 915	3.206 564
340	115 600	39 304 000	18.439 089	6.979 532	4.294 076	3.208 454
341	116 281	39 651 821	18.466 185	6.986 368	4.297 230	3.210 339
342	116 964	40 001 688	18.493 242	6.993 191	4.300 377	3.212 220
343	117 649	40 353 607	18.520 259	7.000 000	4.303 517	3.214 096
344	118 336	40 707 584	18.547 237	7.006 796	4.306 650	3.215 968
345	119 025	41 063 625	18.574 176	7.013 579	4.309 777	3.217 835
346	119 716	41 421 736	18.601 075	7.020 349	4.312 896	3.219 699
347	120 409	41 781 923	18.627 936	7.027 106	4.316 009	3.221 558
348	121 104	42 144 192	18.654 758	7.033 850	4.319 115	3.223 412
349	121 801	42 508 549	18.681 542	7.040 581	4.322 215	3.225 263
350	122 500	42 875 000	18.708 287	7.047 299	4.325 308	3.227 109

A.10 SQUARES, CUBES, AND ROOTS OF N

N = 351–400

N	N^2	N^3	$\sqrt{N}$	$\sqrt[3]{N}$	$\sqrt[4]{N}$	$\sqrt[5]{N}$
351	123 201	43 243 551	18.734 994	7.054 004	4.328 394	3.228 951
352	123 904	43 614 208	18.761 663	7.060 697	4.331 474	3.230 789
353	124 609	43 986 977	18.788 294	7.067 377	4.334 547	3.232 622
354	125 316	44 361 864	18.814 888	7.074 044	4.337 613	3.234 452
355	126 025	44 738 875	18.841 444	7.080 699	4.340 673	3.236 277
356	126 736	45 118 016	18.867 962	7.087 341	4.343 727	3.238 098
357	127 449	45 499 293	18.894 443	7.093 970	4.346 774	3.239 915
358	128 164	45 882 712	18.920 888	7.100 588	4.349 815	3.241 728
359	128 881	46 268 279	18.947 295	7.107 194	4.352 849	3.243 537
360	129 600	46 656 000	18.973 666	7.113 787	4.355 877	3.245 342
361	130 321	47 045 881	19.000 000	7.120 367	4.358 899	3.247 143
362	131 044	47 437 928	19.026 298	7.126 936	4.361 914	3.248 940
363	131 769	47 832 147	19.052 559	7.133 492	4.364 924	3.250 733
364	132 496	48 228 544	19.078 784	7.140 037	4.367 927	3.252 522
365	133 225	48 627 125	19.104 973	7.146 569	4.370 924	3.254 307
366	133 956	49 027 896	19.131 126	7.153 090	4.373 914	3.256 089
367	134 689	49 430 863	19.157 244	7.159 599	4.376 899	3.257 866
368	135 242	49 836 032	19.183 326	7.166 096	4.379 877	3.259 639
369	136 161	50 243 409	19.209 373	7.172 581	4.382 850	3.261 409
370	136 900	50 653 000	19.235 384	7.179 054	4.385 816	3.263 175
371	137 641	51 064 811	19.261 360	7.185 516	4.388 777	3.264 937
372	138 384	51 478 848	19.287 302	7.191 966	4.391 731	3.266 695
373	139 129	51 895 117	19.313 208	7.198 405	4.394 680	3.268 449
374	139 876	52 313 624	19.339 080	7.204 832	4.397 622	3.270 200
375	140 625	52 734 375	19.364 917	7.211 248	4.400 559	3.271 947
376	141 376	53 157 376	19.390 719	7.217 652	4.403 489	3.273 690
377	142 129	53 582 633	19.416 488	7.224 045	4.406 414	3.275 430
378	142 884	54 010 152	19.442 222	7.230 427	4.409 336	3.277 165
379	143 641	54 439 939	19.467 922	7.236 797	4.412 247	3.278 898
380	144 400	54 872 000	19.493 589	7.243 156	4.415 154	3.280 626
381	145 161	55 306 341	19.519 221	7.249 505	4.418 056	3.282 351
382	145 924	55 742 968	19.544 820	7.255 842	4.420 952	3.284 072
383	146 689	56 181 887	19.570 386	7.262 167	4.423 843	3.285 790
384	147 456	56 623 104	19.595 918	7.268 482	4.426 728	3.287 504
385	148 225	57 066 625	19.621 417	7.274 786	4.429 607	3.289 214
386	148 996	57 512 456	19.646 883	7.281 079	4.432 480	3.290 921
387	149 769	57 960 603	19.672 316	7.287 616	4.435 348	3.292 624
388	150 544	58 411 072	19.697 716	7.293 633	4.438 211	3.294 324
389	151 321	58 863 869	19.723 083	7.299 894	4.441 068	3.296 021
390	152 100	59 319 000	19.748 418	7.306 144	4.443 919	3.297 713
391	152 881	59 776 471	19.773 720	7.312 383	4.446 765	3.299 403
392	153 664	60 236 288	19.798 990	7.318 611	4.449 606	3.301 089
393	154 449	60 698 457	19.824 228	7.324 829	4.452 441	3.302 771
394	155 236	61 162 984	19.849 433	7.331 037	4.455 270	3.304 450
395	156 025	61 629 875	19.874 607	7.337 234	4.458 095	3.306 126
396	156 816	62 099 136	19.899 749	7.343 420	4.460 913	3.307 798
397	157 609	62 570 773	19.924 859	7.349 597	4.463 727	3.309 467
398	158 404	63 044 792	19.949 937	7.355 762	4.466 535	3.311 133
399	159 201	63 521 199	19.974 984	7.361 918	4.469 338	3.312 795
400	160 000	64 000 000	20.000 000	7.368 063	4.472 136	3.314 454

A.11 SQUARES, CUBES, AND ROOTS OF N

N	N^2	N^3	$\sqrt{N}$	$\sqrt[3]{N}$	$\sqrt[4]{N}$	$\sqrt[5]{N}$
401	160 801	64 481 201	20.024 984	7.374 198	4.474 928	3.316 110
402	161 604	64 964 808	20.049 938	7.380 323	4.477 716	3.317 762
403	162 409	65 450 827	20.074 860	7.386 437	4.480 498	3.319 411
404	163 216	65 939 264	20.099 751	7.392 542	4.483 275	3.321 057
405	164 025	66 430 125	20.124 612	7.398 636	4.486 046	3.322 699
406	164 836	66 923 416	20.149 442	7.404 721	4.488 813	3.324 338
407	165 649	67 419 143	20.174 241	7.410 795	4.491 574	3.325 974
408	166 464	67 917 312	20.199 010	7.416 859	4.494 331	3.327 607
409	167 281	68 417 929	20.223 748	7.422 914	4.497 082	3.329 237
410	168 100	68 921 000	20.248 457	7.428 959	4.499 829	3.330 863
411	168 921	69 426 531	20.273 135	7.434 994	4.502 570	3.332 486
412	169 744	69 934 528	20.297 783	7.441 019	4.505 306	3.334 106
413	170 569	70 444 997	20.322 401	7.447 034	4.508 037	3.335 723
414	171 396	70 957 944	20.346 990	7.453 040	4.510 764	3.337 337
415	172 225	71 473 375	20.371 549	7.459 036	4.513 485	3.338 948
416	173 056	71 991 296	20.396 078	7.465 022	4.516 202	3.340 555
417	173 889	72 511 713	20.420 578	7.470 999	4.518 913	3.342 160
418	174 724	73 034 632	20.445 048	7.476 966	4.521 620	3.343 761
419	175 561	73 560 059	20.469 489	7.482 924	5.524 322	3.345 360
420	176 400	74 088 000	20.493 902	7.488 872	4.527 019	3.346 955
421	177 241	74 618 461	20.518 285	7.494 811	4.529 711	3.348 547
422	178 084	75 151 448	20.542 639	7.500 741	4.532 399	3.350 136
423	178 929	75 686 967	20.566 964	7.506 661	4.535 081	3.351 723
424	179 776	76 225 024	20.591 260	7.512 572	4.537 759	3.353 306
425	180 625	76 765 625	20.615 528	7.518 473	4.540 433	3.354 886
426	181 476	77 308 776	20.639 767	7.524 365	4.543 101	3.356 463
427	182 329	77 854 483	20.663 978	7.530 248	4.545 765	3.358 038
428	183 184	78 402 752	20.688 161	7.536 122	4.548 424	3.359 609
429	184 041	78 953 589	20.712 315	7.541 987	4.551 078	3.361 178
430	184 900	79 507 000	20.736 441	7.547 842	4.553 729	3.362 743
431	185 761	80 062 991	20.760 539	7.553 689	4.556 374	3.364 306
432	186 624	80 621 568	20.784 610	7.559 526	4.559 014	3.365 865
433	187 489	81 182 737	20.808 652	7.565 355	4.561 650	3.367 423
434	188 356	81 746 504	20.832 667	7.571 174	4.564 282	3.368 976
435	189 225	82 312 875	20.856 654	7.576 985	4.566 909	3.370 527
436	190 096	82 881 856	20.880 613	7.582 787	4.569 531	3.372 076
437	190 969	83 453 453	20.904 545	7.558 579	4.572 149	3.373 621
438	191 844	84 027 672	20.928 450	7.594 363	4.574 762	3.375 164
439	192 721	84 604 519	20.952 327	7.600 139	4.577 371	3.376 703
440	193 600	85 184 000	20.976 177	7.605 905	4.579 976	3.378 240
441	194 481	85 766 121	21.000 000	7.611 663	4.582 576	3.379 744
442	195 364	86 350 888	21.023 796	7.617 412	4.585 171	3.381 306
443	196 249	86 938 307	21.047 565	7.623 152	4.587 763	3.382 834
444	197 136	87 528 384	21.071 308	7.628 884	4.590 349	3.384 360
445	198 025	88 121 125	21.095 023	7.634 607	4.592 932	3.385 883
446	198 916	88 716 536	21.118 712	7.640 321	4.595 510	3.387 404
447	199 809	89 314 623	21.142 375	7.646 027	4.598 084	3.388 921
448	200 704	89 915 392	21.166 010	7.651 725	4.600 653	3.390 436
449	201 601	90 518 849	21.189 620	7.657 414	4.603 218	3.391 949
450	202 500	91 125 000	21.213 203	7.663 094	4.605 779	3.393 458

A.12 SQUARES, CUBES, AND ROOTS OF N

N = 451–500

N	N^2	N^3	$\sqrt{N}$	$\sqrt[3]{N}$	$\sqrt[4]{N}$	$\sqrt[5]{N}$
451	203 401	91 733 851	21.236 761	7.668 766	4.608 336	3.394 965
452	204 304	92 345 408	21.260 292	7.674 430	4.610 888	3.396 469
453	205 209	92 959 677	21.283 797	7.680 086	4.613 437	3.397 971
454	206 116	93 576 664	21.307 276	7.685 733	4.615 980	3.399 470
455	207 025	94 196 375	21 330 729	7.691 372	4.618 520	3.400 966
456	207 936	94 818 816	21.354 156	7.697 002	4.621 056	3.402 460
457	208 849	95 443 993	21.377 558	7.702 625	4.623 587	3.403 951
458	209 764	96 071 912	21.400 935	7.708 239	4.626 144	3.405 439
459	210 681	96 702 579	21.424 285	7.713 845	4.628 638	3.406 925
460	211 600	97 336 000	21.447 611	7.719 443	4.631 157	3.408 408
461	212 521	97 972 181	21.470 911	7.725 032	4.633 671	3.409 889
462	213 444	98 611 128	21.494 185	7.730 614	4.636 182	3.411 367
463	214 369	99 252 847	21.517 435	7.736 188	4.638 689	3.412 842
464	215 296	99 897 344	21.540 659	7.741 753	4.641 192	3.414 315
465	216 225	100 544 625	21.563 859	7.747 311	4.643 690	3.415 785
466	217 156	101 194 696	21.587 033	7.752 861	4.646 185	3.417 253
467	218 089	101 847 563	21.610 183	7.758 402	4.648 675	3.418 719
468	219 024	102 503 232	21.633 308	7.763 936	4.651 162	3.420 182
469	219 961	103 161 709	21.656 408	7.769 462	4.653 645	3.421 642
470	220 900	103 823 000	21.679 483	7.774 980	4.656 123	3.423 100
471	221 841	104 487 111	21.702 534	7.780 490	4.658 598	3.424 555
472	222 784	105 154 048	21.725 561	7.785 993	4.661 069	3.426 008
473	223 729	105 823 817	21.748 563	7.791 488	4.663 535	3.427 459
474	224 676	106 496 424	21.771 541	7.796 974	4.665 998	3.428 907
475	225 625	107 171 875	21.794 495	7.802 454	4.668 457	3.430 352
476	226 576	107 850 176	21.817 424	7.807 925	4.670 913	3.431 795
477	227 529	108 531 133	21.840 330	7.813 389	4.673 364	3.433 236
478	228 484	109 215 352	21.863 211	7.818 846	4.675 811	3.434 674
479	229 441	109 902 239	21.886 069	7.824 294	4.678 255	3.436 110
480	230 400	110 592 000	21.908 902	7.829 735	4.680 695	3.437 544
481	231 361	111 284 641	21.931 712	7.835 169	4.683 131	3.438 975
482	232 324	111 980 168	21.954 498	7.840 595	4.685 563	3.440 404
483	233 289	112 678 587	21.977 261	7.846 013	4.687 991	3.441 830
484	234 256	113 379 904	22.000 000	7.851 424	4.690 416	3.443 254
485	235 225	114 084 125	22.022 716	7.856 828	4.692 837	3.444 676
486	236 196	114 791 256	22.045 408	7.862 224	4.695 254	3.446 095
487	237 169	115 501 303	22.068 076	7.867 613	4.696 767	3.447 512
488	238 144	116 214 272	22.090 722	7.872 994	4.700 077	3.448 927
489	239 121	116 930 169	22.113 344	7.878 368	4.702 483	3.450 339
490	240 100	117 649 000	22.135 944	7.883 735	4.704 885	3.451 749
491	241 081	118 370 771	22.158 520	7.889 095	4.707 284	3.453 157
492	242 064	119 095 488	22.181 703	7.894 447	4.709 679	3.454 562
493	243 049	119 823 157	22.203 603	7.899 792	4.712 070	3.455 965
494	244 036	120 553 784	22.226 111	7.905 129	4.714 458	3.457 366
495	245 025	121 287 375	22.248 595	7.910 460	4.716 842	3.458 765
496	246 016	122 023 936	22.271 057	7.915 783	4.719 222	3.460 161
497	247 009	122 763 473	22.293 497	7.921 099	4.721 599	3.461 555
498	248 004	123 505 992	22.315 914	7.926 408	4.723 972	3.462 947
499	249 001	124 251 499	22.338 308	7.931 710	4.726 342	3.464 337
500	250 000	125 000 000	22.360 680	7.937 005	4.728 708	3.465 724

A.13 SQUARES, CUBES, AND ROOTS OF N

N = 501–550

N	N^2	N^3	$\sqrt{N}$	$\sqrt[3]{N}$	$\sqrt[4]{N}$	$\sqrt[5]{N}$
501	251 001	125 751 501	22.383 029	7.942 293	4.731 071	3.467 109
502	252 004	126 506 008	22.405 356	7.947 574	4.733 430	3.468 492
503	253 009	127 263 527	22.427 661	7.952 848	4.735 785	3.469 873
504	254 016	128 024 064	22.449 944	7.958 114	4.738 137	3.471 252
505	255 025	128 787 625	22.472 205	7.963 374	4.740 486	3.472 628
506	256 036	129 554 216	22.494 444	7.968 627	4.742 831	3.474 002
507	257 048	130 323 843	22.516 660	7.973 873	4.745 172	3.475 374
508	258 064	131 096 512	22.538 855	7.979 112	4.747 510	3.476 744
509	259 081	131 872 229	22.561 028	7.984 344	4.749 845	3.478 120
510	260 100	132 651 000	22.583 180	7.989 570	4.752 176	3.479 478
511	261 121	133 432 831	22.605 309	7.994 788	4.754 504	3.480 841
512	262 144	134 217 728	22.627 417	8.000 000	4.756 828	3.482 202
513	263 169	135 005 697	22.649 503	8.005 205	4.759 149	3.483 561
514	264 196	135 796 744	22.671 568	8.010 403	4.761 467	3.484 918
515	265 225	136 590 875	22.693 611	8.015 595	4.763 781	3.486 273
516	266 256	137 388 096	22.715 633	8.020 779	4.766 092	3.487 626
517	267 289	138 188 413	22.737 634	8.025 957	4.768 400	3.488 977
518	268 324	138 991 832	22.759 613	8.031 129	4.770 704	3.490 326
519	269 361	139 798 359	22.781 571	8.036 293	4.773 044	3.491 672
520	270 400	140 608 000	22.803 508	8.041 452	4.775 302	3.493 017
521	271 441	141 420 761	22.825 424	8.046 603	4.777 596	3.494 359
522	272 484	142 236 648	22.847 319	8.051 748	4.779 887	3.495 700
523	273 529	143 055 667	22.869 193	8.056 886	4.782 175	3.497 038
524	274 576	143 877 824	22.891 046	8.062 018	4.784 459	3.498 374
525	275 625	144 703 125	22.912 878	8.067 143	4.786 740	3.499 708
526	276 676	145 531 576	22.934 690	8.072 262	4.789 018	3.501 041
527	277 729	146 363 183	22.956 481	8.077 374	4.791 292	3.502 371
528	278 784	147 197 952	22.978 251	8.082 480	4.793 563	3.503 699
529	279 841	148 035 889	23.000 000	8.087 579	4.795 832	3.505 025
530	280 900	148 877 000	23.021 729	8.092 672	4.798 096	3.506 349
531	281 961	149 721 291	23.043 437	8.097 759	4.800 358	3.507 671
532	283 024	150 568 768	23.065 125	8.102 839	4.802 616	3.508 992
533	284 089	151 419 437	23.086 793	8.107 913	4.804 872	3.510 310
534	285 156	152 273 304	23.108 440	8.112 980	4.807 124	3.511 626
535	286 225	153 130 375	23.130 067	8.118 041	4.809 373	3.512 940
536	287 296	153 990 656	23.151 674	8.123 096	4.811 619	3.514 252
537	288 369	154 854 153	23.173 260	8.128 145	4.813 861	3.515 563
538	289 444	155 720 872	23.194 827	8.133 187	4.816 101	3.516 871
539	290 521	156 590 819	23.216 374	8.138 223	4.818 337	3.518 178
540	291 600	157 464 000	23.237 900	8.143 253	4.820 571	3.519 482
541	292 681	158 340 421	23.259 407	8.148 276	4.822 801	3.520 785
542	293 764	159 220 088	23.280 893	8.153 294	4.825 028	3.522 085
543	294 849	160 103 007	23.302 360	8.158 305	4.827 252	3.523 384
544	295 936	160 989 184	23.323 808	8.163 310	4.829 473	3.524 681
545	297 025	161 878 625	23.345 235	8.168 309	4.831 691	3.525 976
546	298 116	162 771 336	23.366 643	8.173 302	4.833 906	3.527 269
547	299 209	163 667 323	23.388 031	8.178 289	4.836 117	3.528 560
548	300 304	164 566 592	23.409 400	8.183 269	4.838 326	3.529 849
549	301 401	165 469 149	23.430 749	8.188 244	4.840 532	3.531 136
550	302 500	166 375 000	23.452 079	8.193 213	4.842 735	3.532 422

A.14 SQUARES, CUBES, AND ROOTS OF N $N = 551–600$

N	N^2	N^3	$\sqrt{N}$	$\sqrt[3]{N}$	$\sqrt[4]{N}$	$\sqrt[5]{N}$
551	303 601	167 284 151	23.473 389	8.198 175	4.844 935	3.533 705
552	304 704	168 196 608	23.494 680	8.203 132	4.847 131	3.534 987
553	305 809	169 112 377	23.515 952	8.208 082	4.849 325	3.536 267
554	306 916	170 031 464	23.537 205	8.213 027	4.851 516	3.537 545
555	308 025	170 953 875	23.558 438	8.217 966	4.853 704	3.538 821
556	309 136	171 879 616	23.579 652	8.222 899	4.855 888	3.540 095
557	310 249	172 808 693	23.600 847	8.227 825	4.858 070	3.541 368
558	311 364	173 741 112	23.622 024	8.232 746	4.860 249	3.542 639
559	312 481	174 676 879	23.643 181	8.237 661	8.862 425	3.543 907
560	313 600	175 616 000	23.664 319	8.242 571	4.864 599	3.545 174
561	314 721	176 558 481	23.685 439	8.247 474	4.866 769	3.546 440
562	315 844	177 504 328	23.706 539	8.252 372	4.868 936	3.547 703
563	316 969	178 453 547	23.727 621	8.252 263	4.871 101	3.548 965
564	318 096	179 406 144	23.748 684	8.262 149	4.873 262	3.550 225
565	319 225	180 362 125	23.769 729	8.267 029	4.875 421	3.551 483
566	320 356	181 321 496	23.790 755	8.271 904	4.877 577	3.552 739
567	321 489	182 284 263	23.811 762	8.276 773	4.879 730	3.553 993
568	322 624	183 250 432	23.832 751	8.281 635	4.881 880	3.555 246
569	323 761	184 220 009	23.853 721	8.286 493	4.884 027	3.556 497
570	324 900	185 193 000	23.874 673	8.291 344	4.886 172	3.557 746
571	326 041	186 169 411	23.895 606	8.296 190	4.888 313	3.558 994
572	327 184	187 149 248	23.916 521	8.301 031	4.890 452	3.560 239
573	328 329	188 132 517	23.937 418	8.305 865	4.892 588	3.561 483
574	329 476	189 119 224	23.958 297	8.310 694	4.894 721	3.562 726
575	330 625	190 109 375	23.979 158	8.315 517	4.896 852	3.563 966
576	331 776	191 102 976	24.000 000	8.320 335	4.898 979	3.565 205
577	332 929	192 100 033	24.020 824	8.325 148	4.901 104	3.566 442
578	334 084	193 100 552	24.041 631	8.329 954	4.903 227	3.567 677
579	335 241	194 104 539	24.062 419	8.334 755	4.905 346	3.568 911
580	336 400	195 112 000	24.083 189	8.339 551	4.907 463	3.570 143
581	337 561	196 112 941	24.103 942	8.344 341	4.909 577	3.571 373
582	338 724	197 137 368	24.124 676	8.349 126	4.911 688	3.572 602
583	339 889	198 155 287	24.145 393	8.353 905	4.913 796	3.573 829
584	341 056	199 176 704	24.166 092	8.358 678	4.915 902	3.575 054
585	342 225	200 201 625	24.186 773	8.363 447	4.918 005	3.576 277
586	343 396	201 230 056	24.207 437	8.368 209	4.920 105	3.577 499
587	344 569	202 262 003	24.228 083	8.372 967	4.922 203	3.578 719
588	345 744	203 297 472	24.248 711	8.377 719	4.924 298	3.579 938
589	346 921	204 336 469	24.269 322	8.382 465	4.926 390	3.581 155
590	348 100	205 379 000	24.289 916	8.387 207	4.928 480	3.582 370
591	349 281	206 425 071	24.310 492	8.391 942	4.903 567	3.583 583
592	350 464	207 474 688	24.331 050	8.396 673	4.932 651	3.584 795
593	351 649	208 527 857	24.351 591	8.401 398	4.934 733	3.586 005
594	352 836	209 584 584	24.372 115	8.406 118	4.936 812	3.587 214
595	354 025	210 644 875	24.392 622	8.410 833	4.933 889	3.588 421
596	355 216	211 708 736	24.413 111	8.415 542	4.940 963	3.589 626
597	356 409	212 761 173	24.433 583	8.420 246	4.943 034	3.590 830
598	357 604	213 847 192	24.454 039	8.424 945	4.945 102	3.592 032
599	358 801	214 921 799	24.474 477	8.429 638	4.947 169	3.593 233
600	360 000	216 000 000	24.494 897	8.434 327	4.949 232	3.594 432

A.15 SQUARES, CUBES, AND ROOTS OF N N = 601–650

N	N^2	N^3	$\sqrt{N}$	$\sqrt[3]{N}$	$\sqrt[4]{N}$	$\sqrt[5]{N}$
601	361 201	217 081 801	24.515 301	8.439 010	4.951 293	3.595 629
602	362 404	218 167 208	24.535 688	8.443 688	4.953 351	3.596 825
603	363 609	219 256 227	24.556 058	8.448 361	4.955 407	3.598 019
604	364 816	220 348 864	24.576 411	8.453 028	4.957 460	3.599 212
605	366 025	221 445 125	24.596 748	8.457 691	4.959 511	3.600 403
606	367 236	222 545 016	24.617 067	8.462 348	4.961 559	3.601 592
607	368 449	223 648 543	24.637 370	8.467 000	4.963 605	3.602 780
608	369 664	224 755 712	24.657 656	8.471 647	4.965 648	3.603 966
609	370 881	225 866 529	24.677 925	8.476 289	4.967 688	3.605 151
610	372 100	226 981 000	24.698 178	8.480 926	4.969 726	3.606 334
611	373 321	228 099 131	24.718 414	8.485 558	4.971 762	3.607 516
612	374 544	229 220 928	24.738 634	8.490 185	4.973 795	3.608 696
613	375 769	230 346 397	24.758 837	8.494 807	4.975 825	3.609 874
614	376 996	231 475 544	24.779 023	8.499 423	4.977 853	3.611 051
615	378 225	232 608 375	24.799 194	8.504 035	4.979 879	3.612 227
616	379 456	233 744 896	24.819 347	8.508 642	4.981 902	3.613 401
617	380 689	234 885 113	24.839 485	8.513 243	4.983 923	3.614 573
618	381 924	236 029 032	24.859 606	8.517 840	4.985 941	3.615 744
619	383 161	237 176 659	24.879 711	8.522 432	4.987 957	3.616 914
620	384 400	238 328 000	24.899 799	8.527 019	4.989 970	3.618 081
621	385 641	239 483 061	24.919 872	8.531 601	4.991 981	3.619 248
622	386 884	240 641 848	24.939 928	8.536 178	4.993 989	3.620 413
623	388 129	241 804 367	24.959 968	8.540 750	4.995 995	3.621 576
624	389 376	242 970 624	24.979 992	8.545 317	4.997 999	3.622 739
625	390 625	244 140 625	25.000 000	8.549 880	5.000 000	3.623 898
626	391 876	245 314 376	25.019 992	8.554 437	5.001 999	3.625 057
627	393 129	246 491 883	25.039 968	8.558 990	5.003 995	3.626 215
628	394 384	247 673 152	25.059 993	8.563 538	5.005 989	3.627 371
629	395 641	248 858 189	25.079 872	8.568 081	5.007 981	3.628 525
630	396 900	250 047 000	25.099 801	8.572 619	5.009 970	3.629 678
631	398 161	251 239 591	25.119 713	8.577 152	5.011 957	3.630 830
632	399 424	252 435 968	25.139 610	8.581 681	5.013 942	3.631 980
633	400 689	253 636 137	25.159 491	8.586 205	5.015 924	3.633 128
634	401 956	254 840 104	25.179 357	8.590 724	5.017 904	3.634 276
635	403 225	256 047 875	25.199 206	8.595 238	5.019 881	3.635 421
636	404 496	257 259 456	25.219 040	8.599 748	5.021 856	3.636 566
637	405 769	258 474 853	25.238 859	8.604 252	5.023 829	3.637 708
638	407 044	259 694 072	25.258 662	8.608 753	5.025 800	3.638 850
639	408 321	260 917 119	25.278 449	8.613 248	5.027 768	3.639 990
640	409 600	262 144 000	25.298 221	8.617 739	5.029 734	3.641 128
641	410 881	263 374 721	25.317 978	8.622 225	5.031 697	3.642 266
642	412 164	264 609 288	25.337 719	8.626 706	5.033 659	3.643 401
643	413 449	265 847 707	25.357 445	8.631 183	5.035 618	3.644 536
644	414 736	267 089 984	25.377 155	8.635 655	5.037 574	3.645 668
645	416 025	268 336 125	25.396 850	8.640 123	5.039 529	3.646 800
646	417 316	269 586 136	25.416 530	8.644 585	5.041 481	3.647 930
647	418 609	270 840 023	25.436 195	8.649 044	5.043 431	3.649 059
648	419 904	272 097 792	25.455 844	8.653 497	5.045 378	3.650 186
649	421 201	273 359 449	25.475 478	8.657 947	5.047 324	3.651 312
650	422 500	274 625 000	25.495 098	8.662 391	5.049 267	3.652 436

A.16 SQUARES, CUBES, AND ROOTS OF N

N	N^2	N^3	$\sqrt{N}$	$\sqrt[3]{N}$	$\sqrt[4]{N}$	$\sqrt[5]{N}$
651	423 801	275 894 451	25.514 702	8.666 831	5.051 208	3.653 560
652	425 104	277 167 808	25.534 291	8.671 266	5.053 147	3.654 681
653	426 409	278 445 077	25.553 865	8.675 697	5.055 083	3.655 802
654	427 716	279 726 264	25.573 424	8.680 124	5.057 017	3.656 921
655	429 025	281 011 375	25.592 968	8.684 546	5.058 949	3.658 038
656	430 336	282 300 416	25.612 497	8.688 963	5.060 879	3.659 155
657	431 649	283 593 393	25.632 011	8.693 376	5.062 807	3.660 270
658	432 964	284 890 312	25.651 511	8.697 784	5.064 732	3.661 383
659	434 281	286 191 179	25.670 995	8.702 188	5.066 655	3.662 495
660	435 600	287 496 000	25.690 465	8.706 588	5.068 576	3.663 606
661	436 921	288 804 781	25.709 920	8.710 983	5.070 495	3.664 716
662	438 244	290 117 528	25.729 361	8.715 373	5.072 412	3.665 824
663	439 569	291 434 247	25.748 786	8.719 760	5.074 326	3.666 931
664	440 896	292 754 944	25.768 197	8.724 141	5.076 239	3.668 036
665	442 225	294 079 625	25.787 594	8.728 519	5.078 149	3.669 140
666	443 556	295 408 296	25.806 976	8.732 892	5.080 057	3.670 243
667	444 889	296 740 963	25.826 343	8.737 260	5.081 963	3.671 345
668	446 224	298 077 632	25.845 696	8.741 625	5.083 866	3.672 445
669	447 561	299 418 309	25.865 034	8.745 985	5.085 768	3.673 544
670	448 900	300 763 000	25.884 358	8.750 340	5.087 667	3.674 641
671	450 241	302 111 711	25.903 668	8.754 691	5.089 565	3.675 738
672	451 584	303 464 448	25.922 963	8.759 038	5.091 460	3.676 833
673	452 929	304 821 217	25.942 244	8.763 381	5.093 353	3.677 926
674	454 276	306 182 024	25.961 510	8.767 191	5.095 244	3.679 019
675	455 625	307 546 875	25.980 762	8.772 053	5.097 133	3.680 110
676	456 976	308 915 776	26.000 000	8.776 383	5.099 020	3.681 199
677	458 329	310 288 733	26.019 224	8.780 708	5.100 904	3.682 288
678	459 684	311 665 752	26.038 433	8.785 030	5.102 787	3.683 375
679	461 041	313 046 839	26.057 628	8.789 347	5.104 667	3.684 461
680	462 400	314 432 000	26.076 810	8.793 659	5.106 546	3.685 546
681	463 761	315 821 241	26.095 977	8.797 968	5.108 422	3.686 629
682	465 124	317 214 568	26.115 130	8.802 272	5.110 296	3.687 711
683	466 489	318 611 987	26.134 269	8.806 572	5.112 169	3.688 792
684	467 856	320 013 504	26.153 394	8.810 868	5.114 039	3.689 871
685	469 225	321 419 125	26.172 505	8.815 160	5.115 907	3.690 950
686	470 596	322 828 856	26.191 602	8.819 447	5.117 773	3.692 027
687	471 969	324 242 703	26.210 685	8.823 731	5.119 637	3.693 102
688	473 344	325 660 672	26.229 754	8.828 010	5.121 499	3.694 177
689	474 721	327 082 769	26.248 809	8.832 285	5.123 359	3.695 250
690	476 100	328 509 000	26.267 851	8.836 556	5.125 217	3.696 322
691	477 481	329 939 371	26.286 879	8.840 823	5.127 078	3.697 393
692	478 864	331 373 888	26.305 893	8.845 085	5.128 927	3.698 462
693	480 249	332 812 557	26.324 893	8.849 344	5.130 779	3.669 531
694	481 636	334 255 384	26.343 880	8.853 599	5.132 629	3.700 598
695	483 025	335 702 375	26.362 853	8.857 849	5.134 477	3.701 664
696	484 416	337 153 536	26.381 812	8.862 095	5.136 323	3.702 728
697	485 809	338 608 873	26.400 758	8.866 338	5.138 167	3.703 792
698	487 204	340 068 392	26.419 690	8.870 576	5.140 009	3.704 854
699	488 601	341 532 099	26.438 608	8.874 810	5.141 849	3.705 915
700	490 000	343 000 000	26.457 513	8.879 040	5.143 687	3.706 975

A.17 SQUARES, CUBES, AND ROOTS OF N

N	N^2	N^3	$\sqrt{N}$	$\sqrt[3]{N}$	$\sqrt[4]{N}$	$\sqrt[5]{N}$
701	491 401	344 472 101	26.476 405	8.883 266	5.145 523	3.708 033
702	492 804	345 948 408	26.495 283	8.887 488	5.147 357	3.709 090
703	494 209	347 428 927	26.514 147	8.891 706	5.149 189	3.710 147
704	495 616	348 913 664	26.532 998	8.895 920	5.151 019	3.711 201
705	497 025	350 402 625	26.551 836	8.900 130	5.152 847	3.712 255
706	498 436	351 895 816	26.570 661	8.904 337	5.154 674	3.713 308
707	499 849	353 393 243	26.589 472	8.908 539	5.156 498	3.714 359
708	501 264	354 894 912	26.608 269	8.912 737	5.158 320	3.715 409
709	502 681	356 400 829	26.627 054	8.916 931	5.160 141	3.716 458
710	504 100	357 911 000	26.645 825	8.921 121	5.161 959	3.717 506
711	505 521	359 425 431	26.664 583	8.925 308	5.163 776	3.718 553
712	506 944	360 944 128	26.683 328	8.929 490	5.165 591	3.719 597
713	508 369	362 467 097	26.702 060	8.933 669	5.167 404	3.720 642
714	509 796	363 994 344	26.720 778	8.937 843	5.169 214	3.721 685
715	511 225	365 525 875	26.739 484	8.942 014	5.171 023	3.722 727
716	512 656	367 061 696	26.758 176	8.946 181	5.172 831	3.723 768
717	514 089	368 601 813	26.776 856	8.950 344	5.174 636	3.724 807
718	515 524	370 146 232	26.795 522	8.954 503	5.176 439	3.725 846
719	516 961	371 694 959	26.814 175	8.958 658	5.178 241	3.726 831
720	518 400	373 248 000	26.832 816	8.962 809	5.180 040	3.727 919
721	519 841	374 805 361	26.851 443	8.966 957	5.181 838	3.728 954
722	521 284	376 367 048	26.870 058	8.971 101	5.183 634	3.729 988
723	522 729	377 933 067	26.888 659	8.975 241	5.185 428	3.731 021
724	524 176	379 503 424	26.907 248	8.979 377	5.187 220	3.732 052
725	525 625	381 078 125	26.925 824	8.983 509	5.189 010	3.733 083
726	527 076	382 657 176	26.944 387	8.987 637	5.190 798	3.734 112
727	528 529	384 240 583	26.962 938	8.991 762	5.192 585	3.735 140
728	529 984	385 828 352	26.981 475	8.995 883	5.194 370	3.736 167
729	531 441	387 420 489	27.000 000	9.000 000	5.196 152	3.737 193
730	532 900	389 017 000	27.018 512	9.004 133	5.197 933	3.738 218
731	534 361	390 617 891	27.037 012	9.008 223	5.199 713	3.739 241
732	535 824	392 223 168	27.055 499	9.012 329	5.201 490	3.740 264
733	537 289	393 832 837	27.073 973	9.016 431	5.203 266	3.741 285
734	538 756	395 446 904	27.092 434	9.020 529	5.205 039	3.742 305
735	540 225	397 065 375	27.110 883	9.024 624	5.206 811	3.743 324
736	541 696	398 688 256	27.129 320	9.028 715	5.208 581	3.744 343
737	543 169	400 315 553	27.147 744	9.032 802	5.210 350	3.745 359
738	544 644	401 947 272	27.166 155	9.036 886	5.212 116	3.746 375
739	546 121	403 583 419	27.184 554	9.040 966	5.213 881	3.747 390
740	547 600	405 224 000	27.202 941	9.045 042	5.215 644	3.748 404
741	549 081	406 869 021	27.221 315	9.049 114	5.217 405	3.749 416
742	550 574	408 518 488	27.239 677	9.053 183	5.219 164	3.750 428
743	552 049	410 172 407	27.258 026	9.057 248	5.220 922	3.751 438
744	553 536	411 830 784	27.276 363	9.061 310	5.222 678	3.752 447
745	555 025	413 493 625	27.294 688	9.065 368	5.224 432	3.753 455
746	556 516	415 160 936	27.313 001	9.069 422	5.226 184	3.754 462
747	558 009	416 832 723	27.331 301	9.073 473	5.227 935	3.755 468
748	559 504	418 508 992	27.349 589	9.077 520	5.229 683	3.756 473
749	561 001	420 189 749	27.367 864	9.081 563	5.231 430	3.757 477
750	562 500	421 875 000	27.386 128	9.085 603	5.233 176	3.758 480

A.18 SQUARES, CUBES, AND ROOTS OF N

N	N^3	N^3	$\sqrt{N}$	$\sqrt[3]{N}$	$\sqrt[4]{N}$	$\sqrt[5]{N}$
751	564 001	423 564 751	27.404 379	9.089 639	5.234 919	3.759 482
752	565 504	425 259 008	27.422 618	9.093 672	5.236 661	3.760 482
753	567 009	426 957 777	27.440 845	9.097 701	5.238 401	3.761 482
754	568 516	428 661 064	27.459 060	9.101 727	5.240 139	3.762 481
755	570 025	430 368 875	27.477 263	9.105 748	5.241 876	3.763 478
756	571 536	432 081 216	27.495 454	9.109 767	5.243 611	3.764 474
757	573 049	433 798 093	27.513 633	9.113 782	5.245 344	3.765 470
758	574 564	435 519 512	27.531 800	9.117 793	5.247 075	3.766 464
759	576 081	437 245 479	27.549 955	9.121 801	5.248 805	3.767 457
760	577 600	438 976 000	27.568 097	9.125 805	5.250 533	3.768 450
761	579 121	440 711 081	27.586 228	9.129 806	5.252 259	3.769 441
762	580 644	442 450 728	27.604 347	9.133 803	5.253 984	3.770 431
763	582 169	444 194 947	27.622 455	9.137 797	5.255 707	3.771 420
764	583 696	445 943 744	27.640 550	9.141 787	5.257 428	3.772 408
765	585 225	447 697 125	27.658 633	9.145 774	5.259 148	3.773 395
766	586 756	449 455 096	27.676 705	9.149 758	5.260 865	3.774 381
767	588 289	451 217 663	27.694 765	9.153 738	5.262 582	3.775 366
768	589 824	452 984 832	27.712 813	9.157 714	5.264 296	3.776 350
769	591 361	454 756 609	27.730 849	9.161 687	5.266 009	3.777 333
770	592 900	456 533 000	27.748 874	9.165 656	5.267 720	3.778 315
771	594 441	458 314 011	27.766 887	9.169 623	5.269 429	3.779 296
772	595 984	460 099 648	27.784 888	9.173 585	5.271 137	3.780 276
773	597 529	461 889 917	27.802 878	9.177 544	5.272 843	3.781 254
774	599 076	463 684 824	27.820 855	9.181 500	5.274 548	3.782 232
775	600 625	465 484 375	27.838 822	9.185 453	5.276 251	3.783 209
776	602 176	467 288 576	27.856 777	9.189 402	5.277 952	3.784 185
777	603 729	469 097 433	27.874 720	9.193 347	5.279 651	3.785 160
778	605 284	470 910 952	27.892 651	9.197 290	5.281 349	3.786 133
779	606 841	472 729 139	27.910 571	9.201 229	5.283 046	3.787 106
780	608 400	474 552 000	27.928 480	9.205 164	5.284 740	3.788 078
781	609 961	476 379 541	27.946 377	9.209 096	5.286 433	3.789 049
782	611 524	478 211 768	27.964 263	9.213 025	5.288 125	3.790 019
783	613 089	480 048 687	27.982 137	9.216 950	5.289 814	3.790 987
784	614 656	481 890 304	28.000 000	9.220 873	5.291 503	3.791 955
785	616 225	483 736 625	28.017 851	9.224 791	5.293 189	3.792 922
786	617 796	485 587 656	28.035 692	9.228 707	5.294 874	3.793 888
787	619 369	487 443 403	28.053 520	9.232 619	5.296 557	3.794 853
788	620 944	489 303 872	28.071 338	9.236 528	5.298 231	3.795 817
789	622 521	491 169 069	28.089 144	9.240 433	5.299 919	3.796 780
790	624 100	493 039 000	28.106 939	9.244 335	5.301 598	3.797 742
791	625 681	494 913 671	28.124 722	9.248 234	5.303 275	3.798 703
792	627 264	496 793 088	28.142 495	9.252 130	5.304 950	3.799 663
793	628 849	498 677 257	28.160 256	9.256 022	5.306 624	3.800 622
794	630 436	500 566 184	28.178 006	9.259 911	5.308 296	3.801 580
795	632 025	502 459 875	28.195 744	9.263 797	5.309 967	3.802 537
796	633 616	504 358 336	28.213 472	9.267 680	5.311 636	3.803 493
797	635 209	506 261 573	28.231 188	9.271 559	5.313 303	3.804 448
798	636 804	508 169 592	28.248 894	9.275 435	5.314 969	3.805 402
799	638 401	510 082 399	28.266 588	9.279 308	5.316 633	3.806 356
800	640 000	512 000 000	28.284 271	9.283 178	5.318 296	3.807 308

A.19 SQUARES, CUBES, AND ROOTS OF N

N = 801–850

N	N^2	N^3	$\sqrt{N}$	$\sqrt[3]{N}$	$\sqrt[4]{N}$	$\sqrt[5]{N}$
801	641 601	513 922 401	28.301 943	9.287 044	5.319 957	3.808 259
802	643 204	515 849 608	28.319 605	9.290 907	5.321 617	3.809 210
803	644 809	517 781 627	38.337 255	9.294 767	5.323 275	3.810 159
804	646 416	519 718 464	28.354 894	9.298 624	5.324 931	3.811 108
805	648 025	521 660 125	28.372 522	9.302 477	5.326 586	3.812 055
806	649 636	523 606 616	28.390 139	9.306 328	5.328 240	3.813 002
807	651 249	525 557 943	28.407 745	9.310 175	5.329 892	3.813 947
808	652 864	527 514 112	28.425 341	9.314 019	5.331 542	3.814 892
809	654 481	529 475 129	28.442 925	9.317 860	5.333 191	3.815 836
810	656 100	531 441 000	28.460 499	9.321 698	5.334 838	3.816 779
811	657 721	533 411 731	28.478 062	9.325 532	5.336 484	3.817 721
812	659 344	535 387 328	28.495 614	9.329 363	5.338 128	3.818 662
813	660 969	537 367 797	28.513 155	9.333 192	5.339 771	3.819 602
814	662 596	539 353 144	28.530 685	9.337 017	5.341 412	3.820 541
815	664 225	541 343 375	28.548 205	9.340 839	5.343 052	3.821 479
816	665 856	543 338 496	28.565 714	9.344 657	5.344 690	3.822 417
817	667 489	545 338 513	28.583 212	9.348 473	5.346 327	3.823 353
818	669 124	547 343 432	28.600 699	9.352 286	5.347 962	3.824 289
819	670 761	549 353 259	28.618 176	9.356 095	5.349 596	3.825 223
820	672 400	551 368 000	28.635 642	9.359 902	5.351 228	3.826 157
821	674 041	553 387 661	28.653 098	9.363 705	5.352 859	3.827 090
822	675 684	555 412 248	28.670 542	9.367 505	5.354 488	3.828 021
823	677 329	557 441 767	28.687 977	9.371 302	5.356 116	3.828 952
824	678 976	559 476 224	28.705 400	9.375 096	5.357 742	3.829 882
825	680 625	561 515 625	28.722 813	9.378 887	5.359 367	3.830 812
826	682 276	563 559 976	28.740 216	9.382 675	5.360 990	3.831 740
827	683 929	565 609 283	28.757 608	9.386 460	5.362 612	3.832 667
828	685 584	567 663 552	28.774 989	9.390 242	5.364 232	3.833 594
829	687 241	569 722 790	28.792 360	9.394 021	5.365 851	3.834 519
830	688 900	571 787 000	28.809 721	9.397 796	5.367 469	3.835 444
831	690 561	573 856 191	28.827 071	9.401 569	5.369 085	3.836 368
832	692 224	575 930 368	28.844 410	9.405 339	5.370 699	3.837 290
833	693 889	578 009 537	28.861 739	9.409 105	5.372 312	3.838 212
834	695 556	580 093 704	28.879 058	9.412 869	5.373 924	3.839 133
835	697 225	582 182 875	28.896 367	9.416 630	5.375 534	3.840 054
836	698 896	584 277 056	28.913 665	9.420 387	5.377 143	3.840 973
837	700 569	586 376 253	28.930 952	9.424 142	5.378 750	3.841 891
838	702 244	588 480 472	28.948 230	9.427 894	5.380 356	3.842 809
839	703 921	590 589 719	28.965 497	9.431 642	5.381 960	3.843 726
840	705 600	592 704 000	28.982 753	9.435 388	5.383 563	3.844 642
841	707 281	594 823 321	29.000 000	9.439 131	5.385 165	3.845 557
842	708 964	596 947 688	29.017 236	9.442 870	5.386 765	3.846 471
843	710 649	599 077 107	29.034 462	9.446 607	5.388 364	3.847 384
844	712 336	601 211 584	29.051 678	9.450 341	5.389 961	3.848 296
845	714 025	603 351 125	29.068 884	9.454 072	5.391 577	3.849 208
846	715 716	605 495 736	29.086 079	9.457 800	5.393 151	3.850 118
847	717 409	607 645 423	29.103 264	9.461 525	5.394 744	3.851 028
848	719 104	609 800 192	29.120 440	9.465 247	5.396 366	3.851 937
849	720 801	611 960 049	29.137 605	9.468 966	3.397 926	3.852 845
850	722 500	614 125 000	29.154 759	9.472 682	5.399 515	3.853 752

A.20 SQUARES, CUBES, AND ROOTS OF *N*

N = 851–900

N	N^2	N^3	$\sqrt{N}$	$\sqrt[3]{N}$	$\sqrt[4]{N}$	$\sqrt[5]{N}$
851	724 201	616 295 051	29.171 904	9.476 396	5.401 102	3.854 659
852	725 904	618 470 208	29.189 039	9.480 106	5.402 688	3.855 564
853	727 609	620 650 477	29.206 164	9.483 814	5.404 273	3.856 469
854	729 316	622 835 864	29.223 278	9.487 518	5.405 856	3.857 372
855	731 025	625 026 375	29.240 383	9.491 220	5.407 438	3.858 275
856	732 736	627 222 016	29.257 478	9.494 919	5.409 018	3.859 177
857	734 449	629 422 793	29.274 562	9.498 615	5.410 597	3.860 079
858	736 164	631 628 712	29.291 637	9.502 308	5.412 175	3.860 979
859	737 881	633 839 779	29.308 702	9.505 998	5.413 751	3.861 878
860	739 600	636 056 000	29.325 757	9.509 685	5.415 326	3.862 777
861	741 321	638 277 381	29.342 801	9.513 370	5.416 900	3.863 675
862	743 044	640 503 928	29.359 837	9.517 052	5.418 472	3.864 572
863	744 769	642 735 647	29.376 862	9.520 730	5.420 043	3.865 469
864	746 496	644 972 544	29.393 877	9.524 406	5.421 612	3.866 364
865	748 225	647 214 625	29.410 882	9.528 079	5.423 180	3.867 259
866	749 956	649 461 896	29.427 878	9.531 750	5.424 747	3.868 152
867	751 689	651 714 363	29.444 864	9.535 417	5.426 312	3.869 045
868	753 424	653 972 032	29.461 840	9.539 082	5.427 876	3.869 937
869	755 161	656 234 909	29.478 806	9.542 744	5.429 439	3.870 829
870	756 900	658 503 000	29.495 762	9.546 403	5.431 000	3.871 719
871	758 641	660 776 311	29.512 709	9.550 059	5.432 560	3.872 609
872	760 384	663 054 848	29.529 646	9.553 712	5.434 119	3.873 498
873	762 129	665 338 617	29.546 573	9.557 363	5.435 676	3.874 386
874	763 876	667 627 624	29.563 491	9.561 011	5.437 232	3.875 273
875	765 625	669 921 875	29.580 399	9.564 656	5.438 787	3.876 159
876	767 376	672 221 376	29.597 297	9.568 298	5.440 340	3.877 045
877	769 129	674 526 133	29.614 186	9.571 938	5.441 892	3.877 930
878	770 884	676 836 152	29.631 065	9.575 574	5.443 442	3.878 814
879	772 641	679 151 439	29.647 934	9.579 208	5.444 992	3.879 697
880	774 400	681 472 000	29.664 794	9.582 840	5.446 540	3.880 579
881	776 161	683 797 841	29.681 644	9.586 468	5.448 086	3.881 461
882	777 924	686 128 968	29.698 485	9.590 094	5.449 632	3.882 341
883	779 689	688 465 387	29.715 316	9.593 717	5.451 176	3.883 221
884	781 456	690 807 104	29.732 137	9.597 337	5.452 718	3.884 100
885	783 225	693 154 125	29.748 950	9.600 955	5.454 260	3.884 979
886	784 996	695 506 456	29.765 752	9.604 570	5.455 800	3.885 856
887	786 769	697 864 103	29.782 545	9.608 182	5.457 339	3.886 733
888	788 544	700 227 072	29.799 329	9.611 791	5.458 876	3.887 609
889	790 321	702 595 369	29.816 103	9.615 398	5.460 412	3.888 484
890	792 100	704 969 000	29.832 868	9.619 002	5.461 947	3.889 359
891	793 881	707 347 971	29.849 623	9.622 603	5.463 481	3.890 232
892	795 664	709 732 288	29.866 369	9.626 202	5.465 013	3.891 105
893	797 449	712 121 957	29.883 106	9.629 797	5.466 544	3.891 977
894	799 236	714 516 984	29.899 833	9.633 391	5.468 074	3.892 849
895	801 025	716 917 375	29.916 551	9.636 981	5.469 602	3.893 719
896	802 816	719 323 136	29.933 259	9.640 569	5.471 130	3.894 589
897	804 609	721 734 273	29.949 958	9.644 154	5.472 656	3.895 458
898	806 404	724 150 792	29.966 648	9.647 737	5.474 180	3.896 326
899	808 201	726 572 699	29.983 329	9.651 317	5.475 703	3.897 193
900	810 000	729 000 000	30.000 000	9.654 894	5.477 226	3.898 060

A.21 SQUARES, CUBES, AND ROOTS OF N

N	N^2	N^3	$\sqrt{N}$	$\sqrt[3]{N}$	$\sqrt[4]{N}$	$\sqrt[5]{N}$
901	811 801	731 432 701	30.016 662	9.658 468	5.478 746	3.898 926
902	813 604	733 870 808	30.033 315	9.662 040	5.480 266	3.899 791
903	815 409	736 314 327	30.049 958	9.665 610	5.481 784	3.900 655
904	817 216	738 763 264	30.066 593	9.669 176	5.483 301	3.901 519
905	819 025	741 217 625	30.083 218	9.672 740	5.484 817	3.902 381
906	820 836	743 677 416	30.099 834	9.676 302	5.486 332	3.903 243
907	822 649	746 142 643	30.116 441	9.679 860	5.487 845	3.904 105
908	824 464	748 613 312	30.133 038	9.683 417	5.489 357	3.904 965
909	826 281	751 089 429	30.149 627	9.686 970	5.490 868	3.905 825
910	828 100	753 571 000	30.166 206	9.690 521	5.492 377	3.906 684
911	829 921	756 058 031	30.182 777	9.694 069	5.493 885	3.907 542
912	831 744	758 550 528	30.199 338	9.697 615	5.495 392	3.908 400
913	833 569	761 048 497	30.215 890	9.701 158	5.496 898	3.909 256
914	835 396	763 551 944	30.232 433	9.704 699	5.498 403	3.910 112
915	837 225	766 060 875	30.248 967	9.708 237	5.499 906	3.910 968
916	839 056	768 575 296	30.265 492	9.711 772	5.501 408	3.911 822
917	840 889	771 095 213	30.282 008	9.715 305	5.502 909	3.912 676
918	842 724	773 620 632	30.298 515	9.718 835	5.504 409	3.913 529
919	844 561	776 151 559	30.315 013	9.722 363	5.505 907	3.914 381
920	846 400	778 688 000	30.331 502	9.725 888	5.507 404	3.915 233
921	848 241	781 229 961	30.347 982	9.729 411	5.508 900	3.916 083
922	850 084	783 777 448	30.364 453	9.732 931	5.510 395	3.916 933
923	851 929	786 330 467	30.380 915	9.736 448	5.511 889	3.917 783
924	853 776	788 889 024	30.397 368	9.739 963	5.513 381	3.918 631
925	855 625	791 453 125	30.413 813	9.743 476	5.514 872	3.919 479
926	857 476	794 022 776	30.430 248	9.746 986	5.516 362	3.920 326
927	859 329	796 597 983	30.446 675	9.750 493	5.517 851	3.921 172
928	861 184	799 178 752	30.463 092	9.753 998	5.519 338	3.922 018
929	863 041	801 765 089	30.479 501	9.757 500	5.520 824	3.922 863
930	864 900	804 357 000	30.495 901	9.761 000	5.522 309	3.923 707
931	866 761	806 954 491	30.512 293	9.764 497	5.523 793	3.924 551
932	868 624	809 557 568	30.528 675	9.767 992	5.525 276	3.925 393
933	870 489	812 166 237	30.545 049	9.771 485	5.526 758	3.926 235
934	872 356	814 780 504	30.561 414	9.774 974	5.528 238	3.927 077
935	874 225	817 400 375	30.577 770	9.778 462	5.529 717	3.927 917
936	876 096	820 025 856	30.594 117	9.781 946	5.531 195	3.928 757
937	877 969	822 656 953	30.610 456	9.785 429	5.532 672	3.929 596
938	879 844	825 293 672	30.626 786	9.788 909	5.534 147	3.930 435
939	881 721	827 936 019	30.643 107	9.792 386	5.535 622	3.931 272
940	883 600	830 584 000	30.659 419	9.795 861	5.537 095	3.932 109
941	885 481	833 237 621	30.675 723	9.799 334	5.538 567	3.932 945
942	887 364	835 896 888	30.692 019	9.802 804	5.540 038	3.933 781
943	889 249	838 561 807	30.708 305	9.806 271	5.541 507	3.934 616
944	891 136	841 232 384	30.724 583	9.809 736	5.542 976	3.935 450
945	893 025	843 908 625	30.740 852	9.813 199	5.544 443	3.936 283
946	894 916	846 590 536	30.757 113	9.816 659	5.545 910	3.937 116
947	896 809	849 278 123	30.773 365	9.820 117	5.547 375	3.937 948
948	898 704	851 971 392	30.789 609	9.823 572	5.548 838	3.938 779
949	900 601	854 670 349	30.805 844	9.827 025	5.550 301	3.939 610
950	902 500	857 375 000	30.822 070	9.830 476	5.551 763	3.940 440

A.22 SQUARES, CUBES, AND ROOTS OF N

N = 951–1,000

N	N^2	N^3	$\sqrt{N}$	$\sqrt[3]{N}$	$\sqrt[4]{N}$	$\sqrt[5]{N}$
951	904 401	860 085 351	30.838 288	9.833 924	5.553 223	3.941 269
952	906 304	862 801 408	30.854 497	9.837 369	5.554 682	3.942 098
953	908 209	865 523 177	30.870 698	9.840 813	5.556 141	3.942 926
954	910 116	868 250 664	30.886 890	9.844 254	5.557 598	3.943 753
955	912 025	870 983 875	30.903 074	9.847 692	5.559 053	3.944 579
956	913 936	873 722 816	30.919 250	9.851 128	5.560 508	3.945 405
957	915 849	876 467 493	30.935 417	9.854 562	5.561 962	3.946 230
958	917 764	879 217 912	30.951 575	9.857 993	5.563 414	3.947 054
959	919 681	881 974 079	30.967 725	9.861 422	5.564 865	3.947 878
960	921 600	884 736 000	30.983 867	9.864 848	5.566 315	3.948 701
961	923 521	887 503 681	31.000 000	9.868 272	5.567 764	3.949 523
962	925 444	890 277 128	31.016 125	9.871 694	5.569 212	3.950 345
963	927 369	893 056 347	31.032 241	9.875 113	5.570 659	3.951 166
964	929 296	895 841 344	31.048 349	9.878 530	5.572 105	3.951 986
965	931 225	898 632 125	31.064 449	9.881 945	5.573 549	3.952 806
966	933 156	901 428 696	31.080 541	9.885 357	5.574 992	3.953 625
967	935 089	904 231 063	31.096 624	9.888 767	5.576 435	3.954 443
968	937 024	907 039 232	31.112 698	9.892 175	5.577 876	3.955 260
969	938 961	909 853 209	31.128 765	9.895 580	5.579 316	3.956 077
970	940 900	912 673 000	31.144 823	9.898 983	5.580 755	3.956 893
971	942 841	915 498 611	31.160 873	9.902 384	5.582 192	3.957 709
972	944 784	918 330 048	31.176 915	9.905 782	5.583 629	3.958 524
973	946 729	921 167 317	31.192 948	9.909 178	5.585 065	3.959 338
974	948 676	924 010 424	31.208 973	9.912 571	5.586 499	3.960 151
975	950 625	926 859 375	31.224 990	9.915 962	5.587 933	3.960 964
976	952 576	929 714 176	31.240 999	9.919 351	5.589 365	3.961 776
977	954 529	932 574 833	31.256 999	9.922 738	5.590 796	3.962 588
978	956 484	935 441 352	31.272 992	9.926 122	5.592 226	3.963 399
979	958 441	938 313 739	31.288 976	9.929 504	5.593 655	3.964 209
980	960 400	941 192 000	31.304 952	9.932 884	5.595 083	3.965 018
981	962 361	944 076 141	31.320 920	9.936 261	5.596 510	3.965 827
982	964 324	946 966 168	31.336 879	9.939 636	5.597 935	3.966 636
983	966 289	949 862 087	31.352 831	9.943 009	5.599 360	3.967 443
984	968 256	952 763 904	31.368 774	9.946 380	5.600 783	3.968 250
985	970 225	955 671 625	31.384 710	9.949 748	5.602 206	3.969 056
986	972 196	958 585 256	31.400 637	9.953 114	5.603 627	3.969 862
987	974 169	961 504 803	31.416 556	9.956 478	5.605 047	3.970 667
988	976 144	964 430 272	31.432 467	9.959 839	5.606 467	3.971 471
989	978 121	967 361 669	31.448 370	9.963 198	5.607 885	3.972 275
990	980 100	970 299 000	31.464 265	9.966 555	5.609 302	3.973 078
991	982 081	973 242 271	31.480 152	9.969 910	5.610 718	3.973 880
992	984 064	976 191 488	31.496 031	9.973 262	5.612 133	3.974 682
993	986 049	979 146 657	31.511 903	9.976 612	5.613 546	3.975 483
994	988 036	982 107 784	31.527 766	9.979 960	5.614 959	3.976 283
995	990 025	985 074 875	31.543 621	9.983 305	5.616 371	3.977 083
996	992 016	988 047 936	31.559 468	9.986 649	5.617 781	3.977 882
997	994 009	991 026 973	31.575 307	9.989 990	5.619 191	3.978 680
998	996 004	994 011 992	31.591 138	9.993 329	5.620 599	3.979 478
999	998 001	997 002 999	31.606 961	9.966 666	5.622 007	3.980 275
1 000	1 000 000	1 000 000 000	31.622 777	10.000 000	5.623 413	3.981 072

A.23 COMMON LOGARITHMS OF N*

N = 100–599

N	0	1	2	3	4	5	6	7	8	9	$\bar{\Delta}$
10	0000	0043	0086	0128	0170	0212	0253	0294	0334	0374	43–40
11	0414	0453	0492	0531	0569	0607	0645	0682	0719	0755	39–36
12	0792	0828	0864	0899	0934	0969	1004	1038	1072	1106	36–34
13	1139	1173	1206	1239	1271	1303	1335	1367	1399	1430	34–31
14	1461	1492	1523	1553	1584	1614	1644	1673	1703	1732	31–29
15	1761	1790	1818	1847	1875	1903	1931	1959	1987	2014	29–27
16	2041	2068	2095	2122	2148	2175	2201	2227	2253	2279	27–26
17	2304	2330	2355	2380	2405	2430	2455	2480	2504	2529	26–25
18	2553	2577	2601	2625	2648	2672	2695	2718	2742	2765	24–23
19	2788	2810	2833	2856	2878	2900	2923	2945	2967	2989	22–22
20	3010	3032	3054	3075	3096	3118	3139	3160	3181	3201	21
21	3222	3243	3263	3284	3304	3324	3345	3365	3385	3404	20
22	3424	3444	3464	3483	3502	3522	3541	3560	3579	3598	19
23	3617	3636	3655	3674	3692	3711	3729	3747	3766	3784	18
24	3802	3820	3838	3856	3874	3892	3909	3927	3945	3962	17
25	3979	3997	4014	4031	4048	4065	4082	4099	4116	4133	17
26	4150	4166	4183	4200	4216	4232	4249	4265	4281	4298	16
27	4314	4330	4346	4362	4378	4393	4409	4425	4440	4456	16
28	4472	4487	4502	4518	4533	4548	4564	4579	4594	4609	15
29	4624	4639	4654	4669	4683	4698	4713	4728	4742	4757	15
30	4771	4786	4800	4814	4829	4843	4857	4871	4886	4900	14
31	4914	4928	4942	4955	4969	4983	4997	5011	5024	5038	14
32	5051	5065	5079	5092	5105	5119	5132	5145	5159	5172	13
33	5185	5198	5211	5224	5237	5250	5263	5276	5289	5302	13
34	5315	5328	5340	5353	5366	5378	5391	5403	5416	5428	13
35	5441	5453	5465	5478	5490	5502	5514	5527	5539	5551	12
36	5563	5575	5587	5599	5611	5623	5635	5647	5658	5670	12
37	5682	5694	5705	5717	5729	5740	5752	5763	5775	5786	12
38	5793	5809	5821	5832	5843	5855	5866	5877	5888	5899	11
39	5911	5922	5933	5944	5955	5966	5977	5988	5999	6010	11
40	6021	6031	6042	6053	6064	6075	6085	6096	6107	6117	11
41	6128	6138	6149	6160	6170	6180	6191	6201	6212	6222	10
42	6232	6243	6253	6263	6274	6284	6294	6304	6314	6325	10
43	6335	6345	6355	6365	6375	6385	6395	6405	6415	6425	10
44	6435	6444	6454	6464	6474	6484	6493	6503	6513	6522	10
45	6532	6542	6551	6561	6571	6580	6590	6599	6609	6618	10
46	6628	6637	6646	6656	6665	6675	6684	6693	6702	6712	10
47	6721	6730	6739	6749	6758	6767	6776	6785	6794	6803	9
48	6812	6821	6830	6839	6848	6857	6866	6875	6884	6893	9
49	6902	6911	6920	6928	6937	6946	6955	6964	6972	6981	9
50	6990	6998	7007	7016	7024	7033	7042	7050	7059	7067	9
51	7076	7084	7093	7101	7110	7118	7126	7135	7143	7152	8
52	7160	7168	7177	7185	7193	7202	7210	7218	7226	7235	8
53	7243	7251	7259	7267	7275	7284	7292	7300	7308	7316	8
54	7324	7332	7340	7348	7356	7364	7372	7380	7388	7396	8
55	7404	7412	7419	7427	7435	7443	7451	7459	7466	7474	8
56	7482	7490	7497	7505	7513	7520	7528	7536	7543	7551	8
57	7559	7566	7574	7582	7589	7597	7604	7612	7619	7627	8
58	7634	7642	7649	7657	7664	7672	7679	7686	7694	7701	7
59	7709	7716	7723	7731	7738	7745	7752	7760	7767	7774	7
N	0	1	2	3	4	5	6	7	8	9	$\bar{\Delta}$

*For applications of Tables A.23 to A.24 see Sec. 12.05.

A.24 COMMON LOGARITHMS OF N

N = 600–999

N	0	1	2	3	4	5	6	7	8	9	$\bar{\Delta}$
60	7782	7789	7796	7803	7810	7818	7825	7832	7839	7846	7
61	7853	7860	7868	7875	7882	7889	7896	7903	7910	7917	7
62	7924	7931	7938	7945	7952	7959	7966	7973	7980	7987	7
63	7993	8000	8007	8014	8021	8028	8035	8041	8048	8055	7
64	8062	8069	8075	8082	8089	8096	8102	8109	8116	8122	7
65	8129	8136	8142	8149	8156	8162	8169	8176	8182	8189	7
66	8195	8202	8209	8215	8222	8228	8235	8241	8248	8254	7
67	8261	8267	8274	8280	8287	8293	8299	8306	8312	8319	7
68	8325	8331	8338	8344	8351	8357	8363	8370	8376	8382	6
69	8388	8395	8401	8407	8414	8420	8426	8432	8439	8445	6
70	8451	8457	8463	8470	8476	8482	8488	8494	8500	8506	6
71	8513	8519	8525	8531	8537	8543	8549	8555	8561	8567	6
72	8573	8579	8585	8591	8597	8603	8609	8615	8621	8627	6
73	8633	8639	8645	8651	8657	8663	8669	8675	8681	8686	6
74	8692	8698	8704	8710	8716	8722	8727	8733	8739	8745	6
75	8751	8756	8762	8768	8774	8779	8785	8791	8797	8802	6
76	8808	8814	8820	8825	8831	8837	8842	8848	8854	8859	6
77	8865	8871	8876	8882	8887	8893	8899	8904	8910	8915	6
78	8921	8927	8932	8938	8943	8949	8954	8960	8965	8971	6
79	8976	8982	8987	8993	8998	9004	9009	9015	9020	9025	5
80	9031	9036	9042	9047	9053	9058	9063	9069	9074	9079	5
81	9085	9090	9096	9101	9106	9112	9117	9122	9128	9133	5
82	9138	9143	9149	9154	9159	9165	9170	9175	9180	9186	5
83	9191	9196	9201	9206	9212	9217	9222	9227	9232	9238	5
84	9243	9248	9253	9258	9263	9269	9274	9279	9284	9289	5
85	9294	9299	9304	9309	9315	9320	9325	9330	9335	9340	5
85	9345	9350	9355	9360	9365	9370	9375	9380	9385	9390	5
87	9395	9400	9405	9410	9415	9420	9425	9430	9435	9440	5
88	9445	9450	9455	9460	9465	9469	9474	9479	9484	9489	5
89	9494	9499	9504	9509	9513	9518	9523	9528	9533	9538	5
90	9542	9547	9552	9557	9562	9566	9571	9576	9581	9586	5
91	9590	9595	9600	9605	9609	9614	9619	9624	9628	9633	5
92	9638	9643	9647	9652	9657	9661	9666	9671	9675	9680	5
93	9685	9689	9694	9699	9703	9708	9713	9717	9722	9727	5
94	9731	9736	9741	9745	9750	9754	9759	9763	9768	9773	5
95	9777	9782	9786	9791	9795	9800	9805	9809	9814	9818	5
96	9823	9827	9832	9836	9841	9845	9850	9854	9859	9863	4
97	9868	9872	9877	9881	9886	9890	9894	9899	9903	9908	4
98	9912	9917	9921	9926	9930	9934	9939	9943	9948	9952	4
99	9956	9961	9965	9969	9974	9978	9983	9987	9991	9996	4
N	0	1	2	3	4	5	6	7	8	9	$\bar{\Delta}$

$\log \pi = 0.4972$	$\log \pi/2 = 0.1961$	$\log \sqrt{\pi} = 0.2486$
$\log 2\pi = 0.7982$	$\log \pi/3 = 0.0200$	$\log \sqrt{2\pi} = 0.3991$
$\log 3\pi = 0.9743$	$\log \pi^2 = 0.9943$	$\log \sqrt[3]{\pi} = 0.1657$
$\log 4\pi = 1.0992$	$\log \pi^3 = 1.4915$	$\log \sqrt[3]{4\pi/3} = 0.2074$
$\log e = 0.4343$	$\log e^{\pi} = 1.3644$	$\log \sqrt{e} = 0.2171$
$\log 0 = -\infty$	$\log 1 = 0.0000$	$\log 10 = 1.0000$

A.25 NATURAL LOGARITHMS OF N*

N = 1.00–5.99

N	0	1	2	3	4	5	6	7	8	9	$\bar{\Delta}$
1.0	0.0000	0.0100	0.0198	0.0296	0.0392	0.0488	0.0583	0.0677	0.0770	0.0862	100–92
1.1	0.0953	0.1044	0.1133	0.1222	0.1310	0.1398	0.1484	0.1570	0.1655	0.1740	91–85
1.2	0.1823	0.1906	0.1980	0.2070	0.2151	0.2231	0.2311	0.2390	0.2469	0.2546	83–80
1.3	0.2624	0.2700	0.2776	0.2852	0.2927	0.3001	0.3075	0.3148	0.3221	0.3293	76–72
1.4	0.3365	0.3436	0.3507	0.3577	0.3646	0.3716	0.3784	0.3853	0.3920	0.3988	71–68
1.5	0.4055	0.4121	0.4187	0.4253	0.4318	0.4383	0.4447	0.4511	0.4574	0.4637	66–63
1.6	0.4700	0.4762	0.4824	0.4886	0.4947	0.5008	0.5068	0.5128	0.5188	0.5247	62–59
1.7	0.5306	0.5365	0.5423	0.5481	0.5539	0.5596	0.5653	0.5710	0.5766	0.5822	59–56
1.8	0.5878	0.5933	0.5988	0.6043	0.6098	0.6152	0.6206	0.6259	0.6313	0.6366	55–53
1.9	0.6419	0.6471	0.6523	0.6575	0.6627	0.6678	0.6729	0.6780	0.6831	0.6881	52–50
2.0	0.6931	0.6981	0.7031	0.7080	0.7129	0.7178	0.7227	0.7275	0.7324	0.7372	49
2.1	0.7419	0.7467	0.7514	0.7561	0.7608	0.7655	0.7701	0.7747	0.7793	0.7839	47
2.2	0.7885	0.7930	0.7975	0.8020	0.8065	0.8109	0.8154	0.8198	0.8242	0.8286	45
2.3	0.8329	0.8372	0.8416	0.8459	0.8502	0.8544	0.8587	0.8629	0.8671	0.8713	43
2.4	0.8755	0.8796	0.8838	0.8879	0.8920	0.8961	0.9002	0.9042	0.9083	0.9123	41
2.5	0.9163	0.9203	0.9243	0.9282	0.9322	0.9361	0.9400	0.9439	0.9478	0.9517	40
2.6	0.9555	0.9594	0.9632	0.9670	0.9708	0.9746	0.9783	0.9821	0.9858	0.9895	38
2.7	0.9933	0.9969	1.0006	1.0043	1.0080	1.0116	1.0152	1.0188	1.0225	1.0260	36
2.8	1.0296	1.0332	1.0367	1.0403	1.0438	1.0473	1.0508	1.0543	1.0578	1.0613	35
2.9	1.0647	1.0682	1.0716	1.0750	1.0784	1.0818	1.0852	1.0886	1.0919	1.0953	34
3.0	1.0986	1.1019	1.1053	1.1086	1.1119	1.1151	1.1184	1.1217	1.1249	1.1282	33
3.1	1.1314	1.1346	1.1378	1.1410	1.1442	1.1474	1.1506	1.1537	1.1569	1.1600	32
3.2	1.1632	1.1663	1.1694	1.1725	1.1756	1.1787	1.1817	1.1848	1.1878	1.1909	31
3.3	1.1939	1.1969	1.2000	1.2030	1.2060	1.2090	1.2119	1.2149	1.2179	1.2208	30
3.4	1.2238	1.2267	1.2296	1.2326	1.2326	1.2384	1.2413	1.2442	1.2470	1.2499	29
3.5	1.2528	1.2556	1.2585	1.2613	1.2641	1.2669	1.2698	1.2726	1.2754	1.2782	28
3.6	1.2809	1.2837	1.2865	1.2892	1.2920	1.2947	1.2975	1.3002	1.3029	1.3056	28
3.7	1.3083	1.3110	1.3137	1.3164	1.3191	1.3218	1.3244	1.3271	1.3297	1.3324	27
3.8	1.3350	1.3376	1.3403	1.3429	1.3455	1.3481	1.3507	1.3533	1.3558	1.3584	26
3.9	1.3610	1.3635	1.3661	1.3686	1.3712	1.3737	1.3762	1.3788	1.3813	1.3838	25
4.0	1.3863	1.3888	1.3913	1.3938	1.3962	1.3987	1.4012	1.4036	1.4061	1.4085	24
4.1	1.4110	1.4134	1.4159	1.4183	1.4207	1.4231	1.4255	1.4279	1.4303	1.4327	24
4.2	1.4351	1.4375	1.4398	1.4422	1.4446	1.4469	1.4493	1.4516	1.4540	1.4563	24
4.3	1.4586	1.4609	1.4633	1.4656	1.4679	1.4702	1.4725	1.4748	1.4770	1.4793	23
4.4	1.4816	1.4839	1.4861	1.4884	1.4907	1.4929	1.4951	1.4974	1.4996	1.5019	23
4.5	1.5041	1.5063	1.5085	1.5107	1.5129	1.5151	1.5173	1.5195	1.5217	1.5239	22
4.6	1.5261	1.5282	1.5304	1.5326	1.5347	1.5369	1.5390	1.5412	1.5433	1.5454	21
4.7	1.5476	1.5497	1.5518	1.5539	1.5560	1.5581	1.5602	1.5623	1.5644	1.5665	21
4.8	1.5686	1.5707	1.5728	1.5748	1.5769	1.5790	1.5810	1.5831	1.5851	1.5872	21
4.9	1.5892	1.5913	1.5933	1.5953	1.5974	1.5994	1.6014	1.6034	1.6054	1.6074	20
5.0	1.6094	1.6114	1.6134	1.6154	1.6174	1.6194	1.6214	1.6233	1.6253	1.6273	20
5.1	1.6292	1.6312	1.6332	1.6351	1.6371	1.6390	1.6409	1.6429	1.6448	1.6467	20
5.2	1.6487	1.6506	1.6525	1.6544	1.6563	1.6582	1.6601	1.6620	1.6639	1.6658	19
5.3	1.6677	1.6696	1.6715	1.6734	1.6752	1.6771	1.6790	1.6808	1.6827	1.6845	19
5.4	1.6864	1.6882	1.6901	1.6919	1.6938	1.6956	1.6974	1.6993	1.7011	1.7029	18
5.5	1.7047	1.7066	1.7084	1.7102	1.7120	1.7138	1.7156	1.7174	1.7192	1.7210	18
5.6	1.7228	1.7246	1.7263	1.7281	1.7209	1.7317	1.7334	1.7352	1.7370	1.7387	17
5.7	1.7405	1.7422	1.7440	1.7457	1.7475	1.7492	1.7509	1.7527	1.7544	1.7561	17
5.8	1.7579	1.7596	1.7613	1.7630	1.7647	1.7664	1.7681	1.7699	1.7716	1.7733	17
5.9	1.7750	1.7766	1.7783	1.7800	1.7817	1.7834	1.7851	1.7867	1.7884	1.7901	17
N	0	1	2	3	4	5	6	7	8	9	$\bar{\Delta}$

*For applications of Tables A.25 to A.26, see Sec. 12.06.

A.26 NATURAL LOGARITHMS OF N

N = 6.00–9.99

N	0	1	2	3	4	5	6	7	8	9	$\bar{\Delta}$
6.0	1.7918	1.7934	1.7951	1.7967	1.7984	1.8001	1.8017	1.8034	1.8050	1.8066	16
6.1	1.8083	1.8099	1.8116	1.8132	1.8148	1.8165	1.8181	1.8197	1.8213	1.8229	16
6.2	1.8245	1.8262	1.8278	1.8294	1.8310	1.8326	1.8342	1.8358	1.8374	1.8390	15
6.3	1.8405	1.8421	1.8437	1.8453	1.8469	1.8458	1.8500	1.8516	1.8532	1.8547	15
6.4	1.8563	1.8579	1.8594	1.8610	1.8625	1.8641	1.8656	1.8672	1.8687	1.8703	15
6.5	1.8718	1.8733	1.8749	1.8764	1.8779	1.8795	1.8810	1.8825	1.8840	1.8856	15
6.6	1.8871	1.8886	1.8901	1.8916	1.8931	1.8946	1.8961	1.8976	1.8991	1.9006	15
6.7	1.9021	1.9036	1.9051	1.9066	1.9081	1.9095	1.9110	1.9125	1.9140	1.9155	15
6.8	1.9169	1.9184	1.9199	1.9213	1.9228	1.9242	1.9257	1.9272	1.9286	1.9301	15
6.9	1.9315	1.9330	1.9344	1.9359	1.9373	1.9387	1.9402	1.9416	1.9430	1.9415	14
7.0	1.9450	1.9473	1.9488	1.9502	1.9516	1.9530	1.9544	1.9559	1.9573	1.9587	14
7.1	1.9601	1.9615	1.9629	1.9643	1.9657	1.9671	1.9685	1.9699	1.9713	1.9727	14
7.2	1.9741	1.9755	1.9769	1.9782	1.9796	1.9810	1.9824	1.9838	1.9851	1.9865	14
7.3	1.9879	1.9892	1.9906	1.9920	1.9933	1.9947	1.9961	1.9974	1.9988	2.0001	13
7.4	2.0015	2.0028	2.0042	2.0055	2.0069	2.0082	2.0096	2.0109	2.0122	2.0136	13
7.5	2.0149	2.0162	2.0176	2.0189	2.0202	2.0215	2.0229	2.0242	2.0255	2.0268	13
7.6	2.0282	2.0295	2.0308	2.0321	2.0334	2.0347	2.0360	2.0373	2.0386	2.0399	13
7.7	2.0412	2.0425	2.0438	2.0451	2.0464	2.0477	2.0490	2.0503	2.0516	2.0528	13
7.8	2.0541	2.0554	2.0567	2.0580	2.0592	2.0605	2.0618	2.0631	2.0643	2.0656	13
7.9	2.0669	2.0681	2.0694	2.0707	2.0719	2.0732	2.0744	2.0757	2.0769	2.0782	12
8.0	2.0794	2.0807	2.0819	2.0832	2.0844	2.0857	2.0869	2.0882	2.0894	2.0906	12
8.1	2.0919	2.0931	2.0943	2.0956	2.0968	2.0980	2.0992	2.1005	2.1017	2.1029	12
8.2	2.1041	2.1054	2.1066	2.1078	2.1090	2.1102	2.1114	2.1126	2.1138	2.1150	12
8.3	2.1163	2.1175	2.1187	2.1199	2.1211	2.1223	2.1235	2.1247	2.1258	2.1270	12
8.4	2.1282	2.1294	2.1306	2.1318	2.1330	2.1342	2.1353	2.1365	2.1377	2.1389	12
8.5	2.1401	2.1412	2.1424	2.1436	2.1448	2.1459	2.1471	2.1483	2.1494	2.1506	11
8.6	2.1518	2.1529	2.1541	2.1552	2.1564	2.1576	2.1587	2.1599	2.1610	2.1622	11
8.7	2.1638	2.1645	2.1656	2.1668	2.1679	2.1691	2.1702	2.1713	2.1725	2.1736	11
8.8	2.1748	2.1759	2.1770	2.1782	2.1793	2.1804	2.1815	2.1827	2.1838	2.1849	11
8.9	2.1861	2.1872	2.1883	2.1894	2.1905	2.1917	2.1928	2.1939	2.1950	2.1961	11
9.0	2.1972	2.1983	2.1994	2.2006	2.2017	2.2028	2.2039	2.2050	2.2061	2.2072	11
9.1	2.2083	2.2094	2.2105	2.2116	2.2127	2.2138	2.2148	2.2159	2.2170	2.2181	11
9.2	2.2192	2.2203	2.2214	2.2225	2.2235	2.2246	2.2257	2.2268	2.2279	2.2289	11
9.3	2.2300	2.2311	2.2322	2.2332	2.2343	2.2354	2.2364	2.2375	2.2386	2.2396	10
9.4	2.2407	2.2418	2.2428	2.2439	2.2450	2.2460	2.2471	2.2481	2.2492	2.2502	10
9.5	2.2513	2.2523	2.2534	2.2544	2.2555	2.2565	2.2576	2.2586	2.2597	2.2607	10
9.6	2.2618	2.2628	2.2638	2.2649	2.2659	2.2670	2.2680	2.2690	2.2701	2.2711	10
9.7	2.2721	2.2732	2.2742	2.2752	2.2762	2.2773	2.2783	2.2793	2.2803	2.2814	10
9.8	2.2824	2.2834	2.2844	2.2854	2.2865	2.2875	2.2885	2.2895	2.2905	2.2915	10
9.9	2.2925	2.2935	2.2946	2.2956	2.2966	2.2976	2.2986	2.2996	2.3006	2.3016	10
N	0	1	2	3	4	5	6	7	8	9	$\bar{\Delta}$

$\ln 1 = 0.0000$	$\ln \pi = 1.1447$	$\ln \pi/2 = 0.4516$
$\ln 10 = 2.3026$	$\ln 2\pi = 1.8379$	$\ln \pi/3 = 0.0461$
$\ln 10^2 = 4.6052$	$\ln 3\pi = 2.2433$	$\ln \sqrt{\pi} = 0.5724$
$\ln 10^3 = 6.9078$	$\ln 4\pi = 2.5310$	$\ln \sqrt[3]{\pi} = 0.3816$
$\ln 10^m = 2.3026m$	$\ln e = 1.0000$	$\ln \sqrt{e} = 0.5000$

A.27 TRIGONOMETRIC FUNCTIONS*

$\alpha = 0$–$9°, 81$–$90°$

α	Sine		Cosine		Tangent		Cotangent		
	Nat.	Log.	Nat.	Log.	Nat.	Log.	Nat.	Log.	
0°00′	0.0000	∞	1.0000	0.0000	0.0000	∞	∞	∞	90°00′
10	0.0029	7.4637	1.0000	0.0000	0.0029	7.4637	343.77	2.5363	50
20	0.0058	7.7648	1.0000	0.0000	0.0058	7.7648	171.89	2.2352	40
30	0.0087	7.9408	1.0000	0.0000	0.0087	7.9400	114.59	2.0591	30
40	0.0116	8.0658	0.9999	0.0000	0.0116	8.0658	85.940	1.9342	20
50	0.0145	8.1627	0.9999	0.0000	0.0145	8.1627	68.750	1.8373	10
1°00′	0.0175	8.2419	0.9998	9.9999	0.0175	8.2419	57.200	1.7581	89°00′
10	0.0204	8.3088	0.9998	9.9999	0.0204	8.3089	49.104	1.6911	50
20	0.0233	8.3668	0.9997	9.9999	0.0233	8.3669	42.964	1.6331	40
30	0.0262	8.4179	0.9997	9.9999	0.0262	8.4181	38.188	1.5819	30
40	0.0291	8.4637	0.9996	9.9998	0.0291	8.4638	34.368	1.5362	20
50	0.0320	8.5050	0.9995	9.9998	0.0320	8.5053	31.242	1.4947	10
2°00′	0.0349	8.5428	0.9994	9.9997	0.0349	8.5431	28.636	1.4569	88°00′
10	0.0378	8.5776	0.9993	9.9997	0.0378	8.5779	26.432	1.4221	50
20	0.0407	8.6097	0.9992	9.9996	0.0407	8.6101	24.542	1.3809	40
30	0.0436	8.6397	0.9990	9.9996	0.0437	8.6401	22.904	1.3599	30
40	0.0465	8.6677	0.9989	9.9995	0.0466	8.6682	21.470	1.3318	20
50	0.0494	8.6940	0.9988	9.9995	0.0495	8.6945	20.206	1.3055	10
3°00′	0.0523	8.7188	0.9986	9.9994	0.0524	8.7194	19.081	1.2806	87°00′
10	0.0552	8.7423	0.9985	9.9993	0.0553	8.7429	18.075	1.2571	50
20	0.0581	8.7645	0.9983	9.9993	0.0582	8.7652	17.169	1.2348	40
30	0.0610	8.7857	0.9981	9.9992	0.0612	8.7805	16.350	1.2135	30
40	0.0640	8.8059	0.9980	9.9991	0.0641	8.8067	15.605	1.1933	20
50	0.0669	8.8251	0.9978	9.9990	0.0670	8.8261	14.924	1.1739	10
4°00′	0.0698	8.8436	0.9976	9.9989	0.0699	8.8446	14.301	1.1554	86°00′
10	0.0727	8.8613	0.9974	9.9989	0.0729	8.8624	13.727	1.1376	50
20	0.0756	8.8783	0.9971	9.9988	0.0758	8.8705	13.197	1.1205	40
30	0.0785	8.8946	0.9969	9.9987	0.0787	8.8960	12.706	1.1040	30
40	0.0814	8.9104	0.9967	9.9986	0.0816	8.9118	12.251	1.0882	20
50	0.0843	8.9256	0.9964	9.9985	0.0846	8.9272	11.826	1.0728	10
5°00′	0.0872	8.9403	0.9962	9.9983	0.0875	8.9420	11.430	1.0580	85°00′
10	0.0901	8.9545	0.9959	9.9982	0.0904	8.9563	11.059	1.0437	50
20	0.0929	8.9682	0.9957	9.9981	0.0934	8.9701	10.712	1.0299	40
30	0.0958	8.9816	0.9954	9.9980	0.0963	8.9836	10.385	1.0164	30
40	0.0987	8.9945	0.9951	9.9979	0.0992	8.9966	10.078	1.0034	20
50	0.1016	9.0070	0.9948	9.9977	0.1022	9.0093	9.7882	0.9907	10
6°00′	0.1045	9.0192	0.9945	9.9976	0.1051	9.0216	9.5144	0.9784	84°00′
10	0.1074	9.0311	0.9942	9.9975	0.1080	9.0336	9.2553	0.9604	50
20	0.1103	9.0426	0.9939	9.9973	0.1110	9.0453	9.0098	0.9547	40
30	0.1132	9.0539	0.9936	9.9972	0.1139	9.0567	8.7769	0.9433	30
40	0.1161	9.0648	0.9932	9.9971	0.1169	9.0678	8.5555	0.9322	20
50	0.1190	9.0755	0.9929	9.9969	0.1198	9.0780	8.3450	0.9214	10
7°00′	0.1219	9.0859	0.9925	9.9968	0.1228	9.0801	8.1443	0.9109	83°00′
10	0.1248	9.0961	0.9922	9.9966	0.1257	9.0905	7.9550	0.9005	50
20	0.1276	9.1060	0.9918	9.9964	0.1287	9.1006	7.7704	0.8904	40
30	0.1305	9.1157	0.9914	9.9963	0.1317	9.1194	7.5958	0.8806	30
40	0.1334	9.1252	0.9911	9.9961	0.1346	9.1201	7.4287	0.8709	20
50	0.1363	9.1345	0.9907	9.9959	0.1376	9.1385	7.2687	0.8615	10
8°00′	0.1392	9.1436	0.9903	9.9958	0.1405	9.1478	7.1154	0.8522	82°00′
10	0.1421	9.1525	0.9899	9.9956	0.1435	9.1569	6.9682	0.8431	50
20	0.1449	9.1612	0.9894	9.9954	0.1465	9.1658	6.8269	0.8342	40
30	0.1478	9.1697	0.9890	9.9952	0.1495	9.1745	6.6912	0.8255	30
40	0.1507	9.1781	0.9886	9.9950	0.1524	9.1831	6.5006	0.8169	20
50	0.1536	9.1803	0.9881	9.9948	0.1554	9.1915	6.4348	0.8085	10
9°00′	0.1564	9.1943	0.9877	9.9946	0.1584	9.1997	6.3138	0.8003	81°00′
	Nat.	Log.	Nat.	Log.	Nat.	Log.	Nat.	Log.	α
	Cosine		Sine		Cotangent		Tangent		

*Tables A.27–A.31 adapted by permission from Ross R. Middlemiss, "Analytic Geometry," 2d ed., McGraw-Hill Book Company, New York, 1955, pp. 288–292. For applications see Sec. 12.07.

A.28 TRIGONOMETRIC FUNCTIONS

$\alpha = 9–18°, 72–81°$

α	Sine		Cosine		Tangent		Cotangent		
	Nat.	Log.	Nat.	Log.	Nat.	Log.	Nat.	Log.	
9°00′	0.1564	9.1943	0.9877	9.9946	0.1584	9.1997	6.3138	0.8003	81°00′
10	0.1593	9.2022	0.9872	9.9944	0.1614	9.2078	6.1970	0.7922	50
20	0.1622	9.2100	0.9868	9.9942	0.1644	9.2158	6.0844	0.7842	40
30	0.1650	9.2176	0.9863	9.9940	0.1673	9.2236	5.9758	0.7764	30
40	0.1679	9.2251	0.9858	9.9938	0.1703	9.2313	5.8708	0.7687	20
50	0.1708	9.2324	0.9853	9.9936	0.1733	9.2389	5.7694	0.7611	10
10°00′	0.1736	9.2367	0.9848	9.9934	0.1763	9.2463	5.6713	0.7537	80°00′
10	0.1765	9.2468	0.9843	9.9931	0.1793	9.2536	5.5764	0.7464	50
20	0.1794	9.2538	0.9838	9.9929	0.1823	9.2609	5.4845	0.7391	40
30	0.1822	9.2606	0.9833	9.9927	0.1853	9.2680	5.3955	0.7320	30
40	0.1851	9.2674	0.9827	9.9924	0.1883	9.2750	5.3093	0.7250	20
50	0.1880	9.2740	0.9822	9.9922	0.1914	9.2819	5.2257	0.7181	10
11°00′	0.1908	9.2806	0.9816	9.9919	0.1944	9.2887	5.1446	0.7113	79°00′
10	0.1937	9.2870	0.9811	9.9917	0.1974	9.2953	5.0658	0.7047	50
20	0.1965	9.2934	0.9805	9.9914	0.2004	9.3020	4.9894	0.6980	40
30	0.1994	9.2997	0.9799	9.9912	0.2035	9.3085	4.9152	0.6915	30
40	0.2022	9.3058	0.9793	9.9909	0.2065	9.3149	4.8430	0.6851	20
50	0.2051	9.3119	0.9787	9.9907	0.2095	9.3212	4.7729	0.6788	10
12°00′	0.2079	9.3179	0.9781	9.9904	0.2126	9.3275	4.7046	0.6725	78°00′
10	0.2108	9.3238	0.9775	9.9901	0.2156	9.3336	4.6382	0.6664	50
20	0.2136	9.3296	0.9769	9.9899	0.2186	9.3397	4.5736	0.6603	40
30	0.2164	9.3353	0.9763	9.9896	0.2217	9.3458	4.5107	0.6542	30
40	0.2193	9.3410	0.9757	9.9893	0.2247	9.3517	4.4494	0.6483	20
50	0.2221	9.3466	0.9750	9.9890	0.2278	9.3576	4.3897	0.6424	10
13°00′	0.2250	9.3521	0.9744	9.9887	0.2309	9.3634	4.3315	0.6366	77°00′
10	0.2278	9.3575	0.9737	9.9884	0.2339	9.3691	4.2747	0.6309	50
20	0.2306	9.3629	0.9730	9.9881	0.2370	9.3748	4.2193	0.6252	40
30	0.2334	9.3682	0.9724	9.9878	0.2401	9.3804	4.1653	0.6196	30
40	0.2363	9.3734	0.9717	9.9875	0.2432	9.3859	4.1126	0.6141	20
50	0.2391	9.3786	0.9710	9.9872	0.2462	9.3914	4.0611	0.6086	10
14°00′	0.2419	9.3837	0.9703	9.9869	0.2493	9.3968	4.0108	0.6032	76°00′
10	0.2447	9.3887	0.9696	9.9866	0.2524	9.4021	3.9617	0.5979	50
20	0.2476	9.3937	0.9689	9.9863	0.2555	9.4074	3.9136	0.5926	40
30	0.2504	9.3986	0.9681	9.9859	0.2586	9.4127	3.8667	0.5873	30
40	0.2532	9.4035	0.9674	9.9856	0.2617	9.4178	3.8208	0.5822	20
50	0.2560	9.4083	0.9667	9.9853	0.2648	9.4230	3.7760	0.5770	10
15°00′	0.2588	9.4130	0.9659	9.9849	0.2679	9.4281	3.7321	0.5719	75°00′
10	0.2616	9.4177	0.9652	9.9846	0.2711	9.4331	3.6891	0.5669	50
20	0.2644	9.4223	0.9644	9.9843	0.2742	9.4381	3.6470	0.5619	40
30	0.2672	9.4269	0.9636	9.9839	0.2773	9.4430	3.6059	0.5570	30
40	0.2700	9.4314	0.9628	9.9836	0.2805	9.4479	3.5656	0.5521	20
50	0.2728	9.4359	0.9621	9.9832	0.2836	9.4527	3.5261	0.5473	10
16°00′	0.2756	9.4403	0.9613	9.9828	0.2867	9.4575	3.4874	0.5425	74°00′
10	0.2784	9.4447	0.9605	9.9825	0.2899	9.4622	3.4495	0.5378	50
20	0.2812	9.4491	0.9596	9.9821	0.2931	9.4669	3.4124	0.5331	40
30	0.2840	9.4533	0.9588	9.9817	0.2962	9.4716	3.3759	0.5284	30
40	0.2868	9.4576	0.9580	9.9814	0.2994	9.4762	3.3402	0.5238	20
50	0.2896	9.4618	0.9572	9.9810	0.3026	9.4808	3.3052	0.5192	10
17°00′	0.2924	9.4659	0.9563	9.9806	0.3057	9.4853	3.2709	0.5147	73°00′
10	0.2952	9.4700	0.9555	9.9802	0.3089	9.4898	3.2371	0.5102	50
20	0.2979	9.4741	0.9546	9.9798	0.3121	9.4943	3.2041	0.5057	40
30	0.3007	9.4781	0.9537	9.9794	0.3153	9.4987	3.1716	0.5013	30
40	0.3035	9.4821	0.9528	9.9790	0.3185	9.5031	3.1397	0.4969	20
50	0.3062	9.4861	0.9520	9.9786	0.3217	9.5075	3.1084	0.4925	10
18°00′	0.3090	9.4900	0.9511	9.9782	0.3249	9.5118	3.0777	0.4882	72°00′
	Nat.	Log.	Nat.	Log.	Nat.	Log.	Nat.	Log.	α
	Cosine		Sine		Cotangent		Tangent		

A.29 TRIGONOMETRIC FUNCTIONS

α = 18–27°, 63–72°

α	Sine		Cosine		Tangent		Cotangent		
	Nat.	Log.	Nat.	Log.	Nat.	Log.	Nat.	Log.	
18°00′	0.3090	9.4900	0.9511	9.9782	0.3249	9.5118	3.0777	0.4882	72°00′
10	0.3118	9.4939	0.9502	9.9778	0.3281	9.5161	3.0475	0.4839	50
20	0.3145	9.4977	0.9492	9.9774	0.3314	9.5203	3.0178	0.4797	40
30	0.3173	9.5015	0.9483	9.9770	0.3346	9.5245	2.9887	0.4755	30
40	0.3201	9.5052	0.9474	9.9765	0.3378	9.5287	2.9600	0.4713	20
50	0.3228	9.5090	0.9465	9.9761	0.3411	9.5329	2.9319	0.4671	10
19°00′	0.3256	9.5126	0.9455	9.9757	0.3443	9.5370	2.9042	0.4630	71°00′
10	0.3283	9.5163	0.9446	9.9752	0.3476	9.5411	2.8770	0.4589	50
20	0.3311	9.5199	0.9436	9.9748	0.3508	9.5451	2.8502	0.4549	40
30	0.3338	9.5235	0.9426	9.9743	0.3541	9.5491	2.8239	0.4509	30
40	0.3365	9.5270	0.9417	9.9739	0.3574	9.5531	2.7980	0.4469	20
50	0.3393	9.5306	0.9407	9.9734	0.3607	9.5571	2.7725	0.4429	10
20°00′	0.3420	9.5341	0.9397	9.9730	0.3640	9.5611	2.7475	0.4389	70°00′
10	0.3448	9.5375	0.9387	9.9725	0.3673	9.5650	2.7228	0.4350	50
20	0.3475	9.5409	0.9377	9.9721	0.3706	9.5689	2.6985	0.4311	40
30	0.3502	9.5443	0.9367	9.9716	0.3739	9.5727	2.6746	0.4273	30
40	0.3529	9.5477	0.9356	9.9711	0.3772	9.5766	2.6511	0.4234	20
50	0.3557	9.5510	0.9346	9.9706	0.3805	9.5804	2.6279	0.4196	10
21°00′	0.3584	9.5543	0.9336	9.9702	0.3839	9.5842	2.6051	0.4158	69°00′
10	0.3611	9.5576	0.9325	9.9697	0.3872	9.5879	2.5826	0.4121	50
20	0.3638	9.5609	0.9315	9.9692	0.3906	9.5917	2.5605	0.4083	40
30	0.3665	9.5641	0.9304	9.9687	0.3939	9.5954	2.5386	0.4046	30
40	0.3692	9.5673	0.9293	9.9682	0.3973	9.5991	2.5172	0.4009	20
50	0.3719	9.5704	0.9283	9.9677	0.4006	9.6028	2.4960	0.3972	10
22°00′	0.3746	9.5736	0.9272	9.9672	0.4040	9.6064	2.4751	0.3936	68°00′
10	0.3773	9.5767	0.9261	9.9667	0.4074	9.6100	2.4545	0.3900	50
20	0.3800	9.5798	0.9250	9.9661	0.4108	9.6136	2.4342	0.3864	40
30	0.3827	9.5828	0.9239	9.9656	0.4142	9.6172	2.4142	0.3828	30
40	0.3854	9.5859	0.9228	9.9651	0.4176	9.6208	2.3945	0.3792	20
50	0.3881	9.5889	0.9216	9.9646	0.4210	9.6243	2.3750	0.3757	10
23°00′	0.3907	9.5919	0.9205	9.9640	0.4245	9.6279	2.3559	0.3721	67°00′
10	0.3934	9.5948	0.9194	9.9635	0.4279	9.6314	2.3369	0.3686	50
20	0.3961	9.5978	0.9182	9.9629	0.4314	9.6348	2.3183	0.3652	40
30	0.3987	9.6007	0.9171	9.9624	0.4348	9.6383	2.2908	0.3617	30
40	0.4014	9.6036	0.9159	9.9618	0.4383	9.6417	2.2817	0.3583	20
50	0.4041	9.6065	0.9147	9.9613	0.4417	9.6452	2.2637	0.3548	10
24°00′	0.4067	9.6093	0.9135	9.9607	0.4452	9.6486	2.2460	0.3514	66°00′
10	0.4094	9.6121	0.9124	9.9602	0.4487	9.6520	2.2286	0.3480	50
20	0.4120	9.6149	0.9112	9.9596	0.4522	9.6553	2.2113	0.3447	40
30	0.4147	9.6177	0.9100	9.9590	0.4557	9.6587	2.1943	0.3413	30
40	0.4173	9.6205	0.9088	9.9584	0.4592	9.6620	2.1775	0.3380	20
50	0.4200	9.6232	0.9075	9.9579	0.4628	9.6654	2.1609	0.3346	10
25°00′	0.4226	9.6259	0.9063	9.9573	0.4663	9.6687	2.1445	0.3313	65°00′
10	0.4253	9.6286	0.9051	9.9567	0.4699	9.6720	2.1283	0.3280	50
20	0.4279	9.6313	0.9038	9.9561	0.4734	9.6752	2.1123	0.3248	40
30	0.4305	9.6340	0.9026	9.9555	0.4770	9.6785	2.0965	0.3215	30
40	0.4331	9.6366	0.9013	9.9549	0.4806	9.6817	2.0809	0.3183	20
50	0.4358	9.6392	0.9001	9.9543	0.4841	9.6850	2.0655	0.3150	10
26°00′	0.4384	9.6418	0.8988	9.9537	0.4877	9.6882	2.0503	0.3118	64°00′
10	0.4410	9.6444	0.8975	9.9530	0.4913	9.6914	2.0353	0.3086	50
20	0.4436	9.6470	0.8962	9.9524	0.4950	9.6946	2.0204	0.3054	40
30	0.4462	9.6495	0.8949	9.9518	0.4986	9.6977	2.0057	0.3023	30
40	0.4488	9.6521	0.8936	9.9512	0.5022	9.7009	1.9912	0.2991	20
50	0.4514	9.6546	0.8923	9.9505	0.5059	9.7040	1.9708	0.2960	10
27°00′	0.4540	9.6570	0.8910	9.9499	0.5095	9.7072	1.9626	0.2928	63°00′
	Nat.	Log.	Nat.	Log.	Nat.	Log.	Nat.	Log.	α
	Cosine		Sine		Cotangent		Tangent		

A.30 TRIGONOMETRIC FUNCTIONS

$\alpha = 27\text{–}36°, 54\text{–}63°$

α	Sine Nat.	Sine Log.	Cosine Nat.	Cosine Log.	Tangent Nat.	Tangent Log.	Cotangent Nat.	Cotangent Log.	
27°00′	0.4540	9.6570	0.8910	9.9499	0.5095	9.7072	1.9626	0.2928	63°00′
10	0.4566	9.6595	0.8897	9.9492	0.5132	9.7103	1.9486	0.2897	50
20	0.4592	9.6620	0.8884	9.9486	0.5169	9.7134	1.9347	0.2866	40
30	0.4617	9.6644	0.8870	9.9479	0.5206	9.7165	1.9210	0.2835	30
40	0.4643	9.6668	0.8857	9.9473	0.5243	9.7196	1.9074	0.2804	20
50	0.4669	9.6692	0.8843	9.9466	0.5280	9.7226	1.8940	0.2774	10
28°00′	0.4695	9.6716	0.8829	9.9459	0.5317	9.7257	1.8807	0.2743	62°00′
10	0.4720	9.6740	0.8816	9.9453	0.5354	9.7287	1.8676	0.2713	50
20	0.4746	9.6763	0.8802	9.9446	0.5392	9.7317	1.8546	0.2683	40
30	0.4772	9.6787	0.8788	9.9439	0.5430	9.7348	1.8418	0.2652	30
40	0.4797	9.6810	0.8774	9.9432	0.5467	9.7378	1.8201	0.2622	20
50	0.4823	9.6833	0.8760	9.9425	0.5505	9.7408	1.8165	0.2592	10
29°00′	0.4848	9.6856	0.8746	9.9418	0.5543	9.7438	1.8040	0.2562	61°00′
10	0.4874	9.6878	0.8732	9.9411	0.5581	9.7467	1.7917	0.2533	50
20	0.4899	9.6901	0.8718	9.9404	0.5619	9.7497	1.7796	0.2503	40
30	0.4924	9.6923	0.8704	9.9397	0.5658	9.7526	1.7675	0.2474	30
40	0.4950	9.6946	0.8689	9.9390	0.5696	9.7556	1.7556	0.2444	20
50	0.4975	9.6968	0.8675	9.9383	0.5735	9.7585	1.7437	0.2415	10
30°00′	0.5000	9.6990	0.8660	9.9375	0.5774	9.7614	1.7321	0.2386	60°00′
10	0.5025	9.7012	0.8646	9.9368	0.5812	9.7644	1.7295	0.2356	50
20	0.5050	9.7033	0.8631	9.9361	0.5851	9.7673	1.7090	0.2327	40
30	0.5075	9.7055	0.8616	9.9353	0.5890	9.7701	1.6977	0.2299	30
40	0.5100	9.7076	0.8601	9.9346	0.5930	9.7730	1.6864	0.2270	20
50	0.5125	9.7097	0.8587	9.9338	0.5969	9.7759	1.6753	0.2241	10
31°00′	0.5150	9.7118	0.8572	9.9331	0.6009	9.7788	1.6643	0.2212	59°00′
10	0.5175	9.7139	0.8557	9.9323	0.6048	9.7816	1.6534	0.2184	50
20	0.5200	9.7160	0.8542	9.9315	0.6088	9.7845	1.6426	0.2155	40
30	0.5225	9.7181	0.8526	9.9308	0.6128	9.7873	1.6319	0.2127	30
40	0.5250	9.7201	0.8511	9.9300	0.6168	9.7902	1.6212	0.2098	20
50	0.5275	9.7222	0.8496	9.9292	0.6208	9.7930	1.6107	0.2070	10
32°00′	0.5299	9.7242	0.8480	9.9284	0.6249	9.7958	1.6003	0.2042	58°00′
10	0.5324	9.7262	0.8465	9.9276	0.6289	9.7986	1.5900	0.2014	50
20	0.5348	9.7282	0.8450	9.9268	0.6330	9.8014	1.5798	0.1986	40
30	0.5373	9.7302	0.8434	9.9260	0.6371	9.8042	1.5697	0.1958	30
40	0.5398	9.7322	0.8418	9.9252	0.6412	9.8070	1.5597	0.1930	20
50	0.5422	9.7342	0.8403	9.9244	0.6453	9.8097	1.5497	0.1903	10
33°00′	0.5446	9.7361	0.8387	9.9236	0.6494	9.8125	1.5399	0.1875	57°00′
10	0.5471	9.7380	0.8371	9.9228	0.6536	9.8153	1.5301	0.1847	50
20	0.5495	9.7400	0.8355	9.9219	0.6577	9.8180	1.5204	0.1820	40
30	0.5519	9.7419	0.8339	9.9211	0.6619	9.8208	1.5108	0.1792	30
40	0.5544	9.7438	0.8323	9.9203	0.6661	9.8235	1.5013	0.1765	20
50	0.5568	9.7457	0.8307	9.9194	0.6703	9.8263	1.4919	0.1737	10
34°00′	0.5592	9.7476	0.8290	9.9186	0.6745	9.8290	1.4826	0.1710	56°00′
10	0.5616	9.7494	0.8274	9.9177	0.6787	9.8317	1.4733	0.1683	50
20	0.5640	9.7513	0.8258	9.9169	0.6830	9.8344	1.4641	0.1656	40
30	0.5664	9.7531	0.8241	9.9160	0.6873	9.8371	1.4550	0.1629	30
40	0.5688	9.7550	0.8225	9.9151	0.6916	9.8398	1.4460	0.1602	20
50	0.5712	9.7568	0.8208	9.9142	0.6959	9.8425	1.4370	0.1575	10
35°00′	0.5736	9.7586	0.8192	9.9134	0.7002	9.8452	1.4281	0.1548	55°00′
10	0.5760	9.7604	0.8175	9.9125	0.7046	9.8479	1.4193	0.1521	50
20	0.5783	9.7622	0.8158	9.9116	0.7089	9.8506	1.4106	0.1494	40
30	0.5807	9.7640	0.8141	9.9107	0.7133	9.8533	1.4019	0.1467	30
40	0.5831	9.7657	0.8124	9.9098	0.7177	9.8559	1.3934	0.1441	20
50	0.5854	9.7675	0.8107	9.9089	0.7221	9.8586	1.3848	0.1414	10
36°00′	0.5878	9.7692	0.8090	9.9080	0.7265	9.8613	1.3764	1.1387	54°00′
	Nat.	Log.	Nat.	Log.	Nat.	Log.	Nat.	Log.	α
	Cosine		Sine		Cotangent		Tangent		

A.31 TRIGONOMETRIC FUNCTIONS

α	Sine		Cosine		Tangent		Cotangent		
	Nat.	Log.	Nat.	Log.	Nat.	Log.	Nat.	Log.	
36°00′	0.5878	9.7692	0.8090	9.9080	0.7265	9.8613	1.3764	0.1387	54°00′
10	0.5901	9.7710	0.8073	9.9070	0.7310	9.8639	1.3680	0.1361	50
20	0.5925	9.7727	0.8056	9.9061	0.7355	9.8666	1.3597	0.1334	40
30	0.5948	9.7744	0.8039	9.9052	0.7400	9.8692	1.3514	0.1308	30
40	0.5972	9.7761	0.8021	9.9042	0.7445	9.8718	1.3432	0.1282	20
50	0.5995	9.7778	0.8004	9.9033	0.7490	9.8745	1.3351	0.1255	10
37°00′	0.6018	9.7795	0.7986	9.9023	0.7536	9.8771	1.3270	0.1229	53°00′
10	0.6041	9.7811	0.7969	9.9014	0.7581	9.8797	1.3190	0.1203	50
20	0.6065	9.7828	0.7951	9.9004	0.7627	9.8824	1.3111	0.1176	40
30	0.6088	9.7844	0.7934	9.8995	0.7673	9.8850	1.3032	0.1150	30
40	0.6111	9.7861	0.7916	9.8985	0.7720	9.8876	1.2954	0.1124	20
50	0.6134	9.7877	0.7898	9.8975	0.7766	9.8902	1.2876	0.1098	10
38°00′	0.6157	9.7893	0.7880	9.8965	0.7813	9.8928	1.2790	0.1072	52°00′
10	0.6180	9.7910	0.7862	9.8955	0.7860	9.8954	1.2723	0.1046	50
20	0.6202	9.7926	0.7844	9.8945	0.7907	9.8980	1.2647	0.1020	40
30	0.6225	9.7941	0.7826	9.8935	0.7954	9.9006	1.2572	0.0994	30
40	0.6248	9.7957	0.7808	9.8925	0.8002	9.9032	1.2497	0.0968	20
50	0.6271	9.7973	0.7790	9.8915	0.8050	9.9058	1.2423	0.0942	10
39°00′	0.6293	9.7989	0.7771	9.8905	0.8098	9.9084	1.2349	0.0916	51°00′
10	0.6316	9.8004	0.7753	9.8895	0.8146	9.9110	1.2276	0.0890	50
20	0.6338	9.8020	0.7735	9.8884	0.8195	9.9135	1.2203	0.0865	40
30	0.6361	9.8035	0.7716	9.8874	0.8243	9.9161	1.2131	0.0839	30
40	0.6383	9.8050	0.7698	9.8864	0.8292	9.9187	1.2059	0.0812	20
50	0.6406	9.8066	0.7679	9.8853	0.8342	9.9212	1.1988	0.0788	10
40°00′	0.6428	9.8081	0.7660	9.8843	0.8391	9.9238	1.1918	0.0762	50°00′
10	0.6450	9.8096	0.7642	9.8832	0.8441	9.9264	1.1847	0.0736	50
20	0.6472	9.8111	0.7623	9.8821	0.8491	9.9289	1.1778	0.0711	40
30	0.6494	9.8125	0.7604	9.8810	0.8541	9.9315	1.1708	0.0685	30
40	0.6517	9.8140	0.7585	9.8800	0.8591	9.9341	1.1640	0.0659	20
50	0.6539	9.8155	0.7566	9.8789	0.8642	9.9366	1.1571	0.0634	10
41°00′	0.6561	9.8169	0.7547	9.8778	0.8693	9.9392	1.1504	0.0608	49°00′
10	0.6583	9.8184	0.7528	9.8767	0.8744	9.9417	1.1436	0.0583	50
20	0.6604	9.8198	0.7509	9.8756	0.8796	9.9443	1.1369	0.0557	40
30	0.6626	9.8213	0.7490	9.8745	0.8847	9.9468	1.1303	0.0532	30
40	0.6648	9.8227	0.7470	9.8733	0.8899	9.9494	1.1237	0.0506	20
50	0.6670	9.8241	0.7451	9.8722	0.8952	9.9519	1.1171	0.0481	10
42°00′	0.6691	9.8255	0.7431	9.8711	0.9004	9.9544	1.1106	0.0456	48°00′
10	0.6713	9.8269	0.7412	9.8699	0.9057	9.9570	1.1041	0.0430	50
20	0.6734	9.8283	0.7392	9.8688	0.9110	9.9595	1.0977	0.0405	40
30	0.6756	9.8297	0.7373	9.8676	0.9163	9.9621	1.0913	0.0379	30
40	0.6777	9.8311	0.7353	9.8665	0.9217	9.9646	1.0850	0.0354	20
50	0.6799	9.8324	0.7333	9.8653	0.9271	9.9671	1.0786	0.0329	10
43°00′	0.6820	9.8338	0.7314	9.8641	0.9325	9.9697	1.0724	0.0303	47°00′
10	0.6841	9.8351	0.7294	9.8629	0.9380	9.9722	1.0661	0.0278	50
20	0.6862	9.8365	0.7274	9.8618	0.9435	9.9747	1.0599	0.0253	40
30	0.6884	9.8378	0.7254	9.8606	0.9490	9.9772	1.0538	0.0228	30
40	0.6905	9.8391	0.7234	9.8594	0.9545	9.9798	1.0477	0.0202	20
50	0.6926	9.8405	0.7214	9.8582	0.9601	9.9823	1.0416	0.0177	10
44°00′	0.6947	9.8418	0.7193	9.8569	0.9657	9.9848	1.0355	0.0152	46°00′
10	0.6967	9.8431	0.7173	9.8557	0.9713	9.9874	1.0295	0.0126	50
20	0.6988	9.8444	0.7153	9.8545	0.9770	9.9899	1.0235	0.0101	40
30	0.7009	9.8457	0.7133	9.8532	0.9827	9.9924	1.0176	0.0076	30
40	0.7030	9.8469	0.7112	9.8520	0.9884	9.9949	1.0117	0.0051	20
50	0.7050	9.8482	0.7092	9.8507	0.9942	9.9975	1.0058	0.0025	10
45°00′	0.7071	9.8495	0.7071	9.8495	1.0000	0.0000	1.0000	0.0000	45°00′
	Nat.	Log.	Nat.	Log.	Nat.	Log.	Nat.	Log.	α
	Cosine		Sine		Cotangent		Tangent		

A.32 ELEMENTARY FUNCTIONS*

$x = 0.01 - 0.50$

x, rad	$\sin x$	$\cos x$	$\tan x$	e^x	e^{-x}	$\sinh x$	$\cosh x$	$\tanh x$	x, deg
0.01	0.01000	0.99995	0.01000	1.01005	0.99005	0.01000	1.00005	0.01000	0.57
0.02	0.02000	0.99980	0.02000	1.02020	0.98020	0.02000	1.00020	0.02000	1.15
0.03	0.03000	0.99955	0.03001	1.03045	0.97045	0.03000	1.00045	0.02999	1.72
0.04	0.03999	0.99920	0.04002	1.04081	0.96079	0.04001	1.00080	0.03998	2.29
0.05	0.04998	0.99875	0.05004	1.05127	0.95123	0.05002	1.00125	0.04996	2.86
0.06	0.05996	0.99820	0.06007	1.06184	0.94176	0.06004	1.00180	0.05993	3.44
0.07	0.06994	0.99755	0.07011	1.07251	0.93239	0.07006	1.00245	0.06989	4.01
0.08	0.07991	0.99680	0.08017	1.08329	0.92312	0.08009	1.00320	0.07983	4.58
0.09	0.08988	0.99595	0.09024	1.09417	0.91393	0.09012	1.00405	0.08976	5.16
0.10	0.09983	0.99500	0.10033	1.10517	0.90484	0.10017	1.00500	0.09967	5.73
0.11	0.10978	0.99396	0.11045	1.11628	0.89583	0.11022	1.00606	0.10956	6.30
0.12	0.11971	0.99281	0.12058	1.12750	0.88692	0.12029	1.00721	0.11943	6.88
0.13	0.12963	0.99156	0.13074	1.13883	0.87810	0.13037	1.00846	0.12927	7.45
0.14	0.13954	0.99022	0.14092	1.15027	0.86936	0.14046	1.00982	0.13909	8.02
0.15	0.14944	0.98877	0.15114	1.16183	0.86071	0.15056	1.01127	0.14889	8.59
0.16	0.15932	0.98723	0.16138	1.17351	0.85214	0.16068	1.01283	0.15865	9.17
0.17	0.16918	0.98558	0.17166	1.18530	0.84366	0.17082	1.01448	0.16838	9.74
0.18	0.17903	0.98384	0.18197	1.19722	0.83527	0.18097	1.01624	0.17808	10.31
0.19	0.18886	0.98200	0.19232	1.20925	0.82696	0.19115	1.01810	0.18775	10.89
0.20	0.19867	0.98007	0.20271	1.22140	0.81873	0.20134	1.02007	0.19738	11.46
0.21	0.20846	0.97803	0.21314	1.23368	0.81058	0.21155	1.02213	0.20697	12.03
0.22	0.21823	0.97590	0.22362	1.24608	0.80252	0.22178	1.02430	0.21652	12.61
0.23	0.22798	0.97367	0.23414	1.25860	0.79453	0.23203	1.02657	0.22603	13.18
0.24	0.23770	0.97134	0.24472	1.27125	0.78663	0.24231	1.02894	0.23550	13.75
0.25	0.24740	0.96891	0.25534	1.28403	0.77880	0.25261	1.03141	0.24492	14.32
0.26	0.25708	0.96639	0.26602	1.29693	0.77105	0.26294	1.03399	0.25430	14.90
0.27	0.26673	0.96377	0.27676	1.30996	0.76338	0.27329	1.03667	0.26362	15.47
0.28	0.27636	0.96106	0.28755	1.32313	0.75578	0.28367	1.03946	0.27291	16.04
0.29	0.28595	0.95824	0.29841	1.33643	0.74826	0.29408	1.04235	0.28213	16.62
0.30	0.29552	0.95534	0.30934	1.34986	0.74082	0.30452	1.04534	0.29131	17.19
0.31	0.30506	0.95233	0.32033	1.36343	0.73345	0.31499	1.04844	0.30044	17.76
0.32	0.31457	0.94924	0.33139	1.37713	0.72615	0.32549	1.05164	0.30951	18.33
0.33	0.32404	0.94604	0.34252	1.39097	0.71892	0.33602	1.05495	0.31852	18.91
0.34	0.33349	0.94275	0.35374	1.40495	0.71177	0.34659	1.05836	0.32748	19.48
0.35	0.34290	0.93937	0.36503	1.41907	0.70469	0.35719	1.06188	0.33638	20.05
0.36	0.35227	0.93590	0.37640	1.43333	0.69768	0.36783	1.06550	0.34521	20.63
0.37	0.36162	0.93233	0.38786	1.44773	0.69073	0.37850	1.06923	0.35399	21.20
0.38	0.37092	0.92866	0.39941	1.46228	0.68386	0.38921	1.07307	0.36271	21.77
0.39	0.38019	0.92491	0.41105	1.47698	0.67706	0.39996	1.07702	0.37136	22.35
0.40	0.38942	0.92106	0.42279	1.49182	0.67032	0.41075	1.08107	0.37995	22.92
0.41	0.39861	0.91712	0.43463	1.50682	0.66365	0.42158	1.08523	0.38847	23.49
0.42	0.40776	0.91309	0.44657	1.52196	0.65705	0.43246	1.08950	0.39693	24.06
0.43	0.41687	0.90897	0.45862	1.53726	0.65051	0.44337	1.09388	0.40532	24.64
0.44	0.42594	0.90475	0.47078	1.55271	0.64404	0.45434	1.09837	0.41364	25.21
0.45	0.43497	0.90045	0.48306	1.56831	0.63763	0.46534	1.10297	0.42190	25.78
0.46	0.44395	0.89605	0.49545	1.58407	0.63128	0.47640	1.10768	0.43008	26.36
0.47	0.45289	0.89157	0.50797	1.59999	0.62500	0.48750	1.11250	0.43820	26.93
0.48	0.46178	0.88699	0.52061	1.61607	0.61878	0.49865	1.11743	0.44624	27.50
0.49	0.47063	0.88233	0.53339	1.63232	0.61263	0.50984	1.12247	0.45422	28.07
0.50	0.47943	0.87758	0.54630	1.64872	0.60653	0.52110	1.12763	0.46212	28.65

*For applications of Tables A.32 to A.38 see Sec. 12.08.

x, rad	$\sin x$	$\cos x$	$\tan x$	e^x	e^{-x}	$\sinh x$	$\cosh x$	$\tanh x$	x, deg
0.51	0.48818	0.87274	0.55936	1.66529	0.60050	0.53240	1.13289	0.46995	29.22
0.52	0.49688	0.86782	0.57256	1.68203	0.59452	0.54375	1.13827	0.47770	29.79
0.53	0.50553	0.86281	0.58592	1.69893	0.58860	0.55516	1.14377	0.48538	30.37
0.54	0.51414	0.85771	0.59943	1.71601	0.58275	0.56663	1.14938	0.49299	30.94
0.55	0.52269	0.85252	0.61311	1.73325	0.57695	0.57815	1.15510	0.50052	31.51
0.56	0.53119	0.84726	0.62695	1.75067	0.57121	0.58973	1.16094	0.50798	32.09
0.57	0.53963	0.84190	0.64097	1.76827	0.56553	0.60137	1.16690	0.51536	32.66
0.58	0.54802	0.83646	0.65517	1.78604	0.55990	0.61307	1.17297	0.52267	33.23
0.59	0.55636	0.83094	0.66956	1.80399	0.55433	0.62483	1.17916	0.52990	33.80
0.60	0.56464	0.82534	0.68414	1.82212	0.54881	0.63665	1.18547	0.53705	34.38
0.61	0.57287	0.81965	0.69892	1.84043	0.54335	0.64854	1.19189	0.54413	34.95
0.62	0.58104	0.81388	0.71391	1.85893	0.53794	0.66049	1.19844	0.55113	35.52
0.63	0.58914	0.80803	0.72911	1.87761	0.53259	0.67251	1.20510	0.55805	36.10
0.64	0.59720	0.80210	0.74454	1.89648	0.52729	0.68459	1.21189	0.56490	36.67
0.65	0.60519	0.79608	0.76020	1.91554	0.52205	0.69675	1.21879	0.57167	37.24
0.66	0.61312	0.78999	0.77610	1.93479	0.51685	0.70897	1.22582	0.57836	37.82
0.67	0.62099	0.78382	0.79225	1.95424	0.51171	0.72126	1.23297	0.58498	38.39
0.68	0.62879	0.77757	0.80866	1.97388	0.50662	0.73363	1.24025	0.59152	38.96
0.69	0.63654	0.77125	0.82534	1.99372	0.50158	0.74607	1.24765	0.59798	39.53
0.70	0.64422	0.76484	0.84229	2.01375	0.49659	0.75858	1.25517	0.60437	40.11
0.71	0.65183	0.75836	0.85953	2.03399	0.49164	0.77117	1.26282	0.61068	40.68
0.72	0.65938	0.75181	0.87707	2.05443	0.48675	0.78384	1.27059	0.61691	41.25
0.73	0.66687	0.74517	0.89492	2.07508	0.48191	0.79659	1.27849	0.62307	41.83
0.74	0.67429	0.73847	0.91309	2.09594	0.47711	0.80941	1.28652	0.62915	42.40
0.75	0.68164	0.73169	0.93160	2.11700	0.47237	0.82232	1.29468	0.63515	42.97
0.76	0.68892	0.72484	0.95045	2.13828	0.46767	0.83530	1.30297	0.64108	43.54
0.77	0.69614	0.71791	0.96967	2.15977	0.46301	0.84838	1.31139	0.64693	44.12
0.78	0.70328	0.71091	0.98926	2.18147	0.45841	0.86153	1.31994	0.65271	44.69
0.79	0.71035	0.70385	1.00925	2.20340	0.45384	0.87478	1.32862	0.65841	45.26
0.80	0.71736	0.69671	1.02964	2.22554	0.44933	0.88811	1.33743	0.66404	45.84
0.81	0.72429	0.68950	1.05046	2.24791	0.44486	0.90152	1.34638	0.66959	46.41
0.82	0.73115	0.68222	1.07171	2.27050	0.44043	0.91503	1.35547	0.67507	46.98
0.83	0.73793	0.67488	1.09343	2.29332	0.43605	0.92863	1.36468	0.68048	47.56
0.84	0.74464	0.66746	1.11563	2.31637	0.43171	0.94233	1.37404	0.68581	48.13
0.85	0.75128	0.65998	1.13833	2.33965	0.42741	0.95612	1.38353	0.69107	48.70
0.86	0.75784	0.65244	1.16156	2.36316	0.42316	0.97000	1.39316	0.69626	49.27
0.87	0.76433	0.64483	1.18532	2.38691	0.41895	0.98398	1.40293	0.70137	49.85
0.88	0.77074	0.63715	1.20966	2.41090	0.41478	0.99806	1.41284	0.70642	50.42
0.89	0.77707	0.62941	1.23460	2.43513	0.41066	1.01224	1.42289	0.71139	50.99
0.90	0.78333	0.62161	1.26016	2.45960	0.40657	1.02652	1.43309	0.71630	51.57
0.91	0.78950	0.61375	1.28637	2.48432	0.40252	1.04090	1.44342	0.72113	52.14
0.92	0.79560	0.60582	1.31326	2.50929	0.39852	1.05539	1.45390	0.72590	52.71
0.93	0.80162	0.59783	1.34087	2.53451	0.39455	1.06998	1.46453	0.73059	53.29
0.94	0.80756	0.58979	1.36923	2.55998	0.39063	1.08468	1.47530	0.73522	53.86
0.95	0.81342	0.58168	1.39838	2.58571	0.38674	1.09948	1.48623	0.73978	54.43
0.96	0.81919	0.57352	1.42836	2.61170	0.38289	1.11440	1.49729	0.74428	55.00
0.97	0.82489	0.56530	1.45920	2.63794	0.37908	1.12943	1.50851	0.74870	55.58
0.98	0.83050	0.55702	1.49096	2.66446	0.37531	1.14457	1.51988	0.75307	56.15
0.99	0.83603	0.54869	1.52368	2.69123	0.37158	1.15983	1.53141	0.75736	56.72
1.00	0.84147	0.54030	1.55741	2.71828	0.36788	1.17520	1.54308	0.76159	57.30

A.34 ELEMENTARY FUNCTIONS

x, rad	$\sin x$	$\cos x$	$\tan x$	e^x	e^{-x}	$\sinh x$	$\cosh x$	$\tanh x$	x, deg
1.01	0.84683	0.53186	1.59221	2.74560	0.36422	1.19069	1.55491	0.76576	57.87
1.02	0.85211	0.52337	1.62813	2.77319	0.36059	1.20630	1.56689	0.76987	58.44
1.03	0.85730	0.51482	1.66524	2.80107	0.35701	1.22203	1.57904	0.77391	59.01
1.04	0.86240	0.50622	1.70361	2.82922	0.35345	1.23788	1.59134	0.77789	59.59
1.05	0.86742	0.49757	1.74332	2.85765	0.34994	1.25386	1.60379	0.78181	60.16
1.06	0.87236	0.48887	1.78442	2.88637	0.34646	1.26996	1.61641	0.78566	60.73
1.07	0.87720	0.48012	1.82703	2.91538	0.34301	1.28619	1.62919	0.78946	61.31
1.08	0.88196	0.47133	1.87122	2.94468	0.33960	1.30254	1.64214	0.79320	61.88
1.09	0.88663	0.46249	1.91709	2.97427	0.33622	1.31903	1.65525	0.79688	62.45
1.10	0.89121	0.45360	1.96476	3.00417	0.33287	1.33565	1.66852	0.80050	63.03
1.11	0.89570	0.44466	2.01434	3.03436	0.32956	1.35240	1.68196	0.80406	63.60
1.12	0.90010	0.43568	2.06596	3.06485	0.32628	1.36929	1.69557	0.80757	64.17
1.13	0.90441	0.42666	2.11975	3.09566	0.32303	1.38631	1.70934	0.81102	64.74
1.14	0.90863	0.41759	2.17588	3.12677	0.31982	1.40347	1.72329	0.81441	65.32
1.15	0.91276	0.40849	2.23450	3.15819	0.31664	1.42078	1.73741	0.81775	65.89
1.16	0.91680	0.39934	2.29580	3.18993	0.31349	1.43822	1.75171	0.82104	66.46
1.17	0.92075	0.39015	2.35998	3.22199	0.31037	1.45581	1.76618	0.82427	67.04
1.18	0.92461	0.38092	2.42727	3.25437	0.30728	1.47355	1.78083	0.82745	67.61
1.19	0.92837	0.37166	2.49790	3.28708	0.30422	1.49143	1.79565	0.83058	68.18
1.20	0.93204	0.36236	2.57215	3.32012	0.30119	1.50946	1.81066	0.83365	68.75
1.21	0.93562	0.35302	2.65032	3.35348	0.29820	1.52764	1.82584	0.83668	69.33
1.22	0.93910	0.34365	2.73275	3.38719	0.29523	1.54598	1.84121	0.83965	69.90
1.23	0.94249	0.33424	2.81982	3.42123	0.29229	1.56447	1.85676	0.84258	70.47
1.24	0.94578	0.32480	2.91193	3.45561	0.28938	1.58311	1.87250	0.84546	71.05
1.25	0.94898	0.31532	3.00957	3.49034	0.28650	1.60192	1.88842	0.84828	71.62
1.26	0.95209	0.30582	3.11327	3.52542	0.28365	1.62088	1.90454	0.85106	72.19
1.27	0.95510	0.29628	3.22363	3.56085	0.28083	1.64001	1.92084	0.85380	72.77
1.28	0.95802	0.28672	3.34135	3.59664	0.27804	1.65930	1.93734	0.85648	73.34
1.29	0.96084	0.27712	3.46721	3.63279	0.27527	1.67876	1.95403	0.85913	73.91
1.30	0.96356	0.26750	3.60210	3.66930	0.27253	1.69838	1.97091	0.86172	74.48
1.31	0.96618	0.25785	3.74708	3.70617	0.26982	1.71818	1.98800	0.86428	75.06
1.32	0.96872	0.24818	3.90335	3.74342	0.26714	1.73814	2.00528	0.86678	75.63
1.33	0.97115	0.23848	4.07231	3.78104	0.26448	1.75828	2.02276	0.86925	76.20
1.34	0.97348	0.22875	4.25562	3.81904	0.26185	1.77860	2.04044	0.87167	76.78
1.35	0.97572	0.21901	4.45522	3.85743	0.25924	1.79909	2.05833	0.87405	77.35
1.36	0.97786	0.20924	4.67344	3.89619	0.25666	1.81977	2.07643	0.87639	77.92
1.37	0.97991	0.19945	4.91306	3.93535	0.25411	1.84062	2.09473	0.87869	78.50
1.38	0.98185	0.18964	5.17744	3.97490	0.25158	1.86166	2.11324	0.88095	79.07
1.39	0.98370	0.17981	5.47069	4.01485	0.24908	1.88289	2.13196	0.88317	79.64
1.40	0.98545	0.16997	5.79788	4.05520	0.24660	1.90430	2.15090	0.88535	80.21
1.41	0.98710	0.16010	6.16536	4.09596	0.24414	1.92591	2.17005	0.88749	80.79
1.42	0.98865	0.15023	6.58112	4.13712	0.24171	1.94770	2.18942	0.88960	81.36
1.43	0.99010	0.14033	7.05546	4.17870	0.23931	1.96970	2.20900	0.89167	81.93
1.44	0.99146	0.13042	7.60183	4.22070	0.23693	1.99188	2.22881	0.89370	82.51
1.45	0.99271	0.12050	8.23809	4.26311	0.23457	2.01427	2.24884	0.89569	83.08
1.46	0.99387	0.11057	8.98861	4.30596	0.23224	2.03686	2.26910	0.89765	83.65
1.47	0.99492	0.10063	9.88737	4.34924	0.22993	2.05965	2.28958	0.89958	84.22
1.48	0.99588	0.09067	10.98338	4.39295	0.22764	2.08265	2.31029	0.90147	84.80
1.49	0.99674	0.08071	12.34986	4.43710	0.22537	2.10586	2.33123	0.90332	85.37
1.50	0.99749	0.07074	14.10142	4.48169	0.22313	2.12928	2.35241	0.90515	85.94

A.35 ELEMENTARY FUNCTIONS

$x = 1.51 - 2.00$

x, rad	$\sin x$	$\cos x$	$\tan x$	e^x	e^{-x}	$\sinh x$	$\cosh x$	$\tanh x$	x, deg
1.51	0.99815	0.06076	16.42809	4.52673	0.22091	2.15291	2.37382	0.90694	86.52
1.52	0.99871	0.05077	19.66953	4.57223	0.21871	2.17676	2.39547	0.90870	87.09
1.53	0.99917	0.04079	24.49841	4.61818	0.21654	2.20082	2.41736	0.91042	87.66
1.54	0.99953	0.03079	32.46114	4.66459	0.21438	2.22510	2.43949	0.91212	88.24
1.55	0.99978	0.02079	48.07848	4.71147	0.21225	2.24961	2.46186	0.91379	88.81
1.56	0.99994	0.01080	92.62050	4.75882	0.21014	2.27434	2.48448	0.91542	89.38
1.57	1.00000	0.00080	1255.76559	4.80665	0.20805	2.29930	2.50735	0.91703	89.95
1.58	0.99996	−0.00920	−108.64920	4.85496	0.20598	2.32449	2.53047	0.91860	90.53
1.59	0.99982	−0.01920	−52.06697	4.90375	0.20393	2.34991	2.55384	0.92015	91.10
1.60	0.99957	−0.02920	−34.23253	4.95303	0.20190	2.37557	2.57746	0.92167	91.67
1.61	0.99923	−0.03919	−25.49474	5.00281	0.19989	2.40146	2.60135	0.92316	92.25
1.62	0.99879	−0.04918	−20.30728	5.05309	0.19790	2.42760	2.62549	0.92462	92.82
1.63	0.99825	−0.05917	−16.87110	5.10387	0.19593	2.45397	2.64990	0.92606	93.39
1.64	0.99761	−0.06915	−14.42702	5.15517	0.19398	2.48059	2.67457	0.92747	93.97
1.65	0.99687	−0.07912	−12.59926	5.20698	0.19205	2.50746	2.69951	0.92886	94.54
1.66	0.99602	−0.08909	−11.18055	5.25931	0.19014	2.53459	2.72472	0.93022	95.11
1.67	0.99508	−0.09904	−10.04718	5.31217	0.18825	2.56196	2.75021	0.93155	95.68
1.68	0.99404	−0.10899	−9.12077	5.36556	0.18637	2.58959	2.77596	0.93286	96.26
1.69	0.99290	−0.11892	−8.34923	5.41948	0.18452	2.61748	2.80200	0.93415	96.83
1.70	0.99166	−0.12884	−7.69660	5.47395	0.18268	2.64563	2.82832	0.93541	97.40
1.71	0.99033	−0.13875	−7.13726	5.52896	0.18087	2.67405	2.85491	0.93665	97.98
1.72	0.98889	−0.14865	−6.65244	5.58453	0.17907	2.70273	2.88180	0.93786	98.55
1.73	0.98735	−0.15853	−6.22810	5.64065	0.17728	2.73168	2.90897	0.93906	99.12
1.74	0.98572	−0.16840	−5.85353	5.69734	0.17552	2.76091	2.93643	0.94023	99.69
1.75	0.98399	−0.17825	−5.52038	5.75460	0.17377	2.79041	2.96419	0.94138	100.27
1.76	0.98215	−0.18808	−5.22209	5.81244	0.17204	2.82020	2.99224	0.94250	100.84
1.77	0.98022	−0.19789	−4.95341	5.87085	0.17033	2.85026	3.02059	0.94361	101.41
1.78	0.97820	−0.20768	−4.71009	5.92986	0.16864	2.88061	3.04925	0.94470	101.99
1.79	0.97607	−0.21745	−4.48866	5.98945	0.16696	2.91125	3.07821	0.94576	102.56
1.80	0.97385	−0.22720	−4.28626	6.04965	0.16530	2.94217	3.10747	0.94681	103.13
1.81	0.97153	−0.23693	−4.10050	6.11045	0.16365	2.97340	3.13705	0.94783	103.71
1.82	0.96911	−0.24663	−3.92937	6.17186	0.16203	3.00492	3.16694	0.94884	104.28
1.83	0.96659	−0.25631	−3.77118	6.23389	0.16041	3.03674	3.19715	0.94983	104.85
1.84	0.96398	−0.26596	−3.62449	6.29654	0.15882	3.06886	3.22768	0.95080	105.42
1.85	0.96128	−0.27559	−3.48806	6.35982	0.15724	3.10129	3.25853	0.95175	106.00
1.86	0.95847	−0.28519	−3.36083	6.42374	0.15567	3.13403	3.28970	0.95268	106.57
1.87	0.95557	−0.29476	−3.24187	6.48830	0.15412	3.16709	3.32121	0.95359	107.14
1.88	0.95258	−0.30430	−3.13038	6.55350	0.15259	3.20046	3.35305	0.95449	107.72
1.89	0.94949	−0.31381	−3.02566	6.61937	0.15107	3.23415	3.38522	0.95537	108.29
1.90	0.94630	−0.32329	−2.92710	6.68589	0.14957	3.26816	3.41773	0.95624	108.86
1.91	0.94302	−0.33274	−2.83414	6.75309	0.14808	3.30250	3.45058	0.95709	109.43
1.92	0.93965	−0.34215	−2.74630	6.82096	0.14661	3.33718	3.48378	0.95792	110.01
1.93	0.93618	−0.35153	−2.66316	6.88951	0.14515	3.37218	3.51733	0.95873	110.58
1.94	0.93262	−0.36087	−2.58433	6.95875	0.14370	3.40752	3.55123	0.95953	111.15
1.95	0.92896	−0.37018	−2.50948	7.02869	0.14227	3.44321	3.58548	0.96032	111.73
1.96	0.92521	−0.37945	−2.43828	7.09933	0.14086	3.47923	3.62009	0.96109	112.30
1.97	0.92137	−0.38868	−2.37048	7.17068	0.13946	3.51561	3.65507	0.96185	112.87
1.98	0.91744	−0.39788	−2.30582	7.24274	0.13807	3.55234	3.69041	0.96259	113.45
1.99	0.91341	−0.40703	−2.24408	7.31553	0.13670	3.58942	3.72611	0.96331	114.02
2.00	0.90930	−0.41615	−2.18504	7.38906	0.13534	3.62686	3.76220	0.96403	114.59

x, rad	sin x	cos x	tan x	e^x	e^{-x}	sinh x	cosh x	tanh x	x, deg
2.01	0.90509	−0.42522	−2.12853	7.46332	0.13399	3.66466	3.79865	0.96473	115.16
2.02	0.90079	−0.43425	−2.07437	7.53832	0.13266	3.70283	3.83549	0.96541	115.74
2.03	0.89641	−0.44323	−2.02242	7.61409	0.13134	3.74138	3.87271	0.96609	116.31
2.04	0.89193	−0.45218	−1.97252	7.69061	0.13003	3.78029	3.91032	0.96675	116.88
2.05	0.88736	−0.46107	−1.92456	7.76790	0.12873	3.81958	3.94832	0.96740	117.46
2.06	0.88271	−0.46992	−1.87841	7.84597	0.12745	3.85926	3.98671	0.96803	118.03
2.07	0.87796	−0.47873	−1.83396	7.92482	0.12619	3.89932	4.02550	0.96865	118.60
2.08	0.87313	−0.48748	−1.79111	8.00447	0.12493	3.93977	4.06470	0.96926	119.18
2.09	0.86821	−0.49619	−1.74977	8.08492	0.12369	3.98061	4.10430	0.96986	119.75
2.10	0.86321	−0.50485	−1.70985	8.16617	0.12246	4.02186	4.14431	0.97045	120.32
2.11	0.85812	−0.51345	−1.67127	8.24824	0.12124	4.06350	4.18474	0.97103	120.89
2.12	0.85294	−0.52201	−1.63396	8.33114	0.12003	4.10555	4.22558	0.97159	121.47
2.13	0.84768	−0.53051	−1.59785	8.41487	0.11884	4.14801	4.26685	0.97215	122.04
2.14	0.84233	−0.53896	−1.56288	8.49944	0.11765	4.19089	4.30855	0.97269	122.61
2.15	0.83690	−0.54736	−1.52898	8.58486	0.11648	4.23419	4.35067	0.97323	123.19
2.16	0.83138	−0.55570	−1.49610	8.67114	0.11533	4.27791	4.39323	0.97375	123.76
2.17	0.82578	−0.56399	−1.46420	8.75828	0.11418	4.32205	4.43623	0.97426	124.33
2.18	0.82010	−0.57221	−1.43321	8.84631	0.11304	4.36663	4.47967	0.97477	124.90
2.19	0.81434	−0.58039	−1.40310	8.93521	0.11192	4.41165	4.52356	0.97526	125.48
2.20	0.80850	−0.58850	−1.37382	9.02501	0.11080	4.45711	4.56791	0.97574	126.05
2.21	0.80257	−0.59656	−1.34534	9.11572	0.10970	4.50301	4.61271	0.97622	126.62
2.22	0.79657	−0.60455	−1.31761	9.20733	0.10861	4.54936	4.65797	0.97668	127.20
2.23	0.79048	−0.61249	−1.29061	9.29987	0.10753	4.59617	4.70370	0.97714	127.77
2.24	0.78432	−0.62036	−1.26429	9.39333	0.10646	4.64344	4.74989	0.97759	128.34
2.25	0.77807	−0.62817	−1.23863	9.48774	0.10540	4.69117	4.79657	0.97803	128.92
2.26	0.77175	−0.63592	−1.21359	9.58309	0.10435	4.73937	4.84372	0.97846	129.49
2.27	0.76535	−0.64361	−1.18916	9.67940	0.10331	4.78804	4.89136	0.97888	130.06
2.28	0.75888	−0.65123	−1.16530	9.77668	0.10228	4.83720	4.93948	0.97929	130.63
2.29	0.75233	−0.65879	−1.14200	9.87494	0.10127	4.88684	4.98810	0.97970	131.21
2.30	0.74571	−0.66628	−1.11921	9.97418	0.10026	4.93696	5.03722	0.98010	131.78
2.31	0.73901	−0.67370	−1.09694	10.07442	0.09926	4.98758	5.08684	0.98049	132.35
2.32	0.73223	−0.68106	−1.07514	10.17567	0.09827	5.03870	5.13697	0.98087	132.93
2.33	0.72538	−0.68834	−1.05381	10.27794	0.09730	5.09032	5.18762	0.98124	133.50
2.34	0.78146	−0.69556	−1.03293	10.38124	0.09633	5.14245	5.23878	0.98161	134.07
2.35	0.71147	−0.70271	−1.01247	10.48557	0.09537	5.19510	5.29047	0.98197	134.65
2.36	0.70441	−0.70979	−0.99242	10.59095	0.09442	5.24827	5.34269	0.98233	135.22
2.37	0.69728	−0.71680	−0.97276	10.69739	0.09348	5.30196	5.39544	0.98267	135.79
2.38	0.69007	−0.72374	−0.95349	10.80490	0.09255	5.35618	5.44873	0.98301	136.36
2.39	0.68280	−0.73060	−0.93458	10.91349	0.09163	5.41093	5.50256	0.98335	136.94
2.40	0.67546	−0.73739	−0.91601	11.02318	0.09072	5.46623	5.55695	0.98367	137.51
2.41	0.66806	−0.74411	−0.89779	11.13396	0.08982	5.52207	5.61189	0.98400	138.08
2.42	0.66058	−0.75075	−0.87989	11.24586	0.08892	5.57847	5.66739	0.98431	138.66
2.43	0.65304	−0.75732	−0.86230	11.35888	0.08804	5.63542	5.72346	0.98462	139.23
2.44	0.64543	−0.76382	−0.84501	11.47304	0.08716	5.69294	5.78010	0.98492	139.80
2.45	0.63776	−0.77023	−0.82802	11.58835	0.08629	5.75103	5.83732	0.98522	140.37
2.46	0.63003	−0.77657	−0.81130	11.70481	0.08543	5.80969	5.89512	0.98551	140.95
2.47	0.62223	−0.78283	−0.79485	11.82245	0.08458	5.86893	5.95352	0.98579	141.52
2.48	0.61437	−0.78901	−0.77866	11.94126	0.08374	5.92876	6.01250	0.98607	142.09
2.49	0.60645	−0.79512	−0.76272	12.06128	0.08291	5.98918	6.07209	0.98635	142.67
2.50	0.59847	−0.80114	−0.74702	12.18249	0.08208	6.05020	6.13229	0.98661	143.24

A.37 ELEMENTARY FUNCTIONS

$x = 2.51 - 3.00$

x, rad	$\sin x$	$\cos x$	$\tan x$	e^x	e^{-x}	$\sinh x$	$\cosh x$	$\tanh x$	x, deg
2.51	0.59043	−0.80709	−0.73156	12.30493	0.08127	6.11183	6.19310	0.98688	143.81
2.52	0.58233	−0.81295	−0.71632	12.42860	0.08046	6.17407	6.25453	0.98714	144.39
2.53	0.57417	−0.81873	−0.70129	12.55351	0.07966	6.23692	6.31658	0.98739	144.96
2.54	0.56596	−0.82444	−0.68648	12.67967	0.07887	6.30040	6.37927	0.98764	145.53
2.55	0.55768	−0.83005	−0.67186	12.80710	0.07808	6.36451	6.44259	0.98788	146.10
2.56	0.54936	−0.83559	−0.65745	12.93582	0.07730	6.42926	6.50656	0.98812	146.68
2.57	0.54097	−0.84104	−0.64322	13.06582	0.07654	6.49464	6.57118	0.98835	147.25
2.58	0.53253	−0.84641	−0.62917	13.19714	0.07577	6.56068	6.63646	0.98858	147.82
2.59	0.52404	−0.85169	−0.61530	13.32977	0.07502	6.62738	6.70240	0.98881	148.40
2.60	0.51550	−0.85689	−0.60160	13.46374	0.07427	6.69473	6.76901	0.98903	148.97
2.61	0.50691	−0.86200	−0.58806	13.59905	0.07353	6.76276	6.83629	0.98924	149.54
2.62	0.49826	−0.86703	−0.57468	13.73572	0.07280	6.83146	6.90426	0.98946	150.11
2.63	0.48957	−0.87197	−0.56145	13.87377	0.07208	6.90085	6.97292	0.98966	150.69
2.64	0.48082	−0.87682	−0.54837	14.01320	0.07136	6.97092	7.04228	0.98987	151.26
2.65	0.47203	−0.88158	−0.53544	14.15404	0.07065	7.04169	7.11234	0.99007	151.83
2.66	0.46319	−0.88626	−0.52264	14.29629	0.06995	7.11317	7.18312	0.99026	152.41
2.67	0.45431	−0.89085	−0.50997	14.43997	0.06925	7.18536	7.25461	0.99045	152.98
2.68	0.44537	−0.89534	−0.49743	14.58509	0.06856	7.25827	7.32683	0.99064	153.55
2.69	0.43640	−0.89975	−0.48502	14.73168	0.06788	7.33190	7.39978	0.99083	154.13
2.70	0.42738	−0.90407	−0.47273	14.87973	0.06721	7.40626	7.47347	0.99101	154.70
2.71	0.41832	−0.90830	−0.46055	15.02928	0.06654	7.48137	7.54791	0.99118	155.27
2.72	0.40921	−0.91244	−0.44848	15.18032	0.06587	7.55722	7.62310	0.99136	155.84
2.73	0.40007	−0.91648	−0.43653	15.33289	0.06522	7.63383	7.69905	0.99153	156.42
2.74	0.39088	−0.92044	−0.42467	15.48699	0.06457	7.71121	7.77578	0.99170	156.99
2.75	0.38166	−0.92430	−0.41292	15.64263	0.06393	7.78935	7.85328	0.99186	157.56
2.76	0.37240	−0.92807	−0.40126	15.79984	0.06329	7.86828	7.93157	0.99202	158.14
2.77	0.36310	−0.93175	−0.38970	15.95863	0.06266	7.94799	8.01065	0.98218	158.71
2.78	0.35376	−0.93533	−0.37822	16.11902	0.06204	8.02849	8.09053	0.99233	159.28
2.79	0.34439	−0.93883	−0.36683	16.28102	0.06142	8.10980	8.17122	0.99248	159.86
2.80	0.33499	−0.94222	−0.35553	16.44465	0.06081	8.19192	8.25273	0.99263	160.43
2.81	0.32555	−0.94553	−0.34431	16.60992	0.06020	8.27486	8.33506	0.99278	161.00
2.82	0.31608	−0.94873	−0.33316	16.77685	0.05961	8.35862	8.41823	0.99292	161.57
2.83	0.30657	−0.95185	−0.32208	16.94546	0.05901	8.44322	8.50224	0.99306	162.15
2.84	0.29704	−0.95486	−0.31108	17.11577	0.05843	8.52867	8.58710	0.99320	162.72
2.85	0.28748	−0.95779	−0.30015	17.28778	0.05784	8.61497	8.67281	0.99333	163.29
2.86	0.27789	−0.96061	−0.28928	17.46153	0.05727	8.70213	8.75940	0.99346	163.87
2.87	0.26827	−0.96334	−0.27847	17.63702	0.05670	8.79016	8.84686	0.99359	164.44
2.88	0.25862	−0.96598	−0.26773	17.81427	0.05613	8.87907	8.93520	0.99372	165.01
2.89	0.24895	−0.96852	−0.25704	17.99331	0.05558	8.96887	9.02444	0.99384	165.58
2.90	0.23925	−0.97096	−0.24641	18.17415	0.05502	9.05956	9.11458	0.99396	166.16
2.91	0.22953	−0.97330	−0.23582	18.35680	0.05448	9.15116	9.20564	0.99408	166.73
2.92	0.21978	−0.97555	−0.22529	18.54129	0.05393	9.24368	9.29761	0.99420	167.30
2.93	0.21002	−0.97770	−0.21481	18.72763	0.05340	9.33712	9.39051	0.99431	167.88
2.94	0.20023	−0.97975	−0.20437	18.91585	0.05287	9.43149	9.48436	0.99443	168.45
2.95	0.19042	−0.98170	−0.19397	19.10595	0.05234	9.52681	9.57915	0.99494	169.02
2.96	0.18060	−0.98356	−0.18362	19.29797	0.05182	9.62308	9.67490	0.99464	169.60
2.97	0.17075	−0.98531	−0.17330	19.49192	0.05130	9.72031	9.77161	0.99475	170.17
2.98	0.16089	−0.98697	−0.16301	19.68782	0.05079	9.81851	9.86930	0.99485	170.74
2.99	0.15101	−0.98853	−0.15276	19.88568	0.05029	9.91770	9.96798	0.99496	171.31
3.00	0.14112	−0.98999	−0.14255	20.08554	0.04979	10.01787	10.06766	0.99505	171.89

x, rad	$\sin x$	$\cos x$	$\tan x$	e^x	e^{-x}	$\sinh x$	$\cosh x$	$\tanh x$	x, deg
3.05	0.09146	−0.99581	−0.09185	21.11534	0.04736	10.53399	10.58135	0.99552	174.75
3.10	0.04158	−0.99914	−0.04162	22.19795	0.04505	11.07645	11.12150	0.99595	177.62
3.15	−0.00841	−0.99996	0.00841	23.33606	0.04285	11.64661	11.68946	0.99633	180.48
3.20	−0.05837	−0.99829	0.05847	24.53253	0.04076	12.24588	12.28665	0.99668	183.35
3.25	−0.10820	−0.99413	0.10883	25.79034	0.03877	12.87578	12.91456	0.99700	186.21
3.30	−0.15775	−0.98748	0.15975	27.11264	0.03688	13.53788	13.57476	0.99728	189.08
3.35	−0.20690	−0.97836	0.21148	28.50273	0.03508	14.23382	14.26891	0.99754	191.94
3.40	−0.25554	−0.96680	0.26432	29.96410	0.03337	14.96536	14.99874	0.99777	194.81
3.45	−0.30354	−0.95282	0.31857	31.50039	0.03175	15.73432	15.76607	0.99799	197.67
3.50	−0.35078	−0.93646	0.37459	33.11545	0.03020	16.54263	16.57282	0.99818	200.54
3.55	−0.39715	−0.91775	0.43274	34.81332	0.02872	17.39230	17.42102	0.99835	203.40
3.60	−0.44252	−0.89676	0.49347	36.59823	0.02732	18.28546	18.31278	0.99851	206.26
3.65	−0.48679	−0.87352	0.55727	38.47467	0.02599	19.22434	19.25033	0.99865	209.13
3.70	−0.52984	−0.84810	0.62473	40.44730	0.02472	20.21129	20.23601	0.99878	211.99
3.75	−0.57156	−0.82056	0.69655	42.52108	0.02352	21.24878	21.27230	0.99889	214.86
3.80	−0.61186	−0.79097	0.77356	44.70118	0.02237	22.33941	22.36178	0.99900	217.72
3.85	−0.65063	−0.75940	0.85676	46.99306	0.02128	23.48589	23.50717	0.99909	220.59
3.90	−0.68777	−0.72593	0.94742	49.40245	0.02024	24.69110	24.71135	0.99918	223.45
3.95	−0.72319	−0.69065	1.04711	51.93537	0.01925	25.95806	25.97731	0.99926	226.32
4.00	−0.75680	−0.65364	1.15782	54.59815	0.01832	27.28992	27.30823	0.99933	229.18
4.05	−0.78853	−0.61500	1.28215	57.39746	0.01742	28.69002	28.70744	0.99939	232.05
4.10	−0.81828	−0.57482	1.42353	60.34029	0.01657	30.16186	30.17843	0.99945	234.91
4.15	−0.84598	−0.53321	1.58659	63.43400	0.01576	31.70912	31.72488	0.99950	237.78
4.20	−0.87158	−0.49026	1.77778	66.68633	0.01500	33.33567	33.35066	0.99955	240.64
4.25	−0.89499	−0.44609	2.00631	70.10541	0.01426	35.04557	35.05984	0.99959	243.51
4.30	−0.91617	−0.40080	2.28585	73.69979	0.01357	36.84311	36.85668	0.99963	246.37
4.35	−0.93505	−0.35451	2.63760	77.47846	0.01291	38.73278	38.74568	0.99967	249.24
4.40	−0.95160	−0.30733	3.09632	81.45087	0.01228	40.71930	40.73157	0.99970	252.10
4.45	−0.96577	−0.25939	3.72327	85.62694	0.01168	42.80763	42.81931	0.99973	254.97
4.50	−0.97753	−0.21080	4.63733	90.01713	0.01111	45.00301	45.01412	0.99975	257.83
4.55	−0.98684	−0.16168	6.10383	94.63241	0.01057	47.31092	47.32149	0.99978	260.70
4.60	−0.99369	−0.11215	8.86017	99.48432	0.01005	49.73713	49.74718	0.99980	263.56
4.65	−0.99805	−0.06235	16.00767	104.58499	0.00956	52.28771	52.29727	0.99982	266.43
4.70	−0.99992	−0.01239	80.71276	109.94717	0.00910	54.96904	54.97813	0.99983	269.29
4.75	−0.99929	0.03760	−26.57541	115.58428	0.00865	57.78782	57.79647	0.99985	272.15
4.80	−0.99616	0.08750	−11.38487	121.51042	0.00823	60.75109	60.75932	0.99986	275.02
4.85	−0.99055	0.13718	−7.22093	127.74039	0.00783	63.86628	63.87411	0.99988	277.88
4.90	−0.98245	0.18651	−5.26749	134.28978	0.00745	67.14117	67.14861	0.99989	280.75
4.95	−0.97190	0.23538	−4.12906	141.17496	0.00708	70.58394	70.59102	0.99990	283.61
*5.00	−0.95892	0.28366	−3.38052	148.41316	0.00674	74.20321	74.20995	0.99991	286.48

*For $x > 5.00$ and other numerical data, refer to:

Abramowitz, M., and I. A. Stegun: "Handbook of Mathematical Functions," National Bureau of Standards, Washington, D.C., 1964.

Dwight, H. B.: "Mathematical Tables," 3d ed., Dover, New York, 1961.

Flecher, A., J. C. Miller, L. Rosenhead, and L. J. Comrie: "An Index of Mathematical Tables," 2d ed., Addison-Wesley, Reading, Mass., 1962.

Lebedev, A. V., and R. M. Fedorova: "A Guide to Mathematical Tables," Pergamon, New York, 1960.

A.39 AMOUNT OF 1 AT COMPOUND INTEREST*

$(1+i)^n$

n \ i	1%	2%	3%	$3\frac{1}{2}$%	4%	$4\frac{1}{2}$%	5%	6%	7%	8%
1	1.01000	1.02000	1.03000	1.03500	1.04000	1.04500	1.05000	1.06000	1.07000	1.08000
2	1.02010	1.04040	1.06090	1.07123	1.08160	1.09202	1.10250	1.12360	1.14490	1.16640
3	1.03030	1.06121	1.09273	1.10872	1.12486	1.14117	1.15763	1.19102	1.22504	1.25971
4	1.04060	1.08243	1.12551	1.14752	1.16986	1.19252	1.21551	1.26248	1.31080	1.36049
5	1.05101	1.10408	1.15927	1.18769	1.21665	1.24618	1.27628	1.33823	1.40255	1.46933
6	1.06152	1.12516	1.19405	1.22926	1.26532	1.30226	1.34010	1.41852	1.50073	1.58687
7	1.07214	1.14869	1.22987	1.27228	1.31593	1.36086	1.40710	1.50363	1.60578	1.71382
8	1.08286	1.17166	1.26677	1.31681	1.36857	1.42210	1.47746	1.59385	1.71819	1.85093
9	1.09369	1.19509	1.30477	1.36290	1.42331	1.48610	1.55133	1.68948	1.83846	1.99900
10	1.10462	1.21899	1.34392	1.41060	1.48024	1.55297	1.62889	1.79085	1.96715	2.15892
11	1.11567	1.24337	1.38423	1.45997	1.53945	1.62285	1.71034	1.89830	2.10485	2.33164
12	1.12683	1.26824	1.42576	1.51107	1.60103	1.69588	1.79586	2.01220	2.25219	2.51817
13	1.13809	1.29361	1.46853	1.56396	1.66507	1.77220	1.88565	2.13293	2.40985	2.71962
14	1.14947	1.31948	1.51259	1.61869	1.73168	1.85194	1.97993	2.26090	2.57853	2.93719
15	1.16097	1.34587	1.55797	1.67535	1.80094	1.93528	2.07893	2.39656	2.75903	3.17217
16	1.17258	1.37279	1.60471	1.73399	1.87298	2.02237	2.18287	2.54035	2.95216	3.42594
17	1.18430	1.40024	1.65285	1.79468	1.94790	2.11338	2.29202	2.69277	3.15882	3.70002
18	1.19615	1.42822	1.70243	1.85749	2.02582	2.20848	2.40662	2.85434	3.37993	3.99602
19	1.20810	1.45681	1.75351	1.92250	2.10685	2.30786	2.52695	3.02560	3.61653	4.31570
20	1.22019	1.48595	1.80611	1.98979	2.19112	2.41171	2.65330	3.20714	3.86968	4.66096
21	1.23239	1.51567	1.86029	2.05943	2.27877	2.52024	2.78596	3.39956	4.14056	5.03383
22	1.24472	1.54598	1.91610	2.13151	2.36992	2.63365	2.92526	3.60354	4.43040	5.43654
23	1.25716	1.57690	1.97359	2.20611	2.46472	2.75217	3.07152	3.81975	4.74053	5.87146
24	1.26973	1.60844	2.03279	2.28333	2.56330	2.87601	3.22510	4.04893	5.07237	6.34118
25	1.28243	1.64061	2.09378	2.36324	2.66584	3.00543	3.38635	4.29187	5.42743	6.84848
26	1.29526	1.67342	2.15659	2.44596	2.77247	3.14068	3.55567	4.54938	5.80735	7.39635
27	1.30821	1.70689	2.22128	2.53157	2.88337	3.28201	3.73346	4.82235	6.21387	7.98806
28	1.32129	1.74102	2.28793	2.62017	2.99870	3.42970	3.92013	5.11169	6.64884	8.62711
29	1.33450	1.77584	2.35657	2.71188	3.11865	3.58404	4.11614	5.41839	7.11426	9.31727
30	1.34785	1.81136	2.42726	2.80679	3.24340	3.74532	4.32194	5.74349	7.61226	10.06266
31	1.36133	1.84759	2.50008	2.90503	3.37313	3.91386	4.53804	6.08810	8.14511	10.86767
32	1.37494	1.88454	2.57508	3.00671	3.50806	4.08998	4.76494	6.45339	8.71527	11.73708
33	1.38869	1.92223	2.65234	3.11194	3.64838	4.27403	5.00319	6.84059	9.32534	12.67605
34	1.40258	1.96068	2.73191	3.22086	3.79432	4.46636	5.25335	7.25103	9.97811	13.69013
35	1.41660	1.99989	2.81386	3.33359	3.94609	4.66735	5.51602	7.68608	10.67658	14.78534
36	1.43077	2.03989	2.89828	3.45027	4.10393	4.87738	5.79182	8.14725	11.42394	15.96817
37	1.44508	2.08069	2.98523	3.57103	4.26809	5.09686	6.08141	8.63609	12.22362	17.24563
38	1.45953	2.12221	3.07478	3.69601	4.43881	5.32622	6.38548	9.15425	13.07927	18.62528
39	1.47412	2.16475	3.16703	3.82537	4.61637	5.56590	6.70475	9.70351	13.99482	20.11530
40	1.48886	2.20804	3.26204	3.95926	4.80102	5.81636	7.03999	10.28572	14.97446	21.72452
41	1.50375	2.25220	3.35990	4.09783	4.99306	6.07810	7.39199	10.90286	16.02267	23.46248
42	1.51879	2.29724	3.46070	4.24126	5.19278	6.35162	7.76159	11.55703	17.14426	25.33948
43	1.53398	2.34318	3.56452	4.38970	5.40050	6.63744	8.14967	12.25045	18.34435	27.36665
44	1.54932	2.39005	3.67145	4.54334	5.61652	6.93612	8.55715	12.98548	19.62845	29.55597
45	1.56481	2.43785	3.78160	4.70236	5.84118	7.24825	8.98501	13.76461	21.00245	31.92045
46	1.58046	2.48661	3.89504	4.86694	6.07482	7.57442	9.43426	14.59049	22.47262	34.47409
47	1.59626	2.53634	4.01189	5.03728	6.31782	7.91527	9.90597	15.46592	24.04571	37.23201
48	1.61223	2.58707	4.13225	5.21359	6.57053	8.27146	10.40127	16.39387	25.72891	40.21057
49	1.62835	2.63881	4.25622	5.39606	6.83335	8.64367	10.92133	17.37750	27.52993	43.42742
50	1.64463	2.69159	4.38391	5.58493	7.10668	9.03264	11.46740	18.42015	29.45703	46.90161

*n = number of years (or periods), i = interest. For applications see Sec. 12.11–1.

A.40 PRESENT VALUE OF 1 AT COMPOUND INTEREST* $(1+i)^{-n}$

n \ i	1%	2%	3%	$3\frac{1}{2}$%	4%	$4\frac{1}{2}$%	5%	6%	7%	8%
1	0.99010	0.98039	0.97087	0.96618	0.96154	0.95694	0.95238	0.94340	0.93458	0.92593
2	0.98030	0.96117	0.94260	0.93351	0.92456	0.91573	0.90703	0.89000	0.87344	0.85734
3	0.97059	0.94232	0.91514	0.90194	0.88900	0.87630	0.86384	0.83962	0.81630	0.79383
4	0.96098	0.92385	0.88849	0.87144	0.85480	0.83856	0.82270	0.79209	0.76290	0.73503
5	0.95147	0.90573	0.86261	0.84197	0.82193	0.80245	0.78353	0.74726	0.71299	0.68058
6	0.94205	0.88797	0.83748	0.81350	0.79031	0.76790	0.74622	0.70496	0.66634	0.63017
7	0.93272	0.87056	0.81300	0.78599	0.75992	0.73483	0.71068	0.66506	0.62275	0.58349
8	0.92348	0.85349	0.78941	0.75941	0.73069	0.70319	0.67684	0.62741	0.58201	0.54027
9	0.91434	0.83676	0.76642	0.73373	0.70259	0.67290	0.64461	0.59190	0.54393	0.50025
10	0.90529	0.82035	0.74409	0.70892	0.67556	0.64393	0.61391	0.55839	0.50835	0.46319
11	0.89632	0.80426	0.72242	0.68495	0.64958	0.61620	0.58468	0.52679	0.47509	0.42888
12	0.88745	0.78849	0.70138	0.66178	0.62460	0.58966	0.55684	0.49697	0.44401	0.39711
13	0.87866	0.77303	0.68095	0.63940	0.60057	0.56427	0.53032	0.46884	0.41496	0.36770
14	0.86996	0.75788	0.66112	0.61778	0.57748	0.53997	0.50507	0.44230	0.38782	0.34046
15	0.86135	0.74301	0.64186	0.59689	0.55526	0.51672	0.48102	0.41727	0.36245	0.31524
16	0.85282	0.72845	0.62317	0.57671	0.53391	0.49447	0.45811	0.39365	0.33873	0.29189
17	0.84438	0.71416	0.60502	0.55720	0.51337	0.47318	0.43630	0.37136	0.31657	0.27027
18	0.83602	0.70016	0.58739	0.53836	0.49363	0.45280	0.41552	0.35034	0.29586	0.25025
19	0.82774	0.68643	0.57029	0.52016	0.47464	0.43330	0.39573	0.33051	0.27651	0.23171
20	0.81954	0.67297	0.55368	0.50257	0.45639	0.41464	0.37689	0.31180	0.25842	0.21455
21	0.81143	0.65978	0.53755	0.48557	0.43883	0.39679	0.35894	0.29416	0.24151	0.19866
22	0.80340	0.64684	0.52189	0.46915	0.42196	0.37970	0.34185	0.27751	0.22571	0.18394
23	0.79544	0.63416	0.50669	0.45329	0.40573	0.36335	0.32557	0.26180	0.21095	0.17032
24	0.78757	0.62172	0.49193	0.43796	0.39012	0.34770	0.31007	0.24698	0.19715	0.15770
25	0.77977	0.60953	0.47761	0.42315	0.37512	0.33273	0.29530	0.23300	0.18425	0.14602
26	0.77205	0.59758	0.46369	0.40884	0.36069	0.31840	0.28124	0.21981	0.17220	0.13520
27	0.76440	0.58586	0.45019	0.39501	0.34682	0.30469	0.26785	0.20737	0.16093	0.12519
28	0.75684	0.57437	0.43708	0.38165	0.33348	0.29157	0.25509	0.19563	0.15040	0.11591
29	0.74934	0.56311	0.42435	0.36875	0.32065	0.27901	0.24295	0.18456	0.14056	0.10733
30	0.74192	0.55207	0.41199	0.35628	0.30832	0.26700	0.23138	0.17411	0.13137	0.09938
31	0.73458	0.54125	0.39999	0.34423	0.29646	0.25550	0.22036	0.16425	0.12277	0.09202
32	0.72730	0.53063	0.38834	0.33259	0.28506	0.24450	0.20987	0.15496	0.11474	0.08520
33	0.72010	0.52023	0.37703	0.32134	0.27409	0.23397	0.19987	0.14619	0.10723	0.07889
34	0.71297	0.51003	0.36604	0.31048	0.26355	0.22390	0.19035	0.13791	0.10022	0.07305
35	0.70591	0.50003	0.35538	0.29998	0.25342	0.21425	0.18129	0.13011	0.09366	0.06763
36	0.69893	0.49022	0.34503	0.28983	0.24367	0.20503	0.17266	0.12274	0.08753	0.06262
37	0.69200	0.48061	0.33498	0.28003	0.23430	0.19620	0.16444	0.11579	0.08181	0.05799
38	0.68515	0.47119	0.32523	0.27056	0.22529	0.18775	0.15661	0.10924	0.07646	0.05369
39	0.67837	0.46195	0.31575	0.26141	0.21662	0.17967	0.14915	0.10306	0.07145	0.04971
40	0.67165	0.45289	0.30656	0.25257	0.20829	0.17193	0.14205	0.09722	0.06678	0.04603
41	0.66500	0.44401	0.29763	0.24403	0.20028	0.16453	0.13528	0.09172	0.06241	0.04262
42	0.65842	0.43530	0.28896	0.23578	0.19257	0.15744	0.12884	0.08653	0.05833	0.03946
43	0.65190	0.42677	0.28054	0.22781	0.18517	0.15066	0.12270	0.08163	0.05451	0.03654
44	0.64545	0.41840	0.27237	0.22010	0.17805	0.14417	0.11686	0.07701	0.05095	0.03383
45	0.63905	0.41020	0.26444	0.21266	0.17120	0.13796	0.11130	0.07265	0.04761	0.03133
46	0.63273	0.40215	0.25674	0.20547	0.16461	0.13202	0.10600	0.06854	0.04450	0.02901
47	0.62646	0.39427	0.24926	0.19852	0.15828	0.12634	0.10095	0.06466	0.04159	0.02686
48	0.62026	0.38654	0.24200	0.19181	0.15219	0.12090	0.09614	0.06100	0.03887	0.02487
49	0.61412	0.37896	0.23495	0.18532	0.14634	0.11569	0.09156	0.05755	0.03632	0.02303
50	0.60804	0.37153	0.22811	0.17905	0.14071	0.11071	0.08720	0.05429	0.03395	0.02132

*n = number of years (periods), i = interest. For applications see Sec. 12.11–1.

A.41 AMOUNT OF ANNUITY OF 1*

$$\frac{(1+i)^n - 1}{i}$$

n \ i	1%	2%	3%	3½%	4%	4½%	5%	6%	7%	8%
1	1.00000	1.00000	1.00000	1.00000	1.00000	1.00000	1.00000	1.00000	1.00000	1.00000
2	2.01000	2.02000	2.03000	2.03500	2.04000	2.04500	2.05000	2.06000	2.07000	2.08000
3	3.03010	3.06040	3.09090	3.10622	3.12160	3.13702	3.15250	3.18360	3.21490	3.24640
4	4.06040	4.12161	4.18363	4.21494	4.24646	4.27819	4.31012	4.37462	4.43994	4.50611
5	5.10100	5.20404	5.30914	5.36247	5.41632	5.47071	5.52563	5.63709	5.75074	5.86660
6	6.15202	6.30812	6.40841	6.55015	6.63298	6.71689	6.80191	6.97532	7.15329	7.33593
7	7.21354	7.43428	7.60246	7.77941	7.89829	8.01915	8.14201	8.39384	8.65402	8.92280
8	8.28567	8.58297	8.89234	9.05169	9.21423	9.38001	9.54911	9.89747	10.25980	10.63663
9	9.30853	9.75463	10.15911	10.36850	10.58280	10.80211	11.02656	11.49132	11.97799	12.48756
10	10.46221	10.94972	11.46388	11.73139	12.00611	12.28821	12.57789	13.18079	13.81645	14.48656
11	11.56683	12.16872	12.80780	13.14199	13.48635	13.84118	14.20679	14.97164	15.78360	16.64549
12	12.08250	13.41208	14.19203	14.60196	15.02581	15.46403	15.91713	16.86994	17.88845	18.97713
13	13.80933	14.68033	15.61779	16.11303	16.62684	17.15991	17.71298	18.88214	20.14064	21.49530
14	14.94742	15.97394	17.08632	17.67699	18.29191	18.93211	19.59863	21.01507	22.55049	24.21492
15	16.09690	17.29342	18.59891	19.29508	20.02359	20.78405	21.57856	23.27597	25.12902	27.15211
16	17.25786	18.63929	20.15688	20.97103	21.82453	22.71934	23.65749	25.67253	27.88805	30.32428
17	18.43044	20.01207	21.76159	22.70502	23.69751	24.74171	25.84037	28.21288	30.84022	33.75023
18	19.61475	21.41231	23.41444	24.40969	25.64541	26.85508	28.13238	30.90565	33.99903	37.45024
19	20.81089	22.84056	25.11687	26.35718	27.67123	29.06356	30.53900	33.75999	37.37896	41.44626
20	22.01900	24.29737	26.87037	28.27968	29.77808	31.37142	33.06595	36.78559	40.99549	45.76196
21	23.23919	25.78332	28.67649	30.26947	31.96920	33.78314	35.71925	39.99273	44.86518	50.42202
22	24.47159	27.29898	30.53678	32.32890	34.24797	36.30338	38.50521	43.39229	49.00574	55.45676
23	25.71630	28.84496	32.45288	34.46041	36.61789	38.93703	41.43048	46.99583	53.43614	60.89330
24	26.07346	30.42186	34.42647	36.66653	39.08260	41.08920	44.50200	50.81558	58.17667	66.76476
25	28.24320	32.03030	36.45926	38.94986	41.64591	44.56521	47.72710	54.86451	63.24904	73.10594
26	29.52563	33.67091	38.55304	41.31310	44.31174	47.57064	51.11345	59.15638	68.67647	79.95442
27	30.82089	35.34432	40.70963	43.75906	47.08421	50.71132	54.66913	63.70577	74.48382	87.35077
28	32.12910	37.05121	42.93092	46.29063	49.96758	53.99333	58.40258	68.52811	80.69769	95.33883
29	33.45039	38.79223	45.21885	48.91080	52.96629	57.42303	62.32271	73.63980	87.34653	103.96594
30	34.78489	40.56808	47.57542	51.62268	56.08494	61.00707	66.43885	79.05819	94.46079	113.28321
31	36.13274	42.37944	50.00268	54.42947	59.32834	64.75239	70.76079	84.80168	102.07304	123.34587
32	37.49407	44.22703	52.50276	57.33450	62.70147	68.66625	75.29883	90.88978	110.21815	134.21354
33	38.86901	46.11157	55.07784	60.34121	66.20953	72.75623	80.06377	97.34316	118.93343	145.95062
34	40.25770	48.03380	57.73018	63.45315	69.85791	77.03026	85.06696	104.18375	128.25876	158.62667
35	41.66028	49.99448	60.46208	66.67401	73.65222	81.49662	90.32031	111.43478	138.23688	172.31680
36	43.07688	51.99437	63.27594	70.00760	77.59831	86.16397	95.83632	119.12087	148.91346	187.10215
37	44.50765	54.03425	66.17422	73.45787	81.70225	91.04134	101.62814	127.26812	160.33740	203.07032
38	45.95272	56.11494	69.15945	77.02889	85.97034	96.13820	107.70955	135.90421	172.56102	220.31595
39	47.41225	58.23724	72.23423	80.72491	90.40915	101.46442	114.09502	145.05846	185.64029	238.94122
40	48.88637	60.40198	75.40126	84.55028	95.02552	107.03032	120.79977	154.76197	199.63511	259.05652
41	50.37524	62.61002	78.66330	88.50954	99.82654	112.84669	127.83976	165.04768	214.60957	280.78104
42	51.87899	64.86222	82.02320	92.60737	104.81960	118.92479	135.23175	175.95054	230.63224	304.24352
43	53.39778	67.15947	85.48380	96.84863	110.01238	125.27640	142.99334	187.50758	247.77650	329.58301
44	54.93176	69.50266	89.04841	101.23833	115.41288	131.91384	151.14301	199.75803	266.12085	356.94965
45	56.48107	71.89271	92.71986	105.78167	121.02939	138.84997	159.70016	212.74351	285.74931	386.50562
46	58.04589	74.33056	96.59146	110.48403	126.87057	146.09821	168.68516	226.50812	306.75176	418.42607
47	59.62634	76.81718	100.39650	115.35097	132.94539	153.67203	178.11942	241.09861	329.22439	452.90015
48	61.22261	79.33352	104.40840	120.38826	139.26321	161.58790	188.02539	256.56453	353.27009	490.13216
49	62.83483	81.94059	108.54065	125.60185	145.83373	169.85936	198.42666	272.95840	378.99900	530.34274
50	64.46318	84.57940	112.79687	130.99791	152.66708	178.50303	209.34800	290.33590	406.52893	573.77016

*n = number of years (periods), i = interest. For applications see Sec. 12.11–2.

A.42 PRESENT VALUE OF ANNUITY OF 1*

$$\frac{1-(1+i)^{-n}}{i}$$

n \ i	1%	2%	3%	$3\frac{1}{2}$%	4%	$4\frac{1}{2}$%	5%	6%	7%	8%
1	0.99010	0.98039	0.97087	0.96618	0.96154	0.95694	0.95238	0.94340	0.93458	0.92593
2	1.97040	1.94156	1.91347	1.89969	1.88609	1.87267	1.85941	1.83339	1.80802	1.78326
3	2.94099	2.88388	2.82861	2.80164	2.77509	2.74896	2.72325	2.67301	2.62432	2.57710
4	3.90197	3.80773	3.71710	3.67308	3.62990	3.58753	3.54595	3.46511	3.38721	3.31213
5	4.85343	4.71346	4.57971	4.51505	4.45182	4.38998	4.32948	4.21236	4.10020	3.99271
6	5.79548	5.60143	5.41719	5.32855	5.24214	5.15787	5.07569	4.91732	4.76654	4.62288
7	6.72819	6.47199	6.23028	6.11454	6.00205	5.89270	5.78637	5.58238	5.38929	5.20637
8	7.65168	7.32548	7.01969	6.87396	6.73274	6.59589	6.46321	6.20979	5.97130	5.74664
9	8.56602	8.16224	7.78611	7.60769	7.43533	7.20879	7.10782	6.80169	6.51523	6.24689
10	9.47130	8.98258	8.53020	8.31661	8.11090	7.91272	7.72173	7.30009	7.02358	6.71008
11	10.36763	9.78685	9.25262	9.00155	8.76048	8.52892	8.30641	7.88687	7.49867	7.13896
12	11.25508	10.57534	9.95400	9.66333	9.38507	9.11858	8.86325	8.38384	7.94269	7.53608
13	12.13374	11.34837	10.63496	10.30274	9.98565	9.68285	9.39357	8.85268	8.35765	7.90378
14	13.00370	12.10625	11.29607	10.92052	10.56312	10.22283	9.89865	9.29498	8.74547	8.24424
15	13.86505	12.84926	11.93794	11.51741	11.11839	10.73955	10.37965	9.71225	9.10791	8.55948
16	14.71787	13.57771	12.56110	12.09412	11.65230	11.23401	10.83777	10.10590	9.44665	8.85137
17	15.56225	14.29187	13.16612	12.65132	12.16567	11.70719	11.27407	10.47726	9.76322	9.12164
18	16.39827	14.99203	13.75351	13.18968	12.65930	12.15999	11.68959	10.82760	10.05909	9.37189
19	17.22601	15.67846	14.32380	13.70984	13.13394	12.59329	12.08532	11.15812	10.33560	9.60360
20	18.04555	16.35143	14.87749	14.21240	13.59033	13.00794	12.46222	11.46992	10.59401	9.81815
21	18.85698	17.01121	15.41502	14.69797	14.02916	13.40472	12.82115	11.76408	10.83553	10.01680
22	19.66038	17.65805	15.93692	15.16712	14.45112	13.78442	13.16300	12.04158	11.06124	10.20074
23	20.45582	18.29220	16.44361	15.62041	14.85684	14.14777	13.48857	12.30338	11.27219	10.37106
24	21.24339	18.91393	16.93554	16.05837	15.24696	14.49548	13.79864	12.55036	11.46933	10.52876
25	22.02316	19.52346	17.41315	16.48151	15.62208	14.82821	14.09394	12.78336	11.65358	10.67478
26	22.79520	20.12104	17.87684	16.89035	15.98277	15.14661	14.37519	13.00317	11.82578	10.80998
27	23.55961	20.70690	18.32703	17.28536	16.32959	15.45130	14.64303	13.21053	11.98671	10.93516
28	24.31644	21.28127	18.76411	17.66702	16.66306	15.74287	14.89813	13.40616	12.13711	11.05168
29	25.06579	21.84438	19.18845	18.03577	16.98371	16.02189	15.14107	13.59072	12.27767	11.15841
30	25.80771	22.39646	19.60044	18.39205	17.29203	16.28889	15.37245	13.76483	12.40904	11.25778
31	26.54229	22.93770	20.00043	18.73628	17.58849	16.54439	15.59281	13.92909	12.53181	11.34980
32	27.26959	23.46833	20.38877	19.06887	17.87355	16.78889	15.80268	14.08404	12.64656	11.43500
33	27.98969	23.98856	20.76579	19.39021	18.14765	17.02286	16.00255	14.23023	12.75379	11.51389
34	28.70267	24.49859	21.13184	19.70068	18.41120	17.24676	16.19290	14.36814	12.85401	11.58693
35	29.40858	24.99862	21.48722	20.00066	18.66461	17.46101	16.37419	14.49825	12.94767	11.65457
36	30.10750	25.48884	21.83225	20.29049	18.90828	17.66604	16.54685	14.62099	13.03521	11.71719
37	30.79951	25.96945	22.16724	20.57053	19.14258	17.86224	16.71129	14.73678	13.11702	11.77518
38	31.48466	26.44064	22.49246	20.84109	19.36786	18.04999	16.86789	14.84602	13.19347	11.82887
39	32.16303	26.90259	22.80822	21.10250	19.58448	18.22966	17.01704	14.94907	13.26493	11.87858
40	32.83469	27.35548	23.11477	21.35507	19.79277	18.40158	17.15909	15.04630	13.33171	11.92461
41	33.49969	27.79949	23.41240	21.59910	19.90305	18.56611	17.29437	15.13802	13.39412	11.96723
42	34.15811	28.23479	23.70136	21.83488	20.18563	18.72355	17.42321	15.22454	13.45245	12.00670
43	34.81001	28.66156	23.98190	22.06269	20.37079	18.87421	17.54591	15.30617	13.50696	12.04324
44	35.45545	29.07996	24.25427	22.28279	20.54884	19.01838	17.66277	15.38318	13.55791	12.07707
45	36.09451	29.49016	24.51871	22.49545	20.72004	19.15635	17.77407	15.45583	13.60552	12.10840
46	36.72724	29.89231	24.77545	22.70092	20.88465	19.28837	17.88007	15.52437	13.65002	12.13741
47	37.35370	30.28658	25.02471	22.89944	21.04294	19.41471	17.98102	15.58903	13.69161	12.16427
48	37.97396	30.67312	25.26671	23.09124	21.19513	19.53561	18.07716	15.65003	13.73047	12.18914
49	38.58808	31.05208	25.50166	23.27656	21.34147	19.65130	18.16872	15.70757	13.76680	12.21216
50	39.19612	31.42361	25.72976	23.45562	21.48218	19.76201	18.25593	15.76186	13.80075	12.23348

*n = number of years (periods), i = interest. For applications see Sec. 12.11–2.

A.43 BINOMIAL COEFFICIENTS

$\binom{n}{0} = 1 \qquad \binom{n}{1} = n \qquad \binom{n}{n} = 1$

$$\binom{n}{k} = \frac{n!}{k!\,(n-k)!} = \frac{n(n-1)\cdots(n-k+1)}{k!} = \binom{n}{n-k}$$

n	$\binom{n}{0}$	$\binom{n}{1}$	$\binom{n}{2}$	$\binom{n}{3}$	$\binom{n}{4}$	$\binom{n}{5}$	$\binom{n}{6}$	$\binom{n}{7}$	$\binom{n}{8}$	$\binom{n}{9}$	$\binom{n}{10}$	n
0	1											0
1	1	1										1
2	1	2	1									2
3	1	3	3	1								3
4	1	4	6	4	1							4
5	1	5	10	10	5	1						5
6	1	6	15	20	15	6	1					6
7	1	7	21	35	35	21	7	1				7
8	1	8	28	56	70	56	28	8	1			8
9	1	9	36	84	126	126	84	36	9	1		9
10	1	10	45	120	210	252	210	120	45	10	1	10
11	1	11	55	165	330	462	462	330	165	55	11	11
12	1	12	66	220	495	792	924	792	495	220	66	12
13	1	13	78	286	715	1,287	1,716	1,716	1,287	715	286	13
14	1	14	91	364	1,001	2,002	3,003	3,432	3,003	2,002	1,001	14
15	1	15	105	455	1,365	3,003	5,005	6,435	6,435	5,005	3,003	15
16	1	16	120	560	1,820	4,368	8,008	11,440	12,870	11,440	8,008	16
17	1	17	136	680	2,380	6,188	12,376	19,448	24,310	24,310	19,448	17
18	1	18	153	816	3,060	8,568	18,564	31,824	43,758	48,620	43,758	18
19	1	19	171	969	3,876	11,628	27,132	50,388	75,582	92,378	92,378	19
20	1	20	190	1,140	4,845	15,504	38,760	77,520	125,970	167,960	184,756	20
n	$\binom{n}{0}$	$\binom{n}{1}$	$\binom{n}{2}$	$\binom{n}{3}$	$\binom{n}{4}$	$\binom{n}{5}$	$\binom{n}{6}$	$\binom{n}{7}$	$\binom{n}{8}$	$\binom{n}{9}$	$\binom{n}{10}$	n

Note: For coefficients not given above use $\binom{n}{k} = \binom{n}{n-k}$; for example,

$$\binom{19}{15} = \binom{19}{19-15} = \binom{19}{4} = 3{,}876.$$

Appendix B
CONVERSION TABLES

B.01 DECIMALS OF AN INCH WITH MILLIMETER EQUIVALENTS*

Fraction			Decimal	Millimeters
		1/64	0.015625	0.397
	1/32	2/64	0.031250	0.794
		3/64	0.046875	1.191
1/16	2/32	4/64	0.062500	1.588
		5/64	0.078125	1.984
	3/32	6/64	0.093750	2.381
		7/64	0.109375	2.778
1/8	4/32	8/64	0.125000	3.175
		9/64	0.140625	3.572
	5/32	10/64	0.156250	3.969
		11/64	0.171875	4.366
3/16	6/32	12/64	0.187500	4.763
		13/64	0.203125	5.159
	7/32	14/64	0.218750	5.556
		15/64	0.234375	5.953
1/4	8/32	16/64	0.250000	6.350
		17/64	0.265625	6.747
	9/32	18/64	0.281250	7.144
		19/64	0.296875	7.541
5/16	10/32	20/64	0.312500	7.938
		21/64	0.328125	8.334
	11/32	22/64	0.343750	8.731
		23/64	0.359375	9.128
3/8	12/32	24/64	0.375000	9.525
		25/64	0.390625	9.922
	13/32	26/64	0.406250	10.319
		27/64	0.421875	10.716
7/16	14/32	28/64	0.437500	11.113
		29/64	0.453125	11.509
	15/32	30/64	0.468750	11.906
		31/64	0.484375	12.303
1/2	16/32	32/64	0.500000	12.700

Fraction			Decimal	Millimeters
		33/64	0.515625	13.097
	17/32	34/64	0.531250	13.494
		35/64	0.546875	13.891
9/16	18/32	36/64	0.562500	14.288
		37/64	0.578125	14.684
	19/32	38/64	0.593750	15.081
		39/64	0.609375	15.478
5/8	20/32	40/64	0.625000	15.875
		41/64	0.640625	16.272
	21/32	42/64	0.656250	16.669
		43/64	0.671875	17.066
11/16	22/32	44/64	0.687500	17.463
		45/64	0.703125	17.859
	23/32	46/64	0.718750	18.256
		47/64	0.734375	18.653
3/4	24/32	48/64	0.750000	19.050
		49/64	0.765625	19.447
	25/32	50/64	0.781250	19.844
		51/64	0.796875	20.241
13/16	26/32	52/64	0.812500	20.638
		53/64	0.828125	21.034
	27/32	54/64	0.843750	21.431
		55/64	0.859375	21.828
7/8	28/32	56/64	0.875000	22.225
		57/64	0.890625	22.622
	29/32	58/64	0.906250	23.019
		59/64	0.921875	23.416
15/16	30/32	60/64	0.937500	23.813
		61/64	0.953125	24.209
	31/32	62/64	0.968750	24.606
		63/64	0.984375	25.003
1	32/32	64/64	1.000000	25.400

*Conversion relations presented in this Appendix are based on standards defined by the International Bureau of Weights and Measures (IBWM), Sévres, France; the International Organization for Standardization (IOS), Geneva, Switzerland; the National Bureau of Standards (NBS), Washington, D.C.; and the National Aeronautics and Space Administration (NASA), Washington, D.C. For more detailed information refer to:

IOS Report 31, Part I, "SI Units," 1956, Part II, "Units of Periodic and Related Phenomena," 1958, Part III, "Units of Mechanics," 1960, Part IV, "Units of Heat," 1960, Part V, "Units of Electricity and Magnetism," 1963.

NBS Misc. Publication 286, "Units of Weight and Measure, Definitions and Tables of Equivalents," 1967.

NASA Publication SP-7012, "The International System of Units, Physical Constants and Conversion Factors," C. A. Mechtly, 1964.

B.02 SI SYSTEM—LENGTH, WAVELENGTH, AREA, VOLUME

(1) Length

km = kilometer, m = meter, dm = decimeter, cm = centimeter, mm = millimeter, μm = micrometer

	km	m	dm	cm	mm	μm
km	1	10^{3}	10^{4}	10^{5}	10^{6}	10^{9}
m	10^{-3}	1	10	10^{2}	10^{3}	10^{6}
dm	10^{-4}	10^{-1}	1	10	10^{2}	10^{5}
cm	10^{-5}	10^{-2}	10^{-1}	1	10	10^{4}
mm	10^{-6}	10^{-3}	10^{-2}	10^{-1}	1	10^{3}
μm	10^{-9}	10^{-6}	10^{-5}	10^{-4}	10^{-3}	1

(2) Wavelength

mm = millimeter, μm = micrometer, nm = nanometer
Å = angstrom, pm = picometer, mÅ = milliangstrom

	mm	μm	nm	Å	pm	mÅ
mm	1	10^{3}	10^{6}	10^{7}	10^{9}	10^{10}
μm	10^{-3}	1	10^{3}	10^{4}	10^{6}	10^{7}
nm	10^{-6}	10^{-3}	1	10	10^{3}	10^{4}
Å	10^{-7}	10^{-4}	10^{-1}	1	10^{2}	10^{3}
pm	10^{-9}	10^{-6}	10^{-3}	10^{-2}	1	10
mÅ	10^{-10}	10^{-7}	10^{-4}	10^{-3}	10^{-1}	1

(3) Area

km^2 = square kilometer, m^2 = square meter, dm^2 = square decimeter
cm^2 = square centimeter, mm^2 = square millimeter, μm^2 = square micrometer

	km^2	m^2	dm^2	cm^2	mm^2	μm^2
km^2	1	10^{6}	10^{8}	10^{10}	10^{12}	10^{18}
m^2	10^{-6}	1	10^{2}	10^{4}	10^{6}	10^{12}
dm^2	10^{-8}	10^{-2}	1	10^{2}	10^{4}	10^{10}
cm^2	10^{-10}	10^{-4}	10^{-2}	1	10^{2}	10^{8}
mm^2	10^{-12}	10^{-6}	10^{-4}	10^{-2}	1	10^{6}
μm^2	10^{-18}	10^{-12}	10^{-10}	10^{-8}	10^{-6}	1

(4) Volume

For pure water at 4°C (39.2°F), 1 cubic decimeter = 1 liter = 1 kilogram
hl = hectoliter, l = liter, m^3 = cubic meter
dm^3 = cubic decimeter, cm^3 = cubic centimeter, mm^3 = cubic millimeter

	hl	l	m^3	dm^3	cm^3	mm^3
hl	1	10^{2}	10^{-1}	10^{2}	10^{5}	10^{8}
l	10^{-2}	1	10^{-3}	1	10^{3}	10^{6}
m^3	10	10^{3}	1	10^{3}	10^{6}	10^{9}
dm^3	10^{-2}	1	10^{-3}	1	10^{3}	10^{6}
cm^3	10^{-5}	10^{-3}	10^{-6}	10^{-3}	1	10^{3}
mm^3	10^{-8}	10^{-6}	10^{-9}	10^{-6}	10^{-3}	1

B.03 SI SYSTEM—MASS, FORCE, UNIT WEIGHT, MOMENT

(1) Mass

t = ton, kg = kilogram, Dg = dekagram, g = gram, cg = centigram, mg = milligram

	t	kg	Dg	g	cg	mg
t	1	10^3	10^5	10^6	10^8	10^9
kg	10^{-3}	1	10^2	10^3	10^5	10^6
Dg	10^{-5}	10^{-2}	1	10	10^3	10^4
g	10^{-6}	10^{-3}	10^{-1}	1	10^2	10^3
cg	10^{-8}	10^{-5}	10^{-3}	10^{-2}	1	10
mg	10^{-9}	10^{-6}	10^{-4}	10^{-3}	10^{-1}	1

(2) Force ($\alpha = 9.806\,650$, $\beta = 1.019\,716$)

t_f = ton-force, kg_f = kilogram-force, g_f = gram-force
J/m = joule/meter, N = newton, dyn = dyne

	t_f	kg_f	g_f	J/m	N	dyn
t_f	1	10^3	10^6	$\alpha \times 10^3$	$\alpha \times 10^3$	$\alpha \times 10^8$
kg_f	10^{-3}	1	10^3	α	α	$\alpha \times 10^5$
g_f	10^{-6}	10^{-3}	1	$\alpha \times 10^{-3}$	$\alpha \times 10^{-3}$	$\alpha \times 10^2$
J/m	$\beta \times 10^{-4}$	$\beta \times 10^{-1}$	$\beta \times 10^2$	1	1	10^5
N	$\beta \times 10^{-4}$	$\beta \times 10^{-1}$	$\beta \times 10^2$	1	1	10^5
dyn	$\beta \times 10^{-9}$	$\beta \times 10^{-6}$	$\beta \times 10^{-3}$	10^{-5}	10^{-5}	1

(3) Unit Weight

t_f/m^3 = ton-force/cubic meter, kg_f/m^3 = kilogram-force/cubic meter
kg_f/cm^3 = kilogram-force/cubic centimeter, g_f/cm^3 = gram-force/cubic centimeter
cg_f/cm^3 = centigram-force/cubic centimeter, mg_f/cm^3 = milligram-force/cubic centimeter

	t_f/m^3	kg_f/m^3	kg_f/cm^3	g_f/cm^3	cg_f/cm^3	mg_f/cm^3
t_f/m^3	1	10^3	10^{-3}	1	10^2	10^3
kg_f/m^3	10^{-3}	1	10^{-6}	10^{-3}	10^{-1}	1
kg_f/cm^3	10^3	10^6	1	10^3	10^5	10^6
g_f/cm^3	1	10^3	10^{-3}	1	10^2	10^3
cg_f/cm^3	10^{-2}	10	10^{-5}	10^{-2}	1	10
mg_f/cm^3	10^{-3}	1	10^{-6}	10^{-3}	10^{-1}	1

(4) Moment ($\alpha = 9.806\,650$, $\beta = 1.019\,716$)

$t_f \cdot m$ = ton-force × meter, $kg_f \cdot m$ = kilogram-force × meter
$g_f \cdot cm$ = gram-force × centimeter, J = joule
N·m = newton × meter, dyn·cm = dyne × centimeter

	$t_f \cdot m$	$kg_f \cdot m$	$g_f \cdot cm$	J	N·m	dyn·cm
$t_f \cdot m$	1	10^3	10^8	$\alpha \times 10^3$	$\alpha \times 10^3$	$\alpha \times 10^{10}$
$kg_f \cdot m$	10^{-3}	1	10^5	α	α	$a \times 10^7$
$g_f \cdot cm$	10^{-8}	10^{-5}	1	$\alpha \times 10^{-5}$	$\alpha \times 10^{-5}$	$\alpha \times 10^2$
J	$\beta \times 10^{-4}$	$\beta \times 10^{-1}$	$\beta \times 10^6$	1	1	10^7
N·m	$\beta \times 10^{-4}$	$\beta \times 10^{-1}$	$\beta \times 10^6$	1	1	10^7
dyn·cm	$\beta \times 10^{-11}$	$\beta \times 10^{-8}$	$\beta \times 10^{-1}$	10^{-7}	10^{-7}	1

B.04 SI SYSTEM—VELOCITY, ACCELERATION, STATIC AND INERTIAL MOMENT, PRESSURE

(1) Velocity and Acceleration

km = kilometer, m = meter, cm = centimeter, h = hour, min = minute, s = second

	km/h	m/min	cm/s
km/h	1	$1.666\,667 \times 10$	$2.777\,778 \times 10$
m/min	$6.000\,000 \times 10^{-2}$	1	1.666 667
cm/s	$3.600\,000 \times 10^{-2}$	$6.000\,000 \times 10^{-1}$	1

	km/h^2	m/min^2	cm/s^2
km/h^2	1	$2.777\,778 \times 10^{-1}$	$7.716\,049 \times 10^{-3}$
m/min^2	3.600 000	1	$2.777\,778 \times 10^{-2}$
cm/s^2	$1.296\,000 \times 10^{2}$	$3.600\,000 \times 10$	1

(2) Static and Inertial Moment of Area

m^3 = meter^3, dm^3 = decimeter^3, cm^3 = centimeter^3
m^4 = meter^4, dm^4 = decimeter^4, cm^4 = centimeter^4

	m^3	dm^3	cm^3
m^3	1	10^3	10^6
dm^3	10^{-3}	1	10^3
cm^3	10^{-6}	10^{-3}	1

	m^4	dm^4	cm^4
m^4	1	10^4	10^8
dm^4	10^{-4}	1	10^4
cm^4	10^{-8}	10^{-4}	1

(3) Static and Inertial Moment of Volume

m^5 = meter^5, dm^5 = decimeter^5, cm^5 = centimeter^5

	m^4	dm^4	cm^4
m^4	1	10^4	10^8
dm^4	10^{-4}	1	10^4
cm^4	10^{-8}	10^{-4}	1

	m^5	dm^5	cm^5
m^5	1	10^5	10^{10}
dm^5	10^{-5}	1	10^5
cm^5	10^{-10}	10^{-5}	1

(4) Static and Inertial Moment of Mass

t·m = ton × meter, kg·m = kilogram × meter, kg·cm = kilogram × centimeter
t·m^2 = ton × meter^2, kg·m^2 = kilogram × meter^2, kg·cm^2 = kilogram × centimeter^2

	t·m	kg·m	kg·cm
t·m	1	10^3	10^5
kg·m	10^{-3}	1	10^2
kg·cm	10^{-5}	10^{-2}	1

	t·m^2	kg·m^2	kg·cm^2
t·m^2	1	10^3	10^7
kg·m^2	10^{-3}	1	10^4
kg·cm^2	10^{-7}	10^{-4}	1

(5) Pressure ($\alpha = 9.806\,650$, $\beta = 1.019\,716$)

t_f/m^2 = ton-force/square meter, kg_f/m^2 = kilogram-force/square meter, at = technical atmosphere
N/m^2 = newton/square meter, dyn/cm^2 = dyne/square centimeter, mb = millibar

	t_f/m^2	kg_f/m^2	at	N/m^2	dyn/cm^2	mb
t_f/m^2	1	10^3	10^{-1}	$\alpha \times 10^3$	$\alpha \times 10^4$	$\alpha \times 10$
kg_f/m^2	10^{-3}	1	10^{-4}	α	$\alpha \times 10$	$\alpha \times 10^{-2}$
at	10	10^4	1	$\alpha \times 10^4$	$\alpha \times 10^5$	$\alpha \times 10^2$
N/m^2	$\beta \times 10^{-4}$	$\beta \times 10^{-1}$	$\beta \times 10^{-5}$	1	10	10^{-2}
dyn/cm^2	$\beta \times 10^{-5}$	$\beta \times 10^{-2}$	$\beta \times 10^{-6}$	10^{-1}	1	10^{-3}
mb	$\beta \times 10^{-2}$	$\beta \times 10$	$\beta \times 10^{-3}$	10^2	10^3	1

B.05 SI SYSTEM—ENERGY, WORK, POWER, TEMPERATURE

(1) Energy, Work

J = joule, $m \cdot kg_f$ = meter × kilogram-force, kWh = kilowatt × hour
hp·h = horsepower × hour, kcal = kilocalorie

	J	$m \cdot kg_f$	kWh	hp·h	kcal
J	1	$1.019\,716 \times 10^{-1}$	$2.777\,778 \times 10^{-7}$	$3.776\,765 \times 10^{-7}$	$2.389\,200 \times 10^{-4}$
$m \cdot kg_f$	9.806 650	1	$2.724\,070 \times 10^{-6}$	$3.703\,742 \times 10^{-6}$	$2.343\,005 \times 10^{-3}$
kWh	$3.600\,000 \times 10^{6}$	$3.670\,977 \times 10^{5}$	1	$1.359\,635 \times 10^{-1}$	$8.601\,119 \times 10^{2}$
hp·h	$2.647\,768 \times 10^{6}$	$2.698\,542 \times 10^{5}$	$7.354\,913 \times 10^{-1}$	1	$6.326\,048 \times 10^{2}$
kcal	$4.186\,501 \times 10^{3}$	$4.268\,576 \times 10^{2}$	$1.162\,639 \times 10^{-3}$	$1.580\,766 \times 10^{-3}$	1

(2) Power

W = watt, $m \cdot kg_f/s$ = meter × kilogram-force per second
kW = kilowatt, hp = horsepower, kcal/h = kilocalorie per hour

	W	$m \cdot kg_f/s$	kW	hp	kcal/h
W	1	$1.019\,716 \times 10^{-1}$	10^{-3}	$1.359\,635 \times 10^{-3}$	$8.601\,119 \times 10^{-1}$
$m \cdot kg_f/s$	9.806 650	1	$9.806\,650 \times 10^{-3}$	$1.333\,347 \times 10^{-2}$	8.448 181
kW	10^{3}	$1.019\,716 \times 10^{2}$	1	1.359 635	$8.601\,119 \times 10^{2}$
hp	$7.354\,915 \times 10^{2}$	$7.500\,000 \times 10$	$7.359\,915 \times 10^{-1}$	1	$6.326\,050 \times 10^{2}$
kcal/h	1.162 629	$1.185\,552 \times 10^{-1}$	$1.162\,629 \times 10^{-3}$	$1.580\,752 \times 10^{-3}$	1

(3) Temperature

°C = 1 degree Celsius, °K = 1 degree Kelvin, °F = 1 degree Fahrenheit
°R = 1 degree Rankine, °Re = 1 degree Réaumur

	°C	°K	°F	°R	°Re
1°C	1	1	9/5	9/5	4/5
1°K	1	1	9/5	9/5	4/5
1°F	5/9	5/9	1	1	4/9
1°R	5/9	5/9	1	1	4/9
1°Re	5/4	5/4	9/4	9/4	1
Absolute zero temperature	−273.15	0	−459.67	0	−218.53
Normal freezing point of water	0	273.15	32	491.67	0
Normal boiling point of water	100	373.15	212	671.67	80

B.06 FPS SYSTEM—LENGTH, AREA, VOLUME

(1) Length

	mile	rod	yard	foot	inch
mile	1	$3.200\,000 \times 10^{2}$	$1.760\,000 \times 10^{3}$	$5.280\,000 \times 10^{3}$	$6.336\,000 \times 10^{4}$
rod	$3.125\,000 \times 10^{-3}$	1	5.500 000	$1.650\,000 \times 10$	$1.980\,000 \times 10^{2}$
yard	$5.681\,800 \times 10^{-4}$	$1.818\,182 \times 10^{-1}$	1	3.000 000	$3.600\,000 \times 10$
foot	$1.893\,900 \times 10^{-4}$	$6.060\,606 \times 10^{-2}$	$3.333\,333 \times 10^{-1}$	1	$1.200\,000 \times 10$
inch	$1.578\,282 \times 10^{-5}$	$5.050\,505 \times 10^{-3}$	$2.777\,778 \times 10^{-2}$	$8.333\,333 \times 10^{-2}$	1

(2) Area

	$mile^2$	acre	$yard^2$	$foot^2$	$inch^2$
$mile^2$	1	$6.400\,000 \times 10^{2}$	$3.097\,600 \times 10^{6}$	$2.787\,840 \times 10^{7}$	$4.014\,490 \times 10^{9}$
acre	$1.562\,500 \times 10^{-3}$	1	$4.840\,000 \times 10^{3}$	$4.356\,000 \times 10^{4}$	$6.272\,640 \times 10^{6}$
$yard^2$	$3.228\,300 \times 10^{-7}$	$2.066\,115 \times 10^{-4}$	1	9.000 000	$1.296\,000 \times 10^{3}$
$foot^2$	$3.587\,000 \times 10^{-8}$	$2.295\,684 \times 10^{-5}$	$1.111\,111 \times 10^{-1}$	1	$1.440\,000 \times 10^{2}$
$inch^2$	$2.490\,972 \times 10^{-10}$	$1.594\,225 \times 10^{-7}$	$7.716\,049 \times 10^{-4}$	$6.944\,444 \times 10^{-3}$	1

(3) Volume (dry)

	bushel	$foot^3$	peck	quart	pint
bushel	1	1.244 500	4.000 000	$3.200\,000 \times 10$	$6.400\,000 \times 10$
$foot^3$	$8.035\,354 \times 10^{-1}$	1	3.214 141	$2.571\,314 \times 10$	$5.142\,627 \times 10$
peck	$2.500\,000 \times 10^{-1}$	$3.111\,250 \times 10^{-1}$	1	8.000 000	$1.600\,000 \times 10$
quart	$3.125\,000 \times 10^{-2}$	$3.889\,062 \times 10^{-2}$	$1.250\,000 \times 10^{-1}$	1	2.000 000
pint	$1.562\,500 \times 10^{-2}$	$1.944\,531 \times 10^{-2}$	$6.250\,000 \times 10^{-2}$	$5.000\,000 \times 10^{-1}$	1

(4) Volume (liquid)

	$foot^3$	gallon	quart	pint	gill
$foot^3$	1	7.480 520	$2.992\,209 \times 10$	$5.984\,418 \times 10$	$2.393\,767 \times 10^{2}$
gallon	$1.336\,805 \times 10^{-1}$	1	4.000 000	8.000 000	$3.200\,000 \times 10$
quart	$3.342\,012 \times 10^{-2}$	$2.500\,000 \times 10^{-1}$	1	2.000 000	8.000 000
pint	$1.671\,006 \times 10^{-2}$	$1.250\,000 \times 10^{-1}$	$5.000\,000 \times 10^{-1}$	1	4.000 000
gill	$4.177\,515 \times 10^{-3}$	$3.125\,000 \times 10^{-2}$	$1.250\,000 \times 10^{-1}$	$2.500\,000 \times 10^{-1}$	1

B.07 FPS SYSTEM—WEIGHT, VELOCITY, ACCELERATION, MOMENT, ENERGY, POWER

(1) Weight (avoirdupois)

	ton-force	pound-force	ounce-force	dram-force	grain-force
ton-force	1	$2.000\,000 \times 10^{3}$	$3.200\,000 \times 10^{4}$	$5.120\,000 \times 10^{5}$	$1.400\,000 \times 10^{7}$
pound-force	$5.000\,000 \times 10^{-4}$	1	$1.600\,000 \times 10$	$2.560\,000 \times 10^{2}$	$7.000\,000 \times 10^{3}$
ounce-force	$3.125\,000 \times 10^{-5}$	$6.250\,000 \times 10^{-2}$	1	$1.600\,000 \times 10$	$4.375\,000 \times 10^{2}$
dram-force	$1.953\,125 \times 10^{-6}$	$3.906\,250 \times 10^{-3}$	$6.250\,000 \times 10^{-2}$	1	$2.734\,375 \times 10^{1}$
grain-force	$7.142\,857 \times 10^{-8}$	$1.428\,571 \times 10^{-4}$	$2.285\,714 \times 10^{-3}$	$3.657\,142 \times 10^{-2}$	1

(2) Velocity and Acceleration

	knot	foot/second
mile/hour	$8.684\,000 \times 10^{-1}$	1.466 667
knot	1	1.689 929
foot/minute	$9.868\,234 \times 10^{-3}$	$1.666\,667 \times 10^{-2}$
foot/second	$5.920\,940 \times 10^{-1}$	1
inch/second	$4.934\,117 \times 10^{-2}$	$8.333\,333 \times 10^{-2}$

	foot/second^2	inch/second^2
mile/hour^2	$4.074\,074 \times 10^{-4}$	$4.888\,889 \times 10^{-3}$
mile/minute^2	1.466 667	$1.760\,000 \times 10$
foot/minute^2	$2.777\,778 \times 10^{-4}$	$3.333\,333 \times 10^{-3}$
foot/second^2	1	$1.200\,000 \times 10$
inch/second^2	$8.333\,333 \times 10^{-2}$	1

(3) Static and Inertial Moment of Mass (kip = 1,000 pounds)

	pound-foot	pound-inch
ton-foot	$2.000\,000 \times 10^{3}$	$2.400\,000 \times 10^{4}$
kip-foot	$1.000\,000 \times 10^{3}$	$1.200\,000 \times 10^{4}$
kip-inch	$8.333\,333 \times 10$	$1.000\,000 \times 10^{3}$
pound-foot	1	$1.200\,000 \times 10$
pound-inch	$8.333\,333 \times 10^{-2}$	1

	pound-foot^2	pound-inch^2
ton-foot^2	$2.000\,000 \times 10^{3}$	$2.880\,000 \times 10^{5}$
kip-foot^2	$1.000\,000 \times 10^{3}$	$1.440\,000 \times 10^{5}$
kip-inch^2	$6.944\,444 \times 10$	$1.000\,000 \times 10^{3}$
pound-foot^2	1	$1.440\,000 \times 10^{2}$
pound-inch^2	$6.944\,444 \times 10^{-2}$	1

(4) Energy, Work, and Power (hp = horsepower, Btu = British thermal unit)

	Btu	foot-pound-force
joule	$9.480\,500 \times 10^{-4}$	$7.375\,600 \times 10^{-1}$
hp-hour	$2.547\,000 \times 10^{3}$	$1.980\,000 \times 10^{6}$
Btu	1	$7.780\,000 \times 10^{3}$
foot-pound-force	$1.285\,347 \times 10^{-3}$	1
hp-min	$4.245\,000 \times 10$	$3.300\,000 \times 10^{4}$

	Btu/min	foot-pound/min
watt	$5.688\,282 \times 10^{-2}$	$4.425\,360 \times 10$
hp	$4.245\,000 \times 10$	$3.300\,000 \times 10^{4}$
Btu/min	1	$7.780\,000 \times 10^{2}$
foot-pound-f/min	$1.285\,347 \times 10^{-3}$	1
foot-pound-f/sec	$7.717\,000 \times 10^{-2}$	$6.000\,000 \times 10$

B.08 CONVERSION FACTORS—LENGTH, AREA*

(1) Length

1 meter = 3.280 840 (+00) ft

1 foot = 3.048 000 (−01) meter

angstrom	1.000 000 (−10) meter
astronomical unit	1.495 980 (+11) meter
centimeter	3.937 007 (−01) inch
decimeter	3.937 007 (+00) inch
inch	2.540 000 (−02) meter
kilometer	6.213 711 (−01) mile (U.S. st.)
kilometer	3.280 839 (+03) foot
light year	9.460 550 (+15) meter
light year	6.323 982 (+04) astronomical unit
light year	5.878 512 (+12) mile (U.S. st.)
meter	3.937 007 (+01) inch
mile (U.S. st.)	1.609 344 (+00) kilometer
mile (U.S. naut.)	1.852 000 (+00) kilometer
millimeter	3.937 007 (−02) inch
parsec	3.083 740 (+13) kilometer
parsec	1.916 147 (+13) mile (U.S. st.)
rod	5.029 200 (+00) meter
yard	9.144 000 (−01) meter

(2) Area

1 meter^2 = 1.076 391 (+01) foot^2

1 foot^2 = 9.290 304 (−02) meter^2

acre	4.356 000 (+04) foot^2
acre	4.046 856 (+03) meter^2
acre	1.562 500 (−03) mile^2
are	1.000 000 (+02) meter^2
are	1.076 390 (+03) foot^2
barn	1.000 000 (−28) meter^2
centimeter^2	1.550 003 (−01) inch^2
decimeter^2	1.550 003 (+01) inch^2
circular mil	5.067 075 (−10) meter^2
hectare	1.000 000 (+04) meter^2
hectare	1.076 391 (+05) foot^2
inch^2	6.451 600 (−04) meter^2
kilometer^2	2.471 105 (+02) acre
kilometer^2	3.861 102 (−01) mile^2
meter^2	1.550 000 (+03) inch^2
meter^2	1.195 985 (+00) yard^2
mile^2 (= section)	2.589 988 (+00) kilometer^2
yard^2	8.361 274 (−01) meter^2

All conversion factors are given in the scientific notation of Sec. 12.03–1*c*, where the exponent of 10 is the signed number in parentheses. Example: 1 foot = 3.048 000 (−01) meter = 3.048 000 × 10^{-1} meter = 0.3048 meter.

B.09 CONVERSION FACTORS—VOLUME

(1) Volume (dry)*

1 meter^3 = 3.531 466 (+01) foot^3

1 foot^3 = 2.831 685 (−02) meter^3

board foot	2.359 737 (−03) meter^3
bushel (U.S.)	3.523 907 (−02) meter^3
centimeter^3	6.102 375 (−02) inch^3
cord	3.624 556 (+00) meter^3
decimeter^3	3.531 466 (−02) foot^3
gallon (U.S.)	4.404 883 (−03) meter^3
gallon (U.S.)	4.404 883 (+00) liter
inch^3	1.638 706 (−05) meter^3
inch^3	1.638 706 (+01) centimeter^3
meter^3	1.000 000 (+03) liter
meter^3	1.307 950 (+00) yard^3
peck (U.S.)	8.809 768 (−03) meter^3
pint (U.S.)	5.506 105 (−04) meter^3
quart (U.S.)	1.101 221 (−03) meter^3
yard^3	7.645 549 (−01) meter^3
yard^3	2.700 000 (+01) foot^3

(2) Volume (liquid)

1 liter = 2.641 720 (−01) gallon (U.S.)

1 gallon (U.S.) = 3.785 412 (+00) liter

acre foot	1.233 482 (+03) meter^3
cup	2.365 882 (−04) meter^3
dram	3.696 691 (−06) meter^3
gallon (U.S.)	3.785 412 (−03) meter^3
gallon (U.S.)	4.951 130 (−03) yard^3
gallon (U.S.)	1.336 805 (−01) foot^3
gallon (U.S.)	2.310 000 (+02) inch^3
gallon (Brit.)	4.546 087 (−03) meter^3
gill (U.S.)	1.182 941 (−04) meter^3
liter	1.000 000 (−03) meter^3
liter	1.000 000 (+00) decimeter^3
liter	1.307 950 (−03) yard^3
liter	2.200 000 (−01) gallon (Brit.)
ounce	2.957 353 (−05) meter^3
pint	4.731 765 (−04) meter^3
quart	9.463 530 (−04) meter^3
tablespoon	1.478 676 (−05) meter^3
teaspoon	4.928 922 (−06) meter^3

*Meter^3, decimeter^3, centimeter^3, yard^3, foot^3, and inch^3 are also used as liquid measures.

B.10 CONVERSION FACTORS—MASS, FORCE

(1) Mass

1 kilogram = 2.204 622 (+00) pound

1 pound = 4.535 924 (−01) kilogram

carat (SI)	2.000 000 (−04) kilogram
dram (avoirdupois)	1.771 845 (−03) kilogram
dram (troy)	3.887 935 (−03) kilogram
grain	6.479 891 (−05) kilogram
gram	1.000 000 (−03) kilogram
hundredweight (short)	4.535 923 (+01) kilogram
hundredweight (long)	5.080 235 (+01) kilogram
kilogram	3.527 396 (+01) ounce (avoirdupois)
ounce (avoirdupois)	2.834 952 (−02) kilogram
ounce (troy)	3.110 348 (−02) kilogram
pound (troy)	3.732 417 (−01) kilogram
slug	1.459 390 (+01) kilogram
ton (short)	9.071 847 (+02) kilogram
ton (short)	2.000 000 (+03) pound (avoirdupois)
ton (long)	1.016 047 (+03) kilogram
ton (long)	2.240 000 (+03) pound (avoirdupois)
ton (SI)	1.000 000 (+03) kilogram
ton (SI)	2.204 622 (+03) pound (avoirdupois)

(2) Force

1 newton = 1.019 716 (−01) kilogram-force

1 kilogram-force = 9.806 650 (+00) newton

dyne	1.000 000 (−05) newton
dyne	1.019 716 (−06) kilogram-force
gram-force	1.000 000 (−03) kilogram-force
gram-force	2.204 622 (−03) pound-force
joule/meter	1.019 716 (−01) kilogram-force
joule/meter	2.248 089 (−01) pound-force
kilogram-force	9.806 650 (+00) newton
kilogram-force	2.204 622 (+00) pound-force
kilopound-force (kip)	4.448 222 (+03) newton
kilopound-force (kip)	4.535 923 (+02) kilogram-force
newton	2.248 089 (−01) pound-force
ounce-force	2.780 139 (−01) newton
ounce-force	2.834 952 (−02) kilogram-force
pound-force	4.448 222 (+00) newton
pound-force	4.535 924 (−01) kilogram-force
poundal	1.382 550 (−01) newton
ton-force (SI)	9.806 650 (+03) newton
ton-force (SI)	2.204 622 (+03) pound-force

B.11 CONVERSION FACTORS—UNIT MASS, UNIT WEIGHT, MOMENT

(1) Unit Mass

1 gram/centimeter^3 = 3.612 729 (−02) pound/inch^3

1 pound/inch^3 = 2.767 991 (+01) gram/centimeter^3

gram/centimeter^3	1.000 000 (+03) kilogram/meter^3
kilogram/centimeter^3	1.000 000 (+06) kilogram/meter^3
kilogram/centimeter^3	6.242 800 (+04) pound/foot^3
kilogram/centimeter^3	3.612 720 (+01) pound/inch^3
kilogram/meter^3	6.242 800 (−02) pound/foot^3
kilogram/meter^3	3.612 729 (−05) pound/inch^3
kilopound/foot^3 (kip)	1.601 846 (+04) kilogram/meter^3
pound/inch^3	2.767 991 (+04) kilogram/meter^3
pound/inch^3	1.728 000 (+03) pound/foot^3
pound/foot^3	1.601 846 (+01) kilogram/meter^3
pound/foot^3	5.787 037 (−04) pound/inch^3

(2) Unit Weight*

1 kilogramf/meter^3 = 6.242 800 (−02) poundf/foot^3

1 poundf/foot^3 = 1.601 846 (+01) kilogramf/meter^3

dyne/centimeter^3	1.019 716 (+00) kilogramf/meter^3
dyne/centimeter^3	6.365 883 (−02) poundf/foot^3
kilogramf/centimeter^3	1.000 000 (+06) kilogramf/meter^3
kilogramf/centimeter^3	6.242 800 (+04) poundf/foot^3
kilogramf/meter^3	9.806 650 (+00) newton/meter^3
newton/meter^3	1.019 716 (−01) kilogramf/meter^3
newton/meter^3	6.365 883 (−03) poundf/foot^3
poundf/inch^3	2.767 991 (+04) kilogramf/meter^3
poundf/foot^3	1.570 874 (+02) newton/meter^3
poundf/foot^3	1.601 846 (+01) kilogramf/meter^3
poundf/yard^3	5.932 763 (−01) kilogramf/meter^3

(3) Moment*

1 kilogramf-meter = 7.233 016 (+00) poundf-foot

1 poundf-foot = 1.382 549 (−01) kilogramf-meter

dyne-centimeter	1.000 000 (−07) newton-meter
dyne-centimeter	1.019 716 (−08) kilogramf-meter
kilogramf-meter	9.806 650 (+00) newton-meter
kilogramf-meter	7.233 016 (+00) poundf-foot
newton-meter	1.019 716 (−01) kilogramf-meter
newton-meter	7.375 622 (−01) poundf-foot
poundf-inch	1.152 124 (−02) kilogramf-meter
poundf-inch	1.129 848 (−01) newton-meter
poundf-foot	1.382 549 (−01) kilogramf-meter
poundf-foot	1.355 818 (+00) newton-meter
tonf (SI)-meter	9.806 650 (+03) newton-meter
tonf (SI)-meter	7.233 016 (+03) poundf-foot

*The short forms kilogramf, poundf, and tonf stand for kilogram-force, pound-force, and ton-force, respectively.

B.12 CONVERSION FACTORS—VELOCITY, ACCELERATION, VISCOSITY

(1) Velocity

1 meter/second = 3.280 839 (+00) foot/second

1 foot/second = 3.048 000 (−01) meter/second

centimeter/second	3.600 000 (+01) meter/hour
centimeter/second	3.600 000 (−02) kilometer/hour
centimeter/second	3.937 007 (−01) inch/second
centimeter/second	3.280 839 (−02) foot/second
centimeter/second	1.968 503 (+00) foot/minute
centimeter/second	1.181 102 (+02) foot/hour
foot/minute	5.080 000 (−03) meter/second
foot/hour	8.466 667 (−05) meter/second
inch/second	2.540 000 (−02) meter/second
kilometer/hour	2.777 778 (−01) meter/second
kilometer/hour	6.213 711 (−01) mile (U.S. st.)/hour
knot (international)	1.852 000 (+00) kilometer/hour
knot (international)	1.000 000 (+00) mile (U.S. naut.)/hour
mile (U.S. st.)/hour	1.609 344 (+00) kilometer/hour
mile (U.S. naut.)/hour	1.852 000 (+00) kilometer/hour
mile (U.S. st.)/minute	2.682 224 (+01) meter/second
mile (U.S. st.)/second	1.609 344 (+03) meter/second
yard/second	9.144 000 (−01) meter/second

(2) Acceleration

1 centimeter/second^2 = 3.937 007 (−01) inch/second^2

1 inch/second^2 = 2.540 000 (+00) centimeter/second^2

centimeter/second^2	3.280 839 (−02) foot/second^2
centimeter/minute^2	9.113 414 (−06) foot/second^2
centimeter/minute^2	1.093 613 (−05) inch/second^2
foot/second^2	3.048 000 (−01) meter/second^2
foot/minute^2	8.466 667 (−05) meter/second^2
inch/minute^2	7.055 556 (−06) meter/second^2

(3) Viscosity

1 meter^2/second = 1.076 390 (+01) foot^2/second

1 foot^2/second = 9.290 304 (−02) meter^2/second

centipoise	1.000 000 (−03) newton × second/meter^2
centistoke	1.000 000 (−06) meter^2/second
kilogram/meter × second	6.719 689 (−01) pound/foot × second
pound/foot × second	1.488 164 (+00) kilogram/meter × second
poise	1.000 000 (−01) newton × second/meter^2
stoke	1.000 000 (−04) meter^2/second

B.13 CONVERSION FACTORS—STATIC AND INERTIAL MOMENT, PRESSURE

(1) Static Moment of Volume

1 meter^4 = 1.158 616 (+02) foot^4

1 foot^4 = 8.630 975 (−03) meter^4

centimeter^4 .... 2.402 507 (−02) inch^4
inch^4 .... 4.162 319 (+01) centimeter^4

(2) Inertial Moment of Volume

1 meter^5 = 3.801 233 (+02) foot^5

1 foot^5 = 2.630 725 (−03) meter^5

centimeter^5 .... 9.458 689 (−03) inch^5
inch^5 .... 1.057 228 (+02) centimeter^5

(3) Static Moment of Mass

1 kilogram-meter = 7.233 017 (+00) pound-foot

1 pound-foot = 1.382 549 (−01) kilogram-meter

kilogram-centimeter .... 8.679 614 (−01) pound-inch
pound-inch .... 1.152 125 (−02) kilogram-meter

(4) Inertial Moment of Mass

1 kilogram-meter^2 = 2.373 034 (+01) pound-foot^2

1 pound-foot^2 = 4.214 012 (−02) kilogram-meter^2

kilogram-centimeter^2 .... 3.417 171 (−01) pound-inch^2
pound-inch^2 .... 2.926 397 (+00) kilogram-centimeter^2

(5) Pressure

1 kilogramf/centimeter^2 = 1.422 334 (+01) poundf/inch^2

1 poundf/inch^2 = 7.030 694 (−02) kilogramf/centimeter^2

atmosphere (tech) .... 1.000 000 (+04) kilogramf/meter^2
atmosphere (tech) .... 2.048 161 (+03) poundf/foot^2
bar .... 1.019 716 (+04) kilogramf/meter^2
bar .... 2.088 543 (+03) poundf/foot^2
centimeter of mercury (0°C) .... 1.359 961 (+02) kilogramf/meter^2
centimeter of water (4°C) .... 1.000 000 (+01) kilogramf/meter^2
inch of mercury (0°C) .... 3.453 155 (+02) kilogram/meter^2
inch of water (4°C) .... 2.539 929 (+01) kilogram/meter^2
kilogramf/centimeter^2 .... 1.000 000 (+00) atmosphere (technical)
poundf/foot^2 .... 4.882 423 (+00) kilogramf/meter^2
torr (0°C) .... 1.359 506 (−03) kilogramf/centimeter^2

B.14 CONVERSION FACTORS—ENERGY, WORK, POWER

(1) Energy, Work

1 joule = 1.019 716 (−01) meter-kilogramf

= 7.375 616 (−01) foot-poundf

Btu (British thermal unit-mean)	1.054 800 (+03) joule
Btu (British thermal unit-mean)	1.075 596 (+02) meter-kilogram-force
Btu (British thermal unit-mean)	7.779 799 (+02) foot-pound-force
calorie (technical)	4.186 501 (+00) joule
calorie (technical)	4.269 042 (−01) meter-kilogram-force
calorie (technical)	3.087 802 (+00) foot-pound-force
electron volt	1.602 100 (−19) joule
erg	1.000 000 (−07) joule
erg	1.019 716 (−08) meter-kilogram-force
erg	7.375 616 (−08) foot-pound-force
foot-pound-force	1.355 818 (+00) joule
foot-pound-force	1.382 550 (−01) meter-kilogram-force
horsepower-hour (SI)	2.700 000 (+05) meter-kilogram-force
horsepower-hour (SI)	1.952 913 (+06) foot-pound-force
horsepower-hour (SI)	6.324 603 (+05) calorie (tech)
kilowatt-hour	3.600 000 (+06) joule
kilowatt-hour	3.670 098 (+05) meter-kilogram-force
kilowatt-hour	2.655 222 (+06) foot-pound-force

(2) Power

1 watt = 1.019 716 (−01) meter-kilogramf/second

= 7.375 616 (−01) foot-poundf/second

Btu/second	1.054 800 (+03) watt
Btu/second	1.075 596 (+02) meter-kilogramf/second
Btu/second	7.779 799 (+02) foot-poundf/second
calorie/hour (technical)	1.162 917 (−03) watt
calorie/hour (technical)	1.185 845 (−04) meter-kilogramf/second
calorie/hour (technical)	8.577 228 (−04) foot-poundf/second
foot-poundf/hour	3.766 161 (−04) watt
foot-poundf/minute	2.259 697 (−02) watt
foot-poundf/second	1.355 818 (+00) watt
horsepower (SI)	7.354 988 (+02) watt
horsepower (SI)	5.424 476 (+02) foot-poundf/second
horsepower (FPS)	7.456 999 (+02) watt
horsepower (FPS)	5.500 000 (+02) foot-poundf/second
horsepower (boiler)	9.809 500 (+03) watt
horsepower (boiler)	7.235 111 (+03) foot-poundf/second
horsepower (electric)	7.460 000 (+02) watt
horsepower (electric)	5.502 210 (+02) foot-poundf/second
horsepower (water)	7.460 430 (+02) watt
horsepower (water)	5.502 527 (+02) foot-poundf/second

B.15 CONVERSION FACTORS—ELECTRICITY

(1) Electrostatic Units

1 coulomb = ampere-second $1 \text{ ohm} = \dfrac{\text{volt}}{\text{ampere}}$

$1 \text{ farad} = \dfrac{\text{ampere-second}}{\text{volt}}$ $1 \text{ volt} = \dfrac{\text{watt}}{\text{ampere}}$

$1 \text{ henry} = \dfrac{\text{volt-second}}{\text{ampere}}$ 1 weber = volt-second

ampere	2.997 925 (+09) statampere
coulomb	2.997 925 (+09) statcoulomb
farad	1.000 000 (+00) $\dfrac{\text{ampere}^2\text{-second}^2}{\text{newton-meter}}$
farad	8.987 554 (+11) statfarad
henry	1.000 000 (+00) joule/ampere^2
henry	1.112 650 (−12) stathenry
ohm	1.000 000 (+00) watt/ampere^2
ohm	1.112 650 (−12) statohm
statampere	3.335 640 (−10) ampere
statcoulomb	3.335 640 (−10) coulomb
statfarad	1.112 650 (−12) farad
stathenry	8.987 554 (+11) henry
statohm	8.987 554 (+11) ohm
statvolt	2.997 925 (+02) volt
statweber	2.997 925 (+02) weber
volt	1.000 000 (+00) watt/ampere
volt	3.335 640 (−03) statvolt
weber	3.335 640 (−03) statweber

(2) Electromagnetic Units

maxwell = 1.000 000 (−08) weber

tesla = 1.000 000 (+00) weber/meter^2

abampere	1.000 000 (+01) ampere
abcoulomb	1.000 000 (+01) coulomb
abfarad	1.000 000 (+09) farad
abhenry	1.000 000 (−09) henry
abohm	1.000 000 (−09) ohm
abvolt	1.000 000 (−08) volt
ampere	1.000 000 (−01) abampere
ampere turn	1.256 637 (+00) gilbert
coulomb	1.000 000 (−01) abcoulomb
farad	1.000 000 (−09) abfarad
gamma	1.000 000 (−09) tesla
gauss	1.000 000 (−04) tesla
gilbert	7.957 747 (−01) ampere turn
henry	1.000 000 (+09) abhenry
maxwell	1.000 000 (−08) weber
ohm	1.000 000 (+09) abohm
unit pole	1.256 637 (−07) weber
weber	1.000 000 (+08) maxwell

B.16 CONVERSION FACTORS—ANGLES, TIME

(1) Angles

$$1 \text{ degree} = \pi/180 \text{ radian} = 100/90 \text{ grad}$$

$$1 \text{ radian} = 180/\pi \text{ degree} = 200/\pi \text{ grad}$$

$$1 \text{ grad} = 90/100 \text{ degree} = \pi/200 \text{ radian}$$

1 grad = 1.570 796 (−02) radian
= 9.000 000 (−01) degree
= 5.400 000 (+01) minute
= 3.240 000 (+03) second

1 radian = 6.366 200 (+01) grad
= 5.729 578 (+01) degree
= 3.437 747 (+03) minute
= 2.062 648 (+05) second

1 degree = 1.111 111 (+00) grad
= 1.745 329 (−02) radian
= 6.000 000 (+01) minute
= 3.600 000 (+03) second

1 minute = 1.851 851 (−02) grad
= 2.908 882 (−04) radian
= 1.666 667 (−02) degree
= 6.000 000 (+01) second

1 second = 3.086 419 (−04) grad
= 4.848 137 (−06) radian
= 2.777 778 (−04) degree
= 1.666 667 (−02) minute

(2) Time

1 day = 24 hours = 1,440 minutes = 86,400 seconds

1 year = 3.650 000 (+02) day
= 8.760 000 (+03) hour
= 5.256 000 (+05) minute
= 3.153 600 (+07) second

1 day = 2.739 726 (−03) year
= 2.400 000 (+01) hour
= 1.440 000 (+03) minute
= 8.640 000 (+04) second

1 hour = 1.141 552 (−04) year
= 4.166 667 (−02) day
= 6.000 000 (+01) minute
= 3.600 000 (+03) second

1 minute = 1.902 587 (−07) year
= 6.944 444 (−04) day
= 1.666 667 (−02) hour
= 6.000 000 (+01) second

1 second = 3.170 979 (−08) year
= 1.157 741 (−05) day
= 2.777 778 (−04) hour
= 1.666 667 (−02) minute

B.17 CONVERSION TABLES—ANGULAR MEASURES

(1) Degrees to Radians

Deg.	0	1	2	3	4	5	6	7	8	9	Deg.
0°	0.00000	0.01745	0.03491	0.05236	0.06981	0.08727	0.10472	0.12217	0.13963	0.15708	0°
10°	0.17453	0.19199	0.20944	0.22689	0.24435	0.26180	0.27925	0.29671	0.31416	0.33161	10°
20°	0.34907	0.36652	0.38397	0.40143	0.41888	0.43633	0.45379	0.47124	0.48869	0.50615	20°
30°	0.52360	0.54105	0.55851	0.57596	0.59341	0.61087	0.62832	0.64577	0.66323	0.68068	30°
40°	0.69813	0.71558	0.73304	0.75049	0.76794	0.78540	0.80285	0.82030	0.83776	0.85521	40°
50°	0.87266	0.89012	0.90757	0.92502	0.94248	0.95993	0.97738	0.99484	1.01229	1.02974	50°
60°	1.04720	1.06465	1.08210	1.09956	1.11701	1.13446	1.15192	1.16937	1.18682	1.20428	60°
70°	1.22173	1.23918	1.25664	1.27409	1.29154	1.30900	1.32645	1.34390	1.36136	1.37881	70°
80°	1.39626	1.41372	1.43117	1.44862	1.46608	1.48353	1.50098	1.51844	1.53589	1.55334	80°
90°	1.57080	1.58825	1.60570	1.62316	1.64061	1.65806	1.67552	1.69297	1.71042	1.72788	90°
100°	1.74533										

(2) Minutes to Radians

Min.	0	1	2	3	4	5	6	7	8	9	Min.
0′	0.00000	0.00029	0.00058	0.00087	0.00116	0.00145	0.00175	0.00204	0.00233	0.00262	0′
10′	0.00291	0.00320	0.00349	0.00378	0.00407	0.00436	0.00465	0.00495	0.00524	0.00553	10′
20′	0.00582	0.00611	0.00640	0.00669	0.00698	0.00727	0.00756	0.00785	0.00814	0.00844	20′
30′	0.00873	0.00902	0.00931	0.00960	0.00989	0.01018	0.01047	0.01076	0.01105	0.01134	30′
40′	0.01164	0.01193	0.01222	0.01251	0.01280	0.01309	0.01338	0.01367	0.01396	0.01425	40′
50′	0.01454	0.01484	0.01513	0.01542	0.01571	0.01600	0.01629	0.01658	0.01687	0.01716	50′

(3) Seconds to Radians

Sec.	0	1	2	3	4	5	6	7	8	9	Sec.
0″	0.00000	0.00000	0.00001	0.00001	0.00002	0.00002	0.00003	0.00003	0.00004	0.00004	0″
10″	0.00005	0.00005	0.00006	0.00006	0.00007	0.00007	0.00008	0.00008	0.00009	0.00009	10″
20″	0.00010	0.00010	0.00011	0.00011	0.00012	0.00012	0.00013	0.00013	0.00014	0.00014	20″
30″	0.00015	0.00015	0.00016	0.00016	0.00016	0.00017	0.00017	0.00018	0.00018	0.00019	30″
40″	0.00019	0.00020	0.00020	0.00021	0.00021	0.00022	0.00022	0.00023	0.00023	0.00024	40″
50″	0.00024	0.00025	0.00025	0.00026	0.00026	0.00027	0.00027	0.00028	0.00028	0.00029	50″

(4) Radians to Degrees, Minutes, Seconds

Rad.	1,0000	0.1000	0.0100	0.0010	0.0001	Rad.
1	57°17′44.8″	5°43′46.5″	0°34′22.6″	0°03′26.3″	0°00′20.6″	1
2	114°35′29.6″	11°27′33.0″	1°08′45.3″	0°06′52.5″	0°00′41.3″	2
3	171°53′14.4″	17°11′19.4″	1°43′07.9″	0°10′18.8″	0°01′01.9″	3
4	229°10′59.2″	22°55′05.9″	2°17′30.6″	0°13′45.1″	0°01′22.5″	4
5	286°28′44.0″	28°38′52.4″	2°51′53.2″	0°17′11.3″	0°01′43.1″	5
6	343°46′28.8″	34°22′38.9″	3°26′15.9″	0°20′37.6″	0°02′03.8″	6
7	401°04′13.6″	40°06′25.4″	4°00′38.5″	0°24′03.9″	0°02′24.4″	7
8	458°21′58.4″	45°50′11.8″	4°35′01.2″	0°27′30.1″	0°02′45.0″	8
9	515°39′43.3″	51°33′58.3″	5°09′23.8″	0°30′56.4″	0°03′05.6″	9

B.18 CONVERSION FACTORS—TEMPERATURE, HEAT

(1) Temperature

C = degrees Celsius   R = degrees Rankine

F = degrees Fahrenheit   Re = degrees Reaumur

K = kelvins

Celsius to Fahrenheit degrees	$F = \frac{9}{5}C + 32$
Celsius to Kelvin	$K = C + 273.15$
Celsius to Rankine	$R = \frac{9}{5}C + 491.67$
Celsius to Reaumur	$Re = \frac{4}{5}C$
Fahrenheit to Celsius	$C = \frac{5}{9}(F - 32)$
Fahrenheit to Kelvin	$K = \frac{5}{9}(F + 459.67)$
Fahrenheit to Rankine	$R = F + 459.67$
Fahrenheit to Reaumur	$Re = \frac{4}{9}(F - 32)$
Kelvin to Celsius	$C = K - 273.15$
Kelvin to Fahrenheit	$F = \frac{9}{5}(K - 255.37)$
Kelvin to Rankine	$R = \frac{9}{5}K$
Kelvin to Reaumur	$Re = \frac{4}{5}(K - 273.15)$
Rankine to Celsius	$C = \frac{5}{9}(R - 491.67)$
Rankine to Fahrenheit	$F = R - 459.67$
Rankine to Kelvin	$K = \frac{5}{9}R$
Rankine to Reaumur	$Re = \frac{4}{9}(R - 491.67)$
Reaumur to Celsius	$C = \frac{5}{4}Re$
Reaumur to Fahrenheit	$F = \frac{9}{4}Re + 32$
Reaumur to Kelvin	$K = \frac{5}{4}Re + 273.15$
Reaumur to Rankine	$R = \frac{9}{4}Re + 491.67$

(2) Heat

1 calorie (thermochemical) = 4.184 000 (+00) joule

1 Btu (thermochemical) = 1.054 350 (+03) joule

$\frac{\text{calorie (thermochemical)}}{\text{second}}$	4.184 000 (+00) watt
$\frac{\text{calorie (thermochemical)}}{\text{kilogram}}$	4.184 000 (+00) $\frac{\text{joule}}{\text{kilogram}}$
$\frac{\text{calorie (thermochemical)}}{\text{meter}^2 \times \text{second}}$	4.184 000 (+00) $\frac{\text{watt}}{\text{meter}^2}$
$\frac{\text{Btu (thermochemical)}}{\text{second}}$	1.054 350 (+03) watt
$\frac{\text{Btu (thermochemical)}}{\text{pound}}$	2.324 444 (+03) $\frac{\text{joule}}{\text{kilogram}}$
$\frac{\text{Btu (thermochemical)}}{\text{foot}^2 \times \text{second}}$	1.134 893 (+04) $\frac{\text{watt}}{\text{meter}^2}$
$\frac{\text{Btu (thermochemical)} \times \text{inch}}{\text{foot}^2 \times \text{second} \times {}^\circ\text{F}}$	5.188 732 (+02) $\frac{\text{watt}}{\text{meter} \times \text{K}}$

INDEX

References are to page numbers. In the designation of systems of units the following abbreviations are used:

SI = International system
MKS = Metric system
FPS = English system
AP = Absolute practical system
EM = Electromagnetic system
ES = Electrostatic system

GLOSSARY OF SYMBOLS

Symbol	Meaning	Symbol	Meaning
$=$ or $::$	Equals	$\neq$ or $\neq$	Does not equal
$>$	Greater than	$<$	Less than
$\geqslant$	Greater than or equal	$\leqslant$	Less than or equal
$\equiv$	Identical	$\approx$	Approximately equal
$+$	Plus or positive	$-$	Minus or negative
$\pm$	Plus or minus; Positive or negative	$\mp$	Minus or plus; Negative or positive
$\times$	Multiplied by	$\div$ or $:$	Divided by
a^n	nth power of a	$\sqrt[n]{a}$	nth root of a
log, $\log_{10}$	Common logarithm or Briggs's logarithm	ln, $\log_e$	Natural logarithm or Napier's logarithm
( )	Parentheses	[ ]	Brackets
{ }	Braces		
$\begin{vmatrix} a_1 & a_2 & \cdots \\ b_1 & b_2 & \cdots \\ \cdots & \cdots & \cdots \end{vmatrix}$	Determinant	$\begin{bmatrix} a_1 & a_2 & \cdots \\ b_1 & b_2 & \cdots \\ \cdots & \cdots & \cdots \end{bmatrix}$	Matrix
I	Unit matrix	Adj	Adjoint matrix
A^{-1}	Inverse of the A matrix	A^T	Transpose of the A matrix
$n!$	n factorial	$\binom{n}{k}$	Binomial coefficient
$i = \sqrt{-1}$	Unit imaginary number	$z = x + iy$	Complex variable
sin	Sine	sinh	Hyperbolic sine
cos	Cosine	cosh	Hyperbolic cosine
tan	Tangent	tanh	Hyperbolic tangent
cot	Cotangent	coth	Hyperbolic cotangent
sec	Secant	sech	Hyperbolic secant
csc	Cosecant	csch	Hyperbolic cosecant
vers	Versine	covers	Coversine
$\sin^{-1}$	Inverse sine	$\sinh^{-1}$	Inverse hyperbolic sine
$\cos^{-1}$	Inverse cosine	$\cosh^{-1}$	Inverse hyperbolic cosine
$\tan^{-1}$	Inverse tangent	$\tanh^{-1}$	Inverse hyperbolic tangent
$\cot^{-1}$	Inverse cotangent	$\coth^{-1}$	Inverse hyperbolic cotangent
$\sec^{-1}$	Inverse secant	sech^{-1}	Inverse hyperbolic secant
$\csc^{-1}$	Inverse cosecant	csch^{-1}	Inverse hyperbolic cosecant
$\mathbf{i}, \mathbf{j}, \mathbf{k}$	Unit vectors, cartesian system of coordinates	$\mathbf{v}$	Unit vector in v direction
$\mathbf{r} = \mathbf{i}x + \mathbf{j}y + \mathbf{k}z$	Position vector, cartesian coordinates		
$\mathbf{r}_1 \cdot \mathbf{r}_2$	Scalar product	$\mathbf{r}_1 \times \mathbf{r}_2$	Vector product